Engineering Geology Case Histories 6-10

Edited by George A. Kiersch,
Arthur B. Cleaves, Wm. Mansfield Adams,
Howard J. Pincus, and H.F. Ferguson

Prepared for the
Division on Engineering Geology
of
The Geological Society of America

Library of Congress Catalog Card Number 74-77141
I.S.B.N. 0-8137-4202-1

Published by
The Geological Society of America
3300 Penrose Place
Boulder, CO 80301 USA

Printed in the United States of America
by Edwards Brothers, Inc., Ann Arbor, MI 48104
Type composed by The Geological Society
of America, Boulder, CO 80301

Engineering Geology Case Histories

Number 6

George A. Kiersch
Editor

Prepared for the Division on Engineering Geology
of
The Geological Society of America

1968

Reprinted 1973

Published by

THE GEOLOGICAL SOCIETY OF AMERICA, INC.

3300 Penrose Place

Boulder, Colorado 80301

Printed in the United States of America

CONTENTS

PREFACE

This is the sixth volume in the Case History series of the Division on Engineering Geology of the Geological Society of America, initiated in 1957. Each succeeding volume has enjoyed increasing acceptance as an aid to the practicing geologist and engineer, student, and teacher, alike. This volume is a collection of general case histories on dams, tunnels, highways, and underground construction. Indeed, the Baldwin Hills reservoir failure is another in a long list of cases which demonstrate why the geologic environment, features, and circumstances are of major concern to engineering works.

The preparation of this Case Histories series has been aided by many interested parties. Special thanks are extended to Thomas W. Fluhr who arranged for the papers on "Spruce Run Dam" and the "World's Fair Highway Complex" and gave freely of his time in reviewing both manuscripts. Assistance with the preliminary papers, and with the assembling and typing of the manuscripts, has been provided by several staff members of the Department of Geological Sciences, Cornell University. Manuscripts for this volume were submitted in 1965 and 1966.

George A. Kiersch, Editor
December, 1966

McGraw Hall
Cornell University
Ithaca, New York

FAILURE OF BALDWIN HILLS RESERVOIR, LOS ANGELES, CALIFORNIA

Laurence B. James (State of California, Department of Water Resources, Sacramento, Calif.)

Abstract

The Baldwin Hills reservoir, located 7.5 miles southwest of the Los Angeles City Hall, failed during daylight hours on December 14, 1963. Five lives were lost, scores of homes destroyed, and damages in excess of $15,000,000 incurred.

The reservoir, which was completed in 1951 by the Los Angeles Department of Water and Power, had been hollowed from the head of a ravine on the north slopes of Baldwin Hills. Closure was accomplished by a rolled earth dam approximately 155 feet high across the canyon with five minor embankments placed at low points around the perimeter. The completed structure had a capacity of 897 acre-feet with a water surface area of 19.57 acres.

The foundations of the dams and reservoir consisted of marine sediments of Pliocene and Pleistocene age, comprising largely interbedded sands and silts loosely to moderately consolidated. The site is cut by minor faults associated with the seismically active Newport-Inglewood uplift, source of the catastrophic Long Beach earthquake of 1933 and the seat of moderate late Pleistocene warping and uplift.

Records of periodic spirit levelings in Baldwin Hills since 1917 have disclosed that the slopes, including the reservoir area, are subsiding, and that a maximum subsidence of about 9 feet had occurred by 1962 at a point about a half mile west of the reservoir. Earth deformation was further manifested by a horizontal shift of nearby triangulation stations and by the development of earthcracks up to 2500 feet in length in the region adjoining the reservoir on the southeast. These deformations were probably influenced both by tectonic activities and production from the Inglewood oil field, which flanks the reservoir on the south and west.

The disaster is attributed to the development of an earthcrack similar to those that had earlier appeared southeast of the reservoir. The crack opened along a minor fault mapped by geologists during construction of the project. As a result of this movement, the reservoir floor and the abutment of the main dam were split, creating a narrow passage through erodible sediments in the foundation. Escaping water rapidly enlarged this initial conduit to the point of its collapse. Thus in the final stages of failure Baldwin Hills reservoir emptied through a spectacular breach through the main dam.

CONTENTS

ILLUSTRATIONS

Photo

INTRODUCTION

The Baldwin Hills are situated in Los Angeles County, California, about 4 miles northeast of Los Angeles International Airport. The reservoir was located on the north slope of the hills at the head of a small ravine more than 350 feet above the plain inundated as a result of the failure.

Baldwin Hills reservoir was constructed between 1947 and 1951 to meet inadequacies in supplying peak demands in the growing southern and western parts of Los Angeles. The topography of the hills offered a logical location of sufficient elevation near the area to be served. Certain adverse geologic conditions were recognized, but it appears that there were no desirable alternative sites in the vicinity. A conservative design was adopted for the structure to compensate for the less desirable characteristics. Also, provisions were incorporated which would permit emptying the reservoir in a 24-hour period should an emergency arise.

Photo 1. Baldwin Hills reservoir after failure of December 14, 1963. Explanation: (1) trace of fault 1 (Fig. 1); (2) location of first observed leakage; (3) portal to inlet tunnel; (4) circulator pipes; (5) trace of fault 5 (Fig. 2); and (6) inlet-outlet tower. The drainage inspection chamber is adjacent to the base of this tower 13 feet beneath the floor of the reservoir.

During the period of April 18, 1951 to December 14, 1963, or from dedication of the project to its failure, minor cracking had been observed in the concrete walls of the inspection gallery built beneath the floor of the reservoir; cracks had been noted in the parapet wall of the main dam; subsidence was observed at the reservoir and in the surrounding region; and there had been perceptible horizontal shifting of triangulation stations in the vicinity and movement of survey stations along the perimeter of the reservoir. A slight lengthening of the northeast-southwest diagonal of the reservoir and stretching of the crest of the dam was noted. At the time of observation these phenomena did not appear to be cause for alarm.

At about 11:15 a.m., Saturday, December 14, 1963, the alert caretaker detected an unusually loud sound of running water at the spillway discharge pipe, which was connected to the elaborate underdrainage system of the reservoir. Investigation into the cause of this noise led him to a drainage inspection chamber located under the reservoir floor. Here, he noted three underdrain pipes were "blowing like firehoses," discharging muddy water at a high rate. Authorities of Los Angeles Department of Water and Power were notified, and, at 12:20 p.m., drainage of the reservoir was commenced. At approximately 1:00 p.m., muddy water was seen emerging from the east abutment about 82 feet below dam crest elevation. Drainage was accelerated to meet the emergency, reaching an estimated maximum rate of 450 cubic feet per second. At 1:30 p.m., the Los Angeles Police Department was requested to evacuate the area below the main dam. Up to 3:30 p.m., 8 minutes before the total breach, a detention basin near the downstream toe of the dam accommodated all of the leakage. By 3:38 p.m., the failure was essentially complete, the detention basin had overflowed, and the residential subdivision below the reservoir was flooded. Heavy repair equipment was being moved into the area by 6:50 p.m. Photos 2 through 4 depict the final stages of failure.

Photo 2. Hole in main dam--2:20 p.m., December 14, 1963.

Photo 3. Hole in main dam--3:30 p.m.

PROJECT HISTORY

A need for additional water storage facilities to meet growing demands in the west Los Angeles area was recognized shortly before World War II. Studies by the Los Angeles Department of Water and Power led to the selection of the Baldwin Hills site. It provided the desired elevation and would require a minimum modification of existing facilities. Preliminary boring and soil testing was undertaken in 1941 to determine the characteristics of surface and near-surface materials in the reservoir area. In 1943, additional borings were made which extended into the area occupied by the main dam. Surface reconnaissance accompanied this drilling.

Photo 4. Collapse of hole in dam--3:38 p.m.

The earliest geologic reports presented conflicting conclusions. The first report recommended the site, while the second recommended against construction because of soft and erodible foundation materials (CRA, 1964).

A three-man consulting board was established by the Los Angeles Department of Water and Power to review plans and feasibility of the proposed project. In February 1947 the geologist member of that board reported that the proposed design appeared amply suitable to care for any obvious weaknesses in the foundation, but that it would be well to follow the progress of the excavations closely for evidence of any geological weaknesses not yet disclosed.

Construction of the reservoir began January 13, 1947, and was completed April 18, 1951. During this period further exploration was undertaken and large-scale geologic mapping of the reservoir area compiled. A comprehensive and detailed report, covering site geology, was submitted in February 1949, giving descriptions of geologic formations, faults, and joints. It also discussed the seismicity of the area, and local subsidence occurring in Baldwin Hills at that time (R. R. Wilson, 1949, Geology of the Baldwin Hills reservoir site and immediate vicinity: Los Angeles Department of Water and Power unpublished).

During the life of the project, strict surveillance was maintained. Flow from the elaborate reservoir underdrainage system was inspected and measured daily. Survey crews monitored settlement and horizontal displacements at the reservoir and surrounding area monthly. Also, each month a maintenance group alert for factors relating to safety inspected the facility. Strain gauges, tiltmeters, and seismographs were installed and regularly observed. At no time were conditions regarded as indicative of serious trouble observed during this period.

Following the failure, investigations were undertaken by a special "Blue Ribbon Board" appointed by the Mayor of Los Angeles and by an Engineering Board of Inquiry created by the Director of the Department of Water Resources and consisting of five members--four civil engineers and an engineering geologist. A comprehensive report on the reservoir failure was subsequently published by the State (CRA, 1964). That report has been drawn upon freely in the preparation of this case history.

GEOLOGIC SETTING

The Baldwin Hills rise 510 feet above the coastal plain of Los Angeles County. They are the northernmost topographic expression of the Newport-Inglewood uplift, a chain of structural domes and saddles which extend about 40 miles northwest-southeast between Beverly Hills and Newport Beach. Baldwin Hills reflect the structure of an underlying faulted dome, a prolific source of oil and gas tapped by wells of the Inglewood oil field.

The deepest oil wells in Baldwin Hills penetrate about 12,000 feet of Tertiary marine sediments and bottom in Jurassic (?) schist. Lateral movement along a fault in this basement caused the overlying sediments to wrinkle and pucker, creating a series of en echelon folds, one of which forms Baldwin Hills. This sequence of events explains the creation of the Newport-Inglewood uplift and the ultimate site of the reservoir.

One of the most significant features of the Newport-Inglewood uplift is the Inglewood fault, which passes through Baldwin Hills about 500 feet west of the reservoir (Photo 6). The length of this fault is about 9 miles; however, its continuity is interrupted by cross faults by which it is offset and divided into seven segments. Near the reservoir,

Photo 5. Cracks in reservoir floor following trace of fault 1.

Photo 6. Aerial view of Inglewood fault (fault trace is marked by arrow). Baldwin Hills reservoir lies to the right of the upper arrow.

270 feet of vertical displacement and 1500 feet of right-lateral movement have been measured on the Inglewood fault (Driver, 1943). These deformations affect late Pleistocene formations, and, consequently, it is evident that substantial tectonic movement has occurred during the last 500,000 years. On the other hand, stream channels and fan deposits are not visibly offset where crossed by faults in Baldwin Hills, suggesting that movement on these faults has not persisted throughout Recent geologic time (USGS, 1964).

The Newport-Inglewood uplift remains seismically active. The so-called Inglewood earthquake of 1920, which destroyed a schoolhouse and toppled chimneys near Inglewood, was centered about 4 miles south of the reservoir site. In 1933, the city of Long Beach was partially leveled by shocks originating on an offshore extension of the Newport-Inglewood uplift, the resulting casualties and damage being surpassed in California only by the great San Francisco earthquake of 1906. A plot of epicenters of earthquakes near Los Angeles shows a cluster of points near Baldwin Hills (Richter and others, 1965). Thus, it is obvious that tectonic events are continuing along the uplift. However, records of the Seismographic Laboratory at the California Institute of Technology, 15 miles northeast of the reservoir, indicate that no earthquake of sufficient strength to cause inertial damage to the project had occurred since its construction.

The abutments and foundation for the main dam are Tertiary sediments, mostly of marine origin. The oldest exposed formation is the Pliocene Pico, which consists of silt, sand, and clay, with occasional small lenses of sand and gravel. In general, these materials are moderately consolidated and a few beds appear to be partially cemented.

The Inglewood Formation overlies the Pico and was deposited during Pleistocene time. In the east wall of the breach, it consists mostly of

Photo 7. Zone of fault 1 during reservoir construction.

Photo 8. East wall of breach through reservoir.

nearly flat-lying beds of silt and sand, ranging from half an inch to several feet in thickness (Photo 8). Some of the exposed strata are firm while others contain pockets of sand soft enough to be pulverized by bare hands.

The Palos Verdes Formation is exposed near the reservoir rim. It consists of poorly consolidated silts, sands, and gravels of upper Pleistocene origin.

Several faults were mapped in the vicinity of the reservoir, before and during construction of the project (R. R. Wilson, 1949, Geology of the Baldwin Hills reservoir site and immediate vicinity: Los Angeles Department of Water and Power unpublished). Many of these are no longer discernible, since they are now buried beneath the embankments. The most significant faults had been designated fault 1 and fault 5 by early investigators (Photo 7). Movement along these faults occurred during failure of the reservoir as evidenced by displacement of the asphaltic lining, drains, and observation gallery. The coincidence of the trace of fault 1 with the axis of the breach through the main dam is apparent in Photo 1.

During post-failure investigation, faults 1 and 5 were intensively explored. Thirteen test pits, two shafts, and 256 feet of adit were excavated for this purpose. A geologic section through one of the pits is shown on Figure 1. Some pertinent observations follow.

(1) Both faults 1 and 5 dip steeply westward toward the Inglewood fault.

(2) During reservoir failure both faults experienced normal dip-slip displacement amounting to as much as 7 inches on fault 1.

(3) Seams of loose sand up to 4 inches thick lie between the hanging and footwalls of fault 1 beneath both the reservoir floor and the bottom of the breach. These seams contain occasional voids and a few fragments of asphalt which were apparently derived from the reservoir lining.

(4) Tile drain pipes sheared by the faults 1 and 5 are not offset laterally.

These factors suggest that the faults sprung open during the failure process and the resulting gap was almost, but not entirely, filled with detritus deposited by leakage from the ruptured reservoir.

INGLEWOOD OIL FIELD

Inglewood oil field adjoins Baldwin Hills reservoir on the west and south. Discovered in 1924, the field has produced more than 239 million barrels of oil, 164 billion cubic feet of gas, and 364 million barrels of water (CRA, 1964a). Production is from two unconnected pools separated by the Inglewood fault. Vertically, the field is divided into nine producing zones ranging in depth from 950 to more than 10,000 feet. The Vickers is the most productive of these zones with 294 producing wells which tap sediments described as fine to coarse grained, and generally silty and friable. With the exception of the Investment zone, a minor producer, the Vickers zone is the shallowest source of oil in the Inglewood field. Wells nearest the reservoir enter this zone at elevations between 900 and 1000 feet subsea, as contrasted to the reservoir floor elevation of about 420 feet (Oefelein and Walker, 1963).

The structural dome of Inglewood oil field is intricately faulted, many of the faults joining the Inglewood fault at depth. The plane of one such fault, mapped by subsurface methods by petroleum geologists, was found to trend toward the reservoir area. The plane may be the extension of fault 1.

CHARACTERISTICS OF THE RESERVOIR

Dams

The main dam and five saddle dams were constructed of the sediments excavated during the shaping of the reservoir bowl. These materials were blended and moistened during their emplacement to improve the strength and impermeability of the embankments. The fills were unzoned, and relied upon a surficial blanket of compacted earth to prevent leakage.

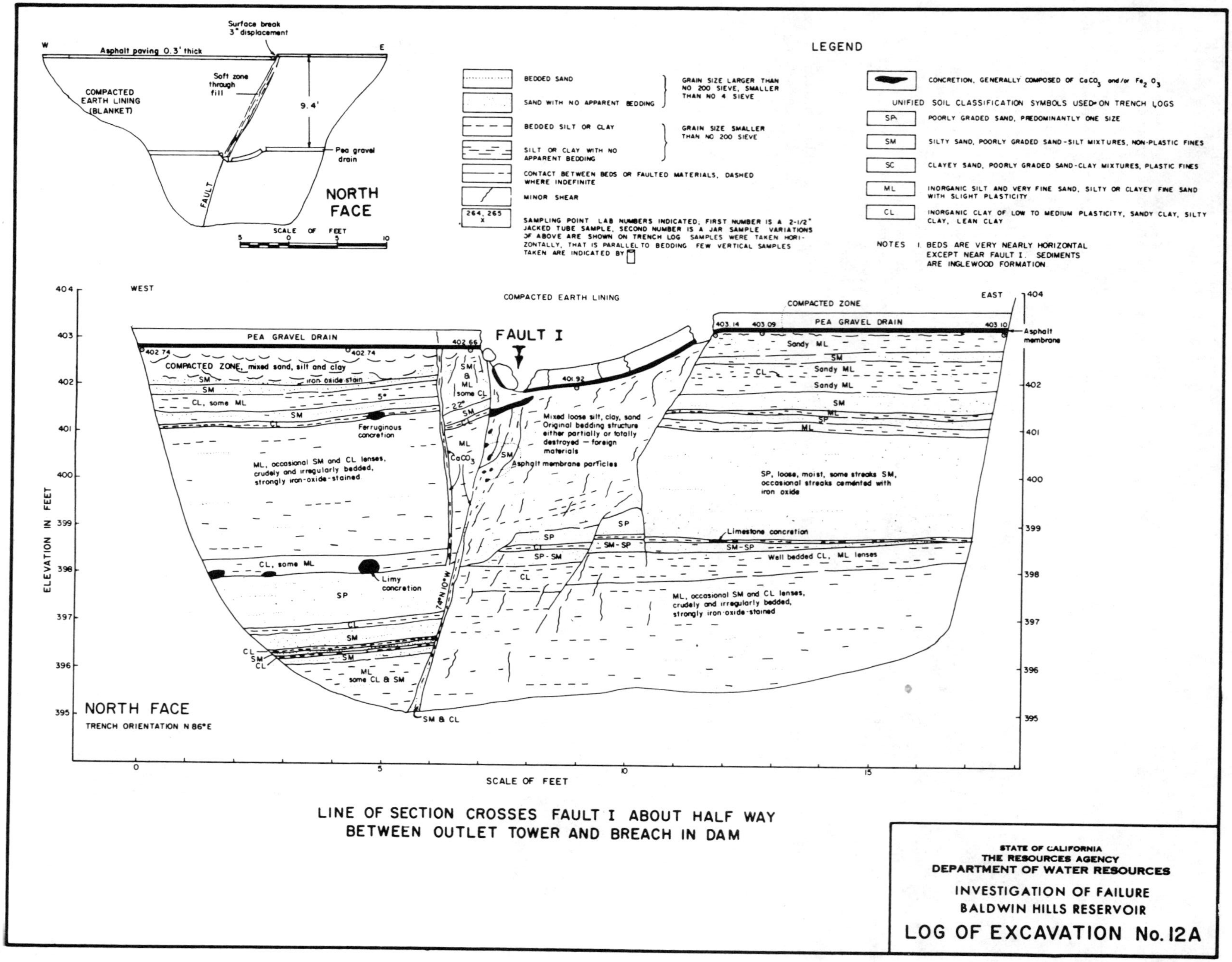

LINE OF SECTION CROSSES FAULT I ABOUT HALF WAY BETWEEN OUTLET TOWER AND BREACH IN DAM

STATE OF CALIFORNIA
THE RESOURCES AGENCY
DEPARTMENT OF WATER RESOURCES

INVESTIGATION OF FAILURE
BALDWIN HILLS RESERVOIR

LOG OF EXCAVATION No. 12A

Figure 1. Log of excavation no. 12A across fault 1.

Lining

A 3-inch layer of porous asphalt lined the reservoir to inhibit weed growth and prevent wave wash. Compacted earth lay 10 feet thick beneath the floor, tapering up the walls to 5 feet in thickness at the reservoir rim. This fill consisted of select materials stockpiled during excavation.

Drainage Facilities

Special precautions were taken to prevent deterioration of the foundation from saturation. These included: (1) an elaborate underdrainage system to divert and remove any seepage penetrating the lining before it contacted the underlying formations; and (2) foundation drains to remove water that might, in some manner, enter these formations.

The underdrain system consisted of a 4-inch pea-gravel layer immediately subjacent to the lining and of a grid of 4-inch clay tile pipes (Photo 9). Seepage through the lining was intercepted by the pea gravel, diverted to the tile drains, and then conducted to a central drainage inspection chamber located near the base of the outlet tower (Photo 10). Here it was measured by weirs, and then conveyed out of the reservoir in a 24-inch blowoff pipe which passed through the main dam.

Photo 9. Reservoir underdrainage system under construction.

The foundation and embankments upon which was placed the pea-gravel drain were first sprayed with a 1/4-inch asphaltic membrane, which provided a water-tight seal at the base of the pea gravel. The security of the reservoir was dependent upon the

Photo 10. Post-failure view of drainage inspection gallery.

integrity of this diaphragm since water penetrating it would enter a foundation susceptible to deterioration.

The facilities for removing water from the foundation included: (1) a 12-inch clay tile pipe placed along the axis of the canyon, and subsequently covered by the main dam; and (2) a number of foundation wells and horizontal drill holes, some of which were connected to the 12-inch canyon drain.

RECENT DEFORMATIONS IN BALDWIN HILLS

Subsidence and Lateral Movement

Periodic-level observations in Baldwin Hills, beginning in 1917, have defined a bowl of subsidence, roughly elliptical in shape and centered about a half mile west of the reservoir (Fig. 2). It was estimated from these levelings that 9.7 feet of subsidence occurred at the center of this bowl from 1917 to 1963, and that 3 feet occurred at the reservoir site during this period. Since the earliest records of usable surveys date from 1917, it was not possible to estimate how much subsidence occurred (if any) prior to that year (CRA, 1964).

Geodetic surveys conducted in 1934, 1961, and 1963 disclosed that first-order triangulation stations in Baldwin Hills are moving laterally and generally toward the trough of the subsidence bowl (Alexander, 1963). One of these stations, designated "Baldwin Aux," located 600 feet south of the

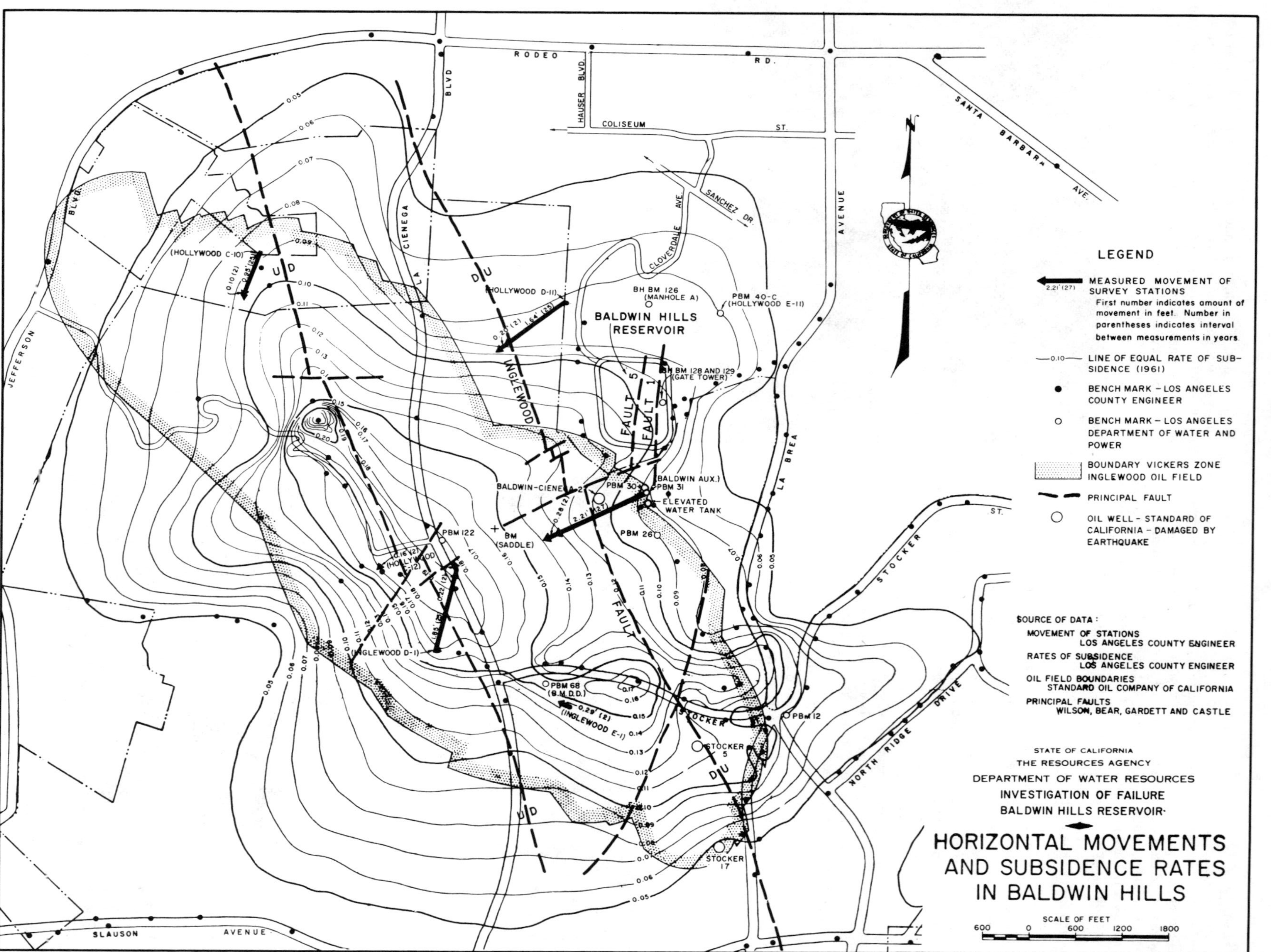

Figure 2. Horizontal movements and subsidence rates in Baldwin Hills.

reservoir rim, moved 2.49 feet during the 29-year period (1934 to 1963). These lateral movements were accompanied by a stretching of the earth's surface near the perimeter of the subsidence bowl, as illustrated by a lengthening of the distance between two survey stations located, respectively, at the northeast and southwest corners of the reservoir. Surveys made in 1947, and repeated in 1962, showed the diagonal line through the reservoir connecting these two stations lengthened 0.4 feet during this interval.

Figure 2 depicts the general outline of the subsidence bowl, the site of the reservoir, lines of equal rates of subsidence for 1961, principal faults in Baldwin Hills, the boundaries of the Vickers oil-bearing zone of Inglewood oil field, and vectors of the lateral movements of triangulation stations. A maximum subsidence rate of almost 3 inches per year is shown to be occurring at a point about 3800 feet west of the reservoir.

Earthcracks

Southeast of the southeast corner of the reservoir, and 2500 to 3800 feet therefrom, a number of open earthcracks were detected as early as May 1957. They measure up to 2500 feet in length, and appear similar in several respects to the cracks which ruptured the reservoir along faults 1 and 5. Some of these similarities, mentioned by the California Department of Water Resources (1964), are:

(1) a general parallelism to pre-existing faults;

(2) a common orientation, generally north-south in direction;

(3) steep dip;

(4) gaps which vary in width along the length of each crack;

(5) no horizontal displacement along these cracks; and

(6) occurrence in areas where rate of subsidence is changing markedly within short distances.

In view of these similarities, it is believed that the cracks southeast of the reservoir and those through the reservoir are related.

Possible Causes of Subsidence

Subsidence is not unique to Baldwin Hills. It occurs elsewhere, and has been attributed to various causes. These causes include: ground-water extraction; densification of loess or mudflow deposits as a result of saturation; mass land movements, such as creep or slides; compaction of soils by earthquake; slow tectonic activity; and oil field production.

The formations underlying Baldwin Hills contain no significant aquifers. Consequently, ground-water extraction can be dismissed as a cause of subsidence.

The reservoir foundation consists of marine sediments deposited underwater and consolidated by the weight of younger overlying strata. Subsequent tectonic events elevated the land mass, and erosion removed former overburden and sculptured present topographic features. Thus, the formations now exposed were originally compressed while submerged. The soil particles were oriented under these conditions, and the present soil structure was formed. The foundation sediments are therefore unlike loess or mudflow, which are deposited in a dry or quasi-saturated state with a resulting open structure that may collapse when saturated. Subsidence in Baldwin Hills is not caused by wetting.

A landslide had occurred in the right abutment of the main dam during construction, and slides were observed after the failure. These factors led to some early speculation that the failure was related to a massive slide or creep in the abutment. This hypothesis was abandoned for the following reasons. First, the cracks in the reservoir extend well beyond the localized area that slid during construction. Second, survey monuments located on the crest and downstream face of the dam did not appreciably shift during the failure. It is highly improbable that any significant mass movement could occur in the abutment without disturbing these monuments. Therefore, the manifestations of sliding observed in the abutment following the failure are believed to have formed after the breach was created, and were attributed to oversteepening of the abutment by the erosive action of the escaping water.

The Seismological Laboratory of the California Institute of Technology is located about 15 miles northeast of the reservoir. The seismographs at this facility are capable of recording tremors which have an energy on the order of 1/100 of that produced by the slightest tremor that could be perceived by a person standing at the site of the reservoir. However, these instruments detected no significant shocks on the day of failure, and a search of past records indicated that during the life of the reservoir no strong earthquakes had occurred. The subsidence in Baldwin Hills persisted throughout this quiescent period, and, consequently, cannot be the result of soil compaction due to earthquakes.

The remaining possible causes of subsidence, namely, tectonic processes other than earthquakes and activities relating to oil and gas production, were considered by the State of California in its investigation. The State comments that any reasonable attempt to estimate the degree to which tectonic or oil field activities are contributing to contemporary subsidence would require greater knowledge of subsurface geology and of earth movements at depth than is now available (CRA, 1964). At present, it is not known whether subsidence in Baldwin Hills is occurring only in or above oil-producing zones, or whether the formations underlying these zones are also affected. Investigation of this important aspect of the problem would involve long and costly subsurface measurements which have not been undertaken.

SEQUENCE OF FAILURE

The final events during rupture of the reservoir eradicated important clues to the failure process. The theoretical failure sequence that will now be reconstructed is therefore, in part, conjectural.

At some unknown depth, and due to a combination of causes, the sediments underlying a portion of Baldwin Hills began to subside. The strata overlying the seat of this subsidence sagged into the growing depression, thereby creating tension in the surface layer--notably near the perimeter of the subsidence bowl. As the stress increased, earthcracks opened southeast of the reservoir, the northeast-southwest diagonal through the reservoir was stretched, and incipient fissures developed at the reservoir--particularly along pre-existing faults 1 and 5, which were planes of weakness. By the spring of 1963, these cracks had opened sufficiently to permit a slight and perceptible increase of flow into the reservoir underdrainage system. This preliminary leakage period is evidenced by the records of flow, measured at the drainage inspection gallery, and by mineral deposits which were found to be coating the sheared surfaces of tile drains cut by fault 1.

Final disruption of the reservoir began early on December 14, 1963, shortly before the caretaker heard the sound of leakage. Beneath the main dam, fault 1 separated as much as 4 inches, and a maximum, normal dip-slip throw of 7 inches occurred (Fig. 1).

The throw on fault 1 was measurable beneath the toe of the main dam where it offset the pea-gravel drain. However, evidence of fault displacement within the dam is lacking. For example, Photos 2 and 3 show the reservoir lining above the hole in the dam is neither sheared nor offset. It appears, therefore, that in the early stages of failure the embankment of the main dam may have "bridged" across the offset in the foundation. Possibly the initial passageway through the reservoir resulted from a combination of fault plane separation and "bridging" action. Post-failure observations of fault 1 suggest that separation along the fault varied from tight cracks at some points to gaps up to 4 inches wide at others. That the escape route was at first tortuous and restricted is suggested by the initial appearance of seepage and progressive increase in flow that followed.

Thus the watertight integrity of Baldwin Hills reservoir was destroyed. Water quickly intruded the underdrainage system under full head and erupted into the drainage inspection gallery, where the drains "blew like firehoses." Near the main dam, water worked northward through the opening along fault 1, emerging at the downstream face of the dam about one hour and forty-five minutes after leakage was first heard. The sediments, through which the fault passed, eroded rapidly, first forming a conduit, and then a breach through the dam.

WATER UNDER THE DAM (Baldwin Hills in Retrospect)

Grunner (1963), in a paper on dam disasters, has observed that, "Every dam which impounds water presents a potential danger which should be neither under- or overestimated. The risk of sudden disaster is forever inescapable, and while knowledge and vigilance may reduce such a risk, it can never be entirely banished."

This sobering truth was well illustrated at Baldwin Hills. Geologic factors had been correctly interpreted and frankly described, after which the reservoir was designed, taking into consideration the environmental conditions. A conservative design was adopted for all embankments; an earthquake safety factor, double that utilized in many other seismic areas, was used; intricate drainage facilities were incorporated to prevent foundation saturation; construction was rigidly controlled; and a strict surveillance program was followed. As an overriding safety measure, the reservoir was provided with extraordinary outlet facilities through which it could be emptied in 24 hours.

This failure has again forcefully demonstrated that a structure is only as strong as its weakest essential component. Geologic elements at the Baldwin Hills site would provide a marginal foundation for any open facility for confinement of water. Sitting on the flank of the sensitive Newport-Inglewood fault system with its associated tectonic restlessness, at the rim of a rapidly depressing subsidence basin, and on a foundation adversely influenced by water, this reservoir was called upon to do more than it was able to do. The rapidity of rupturing on the final day impaired attempts to avert disaster by "dumping" the water in storage. Only four and one-half hours elapsed between the detection of leakage and complete failure of the structure.

ACKNOWLEDGMENTS

Special thanks are due fellow members of the California Department of Water Resources Engineering Board of Inquiry, from whose report, "Investigation of Failure Baldwin Hills Reservoir" (CRA, 1964), most of the information contained in this case history was derived. Members of the Engineering Board of Inquiry were: Robert B. Jansen, chairman; Gordon W. Dukleth, vice chairman (engineering design); Bernard B. Gordon (soil mechanics); Clyde E. Shields (construction); and Laurence B. James (engineering geology).

Consultants to the Board of Inquiry were: J. Barry Cooke (civil engineering); Thomas Leps (soil mechanics); Roger Rhoades (geology); and Pierre St.-Amand (seismology).

Post-failure exploration of the reservoir for the State of California was conducted by crews of engineering geologists and civil engineers from the Department of Water Resources. They were under the supervision of Arthur B. Arnold, senior engineering geologist, and James E. Ley, supervising engineer. Appreciation is expressed to Mrs. Dora Barbee for her efforts and assistance in preparing this manuscript.

REFERENCES CITED

Alexander, I. H., 1963, Horizontal earth movements in the Baldwin Hills, Los Angeles area: Jour. of Geophysical Research, v. 68, no. 11.

CRA, 1964, Investigation of failure Baldwin Hills Reservoir: California Resources Agency, Department of Water Resources, 64 pp., 9 tables, 27 plates.

CRA, 1964a, Data from files: California Resources Agency, Department of Conservation, Div. of Oil and Gas.

Driver, H. L., 1943, Inglewood oil field: California Department Natural Resources, Div. Oil and Gas, Bull. 118.

Grunner, E., 1963, Dam disasters: Institution of Civil Engineers, London, England, February 19, 1963.

Oefelein, F. H., and Walker, J. W., 1963, Field case history--Vickers east zone water flood, Inglewood field, California: Reprint, AIME Paper SPE. 699, October 24, 1963.

Richter, C. F., Allen, C. R., Nordquist, J. M., and St.-Amand, P., 1965, Map of epicenter locations in vicinity of Newport-Inglewood uplift: *in* Allen and others, Bull. Seis. Soc. Amer., v. 55, pp. 753-797.

U. S. Geological Survey, 1964, Preliminary report on recent surface movements through July 1962 in the Baldwin Hills, Los Angeles County, California: Open-file report, January 1964.

ENGINEERING GEOLOGY OF DILLON DAM, SPILLWAY SHAFT, AND DIVERSION TUNNEL, SUMMIT COUNTY, COLORADO

Ernest E. Wahlstrom (Department of Geology, University of Colorado, Boulder, Colorado)
and V. Q. Hornback (Board of Water Commissioners, Denver, Colorado)

Abstract

The Dillon Dam, in Summit County, Colorado, is a large earth-fill dam constructed on a foundation of faulted and extensively jointed Mesozoic sedimentary rocks. Appurtenant features include a diversion tunnel and spillway shaft.

Geologic features determined by preliminary exploratory drilling and by detailed studies during construction permitted close control of foundation excavation and grouting operations. Special problems were created by heavy water flows from pervious gravels up to 80 feet deep that covered bedrock in the valley floor of the Blue River, and by repeated movements of an ancient landslide reactivated during dam construction of loading by a temporary earth embankment and simultaneous excavation near the toe of the slide.

CONTENTS

ILLUSTRATIONS

INTRODUCTION

This report is a condensation of a much more extensive report on the geology of the Dillon Dam site in Summit County, Colorado, prepared for the Board of Water Commissioners, City and County of Denver, after the completion of construction of the dam and appurtenant features.

The location of Dillon Dam in the valley of the Blue River about three-fourths of a mile north of the former location of the town of Dillon, Colorado, was determined largely by the requirements of geography and topography (Fig. 1). Early preliminary geologic investigations of the dam site and the Harold D. Roberts Tunnel location indicated that

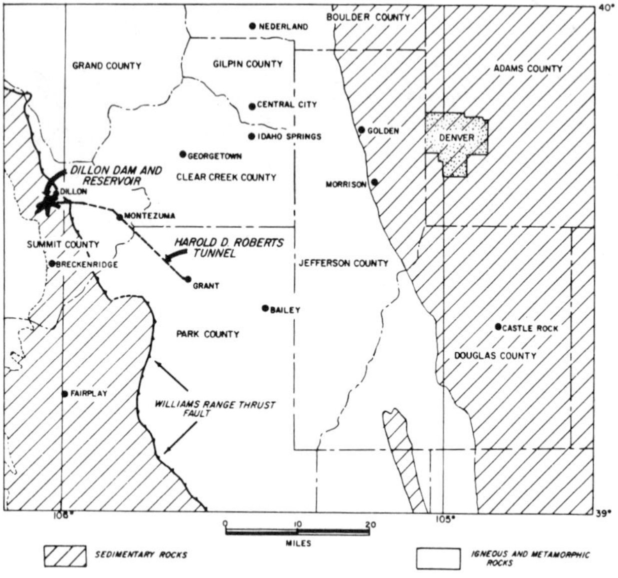

Figure 1. Index map showing location of Dillon Dam and Reservoir.

both were situated in a region of extreme geological complexity and that difficulties probably would be encountered during construction. However, the location of the confluence of the Blue River, Tenmile Creek, and the Snake River, a constriction in the Blue River valley just north of Dillon, and the desirability of gravity flow through the Roberts Tunnel to the Eastern Slope made the selection of the present site of the dam almost mandatory. Geological considerations were, of necessity, of secondary importance.

HISTORY OF GEOLOGIC INVESTIGATIONS

The valley constriction along the Blue River about three-fourths of a mile north of Dillon was regarded as a possible dam site in early investigations of Western Slope water and power development by the United States Bureau of Reclamation. After completion of a topographic map of the area in the vicinity of Dillon, the site was abandoned. More elaborate plans were then made to construct separate dams and interconnecting canals on the Blue River, Tenmile Creek, and Snake River at considerable distances above Dillon. The plan called for diverting water to the Eastern Slope through a tunnel connecting with a reservoir behind the Snake River Dam.

The first significant geological investigation of the Dillon Dam site was made by Ogden Tweto of the United States Geological Survey who on February 1, 1945, submitted a report to the Board of Water Commissioners entitled "Geologic Report on the Dillon Dam Site, Summit County, Colorado." The report included descriptions of the geology, a map, and cross sections. He recommended a dam location very close to the one finally adopted. On November 26, 1955, E. E. Wahlstrom submitted a report to the Board of Water Commissioners entitled "Diamond Drill Exploration of Foundation, Dillon Dam Site, Dillon, Colorado." This study confirmed Tweto's conclusions that the foundation rocks were severely fractured and would require extensive treatment to reduce their permeability. Later, the responsibility for preparing the contract documents, including the specifications and drawings for the construction of Dillon Dam and appurtenant features, was assigned to Tipton and Kalmbach, Inc., of Denver, Colorado. At this time, additional geologic studies were made of the west (left) abutment, and five additional exploratory holes were drilled to assist in the design of the diversion tunnel and spillway shaft. A report on this study by E. E. Wahlstrom in July 1959 was entitled "Supplemental Report on Exploratory Diamond Drilling at Dillon Dam Site, Summit County, Colorado." This report included mention of the distinct possibility of landslides in the west abutment area. A plan of the dam as constructed is shown in Figure 2.

Geologic investigations during construction were made by V. Q. Hornback, resident geologist, and E. E. Wahlstrom, consulting geologist. Detailed maps and sections, on a scale of 1 inch = 20 feet, of the dam foundation, the diversion tunnel, and spillway shaft were prepared as construction progressed. Geologic data were used extensively to determine depth and spacing of grout holes in the dam foundation and the amount of excavation required to obtain suitable foundation rock. Subsidiary investigations were concerned with the location of suitable sources of riprap and the determination of bedrock location and characteristics in an area of earth slides situated upstream from the spillway shaft.

GENERAL GEOLOGY OF THE DAM SITE AND RESERVOIR AREA

The Dillon area is one of extreme geological complexity. A major geological feature is the Williams Range thrust fault which crops out about two miles east of the dam site. This fault, which strikes north-northwest and dips to the northeast, resulted from as much as three miles of displacement of Precambrian igneous and metamorphic rocks over Mesozoic and older sedimentary rocks. Movement was accompanied by extensive folding, faulting, and joint development in the thrust sheet and rocks below.

Bedrock at the dam site and in the spillway shaft and the diversion tunnel consists of sandstone, sedimentary quartzite, shale, siltstone, mudstone and calcareous mudstone. Sills and dikes of felsite porphyry locally are present, especially in the layered quartzites of the Dakota Group. A generalized columnar

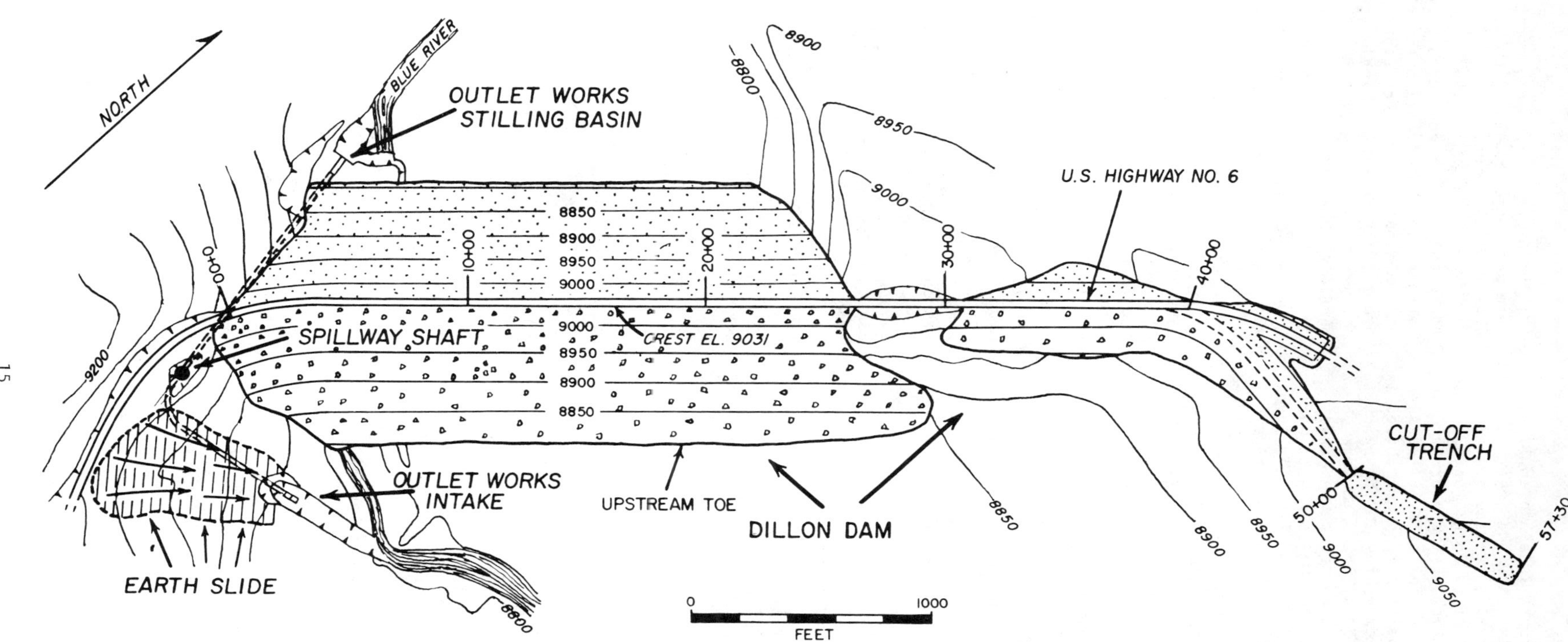

Figure 2. Plan of Dillon Dam and appurtenant structures.

section of the sedimentary rocks is shown in Figure 3. Most of the bedrock in the dam site and reservoir area is buried beneath stream gravels, terrace gravels, glacial deposits, talus, and slide rock. Data regarding the detailed characteristics of the bedrock prior to construction, with the exception of a few measured sections of the Dakota Group, came from core holes in the dam foundation and from exposures in the nearby Harold D. Roberts Tunnel. Additional data were obtained from exposures in the core trench and road cuts as construction progressed. Major faults were identified in a general way during preliminary investigations, but specific locations and characteristics could not be determined until the faults were exposed in excavations during construction. Extensive, closely spaced jointing was noted in rock exposures and drill cores during preliminary investigations,

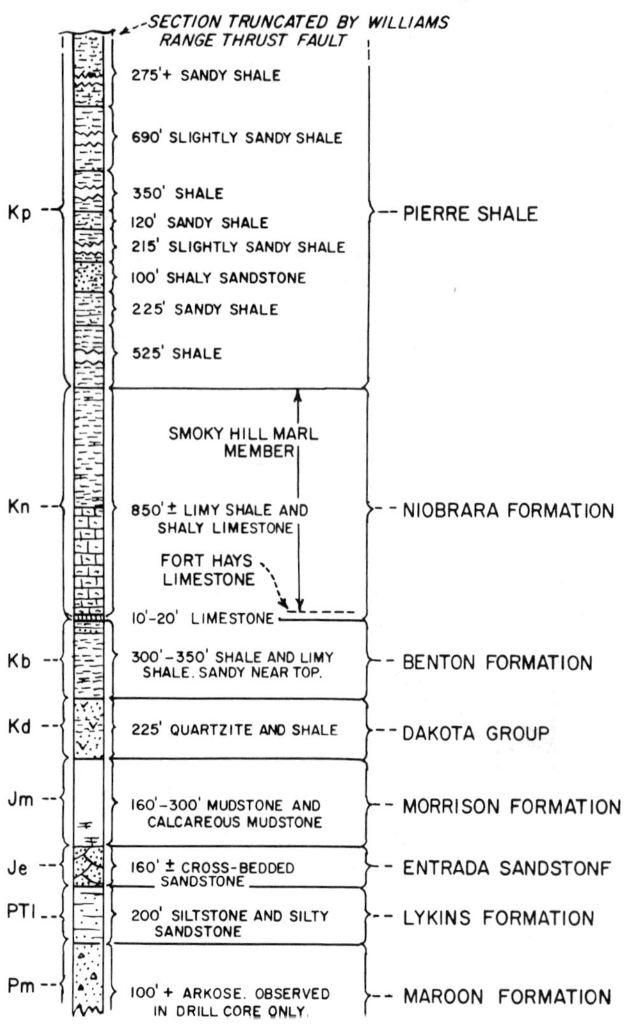

Figure 3. Generalized columnar section.

but the numbers, spacing, and attitudes of the joints were not noted in detail until bedrock was exposed in the core trench.

Rocks and surficial deposits described are those that were encountered in the core trench excavations, in the diversion tunnel and spillway shaft, and in grout holes. Descriptions are given in the order of increasing geologic age of the rocks.

SURFICIAL DEPOSITS

The floor of Blue River valley at the dam site is covered by stream gravels having depths up to 80 feet. The slopes of the dam abutments, before excavation, were covered by slide rock and talus consisting of angular boulders and smaller fragments of Dakota Group rocks intimately mixed with weathered, clayey materials from the Morrison Formation. On the west abutment the talus and slide rock also contained weathered fragments of felsite porphyry from a sill in the Dakota sediments.

The stream gravels on the valley floor are well washed and generally contain only small amounts of clay and silt-size materials. The average permeability is very high, but irregular sorting and stratification produced erratically distributed zones of high permeability. During excavation of the core trench it was noted that water flows into the cut were concentrated mainly in a few gravel layers, especially where these were above a black, manganese-stained zone near the bottom of the gravel deposits.

Terrace gravels up to 50 feet thick are present at the east end of the dam site and rest on sheared black shale. Test holes drilled after the start of dam construction showed that these gravels were highly permeable and that they existed below the high water level of the reservoir. To avoid leakage from the reservoir through the gravel, the cutoff trench was extended to Station 57+60, 730 feet beyond the point originally indicated in the plans and specifications. The shales below the gravel proved to be relatively impermeable and accepted only small amounts of grout during grouting operations.

SEDIMENTARY ROCKS

Pierre Shale

The Pierre Shale, of upper Cretaceous age, underlies much of the eastern portion of the reservoir area, and weathered materials from it provided the main source of impervious material for the core of the dam. The formation is several thousand feet thick, and consists of soft black shales, sandy shales, and shaly sandstones. Exposures of disturbed, weathered Pierre Shale were uncovered in the excavation for the core trench east of a strong fault near Station 47+00, and they extended to the end of the trench at Station 57+60 (Fig. 4).

Niobrara Formation

The Niobrara Formation is about 850 feet thick, as determined from measurements in the Roberts Tunnel, and consists of limestone, limy shale, and shale. The formation is subdivided into two units.

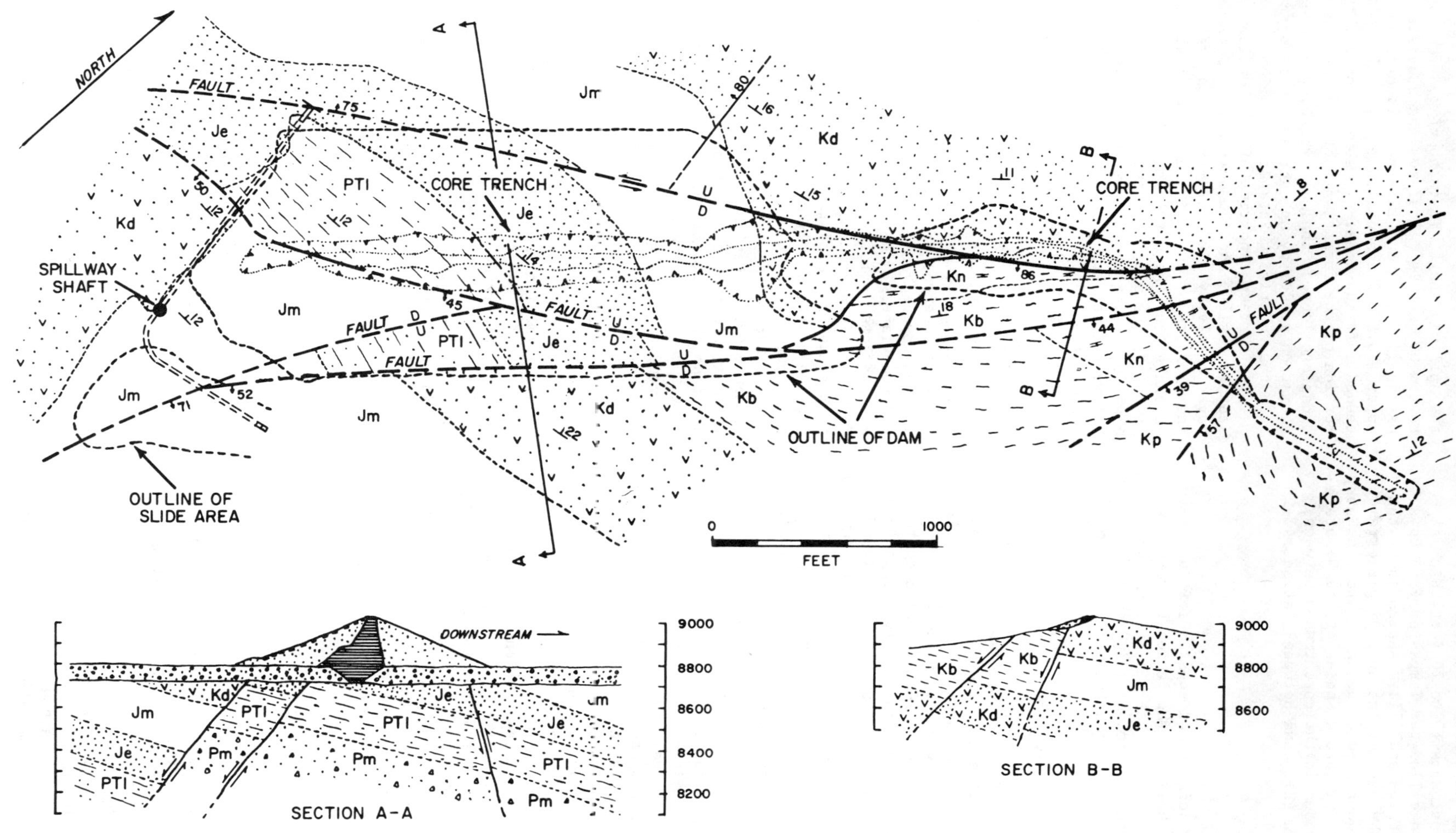

Figure 4. Geologic plan and sections, Dillon Dam.

A massive limestone, some 10 to 20 feet thick at the base, has been called the Timpas or Fort Hays member, and the shales and limy shales above the limestone up to the base of the overlying Pierre Shale have been identified as belonging to the Smoky Hill marl member. Disturbed shales of the Niobrara Formation are exposed in the core trench between Stations 41+50 and 42+15 and between Stations 43+90 and 46+70 (Fig. 4).

Benton Formation

Complexly jointed and sheared soft black shales and sandy shales of the Benton Formation were exposed in the core trench between Stations 42+15 and 43+90 (Fig. 4). The formation is 300 to 350 feet thick, and lies above the quartzite of the Dakota Group.

Dakota Group

The rocks in the Dakota Group constitute a succession of quartzites and black shales approximately 225 feet thick. A massive basal quartzite, 50 to 60 feet thick, is overlain by a brittle siliceous black shale zone 60 to 70 feet thick containing sandstone lenses. The top unit is quartzite containing black shale partings, and is about 100 feet thick. Thicknesses of the shale layers vary considerably from place to place, partly because of thinning and thickening caused by folding and by slippage of the top quartzite over the shale unit, and partly because of the manner of original deposition. The top unit locally is conspicuously cross-bedded, and individual layers tend to be lenticular and laterally discontinuous.

The quartzite layers are very brittle, very hard, and closely jointed. Several sets of joints intersect the rocks and are conspicuous in original outcrops and in core trench exposures. The shale locally is so closely sheared and fragmented that the bedding has been almost destroyed.

Exposures of the Dakota Group are present near the spillway shaft and in the core trench west of Station 0+22, and between Stations 25+00 and 41+70 in the east abutment (Fig. 4). The rocks are also well exposed in the road cut for U. S. Highway 6 in the west abutment.

In several exposures, including that in the top of the hill west of the west abutment, the Dakota Group contains felsite porphyry dikes and sills. However, no porphyry was encountered in excavations for the dam foundation.

Quartzites in the Dakota Group provided the chief source for riprap for the upstream face of the dam.

Morrison Formation

Exposures of the Morrison Formation were encountered in the core trench between Stations 0+25 and 5+80, and between Stations 18+60 and 25+00. The diversion tunnel intersected Morrison sediments between Stations 12+80 and 16+00. The spillway shaft is entirely in the Morrison Formation (Fig. 4).

The formation consists of gray, greenish-gray, and red mudstones and calcareous mudstones. These range in thickness from some 160 to 300 feet. In the spillway shaft, it is about 230 feet thick. The bottom 60 feet is calcareous and consists of alternating lime-rich and lime-poor layers, while the upper 170 feet is noncalcareous, well- to poorly bedded soft mudstone. In some exposures the basal portion of the formation contains Entrada-type sandstone lenses, and, elsewhere, the base of the formation grades by an increase in sandy material into the underlying Entrada Formation.

Surface exposures of the Morrison Formation weather easily, and commonly are concealed by a cover of clayey material mixed with small fragments and boulders of quartzite and with small fragments of shale from the overlying Dakota Group. When wet or moist, the material tends to form earth slides. A slide in overburden over the Morrison Formation became active during construction upstream from the west abutment and threatened burial of the intake structure of the diversion tunnel.

Entrada Sandstone

The Entrada Sandstone consists of a white or light-gray, even grained sandstone containing a calcareous cement or a clayey matrix. The sandstone is cross-bedded and lenticular, and locally contains mudstone layers. Rapid lateral changes in thickness are characteristic. In the vicinity of the dam, the formation is approximately 160 feet thick. Lenses of Entrada-type sandstone are present in the basal portion of the overlying Morrison Formation and in the upper part of the overlying Lykins Formation.

Entrada Sandstone is exposed in the core trench between Stations 12+70 and 18+60 (Fig. 4). A lense in the Lykins Formation is exposed between Stations 12+00 and 12+25. Fairly competent Entrada Sandstone was encountered in the diversion tunnel between Stations 7+80 and 12+80 and in the excavation for the outlet stilling basin.

The sandstone is not as brittle as the quartzites in the Dakota Group, and generally is not as closely fractured.

Lykins Formation

The Lykins Formation is a succession of red and gray well-bedded siltstones, sandy siltstones, and silty sandstones. The formation at the dam site is about 200 feet thick. A few thin layers are dolomitic. The formation was exposed in the core trench between Stations 5+75 and 12+70 and in the diversion tunnel between Stations 17+70 and the outlet portal (Fig. 4).

Maroon Formation

The Maroon Formation, which is of unknown thickness, consists of red and gray arkose and conglomeratic arkose. It was not exposed in the core trench during construction, but from core holes and grout holes is known to occur immediately below the Lykins Formation. The formation was intersected in grout holes between Stations 4+00 and 10+00.

STRUCTURAL GEOLOGY

The sedimentary rocks in the vicinity of the dam site, in general, strike approximately east-west and dip gently to the north at angles commonly less than 25 degrees (Fig. 4). Several faults of large displacement, and numerous minor faults, intersect the bedrocks in the core trench and diversion tunnel. Other faults of large displacement are known to exist in the reservoir area, but no effort was made to locate the faults or to determine their characteristics. Faults encountered in the Roberts Tunnel, some of which cross the reservoir area, are described by E. E. Wahlstrom, C. S. Robinson, and T. C. Nichols in another paper in this Case History volume.

Numerous prominent joint sets were exposed in the core trench excavation. Concentrations of closely spaced joints are associated with faults, but some joints bear no obvious relationship to faulting. Joints are present in great numbers in all types of bedrock, but are especially abundant in the brittle quartzites of the Dakota Group and in shattered rocks adjacent to faults. Particularly noteworthy were several persistent joints crossing the core trench excavation in the west abutment. These joints are in the basal quartzites of the Dakota Group, and opened as a consequence of slight settlement in the relatively incompetent Morrison mudstones below the lower quartzite of the Dakota Group as excavation for the core trench progressed. Repeated attempts to excavate the core trench to an open crack, developed along a particular joint, only caused additional cracks to open uphill from the excavation. As a result it was necessary to extend the excavation for a considerable distance beyond the limit indicated in the construction drawings. Excavation finally was stopped at an arbitrary point, although open cracks were still present in the foundation near the top of the dam. Additional excavation would have required removal of large volumes of rock and would not have resulted in any noteworthy improvement. (One possible explanation is rebound joints due to strain energy release.)

While grouting the foundation in the east abutment beyond Station 26+00, it was noted that the middle black shale unit of the Dakota Group was thicker than normal. In the cliff exposure near Station 26+00, it was observed that slippage of the upper quartzite layers over the shale unit at some time in the past caused the shale to be squeezed in some places and dragged out and attenuated elsewhere. The shale so affected is closely sheared and permeable, and required extensive grouting.

SPILLWAY SHAFT

The spillway shaft is entirely in mudstones and calcareous mudstones of the Morrison Formation. The collar of the shaft is just below the contact of the Morrison Formation with basal quartzites of the Dakota Group. The mudstones exposed in the collar area broke down into clayey material after prolonged exposure. Because of the danger of erosion, the berm around the collar was covered by riprap while a protective concrete structure was placed on the excavation's slopes between the spillway shaft and the nearby access shaft.

The repeated activation of an earth and rock slide upstream from the spillway shaft (Fig. 4) created concern as to the stability of the slope below the spillway shaft. Accordingly, core holes were drilled to test the depth of overburden and to determine the nature of bedrock.

DIVERSION TUNNEL

The diversion tunnel was driven through sedimentary rocks in the Lykins Formation, Entrada Sandstone, and Morrison Formation. Driving of the tunnel required the use of timber lagging and 385, 8-inch wide flange horseshoe steel supports. No unusually bad sections of rock were encountered, and the tunnel appears to be situated in stable, competent rock.

OUTLET WORKS STILLING BASIN

The outlet works stilling basin is located in Entrada Sandstone containing a few faults and strong joints. The foundation is stable, and there should be no settlement. An open cut excavated at the exit portal during construction proved to be unstable, and slide material appeared to threaten the outlet works. Extension of the tunnel and relocation of the stilling basin downstream satisfactorily solved the problems.

EARTH SLIDES

The slope forming the west abutment of the dam was critically examined during preliminary investigations because of the apparent instability of the overburden. Slow movement of earth and rock materials caused maintenance problems along the old location of State Highway 9, and exploratory drill holes in the west abutment area indicated a thick, unstable cover of moist clay and rock overburden. A fortunate consequence of the preliminary investigation was the location of the spillway shaft as far as possible into the hillside and as near as design requirements would permit to the exposures of Dakota quartzite capping the hill above the west abutment.

During construction of the channelway and berms at the intake portal of the diversion tunnel, slip movement in the overburden started and appeared to be the result of loading by a newly constructed haul road embankment, upslope from the intake. The haul road embankment was removed, but subsequent loading and movement of heavy equipment still farther upslope produced repeated additional movements. This danger necessitated removal of a large volume of slide material and construction of a retaining dam to load the toe of the slide and to divert or impound possible additional slides. The slide is in a critical area because of the location of the intake structure for the diversion tunnel. The slide may intermittently continue to be active, especially during rapid drawdown of the reservoir or during periods of heavy precipitation. The responsibility for the study of the slide and recommendations for remedial measures was assigned to W. A. Clevenger of Woodward-Clyde-Sherard and Associates of Denver, Colorado.

GROUTING OPERATIONS

Drilling and grouting of the foundation, tunnel, and spillway shaft were closely correlated with the geology as it was mapped during construction

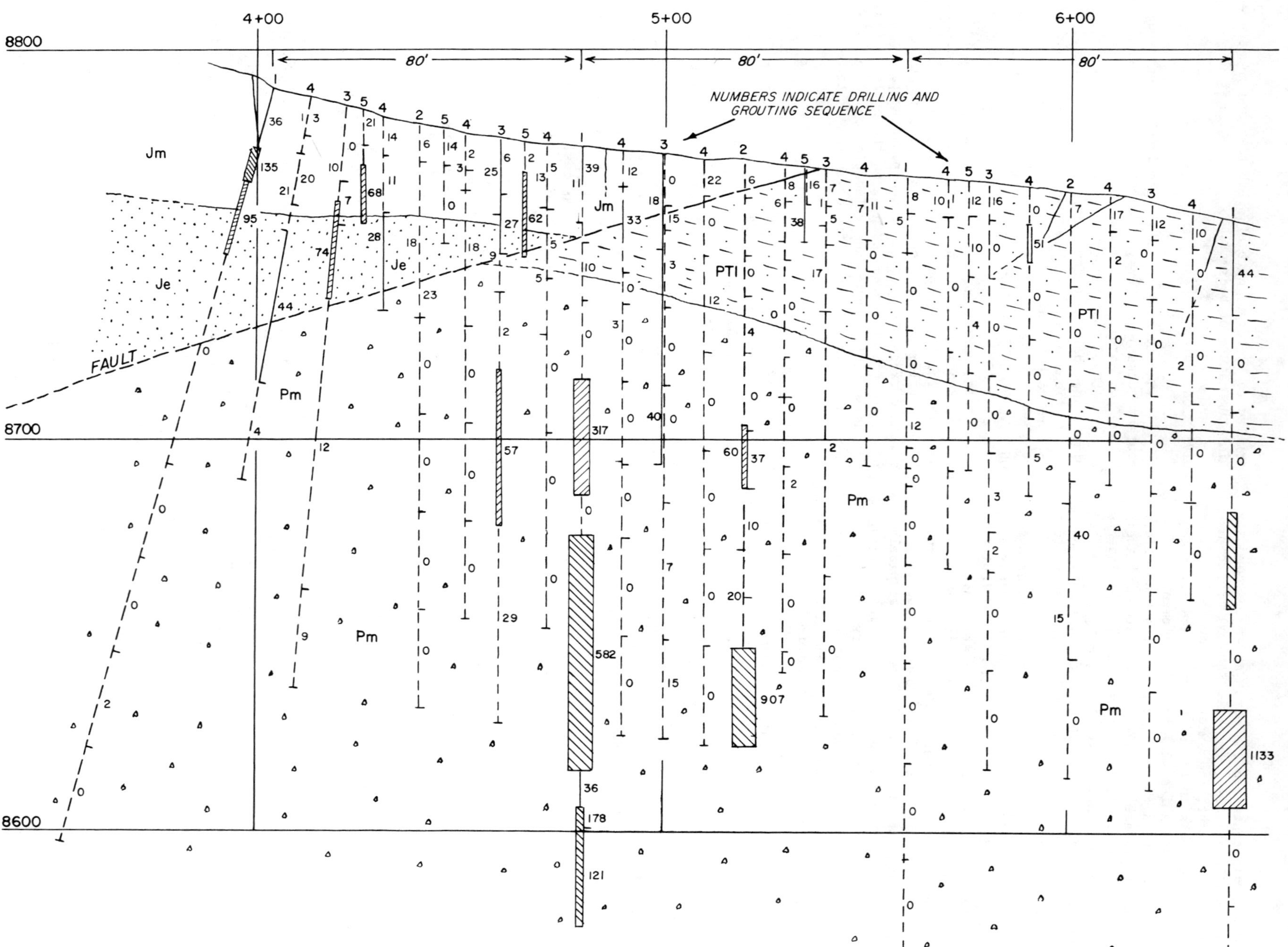

Figure 5. Typical drilling and foundation grouting record. 80-foot closure pattern. (See Fig. 3 for explanation of rock units.)

and with geological data from preliminary investigations. A typical record of the foundation grouting operation is shown in Figure 5.

As was anticipated, fractured brittle rocks, especially sandstone and quartzites in the Dakota Group, the Lykins Formation, the Entrada Sandstone, and the Maroon Formation, accepted large amounts of grout. The amount of grout pumped into the Morrison Formation and the shales of the Dakota Group was somewhat surprising because, initially, it was supposed that these rocks could not maintain open, connecting fractures. Large amounts of grout also were used to seal off faults and fractured rocks associated with the faults.

A total of 630 grout holes, with an aggregate depth of 57,877 feet, was drilled. Two core holes were drilled, one at Station 15+15 to a depth of 145 feet and another at Station 21+25 to a depth of 238 feet, to test the rock at depth after completion of grouting. A total of 175,957 sacks (cubic feet) of cement was placed in the grout holes. Thus the grout take averaged 2.97 sacks per lineal foot of grout hole.

Although extreme precautions were taken to locate the grout holes and drill them to depths indicated by the geological conditions and by the behavior of adjacent grout holes, the complexity of the fracture pattern in the foundation rocks suggested the distinct possibility that all flow channels at depth had not been completely sealed off. There may be some leakage through the foundation rocks as the reservoir is filled. However, it is probable and expected that the load provided by the dam will tend to close many of the fractures not penetrated by the grout.

A favorable aspect of the jointing in the bedrock in the core trench, approximately between Stations 1+00 and 26+00, was the presence of many strong joints subparallel to the grout cap. Field observations indicated that much of the grout traveled at depth along the subparallel joints and probably produced an effective, nearly continuous grout curtain parallel to the dam site.

SUMMARY

(1) Location of the Dillon Dam was mostly determined by geographic and topographic factors which outweighed purely geological considerations.

(2) The geology of the dam site is complex. Sedimentary bedrock is intersected by numerous faults, and brittle rocks, especially sandstones and quartzites, generally contains numerous sets of interconnecting joints. Bedrock is deeply and extensively covered by stream and glacial deposits, terrace gravels, slide rock, and talus.

(3) The spillway shaft, diversion tunnel, and stilling basin appear to be well located in relatively competent rocks.

(4) Earth slides upslope from the diversion tunnels intake structure required considerable excavation and construction of a protective dam above the intake structure. Filling of the reservoir, or drawdown, or periods of excessive rainfall may cause additional movements in the slide rock. Behavior of the slide area should be carefully noted in the future.

(5) Grouting operations in the foundation rocks resulted in the placement of a total of 175,957 sacks of cement in bedrock with an aggregate depth of drill holes of 57,877 feet. The grouting operation was meticulously supervised, and good results should be expected. However, in the event of undesirable leakage through the foundation, detailed geologic maps and sections of the dam site will enable planning of adequate remedial measures.

ENGINEERING GEOLOGY OF SPRUCE RUN DAM AND RESERVOIR, NEW JERSEY*

Cecil B. McGavock, Jr. (Engineering Geologist, Max Meadows, Virginia) and Albert J. Depman (Engineering Geologist, Merchantville, New Jersey)

Abstract

Spruce Run Dam is located near Clinton, Hunterdon County, New Jersey, on Spruce Run, a tributary of South Branch, Raritan River. The dam will retain and supply excess flow for the off-river storage reservoir at Round Valley near Lebanon, New Jersey. The system, constructed by the State of New Jersey, will provide raw water for sale to cities within the State. The dam is earth fill with a select impervious core of weathered glacial drift. Random sections are of weathered shale; riprap is limestone from spillway excavation. The length of the dam, not including the overflow sections and two short dikes is 5400 feet and the height above natural flood plain averages 85 feet. Bedrock is primarily highly faulted, brecciated Kittatinny Limestone bounded on the northeast by older Hardyston Quartzite and on the west by younger Martinsburg Shale. A grout curtain was constructed which spans the limestone belt except for two short stretches where natural water table stands high. The cutoff trench features a concrete grout cap. The upper rock zone beneath much of the fill was blanket-grouted to minimize ground-water movement and sink-hole development. Actual grout quantities were less than estimated. Major construction was completed in 1963.

CONTENTS

---

* Permission to publish granted April 16, 1964 from George R. Shanklin, chief engineer and acting Director, Division of Water Policy and Supply, Department of Conservation and Economic Development, and H. Mat Adams, commissioner, State of New Jersey.

ILLUSTRATIONS

INTRODUCTION

Location

Spruce Run Dam is located on Spruce Run, a short distance below the confluence of Mulhockaway Creek (Fig. 1). The reservoir backs up each stream about two miles. The dam and reservoir lie in and along the foot of the New Jersey Highlands physiographic province. The 40-square-mile watershed is underlain by Precambrian gneiss and granite whereas the dam and much of the reservoir lie in a small isolated area of Paleozoic limestone. The area in Hunterdon County is approximately 45 miles west of Newark, and New York City, and 35 miles north of Trenton. It is less than a mile north of Clinton, New Jersey where Spruce Run flows into South Branch, Raritan River.

Project History and Purposes

The State of New Jersey and affiliated agencies have been active in public water supply under principles established by the 1907 Water Supply Law. A description of the Spruce Run project taken from NJDC (1963) follows.[1]

"The Spruce Run and Round Valley reservoir projects recently constructed by the State are initial steps in the long-range water conservation and development program authorized by the 1958 Water Supply Law and its companion Water Bond Act to protect the availability of New Jersey's abundant natural water resources for use when required by the continued growth of the State. Under this legislation, the Department of Conservation and Economic Development through its Division of Water Policy and Supply as trustee of the water resources of the State is specifically charged with the responsibility for the promotion, construction and operation, on a self-sustaining and self-liquidating basis, of storage reservoir facilities required for the conservation and optimum development of the surface water resources of the State. Through "Stockpiling" of flood waters which otherwise would rush unused to the sea, the two authorized reservoir projects will develop for sale, under contractual terms to repay the cost of construction and operation, the natural surface water resources of the South Branch of the Raritan River for delivery, treatment, and distribution in facilities to be financed, constructed and operated by the purchasers of the water, to meet the present and near future water needs of the northeastern metropolitan area and lower Raritan Valley. In addition, storage is available for the release of water during periods of deficient natural runoff to more or less double the natural minimum daily stream flow available for the South Branch and lower Raritan River for recreation, riparian and other nonconsumptive uses.

The 55 billion gallon Round Valley storage reservoir, to yield 70 million gallons per day, is formed by construction of two dams (125 and 175 feet high) and a dike, closing off gaps in a natural horseshoe shaped rim of diabase (Fig. 1). The valley floor is shale with foundations in the dam areas of gneiss, quartzite and shale. The earth dams and dike, faced with dumped riprap and sod, are of extra width with the clay core "off-center" to permit future raising for possible additional storage of Delaware River water. With no appreciable local drainage area (5 square miles only) Round Valley Reservoir must be filled by pumping from outside sources and no spillway is required for local flood runoff. During construction, local drainage was maintained at both the North and South dams. After construction, the compensatory flows specified by the legislation will be maintained by appropriate release lines through the dams. For the present project to develop 70 m.g.d., water pumped from the South Branch of the Raritan River will enter the reservoir through a tunnel in the west abutment of the South Dam. The supply will be discharged for sale, through two 72-inch conduits and Venturi meters at the North Dam, to pipe lines to be constructed by the purchasers. A limited supply can be released for sale through the South Dam, to be withdrawn from the stream below by the purchasers.

A 350 m.g.d. pumping station at Hamden will maintain the storage via a three-mile force main, located throughout below the hydraulic gradient to assure positive pressure in the event of temporary

[1] State of New Jersey Department of Conservation and Economic Development Division of Water Policy and Supply Information Circular, 1963.

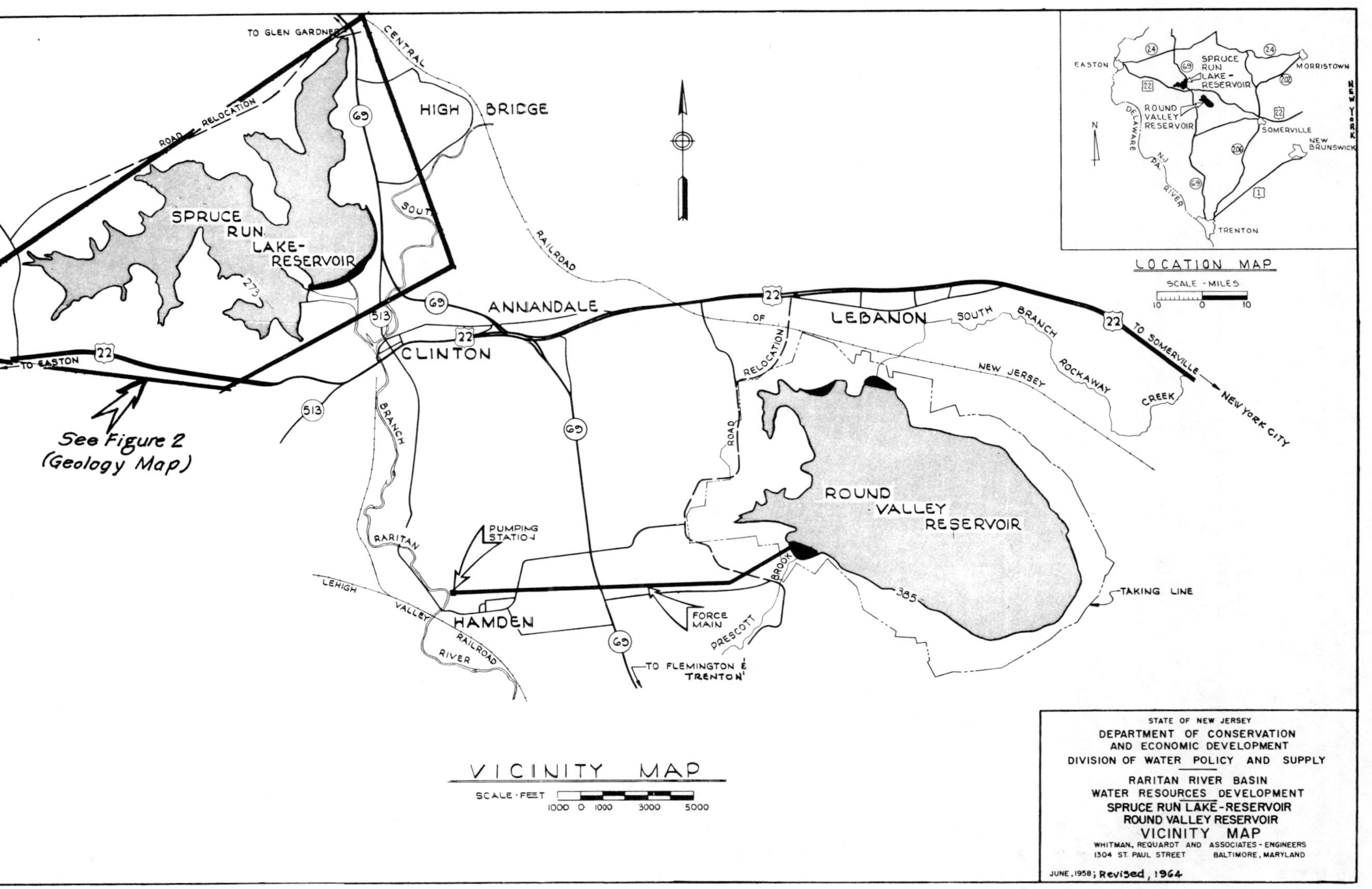

Figure 1. Location map-New Jersey. Spruce Run Reservoir, Round Valley Reservoir.

power failure. There can be no pumping from June 15 to September 15 each year nor at any time when downstream flows are below 40 m.g.d. at Stanton, 70 m.g.d. at Manville, and 90 m.g.d. at Bound Brook.

The 11 billion gallon Spruce Run stream flow regulation reservoir will supply, on behalf of both projects (to save pumping costs at Round Valley), the mandatory stream improvement releases when natural flows are below the required 40, 70 and 90 m.g.d. rates, and in addition, a supply to maintain downstream flow at up to 60 m.g.d. above these statutory minimums, to be sold under contract and withdrawn from the river by the users. Structures include a 6000-foot earth dam (70 feet max. height), two dikes and a spillway with a capacity of 24,000 cubic feet per second, providing 5 feet of flood freeboard (normal 10 feet). Twin 7-foot steel discharge conduits through the dam will also serve for diversion of stream flow during construction. The reservoir and dam are on a limestone formation for which pressure grouting requirements were determined by 220 test borings with an aggregate footage of 25,000 feet. Site experiments developed a successful procedure for sealing voids and fissures by varying the grout density and materials.

Contracts awarded to April 1963 total $19,500,000 covering dams and dikes at both reservoirs, road relocation at Round Valley, force main, administration building, and a relocated bridge at Spruce Run. Agreements have also been executed for a new township road at Spruce Run and relocation of telephone and power lines at both reservoirs. Total cost is estimated at $38,000. Recreational facilities will be provided by the Department's Bureau of Parks and Recreation."

Features of Dam

Spruce Run Dam is in the earth-fill class with an impervious core extending from a concrete grout cap on rock, and composed of local residual limestone soil and weathered glacial soil. The upstream and downstream random sections are constructed of partly weathered local Martinsburg Shale. Riprap is Kittatinny Limestone excavated from the spillway cut. The dam has a maximum height of 91.5 feet and a water depth of 73 feet, with the spillway elevation 273 feet and crest dam elevation 283 feet. The length of the dam is slightly over 6000 feet with two low dikes, 750 and 450 feet long, located in saddles west of the dam. Volume is approximately 1,700,000 cubic yards. The reservoir area at spillway level is 1300 acres with a 17 mile shoreline (Fig. 2).

GEOLOGY OF AREA

General Features and Formations

Spruce Run and its incised valley lie near the center of the Cambro-Ordovician Kittatinny Limestone out crop (see geologic map, Fig. 2). It is bounded on the northeast by the older Cambrian Hardyston Quartzite in stratigraphic, but faulted, relation, and somewhat over a mile southwest of this contact by the younger Ordovician Martinsburg Shale, also in a similar relation. The latter contact is crossed by dike B. The Kittatinny, between the contacts and throughout the cutoff trench and spillway exposure of approximately 6000 feet, has a general northwest strike. Dips are nearly consistently southwest, averaging about 50°, but ranging from 15° to, in rare cases, 70° in some zones. No evidence of folding in the form of synclinal or anticlinal structures, or even reversal of dips, occurs in continuous exposures. Isoclinal folds, such as depicted in the old Raritan USGS Folio in a section across the area, may be present. If so, anticlinal crests have been removed by erosion and synclinal troughs are buried. No recognizable key beds were identified in the Kittatinny. In the cutoff trench and in outcrops, the formation is predominantly medium- to thick-bedded, light- to medium-gray limestone with some dolomitic beds and some thin-bedded carbonaceous zones.

The proximity of the Triassic rocks to the south and the comparatively mild relief in the vicinity of the reservoir strongly suggest that the limestone lowlands were buried by Triassic rocks which were removed by erosion prior to the earliest glaciation (Fig. 2). At present, weathered glacial deposits containing abundant, almost completely weathered granitic and gneissic erratics overlie and intermingle with residual limestone clay in the 260 to 300-foot terrace adjacent to the reservoir and under the dam extensions and dikes. No evidence of Triassic materials has been found above the limestone residuum or beyond the outcrop of the Triassic rocks themselves.

Faulting

Surface evidence of faulting is abundant in the mountains surrounding the limestone valley on the north and northeast in the form of offset alignment of perimeter ridges. Within the limestone terrain, chains of sinkholes, juxtaposition of contrasting topographic features, and straight sections of the stream channel strongly suggest fault zones. Several of these alignments, when projected, traverse the cutoff trench foundation.

In drill cores and the cutoff trench excavation, subsurface evidence of faulting was abundant in the form of brecciation, bedding displacement, slickensides and silicified zones as well as deep solution zones. Many fault zones and displacements are believed to be nearly coincident with bedding planes, and can be identified or suspected only because of concentration of brecciation and solution.

SITE GEOLOGY

Exploration

Diamond drill borings were located on several alternate axes upstream from, as well as on the construction axis. The selected axis is horseshoe-shaped with the dam extensions taking advantage of the high ground along the divides by swinging upstream on both sides. The downstream shifting of the valley portion of the axis was made to avoid, insofar as possible, the crossing of a structurally weak zone in the limestone with known extensive solution cavities and channels in highly brecciated limestone. Location of the axis farther downstream was not feasible because of the lowering of the dividing ridge on the

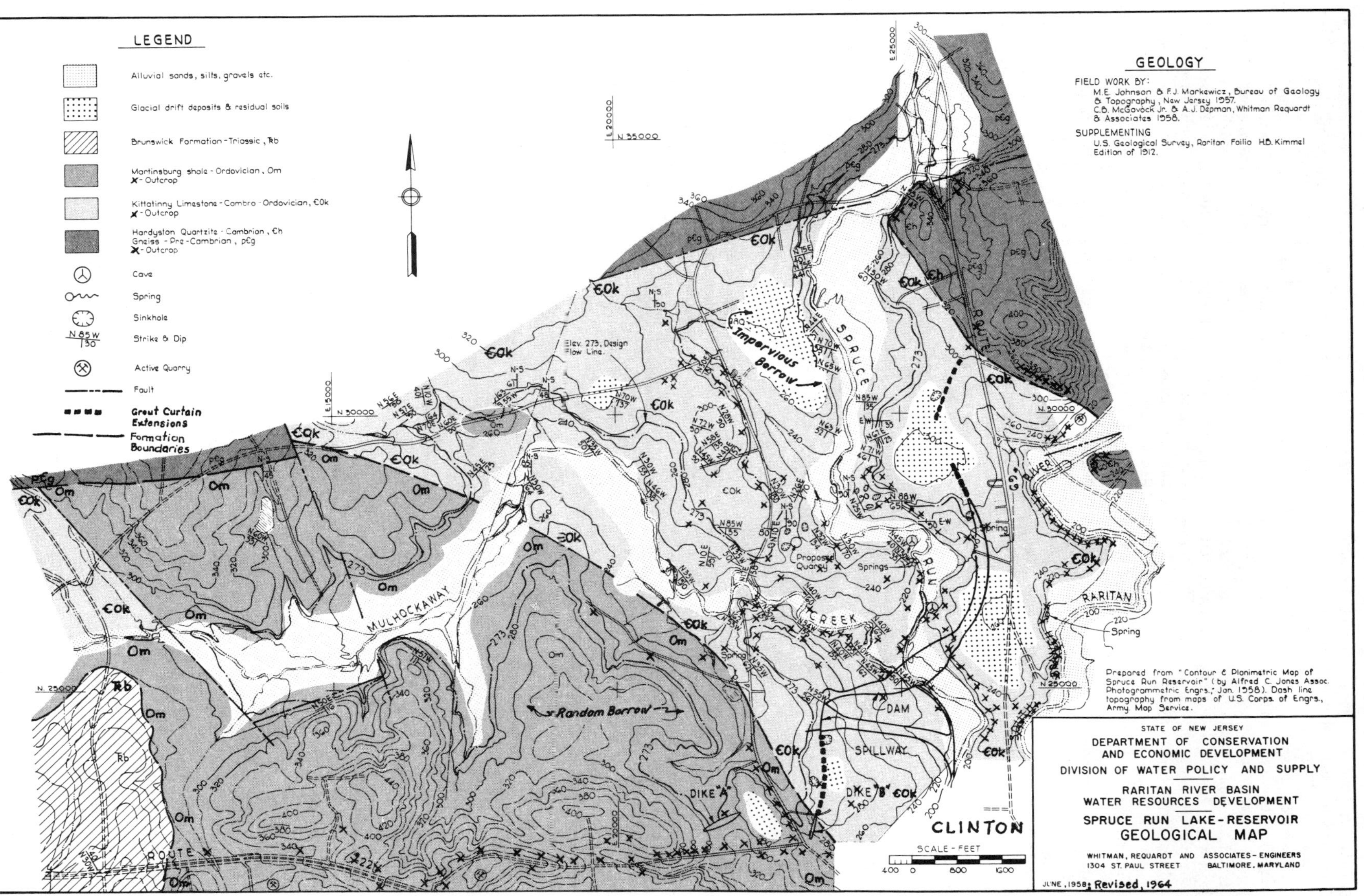

Figure 2. Geological map. Spruce Run Dam, reservoir and environs with dikes, spillway and barrow areas.

east between Spruce Run valley and the south branch of Raritan valley and because of the proximity of Clinton (Fig. 1). Any location upstream of the limestone terrain and on insoluble rocks would have necessitated building two dams with considerably less total drainage and storage areas.

Additional diamond drill core borings were frequently made during the cutoff trench excavation and curtain grouting operations, as a means of checking subsurface details.

Geologic Features Affecting Design and Construction

Essentially all foundation weaknesses were a direct result of the variable, but consistent, solubility of the limestone which underlay all of the dam, part of a dike, and much of the adjacent reservoir rim. This solution of the limestone augmented by the presence of bedding planes and by cross-cutting intersecting joints. Occasional extreme solution action and weathering along the cutoff axis resulted after multiple faulting and massive brecciation. Rarely did these weak zones necessitate excavation to rock lying below the pump-lowered water table, but they did require a triple line of grout holes for the desired cutoff.

Although preconstruction subsurface exploration did not delimit all deeply weathered and dissolved zones in detail, it did indicate the extent of several major zones. This data permitted the design and specifications to adequately cover treatment methods and to anticipate most specialized treatment. In actual construction, the rock surface weathered zone was excavated to a greater depth and extent, and more concrete grout cap was used, than had been anticipated. This not only resulted in use of a smaller total quantity of grout than had been expected, because of less grout leakage and less required small surficial cavity filling, but it also resulted in a more efficient and tighter grout cutoff.

Advantage was taken of acceptable sources of locally available weathered glacial drift and limestone residuum for the impervious core and of partly weathered shale for the random fill (Figs. 2, 3) Riprap materials were locally available from necessary limestone excavations in the spillway and also from nearby gneiss and quartzite deposits.

CUTOFF CONSTRUCTION

Need for Treatment and Trend

Foundation treatment and a good grout curtain are the basic safety features on which a dam is ultimately dependent for integrity and success. Unfortunately, a grout job rarely starts out with more than one qualified man with grouting experience. All too frequently there are none. Grout supervision commonly falls to the engineers, well qualified in other and, to them, more important, phases of construction. Grouting and preparation of rock surface are frequently the most neglected phases of dam construction. This was not the case of Spruce Run.

Personnel

Engineering geology and geology consultants, as well as geologists associated with Spruce Run in other capacities, are too numerous to name. Most prominent of the consulting engineering geologists were E. B. Burwell, Jr., T. W. Fluhr, and Sidney Paige. The senior author was intermittently associated with the project from the earliest stages of exploration in 1957 until final acceptance of the grout curtain in 1963. The junior author was associated with the project during exploration and early construction. Previous papers on Spruce Run Dam have been presented at GSA meetings by Kemble Widmer and F. J. Markewicz of the New Jersey Department of Conservation and Economic Development-Bureau of Geology and Topography. Engineering was by Whitman, Requardt and Associates, Baltimore, Maryland; Justin and Courtney, Philadelphia, Pennsylvania, participated in planning and design.

Inspection of foundation treatment at Spruce Run on behalf of the State was supervised and carried on in detail by civil engineers and other personnel who were initially unfamiliar with grouting and kindred technical work. On-the-job training of engineers and inspectors was conducted in the early stages of drilling and grouting by the senior author. Several of the state engineers developed the rare knacks and "seat of your pants" instinct essential to a good grout job. They became dedicated to the work and to the many small details of which good grout records are composed.

Records

Simultaneous with personnel training, details of record maintenance and methods of keeping the grout chart on a day-to-day development basis were set up. Basic features of the actual grout chart were delineated. Drilling and grout records were added immediately as they became available. The initial outline chart was developed into the final record chart.

Cutoff Requirements

Plans and specifications for Spruce Run Dam called for: a cutoff trench to "rock" for the entire length of dam; relatively short sections of concrete grout cap; a single line of curtain grout holes; and, blanket grout holes under portions of the rolled fill.

Cutoff Trench Variations

Considerably more weathered and broken rock was excavated from the cutoff trench than was anticipated in the design. The grout cap was extended almost the entire length of the core trench rather than placed in a few sections as originally visualized. These two factors tended to offset grout acceptance. Leaks were minimized by the grout cap because much poor, groutable rock was removed by excavation, and this also facilitated placement of cutoff grout. These field modifications resulted in better foundation preparation and treatment.

Cutoff Trench Treatment

The cutoff trench was excavated, in general, down to sound rock with some blasting of pinnacles and considerable hand cleaning with small tools in confined cavities and channels. Little or no

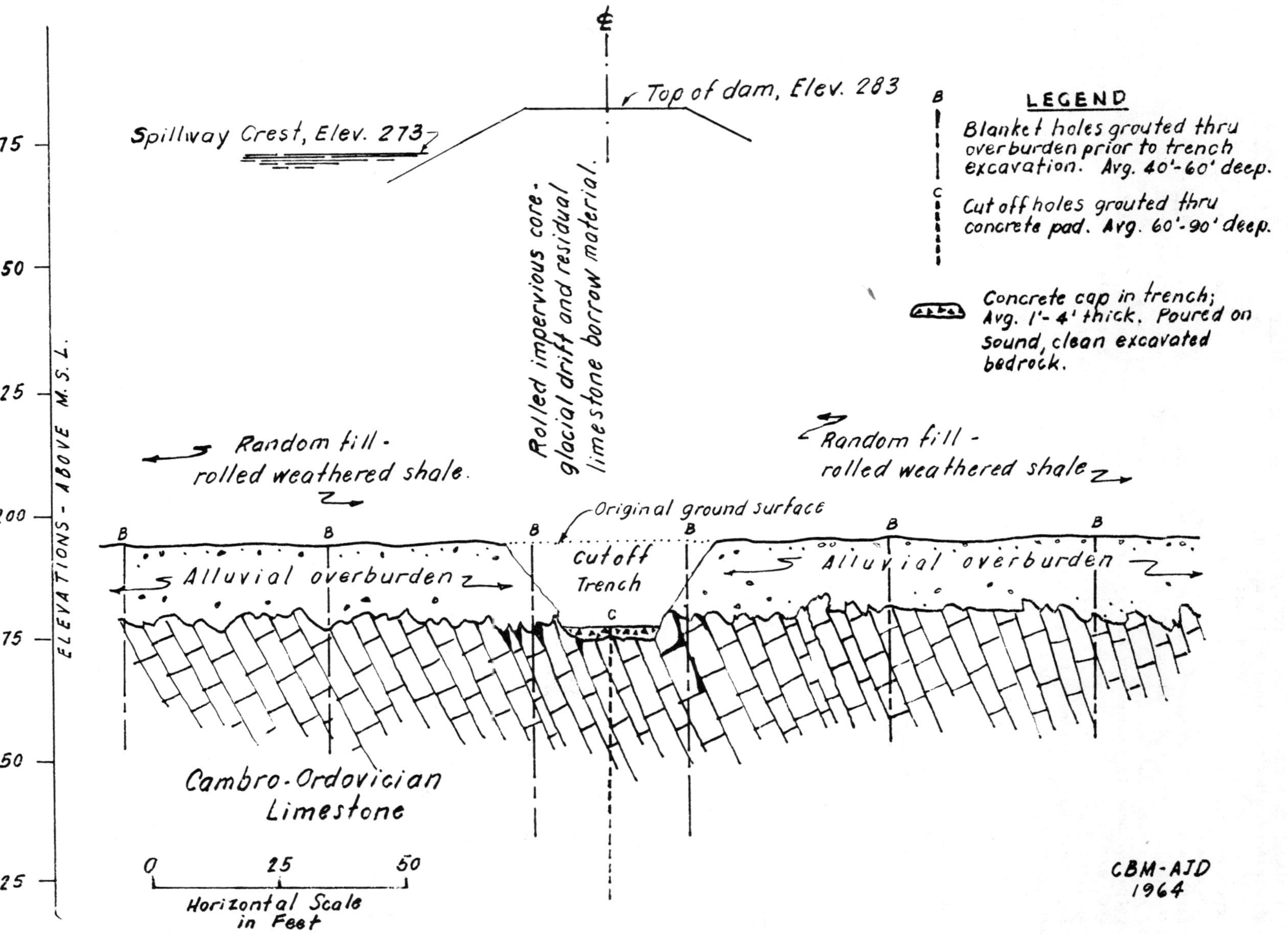

Figure 3. Spruce Run Dam. Typical section through blanket and cutoff grout patterns.

"dental" treatment into caves or under rock projections was necessary. Maximum depth of the trench, about 75 feet, was under the left embankment. Average depth was from 15 feet in the abutments to about 30 feet in the flood plain. The floor of the cutoff trench on which the grout cap was poured generally extended down to sound, although slightly weathered and jointed, rock (Fig. 3). Most of the highly weathered, heavily jointed and blocky rock was removed, except in a few short, excessively deep areas where ground water made clean-up difficult or impossible. These deep areas were treated with triple line grouting and with chemical grout in one short stretch.

Grout Cap

The grout cap was placed for almost the entire length of the cutoff trench, and frequently to its full bottom width, from one to several feet in thickness. It did not resemble a core wall in that its surface was flat to gently undulating. This was to facilitate placement of earth fill upon it. The grout cap greatly facilitated grouting, because grout nipples could easily be set just before placing concrete or drilled into the solid concrete. Surface grout leaks through joints and fractured rock were virtually eliminated. Prior to placing the grout cap concrete, random grout pipes were frequently placed in or aimed to cross known fault or joint zones for later drilling and grouting.

Grouting: What Is It?

Grouting is not an exact science and probably never will be. It is an art based on knowledge, experience, and intuition including: knowledge of rocks, their weaknesses and properties; knowledge of ground water and its movements; knowledge of various grout materials, mixes, pressures, movements, set times, up-lift, mixers, and pumps; and, experience in scheduling activities, in injecting grout, in drilling holes, in hole direction and spacing, in checking results and keeping a workable grout chart, and in locating and patching leaks. To this must be added intuition concerning character of holes, in knowing when to continue and when to stop grouting by sensing when to back check an area, and by knowing when to argue that further grouting is essential to the job, regardless of cost.

Drilling and Grout Quantities

Drilling for grouting consisted of approximately 2810 holes of which 2057 were in the curtain and 753 in blanket areas. Drill footage was approximately 205,100 feet with 153,800 feet in the curtain and 51,300 feet in the blanket. The curtain took 231,350 cubic feet of neat-cement grout and a small amount of chemical grout. The blanket took 341,400 cubic feet of cement-flyash grout. The total grout quantity was 572,750 cubic feet. Drill footage approached the quantity estimated in specifications, although grout quantitites were appreciably lower than was estimated.

Blanket Grouting

Blanket grouting was done first to depths averaging from 60 to 30 feet below top of rock. Spacing was generally on 40-foot centers with 20-foot centers in areas of high takes. Casings were drilled through the overburden and attempts were made to seat and seal them into the top of rock. It is believed the seal was seldom successful and not necessarily desirable. Rock and some overburden was grouted with flyash-cement generally of a 3:1 proportion. The purpose of the blanket grouting was to minimize ground-water flows into the cutoff trench and the subsequent possibilities of sinkholes developing under and in the rolled fill by subsidence or channeling into cavities in the rock. It is believed that this was successful in that flows of ground water into the cutoff trench were nonexistent or minimal. The absence of large inflows was especially evident across the flood plain where flows are normally expected to be large both in weathered limestone and in the base of the valley alluvium. Pumpage from the flood plain cutoff trench was easily handled and was much less than had been anticipated.

Cutoff Grout Curtain Extent

The grout cutoff curtain extends throughout the cutoff trench and spillway, westward across the right abutment divide between the spillway and dike B, and the full length of dike B. Northward from the left end of the dam, the curtain extends to a hill where ground-water levels normally stand above pool level, and thence across a saddle to where high ground water occurs near State Highway 69 (Fig. 2). Except for the two relatively short areas of high ground-water table, the grout curtain extends across the entire Kittatinny Limestone belt from the underlying Hardyston Quartzite of the north into the overlying Martinsburg Shale of the west, a distance of more than 10,000 feet.

Grout Cutoff Methods

For the cutoff grout curtain, in general, nipples were set in the concrete grout cap on 5-foot centers in a single line. In places where it was later found that more closely spaced holes were needed, additional nipples were set as required. The basic drilling-grouting method used in curtain grouting was horizontal split spacing and zones in depth. Primary holes were usually started on 20-foot centers and drilled to depths of 20 to 30 feet. After these were grouted, the space was split by holes drilled to the same depth (10-foot centers and sometimes 5-foot centers), until small water takes under pressure and small or no grout takes indicated a tight section. The method was then repeated in the next lower zone and so on. Upper zone spacing usually ended on 5-foot centers with wider spacing in the lower zones. Exploratory holes were grouted by the packer method. Depths of the curtain in rare instances approached sea level. In a few areas, closure could not be obtained even with minimum spacing of about one foot. In such areas, backup lines of holes within the limits of the grout cap were initiated after much center grouting was done. Consequently, in several areas a triple-line grout curtain was established. Figure 3 shows the generalized cross sectional relations of the blanket grouting to the cutoff grout curtain, cutoff trench, and grout cap.

Triple-Line Cutoff

Effective triple-line grouting is accomplished by determining from exploratory holes and the primary grout holes what areas require intensive treatment. The two outside lines are grouted primarily to serve as plugs or barriers, after which the center line is grouted. In this way, more uniform maximum permissible pressure can be used on the center line holes, and more effective grouting can be done. In general, triple-line grouting is preferable in that it results in a three-dimensional zone, in which there is much less chance of gaps, leakage, or passages than in a single plane of treatment.

Grout Materials

Neat cement was used in the cutoff curtain grout. No flyash was permitted. Chemical grout was used in a short section under the deepest part of the cutoff trench where sand-filled cavities were suspected. Takes were small and closures good.

Cost of neat cement for blanket grouting was considered excessive, and also, the use of neat cement was not thought to be essential. Consequently, experiments were run on use of a soil-cement grout. Local soil as both a wet- and dry-screen was processed and proved acceptable, but treatment and storage required much care. Although use of soil was detailed in the specifications, use of flyash in lieu of soil was agreed on prior to bidding.

Bulk dry flyash was delivered to the grout plants in trailer tanks and pumped to bins for gravity delivery to the scales and batchers. Cement was handled by the same method. Use of flyash in grout proved to be economical and successful in handling, in mix pumpability, in set up time, and in core recovery. Experiments indicated, and field use confirmed, the need for a mix of 3 parts flyash to 1 part cement, with variable amounts of water. Within a few hours, this mixture would set up sufficiently to impede movement. It would also be possible to core it with a diamond drill within a few days.

Drilling and Grouting Equipment

Attempts were made to use percussion drills as provided in specifications. However, removal of cuttings below the water table either by means of air or water, or a combination of the two, was not considered adequate. Increased efficiency in removal of damp percussion cuttings is necessary where percussion drilling is used for grouting. Most of the holes for grouting at Spruce Run were drilled with air-impelled rotary diamond drills.

Large batch plants supplied much of the grout at Spruce Run. Large mixers, larger retention tanks, and frequently, in the early stages, long supply lines between tank and hole were used during the grouting process. Under these conditions, control of pressures and mixes (especially of changes to thicker or thinner mixes) is difficult and slow, while inspection is even more difficult and it also requires more men. Characteristics of the grout acceptance of a hole can change rapidly as final refusal is approached, and mixes need to be changed just as rapidly for good grouting. When oversize equipment was used, travel of some grout beyond the limits of the effective area or the premature refusal of holes necessitated additional drilling.

Preferable Equipment

Small grout units mounted on easily moved trailer beds with mobile material supply units are highly preferable for a more delicate, effective, and economical grout program.

Grout Pump

Upon the grout contractor's suggestion, the moyno-type grout pump was accepted for all grouting, in lieu of piston-type pumps with their undesirable surges. The pump works on the principle of progressive eccentric spirals. It consists of a spiral shaft revolving within a spiral chamber. The shaft is the only moving part; cleaning is by simple flushing. Quantities can be varied from a trickle to several cubic feet per second, viscosities from water to toothpaste range, and pressures from zero to several hundred pounds per square inch. Pressures and volumes can be carefully controlled without harmful surges. This type pump has been specified and used with much success on other grout jobs.

RECOMMENDATIONS

Small Mobile Grout Units

Use of small, highly mobile, complete grout units permits short supply lines with rapid changes in mix viscosity and materials. This permits use of a single inspector with concomitant knowledge, not only of rapidly changing hole characteristics, but also of control of pressures and mixes. A more efficient and, quite probably, a more economical job results in eliminating the excessive travel of the grout. The complete grouting of each hole is also accomplished.

Grout Cap

Use of the concrete grout cap or pad in all but the most "perfect" rock results in a good fill-placement surface and time-saving work surface. It also facilitates placement of grout nipples and minimizes surface leaks, which are time and quantity consuming, because leaks require caulking and necessitate delays for set up or repair of surface loss of grout. Also, somewhat higher pressures are permissible with better rock grouting results. Care must be taken, however, to avoid uplifts of the cap during grouting and damage to the cap during initial fill placement.

The grout cap is particularly effective in minimizing or eliminating the partially grouted, or even ungrouted, shadow zone at the top of the rock working surface in which grout nipples or casings are set. This zone varies from 1 foot to 18 inches in depth in better rock, to 3 feet or more in poor rock. These depths are necessary for anchorage and seal of the nipples. It is physically impossible to subject this zone to desired grout pressures. This shadow zone may well have, and probably has, contributed to the initiation of piping in the top of the rock and the bottom of the impervious earth cores. The result has been partial or total failure of several earth dams. An alternative method of eliminating the shadow zone is to grout from a rock surface several feet higher than the design surface, followed by postgrouting excavation of the shadow zone, with resulting loss of much grout and good rock.

Triple-Line Grouting

Triple-line grouting is accomplished by grouting to refusal, at low to medium pressures, a line of grout holes 5 feet or more from each side of the main cutoff line. Amount of grouting depends upon the width of the cutoff trench and grout cap, which existed prior to any work on the main line. This results in filling of major voids and establishing barriers which restrict the movement of grout away from the main line. More uniform maximum permissible pressures can be built up quickly. The triple-line method constructs a three-dimensional block of grouted rock with little chance of leaving a gap or hole in the grout zone. It also minimizes future difficulties and the possibility of costly remedial treatment through the completed dam. In the case of granular rock- or sand-filled joints, triple-line cement grouting fills both voids and channels, permitting use of chemical grout in a restricted zone where it is most effective.

OPERATION OF RESERVOIR

The construction program for Spruce Run Dam and Reservoir reached a stage where the outlet sluice gates could be closed for initial impoundment of water for release in June 1964 (as originally scheduled). The official impoundment ceremony was held at the dam on September 21, 1963. Water level of the reservoir reached spillway crest in the spring of 1964. There has been no evidence of appreciable water loss through the grout curtain. When the reservoir water was showing a steady rise, the observation wells displayed small irregular rises, considered to be more representative of the general downstream water table, resulting from the winter and spring precipitation or bank storage). A small flow of clear water that discharges from the toe drain of the dam is similarly considered to be primarily due to ground water (Shanklin, personal commun., April 20, 1964).

HYDROGEOLOGY OF THE JOINTED DOLOMITES, GRAND RAPIDS HYDROELECTRIC POWER STATION, MANITOBA, CANADA

R. H. Grice (Department of Geological Sciences, McGill University, Montreal, Canada; formerly engineering geologist, H. G. Acres and Co., Ltd., Niagara Falls, Canada.)

Abstract

The Grand Rapids generating station, completed in 1964, is a 330,000 kilowatt hydroelectric installation located near the mouth of the Saskatchewan River on the west side of Lake Winnipeg, Manitoba, Canada. The operating head is some 130 feet. The 2000-square-mile reservoir is confined at its eastern end by concrete intake and spillway structures and 16 miles of earth dikes with a maximum height of 100 feet. The single-line grout curtain beneath the dike axes is about 18 miles long.

The reservoir area is underlain by Ordovician and Silurian dolomites with a scattered cover of tills and glaciolacustrine silts. There are many sinkholes in the dolomites.

The distribution of joints and joint fillings in the different lithological units were mapped and statistically analyzed. A quantitative ground-water flow technique was devised using the natural vertical water velocity profiles in uncased NX-sized drill holes.

The exploration and research program described was an essential part of the design phase for the grout curtain. Since completion, continuous observations have demonstrated the evolution of the induced ground-water regimen, confirmed the effectiveness of the control measures (grout curtain and pressure relief holes), and provided data for the development and further testing of the analytical procedures for ground-water flow analysis.

CONTENTS

ILLUSTRATIONS

INTRODUCTION

Carbonate rocks may exhibit inherent intercrystalline porosity and secondary, vugular and joint porosity. The primary and vugular porosities are normally of relatively low permeability in the sense of seepage losses from a reservoir. Joint porosity frequently provides appreciable permeability which can increase quickly as ground-water flow is concentrated through a small part of a rock mass.

Thus the characteristics of a ground-water flow system in carbonates are dependent upon the joint pattern and the degree to which some joints can be enlarged by solution. The rate of solution, particularly of dolomites, is slow in terms of the life of an engineering project. However, a new distribution of water may cause the displacement of uncemented joint fillings in an existing system of passages or may promote precipitation.

Studies of the ground-water regimen of the Grand Rapids site (1959-63) proved joint porosity to be the dominant factor in subsurface water movement. Therefore, methods of locating, delineating, and interpreting the characteristics of joints received particular attention.

Location

Grand Rapids, and a settlement of the same name, are situated on the Saskatchewan River near its mouth along the west side of Lake Winnipeg (Fig. 1). The site is some 290 miles northwest of Winnipeg, Manitoba, Canada.

Grand Rapids Generating Station

Since the beginning of the century, there have been proposals for using the Saskatchewan River to generate electric power. The Saskatchewan River flows eastward from the Rocky Mountains and drains many of the interior plains of Canada. The low plateau of the plains is terminated on the east

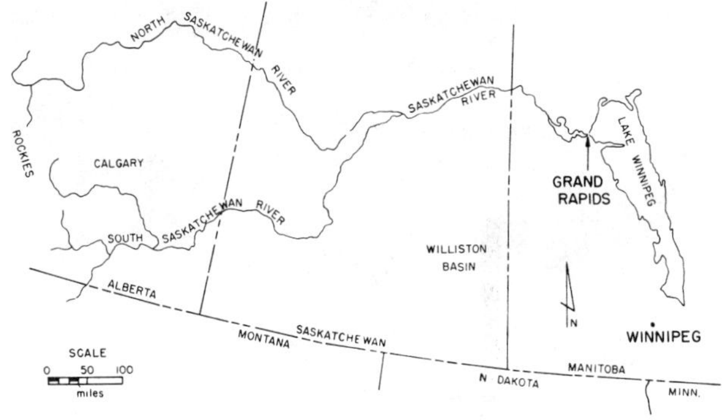

Figure 1. Location of project in the Saskatchewan River basin.

by a raised cliff and beach line parallel to Lake Winnipeg with maximum relief of 180 feet. The river cuts down through the escarpment and drops a total of 100 feet over a series of rapids.

The creation of a reservoir adjacent to and above Lake Winnipeg presented the major problem of appreciable loss from the reservoir into the underlying dolomites which contain many sinkholes and other solution openings. However, after extensive studies in 1959-1960, it was concluded that adequate control of potential seepage was possible and a hydroelectric generating station was feasible at the site.

A reservoir area of 2000 square miles was formed in 1964 by the construction of 16 miles of earth dikes and a concrete spillway structure across the Saskatchewan River (Fig. 2). The original natural fall was increased by more than 20 feet to nearly 130 feet (Fig. 3).

A grout curtain, injected beneath the axes of the dikes, the spillway, and the intake structure for the power house, has a length of 18 miles, and an average depth of 150 feet. It is generally formed of a single line of grout holes and improves the stability of structures as well as controlling seepage losses. The grouting operation has been described by Rettie and others (1962).

The reservoir supplies a flow of 36,600 cubic feet per second, producing 450,000 horsepower.

GEOLOGIC SETTING

The Grand Rapids area is underlain by a maximum thickness of more than 600 feet of Ordovician and Silurian dolomites (Fig. 3) which dip gently to the west. The rocks are jointed and contain many solution features including sinkholes which are typically 20 feet wide below and 5 feet deep with a plugged solution channel or opening beneath.

The dolomites rest on a thin Ordovician sandy siltstone which in turn overlies the Precambrian granodiorite of the western flank of the Canadian Shield. The dolomites are covered locally by a veneer of Pleistocene and recent silty tills, glaciolacustrine silts, and beach gravels.

During and since Paleozoic time, the area was tilted several degrees as the nearby deep Williston basin developed. It is possible that the present jointing and solution features are of Tertiary and Quaternary age, and, consequently, would have been affected greatly by several isostatic adjustments during glacial and postglacial times and the complex of associated proglacial lake and drainage systems (Grice, 1967).

Stratigraphy

The upper 120 feet of the stratigraphic column was exposed in the excavation for the intake structure and power house (Figs. 2, 3 and 4). Each unit was defined lithologically and is described in Table 1. Each unit consists of several beds 0.3 to 4.2 feet thick which can be distinguished over a wide area.

Lithology in the Project Area

Four types of dolomite were exposed in the intake structure of the power house excavation: nodular, pseudobrecciated, bedded, and argillaceous. The first three types range from 6 to 26 feet in thickness, while the argillaceous dolomite do not exceed three feet; their low permeability has an important influence on the ground-water regimen. The only nondolomitic rocks in the area occur at depths below 450 feet.

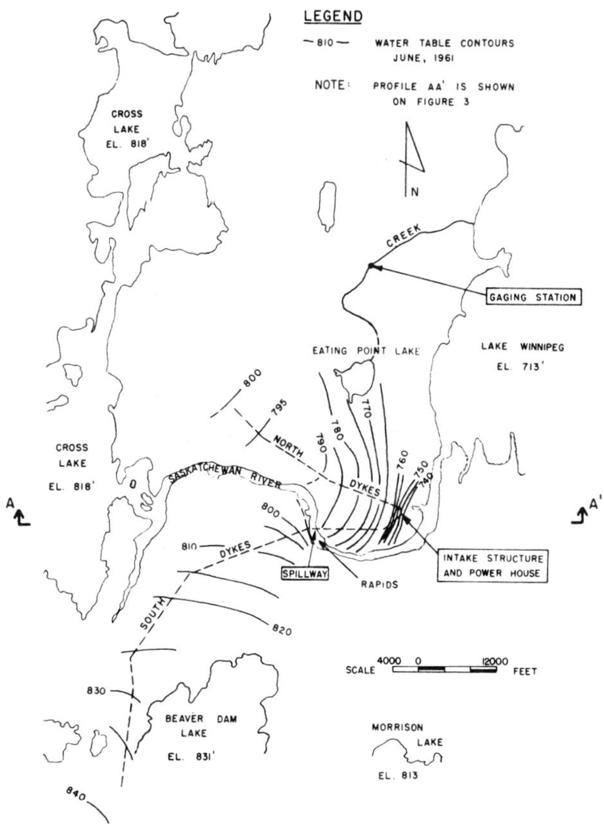

Figure 2. General arrangement of project.

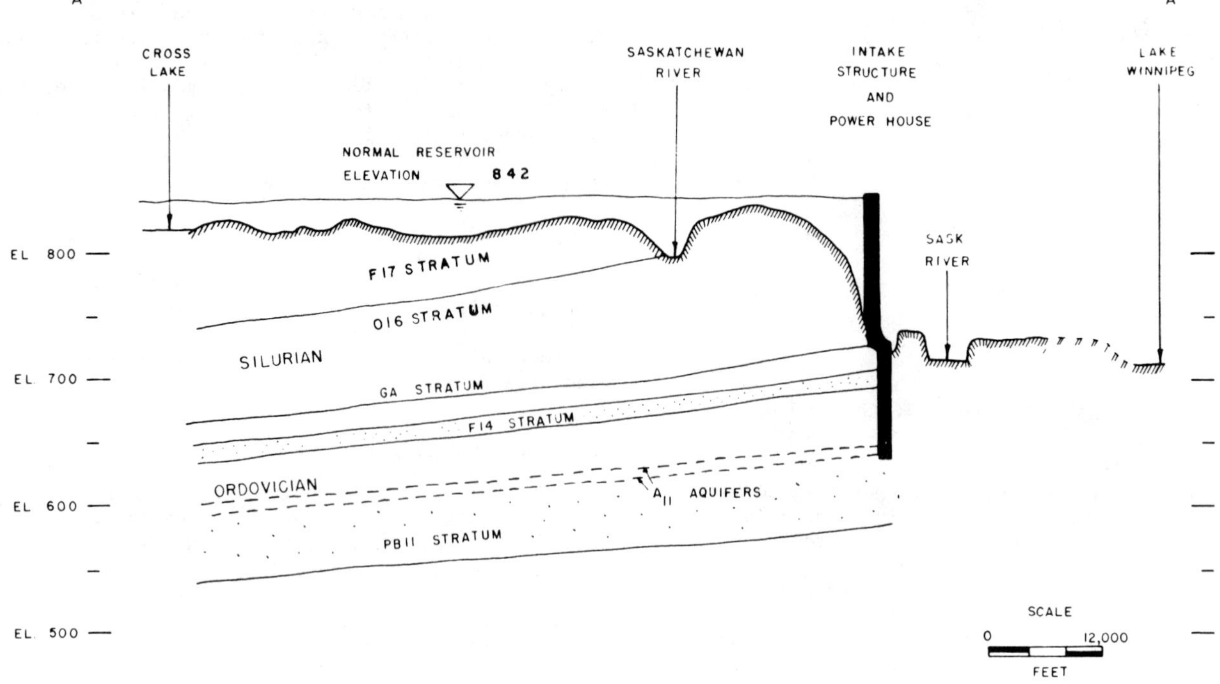

Figure 3. General profile of project.

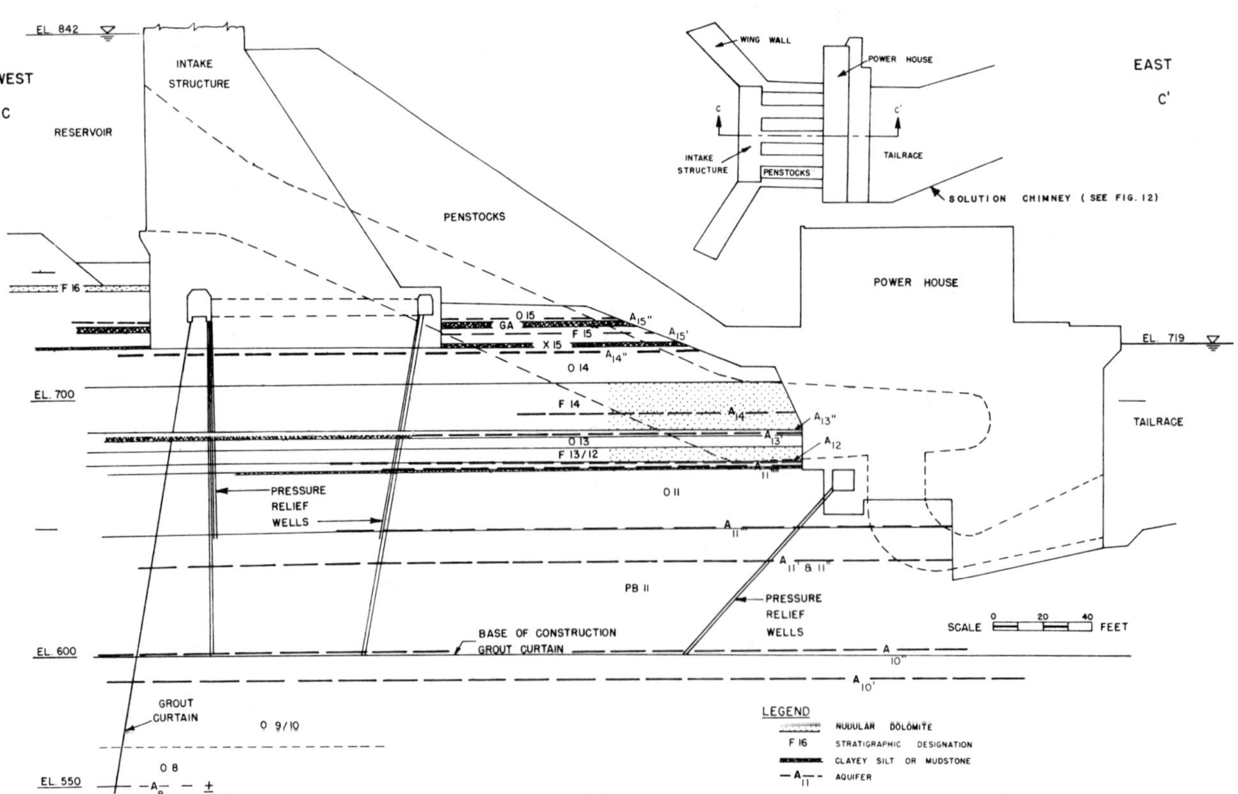

Figure 4. Detailed profile of intake and power house showing hydrogeologic units and installations.

Table 1. Stratigraphic column in intake structure - power house area

Local Units*	Thick-ness (ft)	Bedding-Joint Spacing (ft)	Joint Spacing (ft)	Rock Hard-ness (RH)†	Lithologic Description
F-16	2+	-	3 - 15	4 - 5	DOLOMITE--nodular, pale orange to brown, irregular minor solution, *Virgiana decussata* zone.
O-15	13	0.3 - 3.0	2 - 4	4 - 5	DOLOMITE--bedded, under 1 inch thick, fine grained, negligible solution, unit contains one unconsolidated silt seam less than 1 inch thick.
GA	2	-	Rare	1 - 3	DOLOMITE--argillaceous, arenaceous and conglomeratic locally, green gray or blue gray.
F-15	6	-	2 - 5	3 - 4	DOLOMITE--nodular, gray, interbedded very hard (RH-5) dolomite strata about 2 inches thick, silt laminae less than 1 inch thick in some bedding joints.
X-15	1	-	Rare	1 - 3	DOLOMITE--argillaceous, gray includes soft clayey silt lenses.
O-14	12	0.6 - 2.1	1 - 6	4 - 5	DOLOMITE--bedded, fine grained, very pale orange, generally porous (a dry specimen absorbs water readily).
F-14	18	-	Rare	1 - 5	DOLOMITE--nodular, gray or orange mottled, internodular matrix friable or removed locally, contains irregular stringers less than ¼ inch thick of dark-gray silt.
O-13	6	0.4 - 1.7	0.5 - 3	4 - 5	DOLOMITE--bedded, fine grained, very pale orange, locally moderately porous, unit includes an argillaceous stratum 1.5 feet thick, RH 2-4, joint spacing 5 to 10 feet, open bedding joints above and below this stratum.
F-13/12	10	-	Rare	2 - 5	DOLOMITE--nodular, orange to brown mottled, minor solution of matrix.
O-11	26	0.5 - 4.2	3 - 10	4	DOLOMITE--bedded, fine grained, yellow-gray, highly porous and strong solution locally, argillaceous stratum, 0.5 feet thick, RH1.
PB-11	21	1.4 - 3.0	5 - 15	5	DOLOMITE--pseudobrecciated, blue-gray and olive-gray mottled, negligible solution.

* The alphabetical prefix of the unit indices indicates the principal macroscopic characteristic, textural or color.

† See Appendix A for description.

Nodular Dolomites. The Nodular Dolomite is a mottled rock of aphanitic nodules or sponge-like complexes in a matrix of silty dolomite. Texturally, the dolomite of both the nodules and the matrix is composed of silt-size grains. The colors are tones of either orange or gray. The nodules and complexes are very hard (RH-5 on scale Appendix A). The matrix ranges in hardness from hard (RH-4) to friable (RH-1). Where exposed, and in some zones below the surface, the matrix has been removed by solution leaving a more or less stable mass of honeycombed dolomite (Fig. 5). Joints and bedding-plane joints are rare or inconspicuous except where accentuated adjacent to open cuts (Fig. 6). Nevertheless, bedding-plane aquifers have developed locally and exceptional solution activity has occurred in and around favositid colonies, which are common in some zones.

Pseudobrecciated Dolomite. The pseudobrecciated dolomite is a blue-gray and olive-gray mottled aphanitic to fine-grained rock containing contorted stringers and laminae of brown shaly dolomite. It has a hardness of RH-5 and is stronger than the nodular dolomite. The tight bedding-plane joints are spaced from 1 to 3 feet apart, and the crosscutting joints are spaced from 3 to 10 feet. Joints which have been opened by solution are over 10 feet apart.

Bedded Dolomite. The bedded dolomite ranges in texture from a very fine-grained, medium-hard (RH-3) chalky dolomite to aphanitic very hard (RH-5) rock. It is commonly light orange in color and jointed at spacings between a few inches and a few feet. Stromatolites are common and form domed, laminated structures. Many of the contacts between the bedded dolomite and the nodular and pseudobrecciated dolomites are gradational.

Some zones of the aphanitic dolomite have been softened adjacent to joints closely spaced at an inch or two. Solution cavities have formed at the junctions of joints and in the cores and between stromatolitic domes (Fig. 7).

Argillaceous Dolomite. The argillaceous dolomite is blue-gray and green-gray uncemented mud-

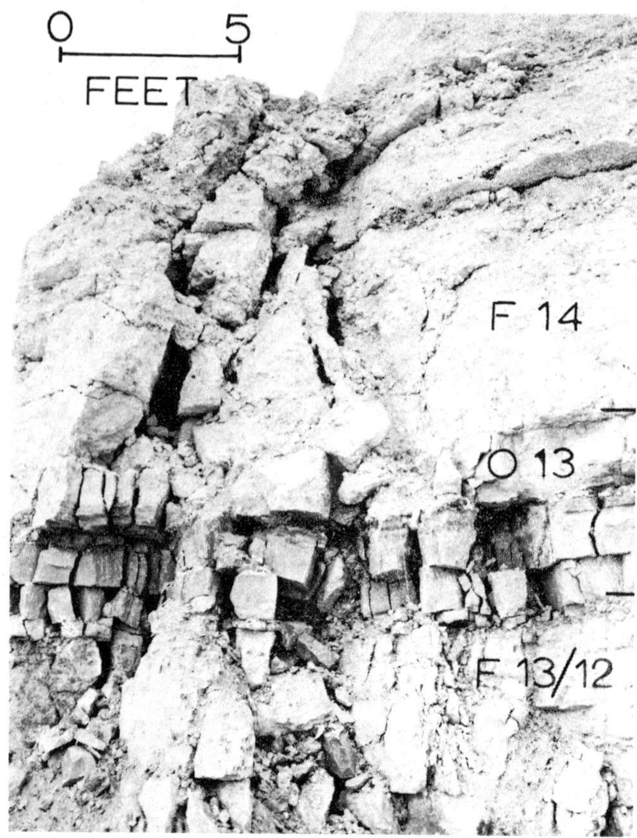

Figure 6. Nodular and bedded dolomites, opened joints in face of penstock excavation. "Slumping and excessive fracturing due to joints and rebound along fresh cut, steep slope."

Figure 5. F-14-- nodular dolomite, rubbly texture.

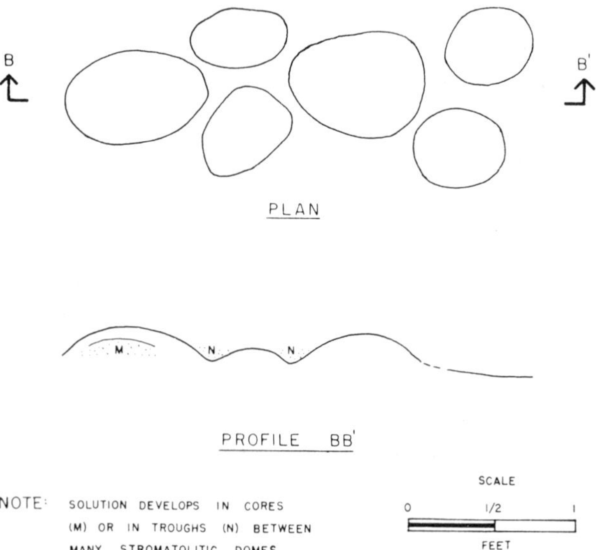

Figure 7. Stromatolitic bedding surface.

stone with distinct lower contacts and gradational upper limits. Although predominantly medium hard (RH-3) with sandy zones, this facies is interbedded with discontinuous, soft plastic seams of clayey silt. Illite is the principal clay mineral. Joints are spaced from 15 to 80 feet apart and are tight.

STRUCTURAL FEATURES

The regional and local structure has been described by Baillie (1951, 1952) and Stearn (1956). The Paleozoic rocks which crop out along the western side of Lake Winnipeg dip west to southwest at a low angle toward the Williston basin (Fig. 1). The average dip of these rocks is 10 feet to a mile, while locally at Grand Rapids it is 5 feet to a mile (Fig. 3). Only minor folding which occurs as a local flattening of dip has been observed. The "amplitudes" are less than 5 feet.

One possible minor thrust fault was observed in the side of an excavation within 10 feet of the bedrock surface. It could have been produced by natural pressure relief or blasting. Displacements of vertical joints not exceeding 6 inches were seen in the power house excavation.

Secondary Structural Features

Joints provide paths for the movement of fluids and for the promotion of subsurface alteration in rock masses. They are of major importance in the control of ground water in carbonate rocks.

Origin of Joints. Many joints appear to have been produced by the horizontal tensions related to pressure relief when vertical stresses are reduced by erosion. The unloading effect is accentuated in glaciated areas, such as Grand Rapids, by intensive erosion and by oscillatory loading of a succession of continental ice sheets.

Very few joint surfaces are mathematical planes for appreciable areas. Some joints were observed to be almost planar within lithologic units, but their continuations were diffracted across the boundaries. In other examples, the changes in orientation of the joint surfaces are gradual, caused by a combination of the effects of variations in facies and stress fields. Some joints are of limited extent individually, but occur en echelon and form distinct zones of discontinuities.

Origin of Bedding-Plane Joints. Many depositional bedding planes are welded by consolidation and cementation, but they remain discontinuities with respect to particle shape and size, mineral content, cementing agents, density, porosity, permeability, strength, and deformation characteristics.

Most bedding surfaces are undulatory in some degree. The smallest undulations are suggestive of ripple marks with a wave length of a few inches and amplitudes of small fractions of an inch (Pettijohn, 1957, p. 185). At Grand Rapids, there are also larger undulations with wave lengths on the order of 6 inches to 10 feet which appear to correspond to the upper surfaces of colonies of stromatolites (Fig. 7).

Bedding-plane joints are produced by the reduction of vertical stress (rebound) and are closer together near the surface. At Grand Rapids, the most intensive development has occurred above the PB-11 stratum (Fig. 4).

Characteristics of Joints. The importance of joints in the movement of ground water is well known since the first discussions by Meinzer (1923). The important characteristics of joints include: (1) spacing or frequency; (2) extent or area of continuity; (3) orientation; (4) geometry of the surfaces; and (5) distribution of sets and the size of the individual rock block bounded by joint surfaces.

The associated characteristics are: (1) nature of joint fillings; (2) potential for fluid distribution in openings; (3) chemical composition of joint water; and (4) temperature distribution in rock adjacent to the joints and in the mineral or fluid fillings.

Terzaghi (1963), like many others, has described the practical difficulties in evaluating these characteristics; unless limits can be ascertained, realistic ground-water movement surveys cannot be made.

Development of Joint Openings by Construction Operations. Construction operations affect the characteristics of joints in one of two ways: (1) excavation, blasting (Fig. 6), and grouting may open the joints; and (2) placing of concrete and other mass materials, or grouting, closes or plugs the joints. Either opening or closing can have serious effects on adjacent areas, because of changes caused in the ground-water regimen. The potential effects of grouting beyond the site limits requires, therefore, that constant attention be given to the reactions caused by grout.

JOINT ANALYSIS

Field Observation

The Joint system at Grand Rapids was observed in: (1) outcrops; (2) drill cores; (3) excavation exposures; and (4) drill holes by measuring the natural vertical water velocity.

(1) Outcrops afforded an accurate measurement of the orientation, spacing or frequency, persistence, and nature of the joint surfaces. However, the near-surface rock was weathered and an assessment of a joint system near the surface usually differs from conditions at depth.

(2) NX-sized drill cores provided an inexpensive method for locating the subsurface discontinuities whenever core recovery was high. However, drilling operations produced many mechanical breaks and grinding of fracture surfaces because the core holes were normal to the flat bedding.

(3) Some of the excavations penetrated below the near-surface zone of stress relief and weathering and thus enabled "unaltered" rock to be examined in three dimensions. Unfortunately, the excavations were not available until after the major design decisions and construction had begun.

The three-dimensional examination provided a full appreciation of the extent, persistence, spacing, and orientation of different joint sets. Usually natural and artificial fractures could be distinguished in new excavations, although severe mechanical disturbance could occur during excavation (Fig. 6).

Unfortunately, significant variations in the joint pattern and associated alteration of the rock could exist just beyond the excavation face. This occurred beneath the intake structure where a zone of friable dolomite was encountered within a normally hard, fine-grained dolomite. The normal medium-spaced joint pattern occurred everywhere in the hard dolomite unit to within 2 feet of the soft friable zone.

(4) Natural vertical water flows occur in many drill holes that intersect a series of aquifers. If so, each zone usually has its own potential for fluid distribution which establishes a vertical potential gradient between one aquifer zone and the others cut by the hole. At Grand Rapids, an aquifer was defined as either a single joint, a set of closely spaced joints, or open bedding joints. The profiles of the natural vertical flows in adjacent NX-sized drill holes (Fig. 8) diagrammatically illustrate an aquifer. Observations were repeated periodically to determine changes in potential distribution. A 1 3/8 inch diameter electric flowmeter with a vane was used for measurements in uncased drill holes. Such holes, if drilled without caving, have remained in satisfactory condition four years (to this date).

Correlation of the elevations of the changes in velocity (Fig. 8) indicated the widespread extent of a number of thin aquifers or open bedding joints. The positions of the aquifers correlate with lithologic and stratigraphic horizons. The velocity changes and the shape of respective profiles demonstrated the dimensions of the water-bearing fissures, and their relative permeability and fluid potential distribution. A sharp change in vertical velocity in a drill hole was indicative of an isolated open fracture. More gradual changes in velocity suggested either: (1) a system of closely spaced, more or less horizontal fissures, that possibly intersect one or more steeply inclined joints; or (2) a zone of high primary porosity and permeability. Experience showed that certain widespread joints could not be detected in all intersecting holes by natural velocity observations, owing to a lack of local fluid potential differences. When nonroutine tests were made by measuring velocities at the same time as water was pumped into drill holes, the presence of such open joints was proved.

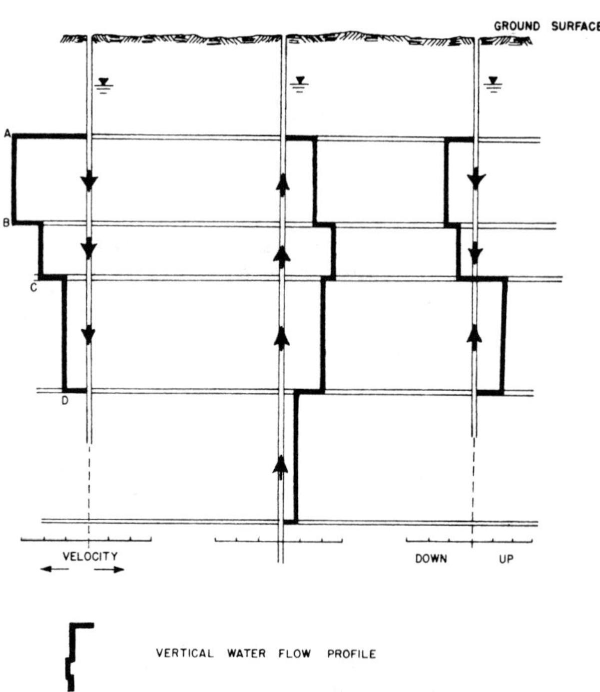

Figure 8. Natural vertical water velocity profiles in adjacent drill holes.

Variations in the velocity profiles have been consistent with seasonal hydrologic changes and with constructional events such as grouting. In the latter case, a new fluid potential distribution was developed in the aquifers intersected by a grout curtain and those below it. As a result, water moved downward through the aquitards on the upstream side of the curtain and passed beneath it along ungrouted aquifers. (Aquitard denotes a rock unit of low, but significant, permeability relative to the nearby aquifers).

An example of the changes of natural water velocities in a drill hole located inside the reservoir area before impounding and adjacent to the grout curtain is shown on Figure 9. The exceptionally large desçending velocities (profile B) were observed at the time when the grout curtain did not intercept aquifer A-10. This aquifer drained beneath the curtain and upward through joints into the deep power house excavation (Fig. 4). As the discharge path was short and unobstructed, the fluid potential at the intersection of the drill hole and aquifer was relatively low, and the potential gradient between the upper aquifers, A-13, A-12, and A-11, and the lower aquifer, A-10, was relatively high. When the grout curtain was deepened to intersect aquifer A-10, the discharge was obstructed, the potential at the lower intersection increased, the potential gradient between upper and lower aquifers was reduced, and the natural vertical water velocity in the drill hole became smaller.

The most detailed structural and lithologic mapping was carried out in the 120-foot-deep intake structure and power house excavation (Fig. 4) as follows.

(1) Horizontal bedding joints were mapped, the discontinuities described, and the fillings sampled. All the horizontal discontinuities which extended around the 1000-foot periphery of the excavation were used to define the subunits of the stratigraphic column (Tables 1 and 2). Only a small number of the subunit limits are lithologic contacts without mechanical discontinuity.

(2) The vertical and inclined joints were mapped with particular attention to their vertical limits and horizontal displacements along bedding joints (Fig. 10). Visual estimates of the orientations of the joints were made within 5 degrees by using the direction of the excavated faces as reference planes.

Table 2. Apparent average joint spacing in intake structure excavation in 0-15 unit

Sub-unit	Vertical Thickness (ft)	Joint Spacing West Wall (ft)	South Wall (ft)	West & South Wall (ft)
F-16	A 2	12.2 (23)	-	-
	H 1.6	10.3 (27)	-	-
	G 1.7	8.3 (34)	-	-
	F 2.2	4.2 (66)	5.0 (28)	4.4 (94)
	E 3.0	4.0 (70)	4.5 (31)	4.2 (101)
0-15	D 0.4±	7.8 (36)	23.2 (6)	10.0 (42)
	C 0.9±	8.0 (35)	17.5 (8)	9.8 (43)
	B 1.0±	9.6 (29)	17.5 (8)	11.3 (37)
	A 2.1	10.3 (27)	14.0 (10)	11.3 (37)
	C 0.1	110.0 (3)	17.5 (8)	43.0 (11)
GA	B 0.7	110.0 (3)	143.0 (1)	118.0 (4)
	A 0.7	110.0 (3)	34.4 (4)	65.8 (7)
	E 1.3	-	17.5 (8)	-
F-15				
	D 1.8	-	23.2 (6)	-
Total	19.5			

Notes: Total number of joints observed in each subunit is shown in brackets. The length of the west wall was 240 feet and of the south wall 140 feet. For location, see Figure 4.

Although it is common practice to record the strike and dip of joints as N. 30 E. and 35 SE., the writer finds it more convenient to note the azimuth of the dip direction and the dip angle. Such a reading as 120/35 can be plotted directly on a joint frequency diagram (Fig. 11).

Results

Examples of the distribution of joints on two mutually perpendicular faces of an excavation are shown on Figure 10. There is an appreciable difference in the patterns. This is to be expected wherever any one set of vertical or inclined joints strikes at a small angle to the orientation of one face normal to a mutually perpendicular pair. It is preferable to map all four sides of an excavation and the floor as the excavation is deepened. In this way, with additional data from drill holes, optimum extrapolations of joint patterns can be made in the rock mass around an excavation.

A tabulation in Table 2 of the average apparent joint spacings in the strata exposed in the intake structure excavation emphasizes the differences of spacings in the subunits which together form a number of more or less homogeneous zones. The true average spacings of the joints in each of the sets in the 0-15-E and F subunits are listed in Table 3. These values were obtained by plotting joints on a plan of the excavation, and correlations

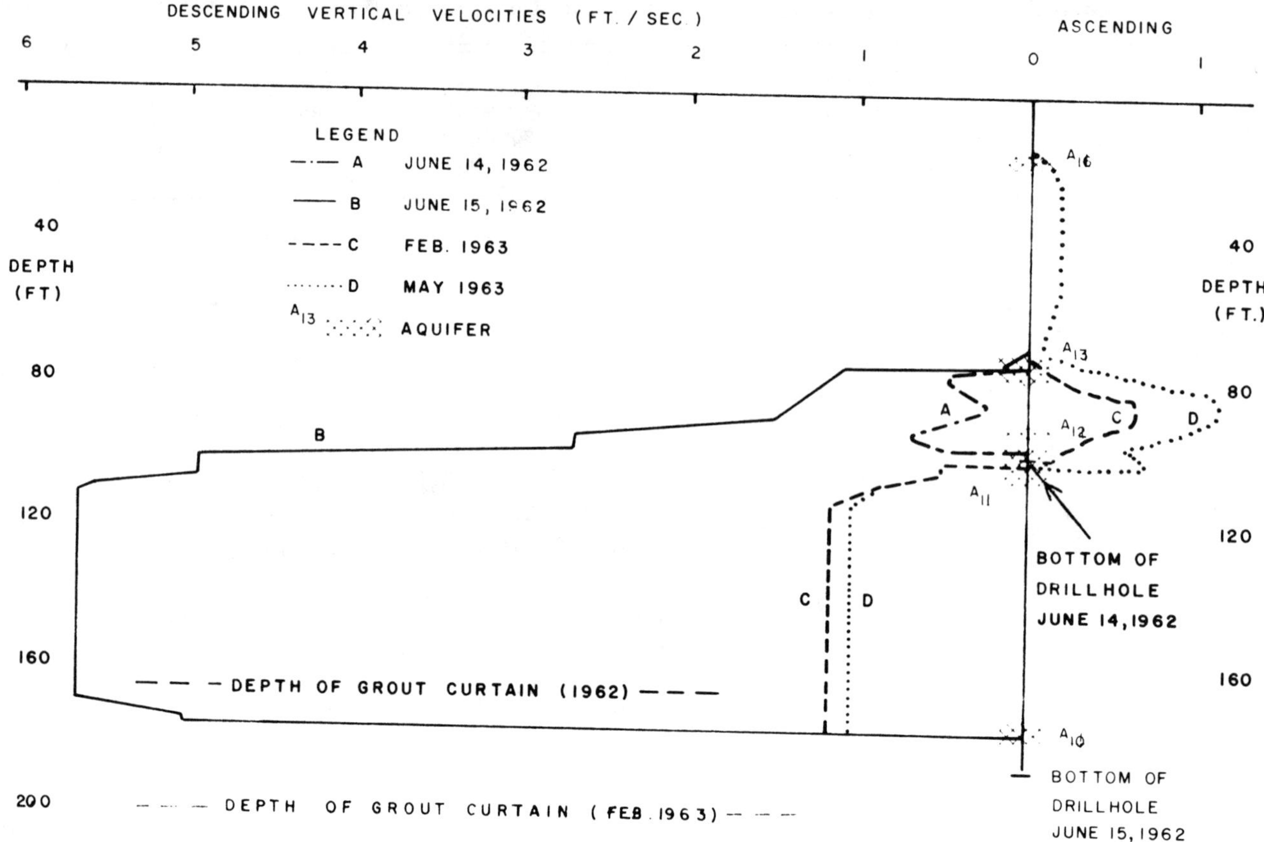

Figure 9. Changes in the vertical water velocities in a drill hole due to construction activities.

were made from data which had been observed on the west and south walls of the excavation. The average joint block plan area was about 37 square feet.

The orientations and inclinations of the dips of the joints were plotted on a contoured frequency diagram (Fig. 11) or equal area polar plot in which the poles have been projected from the upper hemisphere. This method differs from existing fabric and mineralogical analysis practice; however, the method is open to less confusion and error as it follows the topographic convention of direction of north to the top and east to the right of a plot.

GROUND WATER

Solution of Carbonates along Joints

Solution of carbonates has been discussed by Davis (1930), Bretz (1942, 1956), Hamilton (1948), and Thornbury (1954). In addition to the requirements of climate and a soluble, well-jointed rock, an active ground-water system is necessary. The general recharge-discharge ground-water pattern of Hubbert (1940) is as effective for vigorous solution as the special case of a major entrenched valley in an upland area. The relief in the vicinity of Grand Rapids is gentle, but there are many springs in the marshy strip adjacent to Lake Winnipeg confirming a zone of discharge along the toe of the escarpment. Other discharge zones probably existed in the past, particularly during the various proglacial phases of the Pleistocene epoch (Grice, 1967).

Many sinkholes and other features, including solution chimneys, were observed to have developed at the intersections of the major vertical joint

Table 3. Joint spacing in intake structure excavation in O-15-E and -F strata

Strike of Joint Set	No. of Joints in 120-foot-square area	True Average Joint Spacing (ft)*
NE	45	170/45 = 3.8
NW	10	170/10 = 17
N	6	120/6 = 20
E	1	120/1 = 120

Notes: For location of intake structure see Figure 4.

For stratigraphic succession see Tables 1 and 2.

* The length of the 45° diagonal across the square test area is 170 feet.

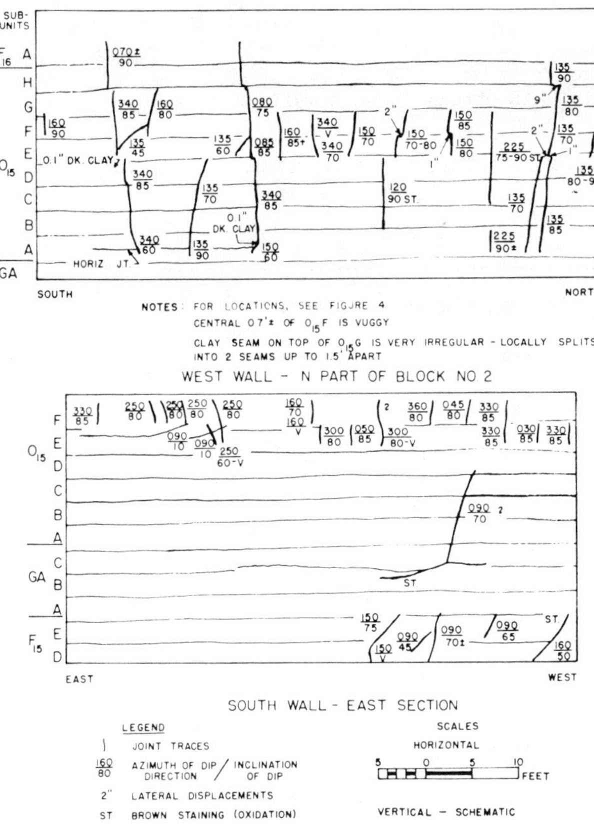

Figure 10. Schematic distribution of joints in the intake structure excavation.

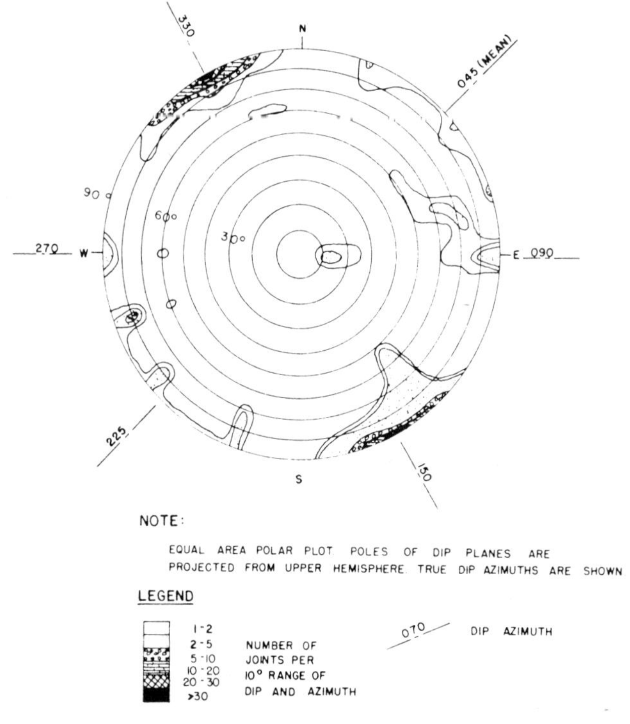

Figure 11. Joint frequency diagram for the intake structure area.

sets. The major axes of elongated sinkholes are orientated along the principal joint strikes. Two excellent examples of discharge solution chimneys were exposed in the north and south walls of the tailrace excavation, which was located in an active spring area. Each chimney was about 30 feet high and 3 feet wide and traversed several lithologic units (Fig. 12).

In general, both vertical and bedding joints are more closely spaced in the harder and more brittle strata. Thus, where the joint blocks are very small, the permeability is relatively greater so that the harder, less soluble dolomites are dissolved at a more rapid rate than the more soluble but less jointed units.

There are two major questions concerning the solution cavities along joint planes which have developed principally in the upper 100 feet, and, to a lesser extent, in the upper 200 feet of the bedrock, owing to rebound and stress relief. First, why has solution activity been concentrated along a number of preferred planes? Second, are there any definable types of plugged openings which may open when subject to new hydrologic conditions? To answer these questions, detailed examinations were made of the lithologic divisions, of the geometry of the discontinuities, and of the joint fillings.

The different characteristics of the bedding-plane discontinuities of a 13-foot-thick lithologic unit, 0-15, are shown in Table 4. The outcrop was located in the upper part of the intake structure excavation (Fig. 4). Several joints were tight but contained thin seams of clayey silt, and two exhibited marginal decomposition of the rock. The 0-15 B/A joint zone was stained brown from recent groundwater movements. Vertical water velocity profiles in nearby drill holes measured prior to the excavation confirmed this. Above-average marginal decomposition had occurred along the 0-15 E/D joint, which was characterized also by above-average undulation. This is a typical condition for the evolution of a network of solution channels (Fig. 7).

The most prominent aquifer exposed in the power house excavation, the A-11' aquifer (Fig. 4), consists of a number of openings about 0.5 feet wide and 0.2 feet high, which together form a network. Many of the openings are located at the intersection of near-vertical joints and the undulating bedding-plane surface. However, there were no macro- or microscopic indications of why the aquifer developed there as opposed to along several other undulating discontinuities. On the other hand, the locations of some other aquifers were controlled by recognizable lithologic features. For instance, the aquifer A-11''' at the contact of the PB-11 and 0-11 units developed in the latter unit which is more jointed and soluble. Aquifer A-13" evolved immediately above a relatively impermeable argillaceous dolomite, and aquifer A-15' below a similar stratum.

It can be concluded that joint aquifers are likely to form at the junction of two lithologies, particularly when one is relatively impermeable and each has different strength and deformation properties. Other joint aquifers are located at apparently random positions in homogeneous lithologic units, although a bedding joint with above-average amplitudes of undulations may be a preferred "plane".

Joint and Cavity Fillings

Many of the joints and cavities in the bedrock at Grand Rapids have been partially or completely filled with unconsolidated material ranging from

Table 4. Bedding plane discontinuities in 0-15 strata in west wall of intake structure excavation

Discontinuity	Amplitude of Undulation	Comment
0-15-H/G	-	½ inch clay-silt seam.
0-15-G/F	0.2"	Tight.
0-15-F/E	0.2"	Tight.
0-15-E/D	0.4" - 0.8"	4 x 0.4 inches thick laminae with silt partings, lower 0.1 inch of overlying 0-15-E softened, upper 0.2 inch of underlying 0-15-D softened.
0-15-D/C	0.1"	Tight persistent joint pair spaced at 0.8 inch.
0-15-C/B	0.1"	Tight.
0-15-B/A	0.2"	Stained nodular surface overlying 0-15-B apparently unaltered, upper 0.1 inch of underlying 0-15-A brown stained and softened.
0-15-A/GA (base of 0-15 stratum)	-	Sharp lithologic break from a dolomite to an arenaceous dolomite with no mechanical break.

Note: For location see Figure 4.

clay- to gravel-size particles. In many joints the passageways from the surface are small and the coarser material is residual. In a typical filled horizontal opening about a quarter of an inch high, a thin clay lamina overlies an equally thin silt lamina. In contrast, some horizontal pipe-like cavities a foot or two in diameter are filled by complex stratified deposits, including cut-and-fill and slump structures (Fig. 13). The results of X-ray analysis of the components in this example are shown in Table 5. The components containing kaolinite-chlorite, illite, and degraded micas (not shown in table) are mostly transported plastic clays, while the clay mineral-free dolomite-quartz-feldspar sludge is probably residual.

The filling in the lowest elevation joint, sampled at a depth of about 100 feet below surface, included chlorite and illite and had a composition similar to a till in the area. There was no evidence that the filling was a fractionation from any of the local bedrocks, while the argillaceous dolomites contain illite, they do not include chlorite.

No quantitative tests were made on the permanence of joint and cavity fillings under sustained prototype pressures because it was not possible to (1) select a large enough number of test zones below the ground surface for statistical analysis; and (2) maintain test pressures for more than a few hours or days. However, knowledge of the nature and location of representative openings and their fillings, as well as of soft strata, was essential for the interpretation of subsurface hydrologic observations

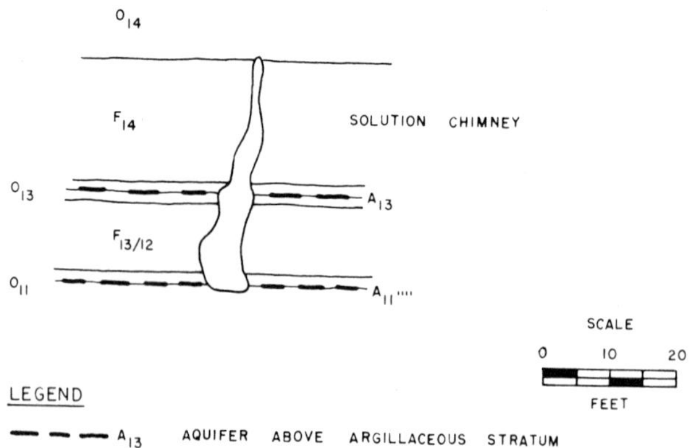

Figure 12. Solution chimney in dolomites exposed on south side of tailrace excavation.

Table 5. X-ray analyses of cavity fillings in 0-14-E stratum

Lattice Spacing (Å)	Illite 10.16	Kaolinite-Chlorite 7.23	Quartz 4.29	Dolomite 4.04	Feldspar 3.21	Calcite 3.03
Sample No. GS	Peak Intensities (%)					
94	11	27	28	10	14	10
95	9	20	42	11	12	11
90	6	59	13	13	4	5
93	10	73	7	7	3	-
92	17	58	13	5	7	-
91	-	-	7	67	26	-

Note: The percentage peak intensities are for comparison only as their quantitative significance has not been checked against known mixtures.

The 0-14-E stratum is a subunit of the 0-14 unit.

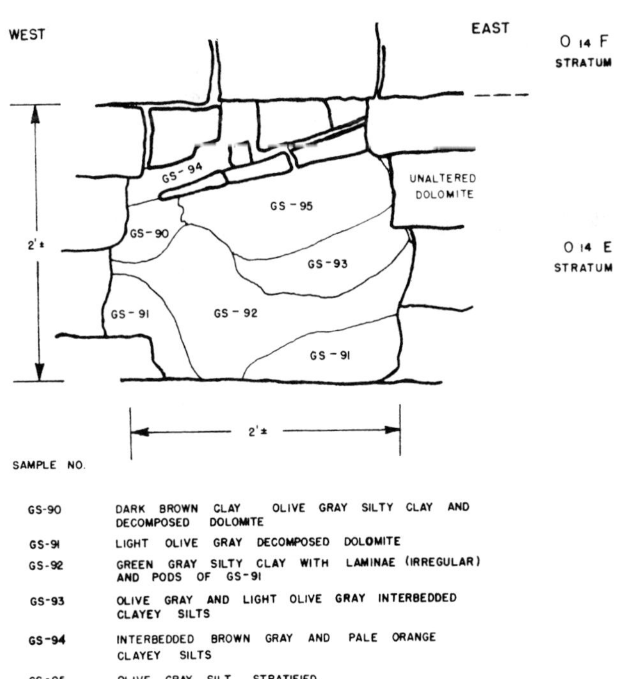

Figure 13. Cut-and-fill plugging of a solution cavity in the 0-14-E stratum.

and for the design and control of grouting programs. The semiquantitative information suggested which zones might fail hydraulically. These zones were either replaced by concrete or grout, or instrumented with point or open pipe piezometers (Appendix B) for observation as the full reservoir head was progressively approached and maintained.

Ground-Water Flow Patterns

The natural ground-water flow pattern in the project area prior to construction is indicated by the water table contours on Figure 2. The flow to the east was modified by local discharge into the Saskatchewan River. Regional discharge occurs in the marshy area adjacent to Lake Winnipeg.

Prior to the deep excavations, standard water pressure tests and natural vertical water velocity measurements in drill holes had proved that horizontal water flow is concentrated along a number of joint aquifer zones, each of limited vertical, but of great horizontal, extent. This was confirmed in the excavations and most aquifers were seen to consist of only one or two open bedding-plane joints totalling about an inch in height. The aquifers are joined vertically by joints which are generally tighter than the horizontal openings. There are large vertical openings which include sinkholes, and which are several feet or tens of feet in horizontal and vertical extent.

Design of Depth of Grout Curtain. There are no aquitards of appreciable thickness beneath the reservoir area and therefore there is no obvious base for a grout curtain around the periphery. Therefore, the positions of the aquifer zones were used as design controls for the depths of the curtain beneath various parts of the dike axes (Fig. 14).

CALCULATION OF HORIZONTAL AND VERTICAL GROUND-WATER FLOWS

Shea and Whitsett (1958) have discussed the prediction of seepage under dams on multilayered foundations from the results of pumping tests. At Grand Rapids, flow was calculated from fluid potential and velocity measurements in drill holes, generally without recourse to pumping. It was assumed that all horizontal flow is confined to the aquifers and all vertical flow to the aquitards, namely the strata of relatively low permeability between the aquifers. For analytical purposes, the upper 200 feet of the bedrock in the project area was divided into a series of columnar zones. The horizontal and vertical flows in each column were calculated by Darcy's Law applying the fluid potentials and transmissibilities (Todd, 1959, p. 50) of each aquifer and the thickness and permeability of each aquitard.

Vertical water velocities and water level data from each drill hole were used to calculate the potentials in the aquifers. As virtually no changes in flow patterns were caused by the observational procedures used, in contrast to the changes produced by the standard pumping tests, equilibrium was maintained. Flows were assumed to be laminar, although turbulence must occur in most cases when water passes between an aquifer and a hole. With these simplifying assumptions, the potential and flow relationships in drill holes were derived from the empirical formula of Hazen and Williams (Davis, 1952, pp. 1209-1210). The potential values were corrected as the effect of flow between the aquifers and holes produced cones of depression or elevation. Thiem's equation (1906) was used as suggested by Sokol (1963).

An average permeability was estimated by the routine pumping from the power house excavation, which was treated as a well. The values for each aquifer were checked independently by comparison of calculated and observed potentials in individual aquifers.

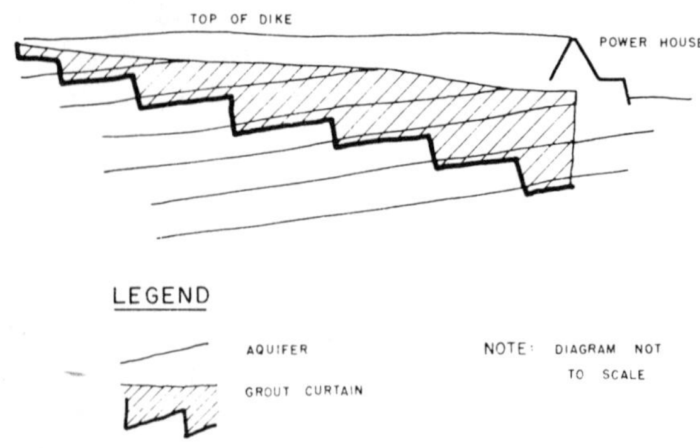

Figure 14. Generalized profile of grout curtain showing the position of the aquifers.

Vertical Permeability

No direct measurements of vertical permeability could be made, and estimates were necessary. These were based on qualitative geologic and hydrologic information. For instance, the open joints were more common in the 0-11 stratum than in the underlying PB-11 stratum. Furthermore, the hydrologic observations, summarized in Figure 9 and shown by Profile B, indicate that a large potential difference was maintained across the PB-11 stratum before the A-10 aquifer was grouted.

The estimated values of permeability were checked for the flow continuity requirements in the system of contiguous analytical columns described above.

Horizontal Ground Water Velocities

The velocities in the A-10 aquifers beneath the reservoir near the intake structure (Fig. 4),

Table 6. Estimated effect of grout curtain on ground water flow with full reservoir

	Flow (cubic feet per second)	
	Natural*	After Grouting
Horizontal flow in upper aquifers (aquifers above base of grout curtain)	100	17
Vertical flow from upper aquifers into lower aquifers and out of reservoir area	-	.20
Total seepage from reservoir area	100	37

*If a grout curtain had not been injected.

after the completion of the grout curtains but before impounding, was about 20 feet per day, assuming an average total height of the openings as one inch. Earlier in the same area, before construction began, dye tests had indicated local velocities as high as 14,000 feet per day.

Over-all Rate of Ground Water Flow

The calculated natural flow eastward across a north-south line joining the western extremities of the dike lines (Fig. 2), through the A-11 and aquifers at higher elevations (Fig. 3), was about five cubic feet per second. This value can be compared to the dry weather flow of 10 cubic feet per second in the creek flowing northward from Eating Point Lake. This flow was derived from springs and represented, before the reservoir was filled, much of the flow in and above the A-11 aquifers from a similar north-south "front".

Predicted Losses from the Reservoir

The predicted loss with a full reservoir is 37 cubic feet per second (17,000 gpm). The estimated effect on the ground-water flow and its pattern is summarized in Table 6, which shows that while the curtain reduces losses through the grouted aquifers by 80 percent, it increases losses below the base of the curtain. This deflection reduces the possibilities for piping and the loss of stability of structures.

Effect of Impounding

Impounding began progressively in June 1964. Observations since then indicate that the seepage control is satisfactory. Analysis of new subsurface data is continuing.

SUMMARY AND CONCLUSIONS

Surface mapping, examination of NX-sized drill cores, and observations in construction excavations established that the Grand Rapids generating station area is underlain by almost horizontal bedded dolomites with significantly different textures. Joint and bedding-joint patterns, as well as the degree of solution along the joints, can usually be correlated with observable lithologic variations.

Ground-water flow is concentrated along a limited number of thin joint aquifer zones whose widespread horizontal extent was demonstrated by observations of water levels and natural water flow velocities in NX-sized drill holes. Typical horizontal flow velocities in the aquifers are of the order of 20 feet per day. The total calculated escape flow beneath the reservoir limits in and above the A-11 aquifers (the upper 200 feet), increased from 5 to about 40 cubic feet per second (1/2 to 4 million cubic feet per day) when the reservoir was filled (before grouting).

Geologic data on joint frequencies, size of opening, and nature of the filling provided essential background for the planning of the grout curtain and for the interpretation of the hydrologic observations. The grouting program, as usual, was selective because of economic reasons.

Analysis of the continuing observations program is permitting a close watch on the evolution of the new ground water regimen, a confirmation that the control measures (grout curtain and pressure relief holes) are adequate, and the development and proving of analytical procedures for ground-water flow analysis in similar hydrogeological environments.

ACKNOWLEDGMENTS

The great assistance of numerous members of the Manitoba Hydro Electric Board, H. G. Acres & Co. Ltd., Niagara Falls, Ontario, Canada, and the University of Illinois, Urbana, Illinois, is gratefully acknowledged.

REFERENCES CITED

Baillie, A. D., 1951 Silurian geology of the Interlake area, Manitoba: Dept. Mines and Natural Resources, Mines Branch, Prov. of Manitoba, Pub. 50-1., 82 pp.

-----1952, Ordovician geology of Lake Winnipeg and adjacent areas: Dept. Mines and Natural Resources, Mines Branch, Prov. of Manitoba, Pub. 51-6., 64 pp.

Bretz, J. H., 1942, Vadose and phreatic features of limestone caverns: Jour. Geol. v. 50, pp. 675-811.

-----1956, Caves of Missouri: Div. Geol. Survey and Water Resources, State of Missouri, V. XXXIX, 2nd ser., 490 pp.

Davis, C. V., 1952, Handbook of applied hydraulics: New York: McGraw-Hill, 2nd ed., 1272 pp.

Davis, W. M., 1930, The origin of limestone caverns: Geol. Soc. America Bull. v. 41, pp. 475-628.

Grice, R. H., 1967, Quaternary geology of the Grand Rapids area, Manitoba.

Hamilton, D. K., 1948, Some solutional features of the limestone near Lexington, Kentucky: Econ. Geol., v. 43, no. 1, pp. 39-52.

Meinzer, O. E., 1923, The occurrence of ground water in the United States: U. S. Geol. Survey Water-Supply Paper 489, Washington, D. C., 71 pp.

Pettijohn, F. J., 1957, Sedimentary rocks: New York: Harper & Brothers, 2nd ed., 718 pp.

Rettie, J. R., Koropatnick, A., and Isom, W. S., 1962, Foundation treatment and construction of sixteen miles of dykes at Manitoba's Grand Rapids project: Western Zone Technical Conference, Eng. Inst. Canada, Edmonton, Alberta.

Shea, P. H., and Whitsett, H. E., 1958, Predicting seepage under dams on multilayered foundations: Jour. Soil Mech. Found. Div., Am. Soc. Civil Engineers Proc., v. 84, No. SM3, paper 1727, 41 pp.

Sokol, D., 1963, Position and fluctuations of water levels in wells perforated in more than one aquifer: Jour. Geophys. Research, v. 68, no. 4, pp. 1079-1080.

Stearn, C. W., 1956, Stratigraphy and paleontology of the Interlake group and Stonewall formation of southern Manitoba: Canada Geol. Survey, Memoir 281, Dept. Mines and Tech. Survey, Ottawa, 162 pp.

Terzaghi, K., 1963, Discussion on K. W. John's paper "An approach to rock mechanics": Jour. Soil Mech. Found. Div., Am. Soc. Civil Engineers Proc., v. 89, No. SM1, pp. 295-300.

Thiem, G., 1906, Hydrologische methoden: Gebhardt, Leipzig, 56 pp.

Thornbury, W. D., 1954, Principles of geomorphology: New York: John Wiley, 618 pp.

Todd, D. K., 1959, Ground water hydrology: New York: John Wiley, 336 pp.

Trainer, F. W. and Eddy, J. E., 1964, A periscope for the study of borehole walls, and its use in ground-water studies in Niagara County, New York: U. S. Geol. Survey Prof. Paper 501-D, pp. D203-D-206.

APPENDICES

Appendix A. Scale of Rock Hardness

RH-1	-	Soft	Slightly harder than very hard overburden, rocklike character but crumbles or breaks easily by hand. (Some clay-shales and uncemented sandstones and agglomerates).
RH-2	-	Medium Soft	Cannot be crumbled between fingers but can be easily picked with light blows of the geology hammer. (Some shales and slightly cemented sandstones and agglomerates).
RH-3	-	Medium Hard	Can be picked with moderate blows of geology hammer. Can be cut with knife.
RH-4	-	Hard	Cannot be picked with geology hammer but can be chipped with moderate blows of the hammer.
RH-5	-	Very Hard	Chips can be broken off only with heavy blows of the geology hammer.

(after Panama Canal Co., Deere, 1962, Personal Commun., Talbot Univ., Univ. Illinois, Urbana, Ill.)

Appendix B. Observation Holes and Piezometers

In limestone and dolomite terrain, observation holes with vertical water velocity measurements are better than piezometers on two counts.

(1) Measurements on a drill hole meter the potentials in all aquifers intersected by the hole, including those which were undeveloped at the time of initial exploration.

(2) A single open hole is cheaper than even a single hole with multiple piezometer installations.

Permanent observation holes must be available to measure the potential of at least one aquifer below grout curtains so that the seepage losses by downward deflection can be calculated and, if need be, the value of deepening a curtain can be assessed.

Other extensive ground water studies have been made in dolomites including the Lockport dolomite, Niagara County, New York (Trainer and Eddy, 1964).

ENGINEERING GEOLOGY OF INTERSTATE HIGHWAY 94 UNDERPASS AT NORTHERN PACIFIC RAILWAY, FARGO, NORTH DAKOTA

Gordon L. Bell (Geologist, North Dakota State Highway Department, Bismarck, North Dakota)

Abstract

Construction of four-lane Interstate Highway 94, at the Streeter Branch of Northern Pacific Railway south of Fargo, North Dakota, encountered excessive ground water in silty sand. The underpass required special excavation with two stages of well points where "quicksand" was encountered in an abandoned stream channel within sediments of ancient Lake Agassiz.

Exploration drilling of 11 test holes indicated as much as 78 feet of "quicksand" under the structure site. Discovery of these subsurface conditions resulted in several construction changes including the use of well points, vertical sand drains, and special filter-sand-encased subdrains discharging into a lift station.

Special excavation during abnormally rainy weather was accomplished with the aid of a second well-point system along the median below the 896-foot level in the buried channel system.

The area was successfully excavated and drained to a lift station. The Northern Pacific Railway bridge and the twin highway were completed. The concrete pavement has remained stable under constant use since November 1960.

This is an example of the use of well points in stabilizing saturated soils for highway foundations.

CONTENTS

ILLUSTRATIONS

INTRODUCTION

Fargo, North Dakota, is situated on the Red River of the North. Here, on the broad plain of ancient Lake Agassiz, man has experienced a wide range of interesting and unusual construction problems. Many of the problems concern foundations in the various glacial and lake deposits.

The extensive Interstate Highway program required new routes and new approaches to accommodate modern four-lane highways for local and interstate travel. The Fargo Bypass segment of Interstate Highway 94 was planned south of the city where it was necessary to go over or under two railways: the Northern Pacific, Streeter Branch line, and the Chicago, Milwaukee, St. Paul, and Pacific Railway. It was decided to construct the twin highway under the railways.

The North Dakota State Highway Department coordinated its planning and construction with the railways for uninterrupted rail traffic, and for satisfactory completion of the projects. Although both projects required special consideration of their geologic soil and hydrologic conditions, it is the purpose of this report to discuss only the engineering geology at the Northern Pacific Railway separation or underpass.

The project included two primary construction phase contracts, the structural contract for the railway bridge and the grading contract.

During the planning stages in September 1955, four core drill holes were made to determine the soil conditions at the site. The borings were made at the proposed bridge site, on or near the centerline of the highway. The deepest hole drilled at that time was 58 feet at highway survey station 521+00 on center line. The water table was reported to be at elevation 900.3 feet, or 7.7 feet below the land surface. The water was contained in silty sand and clay. The high water table is common in the area and nothing unusual was reported.

CONSTRUCTION PROBLEMS

Construction was started in June 1958, and unusual soil conditions and pile reactions developed after preliminary excavation for the bridge. During the initial piling operations for the north abutment and piers, the Resident Engineer and Inspectors reported unusual behavior of the treated timber piles. As pile driving continued, the action caused some piles to rise, and water with silty fine-grained sand began to appear in neighboring shallow excavations, as much as 5 feet above pier base and pile cutoff. During driving with a gravity hammer, a few piles ceased movement as if they had struck boulders.

The north abutment No. 1, elevation 896.9 feet, and south abutment No. 5, elevation 896.9 feet, were completed. Pile driving was continued for Pier No. 2, elevation 885.5 feet, and the unusual behavior of the piles was again noticed.

Treated timber piles were used with a design bearing of 16 tons to the pile for the piers and 14 tons to the pile for the abutments, all on 3-foot centers. Table 1 gives the pile data.

The contractor attempted to use steel sheet piling for a cofferdam to protect the excavation for pier piling and pier base construction at Pier No. 3, elevation 882.5 feet. The general water level in the ridge was elevation 901 feet, and this required lowering the water about 18 feet. Sump pumps were used to dewater the cofferdam. Outside pressure increased as the load of sand and water was decreased in the cofferdam, and "blow-ins" resulted. The sheet piling was displaced as sand and water flowed in the bottom and sides.

This apparent great volume of "quicksand" was unusual in the area and it was decided to explore the site in greater detail. Construction work was temporarily suspended September 20, 1958 until results of the exploration could be obtained.

EXPLORATION

In view of the complex geologic setting, a new program of exploration, including surface geology and drilling, was authorized. Eleven test holes were drilled in the vicinity of the bridge site and along the center line of the proposed excavation for the underpass (Fig. 1). The first ten holes were drilled with a truck-mounted continuous flight 4-inch diameter spiral auger drill. A stand-by truck crane was used as required to help pull the auger from the hole. This worked very well, and none of the 5-foot-long spiral sections were lost. Test hole No. 11, drilled sometime after the 10 auger holes, was drilled with a core drill. The holes ranged in depth from 44 feet in holes No. 6 and No. 7, to 80 feet in hole No. 2, to 89.5 feet in the core drill hole No. 11.

The first five test holes were drilled along a northeast line on the west side of the railway detour or "shoofly." The remaining six holes were drilled, as numbered, at intervals along the center line of the proposed excavation for the underpass, from survey station 512 to 528.

The eleven test holes were drilled in the knowledge that the railway was constructed along an extensive ridge with possible connections with the "Fargo Ridge" and the "West Fargo Ridge" one mile west of the project. Parts of these water-bearing broad ridges had previously been described by Upham (1895) and Dennis and others (1949). These broad ridges rise from 5 to 15 feet above the general plain. It is thought that early railway engineers used these prominences for location above spring floods. They certainly served the purpose.

The general area is flat and nearly featureless, on casual observation. However, there are features, including the ridges, that are of local significance. The Fargo Ridge and West Fargo Ridge distributary, from the Horace Ridge two miles south of the underpass, comprise a ground-water system in filled and abandoned channels of an earlier Red River that apparently kept pace with the alternately filled and dried Lake Agassiz. These filled and abandoned stream channels with their natural levees, and interbedded and associated lake sediments, are significant and common structures in the Lake Agassiz basin of sediments. The ridges are surface drainage barriers and important aquifers, and with their water-charged sand lenses, they are unstable and weak foundation areas.

This report deals with the Fargo Ridge of this system, its geology, and the construction practices required to complete the highway underpass and railway bridge in the wet sand ridge.

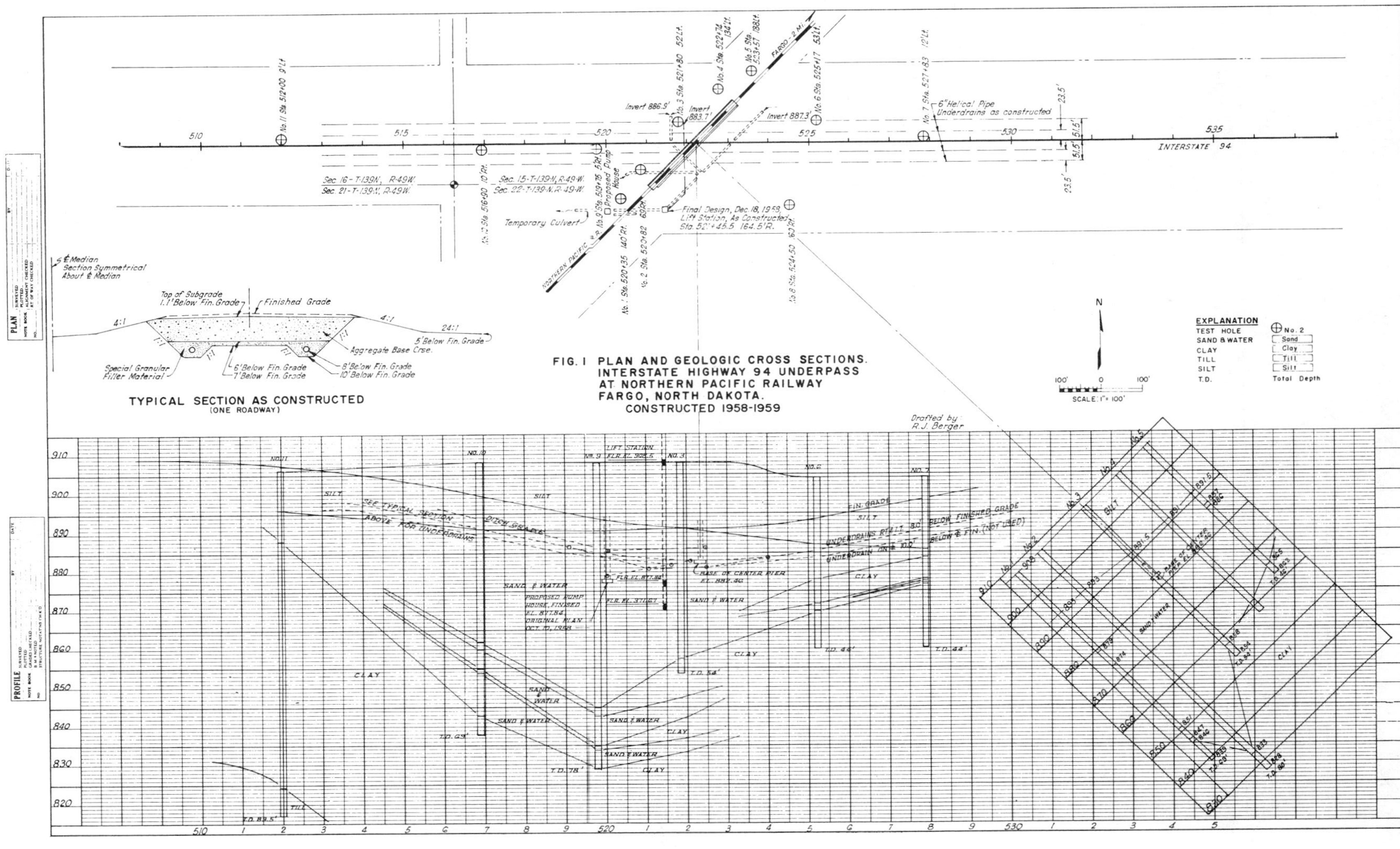

Figure 1. Plan and geologic cross sections, interstate highway 94 underpass at Northern Pacific Railway Fargo, North Dakota.

Table 1. Pile record, Northern Pacific Railway bridge, Interstate Highway 94, Fargo, North Dakota

	Elevation	Number of Piles	Tons Average Bearing	Average Length Feet
Abutment 1	896.9	126	78	74
Pier 2	885.5	70	47	73
Center Pier 3	882.5	96	39	52
Pier 4	885.5	70	66	29
Abutment 5	896.9	126	53	35

GEOLOGY

To better understand the geology of the ridge, it seems pertinent to consider more of the regional physiographic features. The region, at an altitude of 905 feet, is drained by the Sheyenne River two miles west, and the master stream, the Red River of the North, four miles east of the structure. Both rivers flow north in meandering channels 15 to 30 feet deep in tan to gray silt, silty sand, and clay, all in part of the upper deposits of Lake Agassiz.

Lesser intermittent drainage is developed along each ridge, and local drainage ditches contain water, at places, during the driest summer months. This reflects the high water table characteristic of the central lake plain. The water is held by the silt with its low specific yield. The permeability is so low that the water cannot flow to lower elevations rapidly enough to lower the water table. At places, silt and sand lenses in clay support perched water tables of little yield and slow recharge.

At the structure, the northeast-trending ridge is about 800 feet wide at the surface and 1500 feet wide at elevation 895 feet. The center of the ridge is considered to be at highway survey station 520+00 (Fig. 1, cross sections). The ridge rests on upper Lake Agassiz sediments that are composed of 80 or more feet of plastic blue clay covered by tan to gray thin-bedded silt and clay. Core drill hole No. 11, at the west edge of the ridge, entered glacial till at a depth of 83 feet, which corresponds to elevation 825 feet.

The glacial deposits, locally referred to as till, with included glacial stream deposits of sand and gravel, underlie the blue-gray lake clay. The till may be 100 feet thick at the project site where the till was accummulated unconformably on granitic bedrock of the preglacial Red River Valley. The glacial stream deposits occur at various depths in the basal till and are the principal aquifers in the region. West Fargo obtains industrial and municipal supplies of water from these lower deposits of sand and gravel. Artesian conditions exist in some of these confined aquifers, and it was important to inspect farm wells in the area before drilling at the construction site. It was found that the artesian water at the farm in the southwest corner of the area rises to elevation 868 feet, or within 40 feet of the surface of the land.

Two possible problems were indicated by the artesian head. It was important to avoid recharging the sand ridge with artesian water or losing water to lower permeable layers with consequent settlement at the construction site. Fortunately, it was not necessary to drill deeper than 100 feet.

CONSTRUCTION

After the investigative work described above was completed, the Materials Department retained D. J. Henkel, soils technician, to appraise the conditions at the project. He concluded that the underpass design was feasible, provided certain construction changes were made. He recommended that a well-point system be installed to lower the water table in the area and to determine the amount of dewatering required for construction and maintenance of the highway.

The initial well-point system was installed early in October 1958 by the John W. Stang Corporation, under the direction of H. J. Hoody. Seventy-one sandpoints and two observation wells were installed around the site of Pier No. 4. The dimension of the header pipe layout was 40 feet by 80 feet. The well points were spaced on 3-foot centers and jetted 24 feet below their connections with the header pipe. Twenty additional observation wells, of well-point pipe 35 feet long, were installed on a quadrant extending approximately 400 feet from the center of Pier No. 4. The quadrant was composed of coordinates of 50 feet, 75 feet, and 100 feet for spacing the pipe observation wells. Each well point and observation well was enclosed in filter sand (concrete sand).

Pumping started October 22, 1958 at 8:00 A.M. After 16 hours of pumping at 140 gallons per minute the drawdown at the header pipe was 19 feet. This was maintained and increased temporarily as the quantity pumped decreased to a fairly uniform rate of 70 gallons per minute.

A possible area of influence in the area being dewatered was calculated using 898.0 feet as the initial ground water elevation and an average measured well-point discharge of 70 gallons per minute. Dr. Henkel (1959) reported and recommended the following.

(1) The maximum drawdown was 14.8 feet from the original level at a radius of 50 feet, and a total range of influence of 9600 feet.

(2) Permeability of the sand as 0.017 feet per minute, using the measured yield of 70 gallons or 9 cubic feet per minute.

(3) The ground-water level was to be lowered to the 886.7-foot level.

(4) Installation of perforated pipes 8 feet below finished grade line with suitable filter layers to prevent internal erosion of the silty sand and clogging of the drains.

(5) Installation of relief wells of filter sand 6 inches in diameter and to extend 16 feet below pavement level. The relief wells to be spaced 20 feet apart along the line of subdrains.

All construction and pumping was suspended in December 1958 when the ground froze to depths of 6 to 8 feet.

The State Highway Department and the U. S. Bureau of Public Roads reviewed all recommendations, tests, and records, including an estimate from the Northern Pacific Railway for raising their grade 4.5 feet. The following measures were approved:

(1) To construct the underpass to control the water because a certain amount of water had to be pumped in any alternate plan involving an underpass.

(2) To construct a subdrain system of perforated subdrain pipes 8 feet below finished grade (Fig. 1).

(3) It was apparent that it would be necessary to pump water for the life of the underpass. This required an underground drainage system to handle the water the year round. The ditches in the underpass were also drained to the lift station (Fig. 1).

(4) To construct a two-level lift station in more stable ground on the east side of the Railway.

(5) To install relief wells of filter sand below the subdrain pipe that in turn would be encased in filter material.

(6) To have a grain-size analysis made of the sand exposed in the bottom of the excavation.

The criterion used by the State Highway Laboratory for determining the suitability of the filter material is that developed by Terzaghi and Peck (1948) and recommended by the U. S. Waterways Experiment Station. The formula may be expressed as follows:

$$\frac{D\ 15(\text{of filter})}{D\ 85(\text{of foundation})} < 4 < \frac{D\ 15(\text{of filter})}{D\ 15(\text{of foundation})}.$$

To prevent the foundation material from passing through the pores of the filter, the 15 percent size of the filter material must not exceed 4 times the 85 percent size of the foundation material. To maintain minimum seepage forces within the filter material, the ratio of 15 percent sizes of filter and foundation material must be greater than 4. Generally, washed concrete sand will meet the requirements for filter material about perforated helical metal pipe subdrains.

Construction of the underpass highway was resumed in the spring of 1959. The grading contractor installed a well-point system consisting of several well points remaining under the bridge and 88 well points at 5-foot intervals along the centerline of the median starting at station 521+00 and terminating at station 525+40. The area was dewatered by this system after excavation to elevation 896.0 feet in June 1959. The excavation work continued through September 1959. More than 200,000 cubic yards of the silty sand were removed for the underpass, including some 10,000 cubic yards for the underdrains.

Excessive rain during June and July made haul roads impassable. Some of the surface water that entered the excavation was removed by suction pumps operating on the surface until the well-point system could be installed along the median. The surface water and quicksand condition below elevation 896.0 feet made it impossible to operate self-propelled scrapers in much of the underpass area. The soil material (quicksand), below 896.0 feet, is 94 percent fine-grained spherical sand and 6 percent silt and clay-size material with interbedded lenses of silt and clay. This near-fluid material was difficult to handle. The contractor used draglines for loading and bottom dump trailer units for hauling. The trucks were assisted by a D-8 caterpillar tractor where necessary.

The excavated material was placed in a test fill at a site 1.7 miles from the excavation. The material lacked sufficient stability for geometric placement, and the result was a large stockpile of material for later use on north-south Interstate Highway 29 projects in this area.

The underpass was satisfactorily completed, paved, and included in the Interstate Highway 94 system in November 1960. The concrete pavement was undisturbed, and the subdrains continued to function four years after completion.

REFERENCES CITED

Dennis, P. E., Akin, P. D., and Worts, G. F., Jr., 1949, Geology and groundwater resources of parts of Cass and Clay Counties, North Dakota and Minnesota: North Dakota Groundwater Studies No. 11; Minnesota Groundwater Studies No. 1; North Dakota State Water Commission, Bismarck, North Dakota, p. 17.

Henkel, D. J., 1959, Report on the soils and groundwater problems associated with the construction and operation of the Northern Pacific Railroad Underpass on Route 94 near Fargo, North Dakota: (Unpublished).

Taylor, D. W., 1948, Fundamentals of soil mechanics: New York, John Wiley and Sons, p. 135.

Terzaghi, Karl and Peck, Ralph B., 1948, Soil mechanics in engineering practice: John Wiley and Sons, Inc., New York, p. 50.

Upham, Warren, 1895, The glacial Lake Agassiz: U. S. Geol. Survey Mon. 25.

GEOLOGY AND FOUNDATION TREATMENT OF SENSITIVE SEDIMENTS--WORLD'S FAIR HIGHWAY COMPLEX, NEW YORK

Charles A. Baskerville (Department of Geology, The City College, New York, N.Y.)

Abstract

The highways and 1964 World's Fair site are located in a tidal valley in the Flushing Meadows of Queens County, New York. Highway construction costs were about $120,000,000.

The Meadows contains the Flushing River, which connects Flushing Bay with Meadow and Willow Lakes to the south. These waters receive fresh water from surface runoff and salt water from the East River.

Deposition in the valley is believed by others to be due to submergence. Evidence encountered during highway construction suggests that it was formed by the breaching of an older terminal moraine and by subsequent rising sea level. The Meadows is close to sea level. The valley floor deposits are composed of plastic organic silt and clay overlain by peat. Peat layers cored in borings at the bottom of the organic silts overlie a clayey till. Radiocarbon methods indicate a bottom-peat age of 7000 to 9000 years before the present.

These sensitive organic sediments have average moisture contents (water weight/dry weight) from 90 to 100 percent, plastic limit 40 percent, and specific gravity 2.62. This specific gravity is consistent with the granitic material from which the sediments were derived. This material has a low shear strength, which increases slightly with depth because of compaction.

Special foundation treatments to reduce postconstruction settlements included: relieving structures over drainage lines and light-weight fill embankments with counter-weight berms; and sand drains to increase consolidation rate with accompanying shear-strength increase.

CONTENTS

ILLUSTRATIONS

INTRODUCTION

Approximately $120,000,000 was spent on highway construction in the Flushing Meadows area of Queens County, New York (Fig. 1). This road network surrounds the location of the 1964-1965 World's Fair, and was built to help accommodate the expected excessive traffic influx. The entire highway system and the Fair are located in a tidal valley which was probably formed by glaciation.

When marsh or swamp lands are selected for a highway right-of-way, it is generally to avoid the costs of acquiring right-of-way through expensive residential areas, heavily built-up industrial areas and public institutions, and to eliminate the need to build extra overpass or underpass bridges. This was the case in the selection of the route for the New Jersey Turnpike through the New Jersey Meadows, and also the proposed West Shore Expressway on Staten Island through the tidal flats parallel to Arthur Kill, which separates Staten Island from New Jersey.

Flushing Meadows, formerly known as the Corona Dumps, is an area about 3.6 miles north-south and approximately 0.8 miles wide across its midpoint (Jewel Ave.). It extends from Flushing Bay, which is an embayment of the East River, to Kew Gardens, Queens County, New York. The Flushing River, which is partially navigable to the north, empties into Flushing Bay on its north end and connects with Meadow and Willow Lakes through tide gates to the south (Fig. 2). These lakes were created for the 1939 World's Fair and occupy a large area of the south half of the Meadows. Prior to the 1939 Fair, the original Flushing River occupied practically the entire width of this tidal valley with broad meanders and numerous tributaries.

This paper describes the geology of the Flushing Meadows area, foundation problems encountered in the construction of the World's Fair highway complex, and how the sensitive soil-rock units were stabilized for heavy construction.

GEOLOGY OF THE AREA

Flushing Bay and Flushing River are brackish tidal estuaries receiving fresh water from the south by way of storm sewers and surface runoff from high ground. The salt water enters from the East River to the north, which also contains, at this point, a mixture of fresh water derived from the Bronx River which enters the East River just opposite Flushing Bay. This water is turbid, at times, which is effective in blocking light penetration and subsequently photosynthesis, which thereby eliminates normal marine organisms.

Most of the central "valley" of the Meadows is close to sea level. Surrounding the valley to the east, west and south are glacier deposited ridges which range from 50 to more than 100 feet higher. They are composed of sediments that range from well-sorted fluvioglacial channel fillings to poorly sorted sands, gravels, and boulder tills.

Lowest Sand and Gravel Unit

The valley bottom sediments consist of a brown micaceous sand and gravelly till which is overlain by peat at the base, then a plastic organic silt and clay, followed by a surface peat and the dump fill (geological section shown on Figure 5). The author observed traces of vegetation and mollusk and Foraminifera shells in the organic silt from borings taken to depths of more than 110 feet for bridge footings in this area. Numerous discontinuous large lenses of brown laminated silts occur in the bottom till. The lower part of these beds is often gravelly. None of these laminated lenses are known to contain any marine organisms. This evidence suggests that these lower deposits are glacial outwash from an earlier glacial advance than the Harbor Hill.

Bottom Peat Unit

The peat layer overlying the gravelly till would indicate that during a previous interglacial or interstadial period, the sea level was nearly 100 feet lower than present. Based on carbon-14 dates on samples of the bottom peat, an age of 7000 to 9000 years before present is given by Newman (1965, Personal Commun., Queens College).

Glacial History and Effect on Foundation Characteristics

Generally, the peat-organic silt and clay strata range from about 25 feet thick near the valley borders to 70 or more feet at the center (some structural piles have penetrated to 185 feet before fetching up).

The structure of the original valley has led some to believe that the deposition is indicative of submergence. However, based on evidence encountered during construction, the author believes that the valley was formed by breaching of an older terminal moraine along the northeast coast of Long Island and subsequent gouging of the valley by an ice lobe that was thicker at this point than elsewhere. In addition, the old moraine may have been lower and probably weaker at this point.

Evidence for this origin of Flushing Meadows is based on a cut for an entrance ramp (designated 'M') and an associated retaining wall (designated as 'WM' on the construction plans) located behind the LaGuardia Motel across Grand Central Parkway from the LaGuardia Airport (① on Fig. 2). A reddish-brown fine till, composed of shale and sandstone particles, was encountered, containing a thick lens of cross-bedded sand that grades, top to bottom, from fine- to medium- to coarse-grained. The material, derived from erosion of the Triassic series of New Jersey, indicates that, at the time of deposition, a glacial lobe passed over New Jersey and entered this area from the northwest. Similar material occurs on Staten Island and Lower Manhattan which lie in this path and supports this interpretation of the source.

The material in the vicinity of this 'WM' wall has been overridden by a still more recent glacial advance, probably from the north to northeast. This source direction is indicated by the older reddish-brown till deposits which contain shear zones inclined to the south that parallel the axes of overturned anticlinal folds in the sediments, also overturned to the south. The overriding mass,

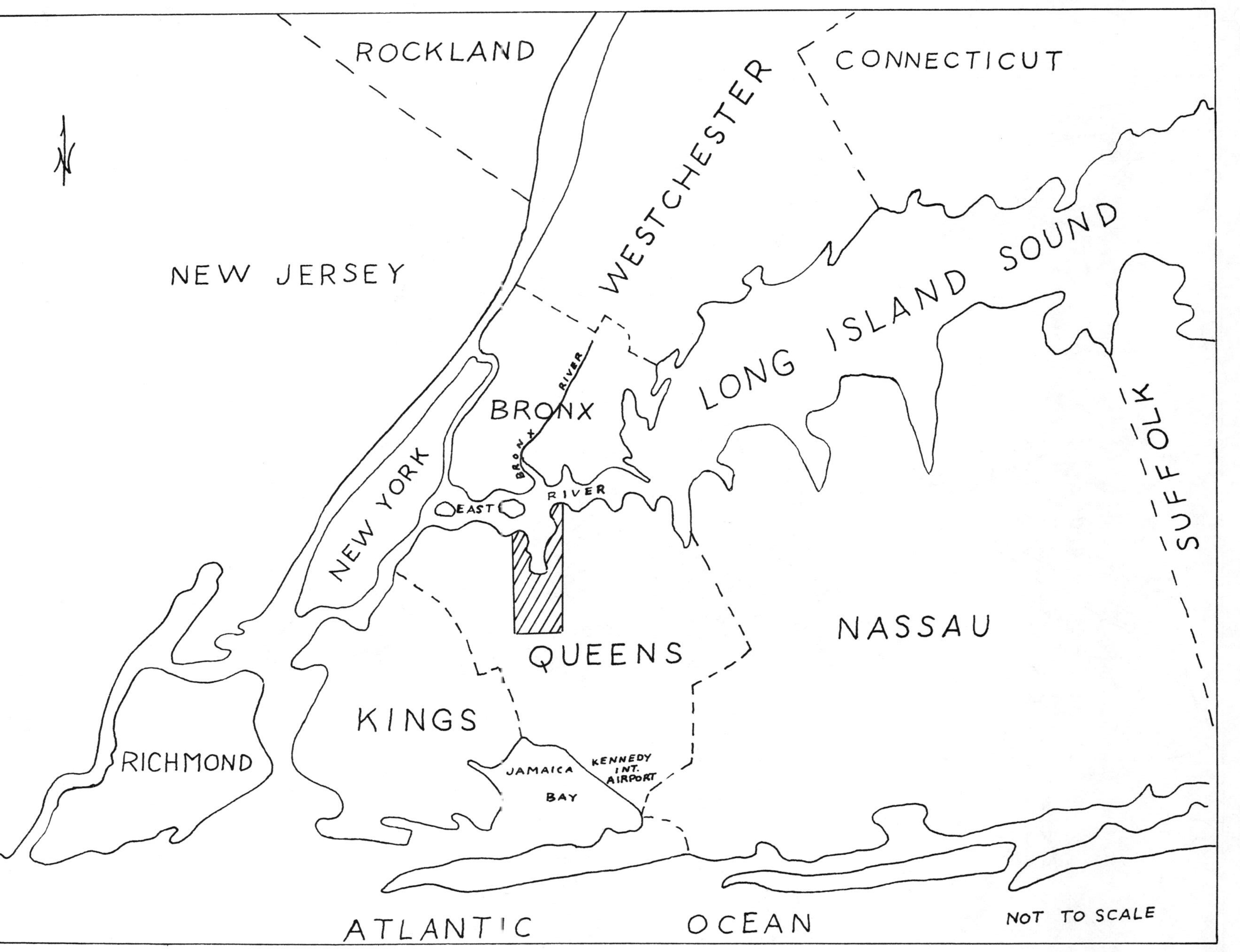

Figure 1. Metropolitan New York area. [hatched box] Flushing Meadows area of Queens County, New York.

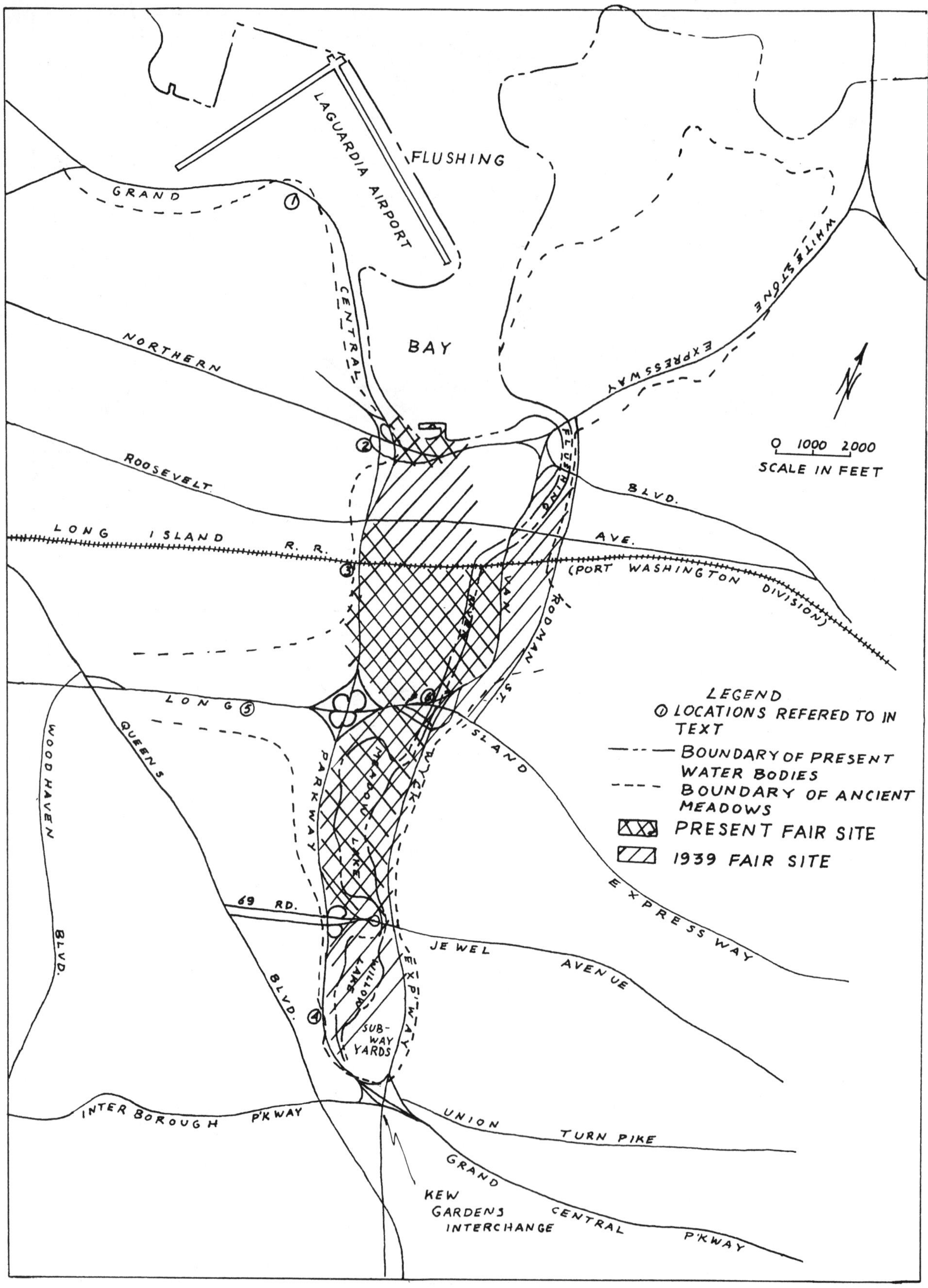

Figure 2. Location map showing World's Fair highway network in Queens County, New York.

still in evidence, contains a heavy boulder layer 6 or more feet in thickness at its base. This overriding mass is made up of granitic and metamorphic detritus, which reinforces the contention of a north to northeast direction from which the more recent ice advanced, since bedrock to the north and northeast are of this type.

Sequence of Events. The Flushing Meadows area has been modified by the following geologic events that affected the sediments involved with the highway structures.

(1) When the ice retreated from this area some 10,000 to 11,000 years before present, the sea level apparently remained low, allowing vegetation to flourish. Such a condition would explain the bottom peat layer. This layer helps to maintain the saturated condition of the overlying organic silt by reducing the vertical permeability.

(2) The granitic-metamorphic glacial till deposits, of the latest glacial advance, are micaceous like the organic silt filling the main valley. This suggests that the valley-filling, organic silt-clay material was probably derived from glacial rock flour leached from the till by meltwater streams. Melting from the ice front and stagnant blocks of ice left on the surrounding high ground probably led to the formation of small streams draining into the main Meadows valley (then a swamp) which carried rock flour in suspension. Many good-sized kettles have been encountered on surrounding high ground and the terminal moraine to the south, confirming that many large blocks of ice were left behind at the time of final retreat. Some of these kettles contain from 1 to 12 feet of peat underlain by a gray clayey till (Fig. 3).

The existence of melt-water streams is evidenced by a large cross-bedded delta trending to the southeast about a mile and a quarter south of wall 'WM' and in the vicinity of Northern Boulevard on the valley's west side (② on Fig. 2). The cross-strata of this delta dip 20 to 30 degrees southeastward. About a mile farther south, along the Grand

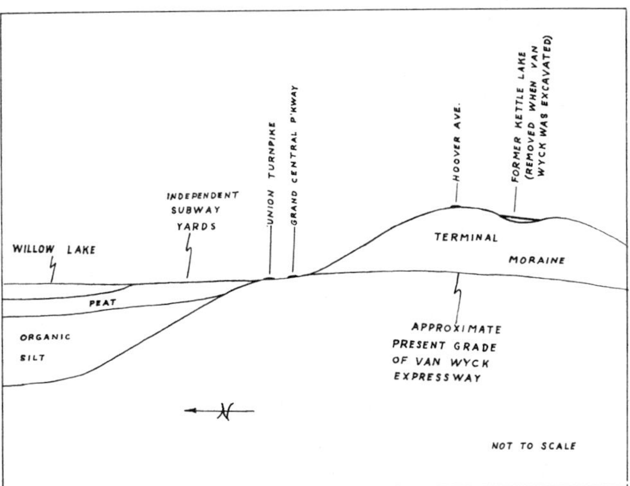

Figure 3. Generalized north-south sketch of the relationship of a typical terminal moraine kettle to the Meadows Valley.

Central Parkway at the Long Island Railroad overpass (③ on Fig. 2), there is another deltaic tongue of smaller dimensions at a depth of from 20 to 23 feet below the present roadway. This delta was of rather short duration in development, and probably originated when the main valley had already been partially filled with fine silt-clay sediments from other streams. Borings show organic silt below this delta in contrast to the Northern Boulevard delta in which borings penetrated 100 or more feet and were in sand throughout.

Valley filling was quite slow, as suggested by the depth of organic silt with roots and stems of vegetation throughout. Apparently, vegetation was able to take hold during about one season each year. Furthermore, the rising sea level was keeping pace with stream deposition, as indicated by marine shells located throughout the depth of the borings. This sequence would indicate that the valley bottom was always slightly below sea level (a large estuarine embayment). The large particles coming down the streams at some velocity would be deposited at the shoreline due to the braking action of this large body of water. The finer silts and clays would be carried out into the water and settle out by flocculation. The organic content tends to cut down shear strength, as does the mica, which tends to establish shear planes in the clay-silt, both of which, with the high water content, cause an increase in compressibility ($\emptyset = 12^{\circ}$, shear strength C= 300 lbs/sq ft).

As mentioned previously, deep borings in the Northern Boulevard delta have indicated no change in materials. Some of these borings were taken below the bottom of the valley. This would indicate that this delta, whose top is about 20 feet above sea level, was in existence throughout the most recent glacial retreat. A similar, large deltaic-till structure was observed near the south end of the Meadows valley, again on the west side, in the vicinity of wall "6" on the last East Bound entrance to the Grand Central Parkway (④ on Fig. 2).

No such structures were observed on the east side of the valley since construction of the Van Wyck Expressway was carried out on the valley surface proper. But it is quite possible that such deltas exist on the east side of the valley.

(3) The top peat zone across the valley surface is quite irregular, due to recent overloading. Although this layer probably covers the last thousand years, no carbon date is available. Its beginning probably represents a stabilization of sea level rise at near-present state and a cessation, or nearly so, of meltwater streams. This layer, when directly beneath proposed structures such as pavement, will compress under load and cause detrimental differential settlement of engineering structures due to the variable thickness (material has a $\emptyset$ of 12°, and C of 150 to 200 lbs/sq ft).

At the turn of the century, there were in existence stream networks flowing into the main valley, especially from the west in a low valley from Queens Boulevard along the present Long Island Expressway route (Fig. 4 and ⑤ on Fig. 2). As recent as about 1940, there was water in kettles located in Briarwood on top of the terminal moraine at the south end of the Meadows valley (Fig. 3). The kettles have since been removed by excavation for construction of

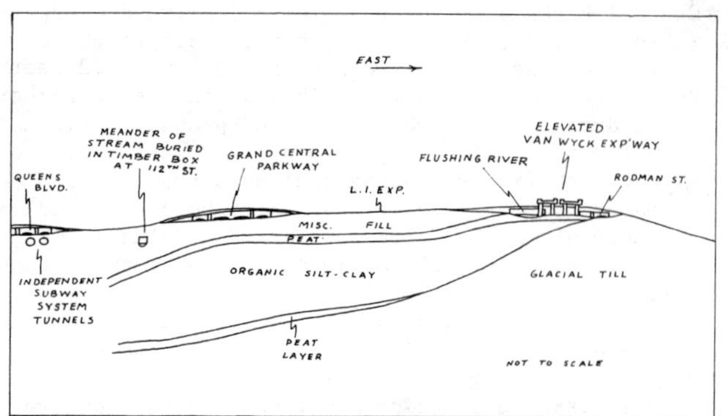

Figure 4. Generalized east-west cross section along the route of the Long Island Expressway.

the Van Wyck Expressway at the Kew Gardens Interchange. These features indicate that the original valley configuration remained unchanged until recently when changes were brought about by man's interference.

ENGINEERING CHARACTERISTICS OF ORGANIC CLAY-SILT

The organic silt unit is actually a very plastic, organic clay-silt. This organic material consists of macro- and micro-organisms deposited with clay and silt as the tidal salt water and fresh water mixed in the shallow water basin. (True marine and fresh water organisms die when introduced into a brackish environment).

This material has a low shear strength which increases slightly with depth due to compaction. An embankment with steep side slopes may be built to a maximum height of 8 feet before overstressing and displacing the organic silt.

This sediment is very compressible, and compresses quite slowly under embankment loads due to its extremely low permeability. For example, in an area where 15 feet of fill was carefully placed in 1938, the surface of the ground has settled approximately 6 feet in 20 years due to the consolidation of the 40 feet of organic silt beneath (Appendix).

CONSTRUCTION OF HIGHWAY COMPLEX

History

Prior to the 1939 World's Fair, held in the same Meadows (Fig. 2), this was a dump area for New York City. In the area of Grand Central Parkway, there were 120 feet of ashes, 80 feet above ground and 40 feet below. This dump material was used to grade the 1939 Fair site. Surface drainage was also used to grade the Fair site. This drainage was poor due to differential settlement. The existing parkways were closed at times after heavy rains.

In 1957 the Long Island Expressway was built across these Meadows. All highway drainage structures were placed on piles or concrete rubble mats. For a north service road bridge over the Flushing River near Rodman Street (east side of the Meadows north of the Long Island Expressway and ⑥ on Fig. 2), two methods were used to stabilize the approaches. On the east side of this structure, the abutment and approach areas were mucked out since the organic silt depth was less than 25 feet. On the west side, the abutment and approaches were treated by use of vertical sand drains. The west approach embankment was to be 15 feet high and the sand drains were used to increase the rate of settlement. Six feet of primary consolidation occurred in one year. The rate of secondary settlement after construction exceeded that predicted due to the disturbance of the soil caused by the driving of the sand drains.

Completed Construction

All construction was scheduled for completion in time for the World's Fair opening in April 1964. Most of the construction on the World's Fair highway complex involved widening and reconstruction of existing Grand Central Parkway, Whitestone Expressway, Northern Boulevard and 69th Road-Jewel Avenue. New construction consisted of the Van Wyck Expressway extension to connect the existing Van Wyck Expressway to the Whitestone Expressway. The goal was to minimize postconstruction settlements as much as time and money would allow.

FOUNDATION TREATMENT FOR SENSITIVE SOIL-ROCK UNITS

The requirements of a soil as a highway foundation are: (1) sufficient strength to support embankments, traffic, and structure loads without becoming overstressed and displaced; and (2) the ability to support these loads without undergoing excessive consolidation that results in detrimental structural settlements and poor riding qualities of the pavement.

The methods used to attain these conditions were the excavation of "unsuitable material" and its replacement with suitable granular material. The method of elimination of post construction settlements in areas of shallow peat and organic silt was by preloading with a surcharge embankment which remained in place for a number of months before actual road construction (Van Wyck Expressway). Pile structures at grade were used to support the roadway where fill heights were very critical in relation to stability (both Van Wyck and Whitestone expressways). Extended piles support viaduct structures across sensitive areas to limit embankments to 5 foot heights (portions of Northern Boulevard), and use specially designed heavily reinforced approach slabs for structures. To reduce abrupt settlement bumps at structure abutments, a light-weight fill was used which reduced settlement and achieved embankment stability. The light-weight fill is black anthracite ash from the New Jersey Zinc Company smelter at Palmerton, Pennsylvania. Its unit weight compacted is 67 to 75 pounds per cubic foot compared with some 130 pounds per cubic foot for regular embankment soil. With this light-weight fill embankment, stabilizing counterbalance berms have also been employed at the Jewel Ave.-Grand Central Parkway Interchange. Floating cover or relieving structures, over pile-supported pipes across paved areas, were used to eliminate differential settlement on the Van Wyck Expressway. This consisted of placing a semicircular corrugated steel arch supported on 12 by 12-inch creosoted timbers over the pipe with a void between the arch and the pipe before backfilling the pipe trench. Also, to eliminate differential settlement, old piles remaining from the 1939 World's Fair structures were removed (Van Wyck Expressway). And, last but not least, sand drain treatment was used to increase the rate of consolidation with an accompanying increase in shear strength (Whitestone Expressway).

In selecting each of these foundation treatments for particular structures, it was necessary to set up an extensive exploration program with which the author was associated. The Bureau of Soil Mechanics Laboratory, of the New York State Department of Public Works, performed numerous strength, consolidation, and classification tests. This data was interpreted by state soils engineers, in conjunction with consulting engineers employed to design the projects, in order that procedures, construction schedules, design requirements, and specifications could be coordinated and correlated. In addition, an evaluation of the performance of previous highways and structures in the area was made. In the last analysis, an appreciation of the economics of the various methods of treatment available had to be considered in order to choose the most economical treatment that would result in a satisfactorily finished highway system.

Vertical Sand Drains

One of the most successful methods for stabelizing sensitive sediments is the use of sand drains, although further research and improvement is needed. The first sand drain installation in the Flushing Meadows area was in 1957 in conjunction with the construction of the Long Island Expressway.

O. J. Porter (1952) describes the design of sand drains as causing the interstitial water to escape from saturated materials as an overlying fill or load is applied (amount of settlement increases with an increase in the load applied), thus consolidating the soft material in place without special treatment. Also, the shearing strength of the soil increases with consolidation. It is essential that a construction procedure be provided that is safe against a possible sliding failure.

The time required for this stabilization process varies according to the square of the distance the water must travel through the muck to escape. In other words, where a 5-foot layer of soft material under a previous embankment or base will stabilize during a construction period of, say, one year, a 50-foot layer, without vertical drains, requires approximately one hundred years to reach the same state of equilibrium.

In deep layers of muck, vertical sand drains are installed on spacings which will provide drainage channels and drainage conditions comparable to that for thin layers. The greater the spacing, the slower the consolidation. This analysis assumes that the foundation material is homogeneous. Fortunately, this is not generally true. Fine sediments, when laid down in water, usually segregate and grains are sorted, resulting in greater permeability in a horizontal than in a vertical direction. Sand drains are designed to take full advantage of this natural condition. Consequently, it is possible to stabilize thick layers of compressible foundation sediments in three months to a year, and thus obtain most of the settlement during the construction period.

Description of Sand Drain Placement

(1) Long Island Expressway 1957. Settlement platforms were placed first, followed by a granular drainage blanket. The specifications for size gradation were important in order to ensure that the drainage blanket had adequate permeability. Care was taken so that the contractor did not contaminate the granular blanket during its placing. The drains were placed by driving a large diameter steel displacement mandrel using a pile hammer of special design. A record was kept of the driving of each drain. It was important that the sand be placed in the mandrel in sufficient quantity, and with required air pressure (100 psi, minimum specified), in order to fill the hole with sand before its diameter decreased. Therefore, by prior computation it was possible to check the operation by comparing the volume of sand going into the hole with the theoretical volume of the hole. The blows per foot on the mandrel were recorded for sand drains installed adjacent to original exploratory drill hole locations for penetration comparisons.

After completion of the sand drain installation, piezometers were installed. Proper installation procedure is essential for the proper operation of the Bourdon gauges during construction. After installation of the piezometer systems, records were kept of the gauge readings. The critical periods were immediately after installation when the frequency of readings was once a day until the readings stabilized; then once a week until surcharge operations started.

During the placing of the surcharge on the drainage blanket, readings were taken on the settlement platforms once a week to record the rate of settlement in the sand drain area. Also, water readings were taken in the platform standpipes and checked against that observed in observation wells dug adjacent to the fill. Control stakes were also set. Purpose of the stakes is to help detect any evidence of an impending shear failure in the embankment. General shear failure of an embankment is usually preceded by the following warning signs: cracks appear in the fill, settlement occurs in the fill, and heaving of the ground takes place beyond the toe of the embankment. The embankment control stakes are installed in the expected critical areas in order to measure the vertical and horizontal magnitude of any heaving of the ground.

Any failure will reduce the strength of the soil and will shear off the sand drains, thereby disrupting the drainage of the foundation. Experience has shown that expensive corrective measures are necessary to repair such failures and that settlement will continue for long periods of time in slide areas. Besides control stake and settlement platform readings, information concerning impending failures should be evident from piezometer records (excessive pore pressure readings on the Bourdon gauges relative to specified pressures for the area during embankment construction). Figure 5 shows a typical sand drain installation.

Postconstruction studies were made by the Bureau of Soil Mechanics on secondary settlements that occurred in this project by use of in-place vane shear tests in bore holes and triaxial tests on undisturbed piston samples. The increase in shear

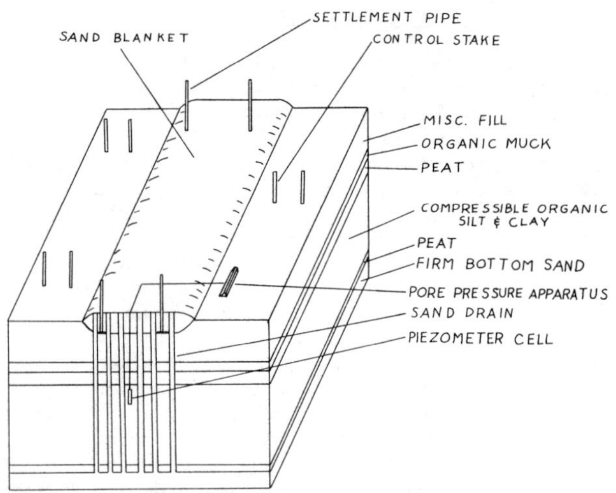

Figure 5. Block diagram illustrating a typical sand drain installation and geological section.

strength expected had occurred, but secondary settlements detrimental to the roadway were also taking place. These settlements were found to be caused by the use of the displacement mandrel in placing the drains, which caused "smear" along the walls of the drain (a compacted layer of the displaced material coats the outside of the drain and decreases permeability).

(2) Whitestone Expressway 1961. A nondisplacement method of placing drains can eliminate the "smear" and was developed by R. Landau of Brill Engineering ('Landau Method'). Based on experience with the Long Island Expressway, this method will not cut the permeability of the compressible strata which would lead to secondary settlement.

To summarize the theoretical considerations: when an embankment load is placed on a soil of very low permeability such as organic silt, the pressure in the soil mass caused by the load is carried by the water between soil particles. This pore water under pressure tends to drain to points of lower pore pressure, in this case at the surface of the water table. As the pore water drains from the soil, the soil particles are able to compress into a smaller volume. This is the process of consolidation of the organic silt layer and the visible evidence is the settlement of the embankment.

In order to hasten the consolidation process, it is necessary to decrease the distance the pore water has to travel through the organic silt of low permeability in order to reach points of lower pore pressure. This is accomplished by driving sand drains into the silt and filling them with very permeable sand. This decreases greatly the time required for consolidation of the organic silt to take place. The pore water flows to the surface of the ground under a pressure head supplied by the weight of the surcharge embankment.

As the consolidation process progresses, an increasing amount of the embankment load is carried by the soil particles and a decreasing amount of the embankment load by the pore water. When consolidation is 100 percent, all the embankment weight is carried by soil particles and there is no pressure from the embankment load in the pore water. Along with consolidation of the soil, there is also an increase in the shearing strength as the soil particles are knit closer together with greater surface contact.

GEOLOGIC CONDITIONS RELEVANT TO HIGHWAY LOCATION AND CONSTRUCTION

Of course the ideal right-of-way for highway location would be along a glacial terminal moraine or solid, uniform bedrock where construction would involve simple cuts and fills. Unstable soil conditions, which add expense to projects, would be eliminated. Unfortunately, existing moraines do not connect leading centers of transportation in every case. One exception is the Interborough-Grand Central-Northern State Parkway system from Brooklyn to eastern Suffolk County Long Island, which follows the gap of the Harbor Hill moraine.

Construction of express highways in urban areas usually calls for skirting the central city, and especially port city sites, which means construction on unstable soil profiles. Or, based on traffic origin and destination surveys between large transportation hubs, routes have to be selected that cover, generally, the shortest distance, which along coasts also tends to cross unstable ground such as river mouths, tidal flats, or old municipal land fills.

To determine construction methods, the results of preliminary subsurface explorations must be utilized to the fullest extent. Profiles must be drawn to determine the total transverse and longitudinal extent of organic soils. Tests must be performed on disturbed and undisturbed samples of this material, such as triaxial shear and bearing ratios, to determine their ultimate strength.

If it is economically feasible, and depth and lateral extent of the entire mass of unstable material is not great, excavation and replacement of this soil eliminates special treatments which can run costs high. Otherwise, amounts of settlements of each unstable soil type must be computed along with the time-rate of settlement so that proper treatments should also take into consideration the reduction of long range maintenance on these highways.

REFERENCES CITED

Porter, O. J., and L. C. Urquhart, 1952, Sand drains expedited stabilization of marsh section: Civil Engineering, v. 22, no. 1, p. 51.

Spangler, Merlin G., 1951, Soil engineering: Scranton, Pennsylvania, Int. Textbook Co., Chaps. 18, 19.

Taylor, Donald W., 1948, Fundamentals of soil mechanics: New York, John Wiley.

APPENDIX A

1. Laboratory tests.

Tests on organic silts indicate 5 percent organic material by weight. The average moisture content (wt. of water/dry wt.) ranges from 90 to 100 percent, the liquid limit is 90 percent, plastic limit is 40 percent, and specific gravity 2.62. The specific gravity is similar to that of granitic material from which the silt was derived.

2. Computation of settlements in unstable soil-rock units.

Engineering-wise, settlement of a structure is not considered detrimental, if uniform. Settlement becomes a factor in foundation treatment when it is differential, and not uniform, throughtout. Primary consolidation of soil masses causes settlements due to a decrease in volume of void spaces. There is a time lag in this type of settlement because water contained in the voids of the soil must be eliminated before the volume decreases. This depends on permeability. Permeability in organic silts varies vertically due to the fact that these sediments are deposited horizontally. Each layer has a different permeability. Therefore, the pore water has to travel laterally at different rates in different layers, and much slower vertically, as it crosses interfaces of varying permeability. Thus, structural settlements can continue over a number of years.

To compute the probable settlement of a proposed structure, two quantitites are needed: (1) the total settlement or total soil-rock unit compression; and (2) the time-rate of this settlement or compression. The data for these computations are obtained by subjecting an undisturbed sample of the soil to consolidation tests in the laboratory (Spangler, 1951). To calculate the total compression (S), the following formula is used:

$$s = d \frac{v_1 - v_2}{1 + v_1}$$

where s =total compression of soil layer for a long time span
d =compressible layer thickness
v_1=initial void ratio (porosity) before applying load
v_2=final void ratio after applying load for some time.

To calculate the total settlement of the compressible soil-rock unit, see formulas and explanation for this in Spangler (1951) and Taylor (1948).

SUBSIDENCE AT SAN MANUEL COPPER MINE, PINAL COUNTY, ARIZONA

Allen W. Hatheway (Department of Mining and Geological Engineering, and the Lunar and Planetary Laboratory, University of Arizona, Tucson, Arizona)

Abstract

The phenomenon of subsidence has plagued the mining industry for centuries. Although subsidence is directly caused by withdrawal of ore from underground workings, the particular combination of rock type, geologic structure, and method of mining at the site is responsible for the physical character of subsidence appearing at the surface.

Continued mining of porphyry copper ores at the San Manuel Mine for ten years has resulted in the formation of three large subsidence pits in a wedge-shaped caprock of Tertiary conglomerates. The pits have a total surface area exceeding 160 acres. Peripheral growth of the pits is controlled by the geologic structure and engineering properties of the caprock and of the ore body, as well as by the relative location of caved blocks and rate of withdrawal of ore.

Specific items of interest at the site are numerous subparallel faults, a definite pattern of advancing ground fractures, and well-defined units of mass wastage within the pits. The study has shown that there are definite relationships between geologic structure in the area and the growth and configuration of the subsidence pits.

CONTENTS

ILLUSTRATIONS

INTRODUCTION

The three subsidence pits shown in the stereo-triplet of Figure 1 are the direct result of the removal of nearly 100 million tons of copper ore from beneath the surface of the ground. The San Manuel deposit is one of the country's major low-grade porphyry copper discoveries and lies between the towns of Oracle and Mammoth, approximately 50 airline miles northeast of Tucson, Arizona.

Active mining by the block-caving method has been going on at the property since 1956. This method of mining requires construction of two parallel levels of drifts, 75 feet apart, below the portion of the ore body to be mined. The lower of the two drifts is called the haulage level and is utilized solely for that purpose. From the uppermost (or grizzly) drifts, finger raises are driven up into the ore body. Since block caving depends upon weathered, altered, and fractured rock, initial blasting is all that is required to begin an even flow of rock into the finger raises. The grizzly level serves as a working access to the finger raises. Broken ore passes directly from the ore body, through steel grizzlies (rails placed about 12 inches apart), to the haulage level, where it is loaded on ore cars and taken to raise shafts for transportation to the surface. Miners with "mucking" duties hand break all boulders retained by grizzlies. An important economic factor is that the ore can be crushed as it moves over several hundreds of feet before arriving at the ore shoots. The primary crusher located at the surface will accept all fragments of 12 inches, or less, in diameter.

Block caving requires no attempt to fill, or otherwise shore up, any portion of the mine which has been drawn. The logical result of this practice is continuing failure of the "roof" of the drawn area until breakage reaches the ground surface, and a pit is eventually formed.

PHYSICAL CHARACTER OF THE SUBSIDENCE PITS

Conditions of subsidence at the surface of the mined area during 1965 had become stabilized, and trends and characteristics of the process were clearly represented. Two portions of the mine were inactive during this period, and the two related subsidence pits had become quiescent. These areas were over the North Ore Body and are seen as the two smaller pits in Figure 1.

The largest of the pits is that corresponding to the South Ore Body; during 1965 it had attained a depth of over 500 feet. Maximum dimensions in plan are 3000 feet in length by 2000 feet in width (Fig. 2). Each pit is ringed by a series of nearly vertical peripheral tension fractures which have marked the advance of the effects of subsidence. The most important of these fractures are shown in Figure 3. Downward movement of blocks of conglomerate occurs at the surface in response to a general lowering along the center line of the pits as ore is removed through draw raises many hundreds of feet below. This movement is marked by an advancing pit escarpment and the formation of new tension fractures as mining stresses are concentrated against the boundary areas of the pits.

GEOLOGIC SETTING

Lithology

The granitic rocks comprising the ore body form part of a basement complex that is exposed at the surface in a few places, but which is mostly buried by more than a thousand feet of conglomerate. It is this wedge-shaped caprock of conglomerate that figures most prominently at the surface in the expression of subsidence. Geologic conditions before mining are shown in Figure 4, where the caprock is represented by the Cloudburst, San Manuel, and Quiburis formations of early Tertiary to Pliocene age.

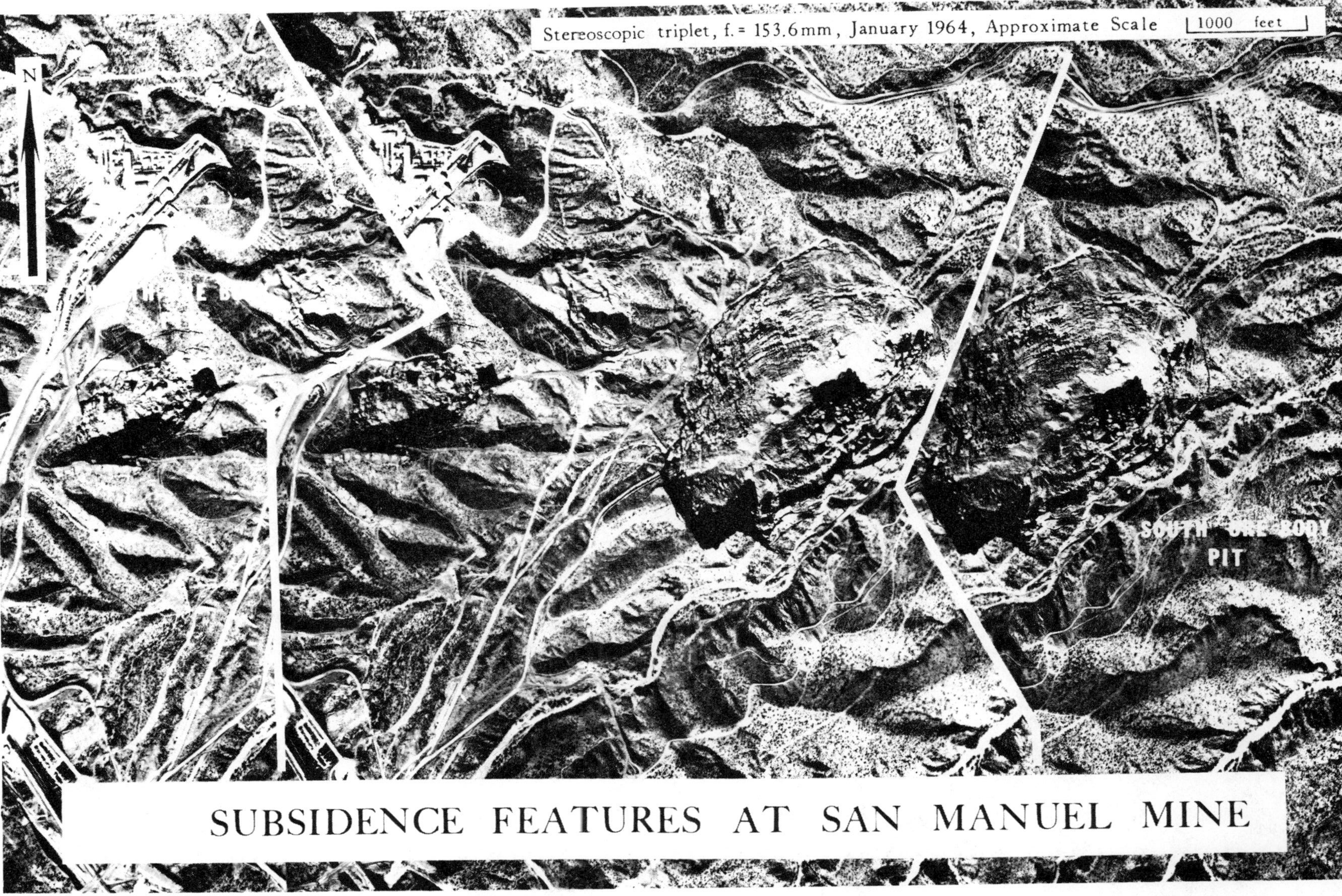

Figure 1. Subsidence at San Manuel Mine

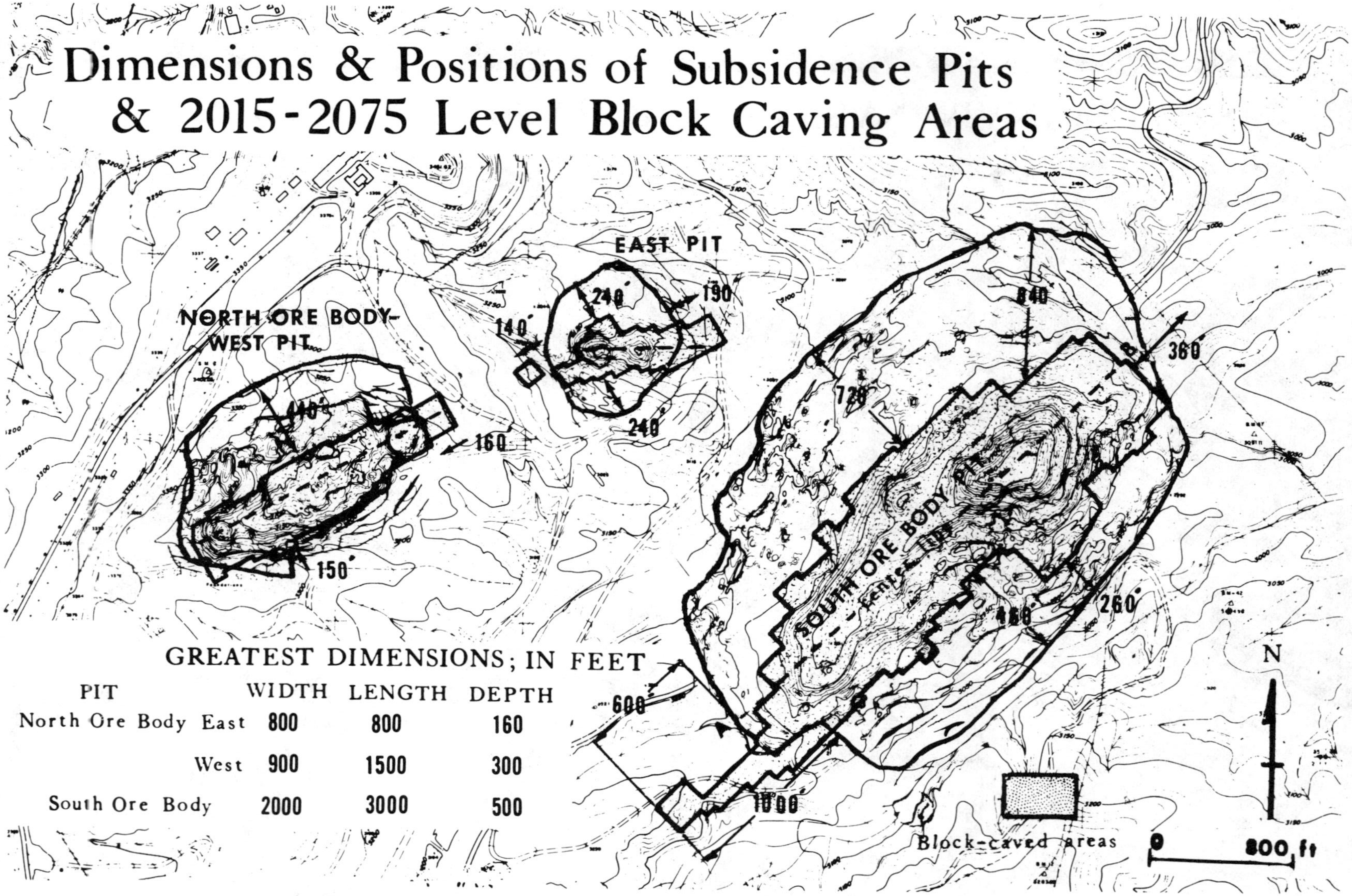
Dimensions & Positions of Subsidence Pits
& 2015-2075 Level Block Caving Areas
NORTH ORE BODY
WEST PIT
EAST PIT
SOUTH ORE BODY PIT
140
150
160
140
240
190
240
720
840
360
260
460
600
1000
GREATEST DIMENSIONS; IN FEET
PIT | WIDTH | LENGTH | DEPTH
North Ore Body East | 800 | 800 | 160
West | 900 | 1500 | 300
South Ore Body | 2000 | 3000 | 500
Block-caved areas
N
0
800 ft

Figure 2. Dimensions and positions of subsidence pits.

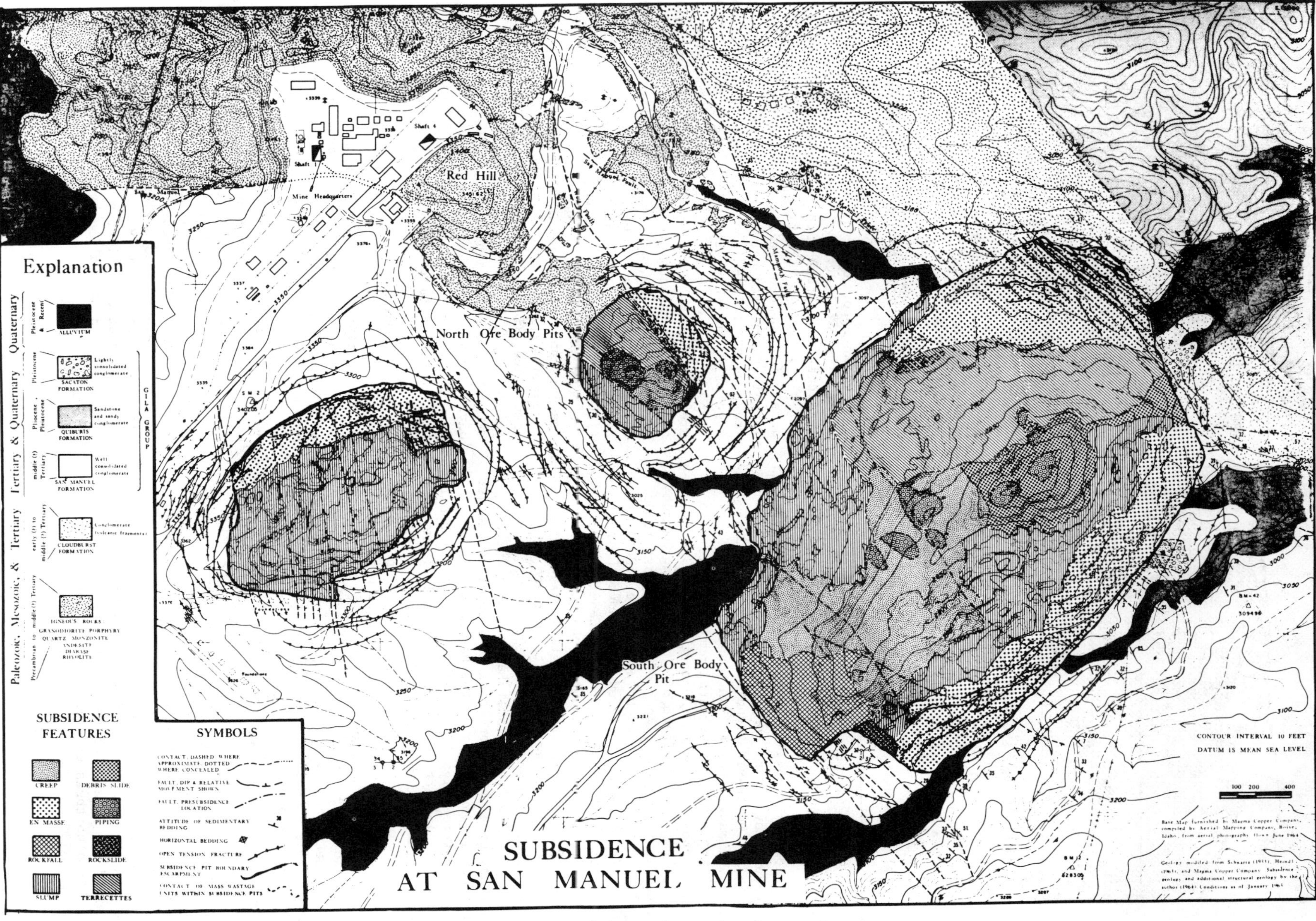

Figure 3. Faulting and tension fracturing.

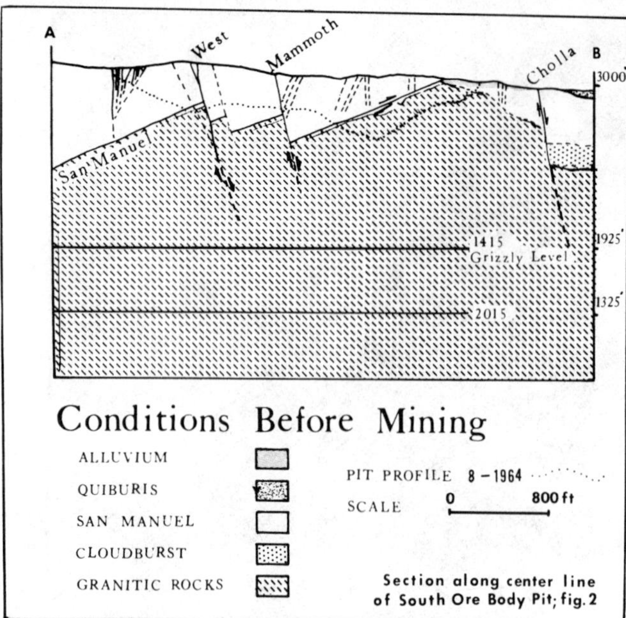

Figure 4. Conditions before mining.

Geologic Structure

Both the massive rocks of the ore body and the beds of overlying conglomerate have been cut by three major northwest-trending faults. Associated with these major faults (San Manuel, Mammoth, and Cholla) are many near-vertical normal faults that exhibit opposing dips to the northeast and the southwest. The greatest effect of this faulting on subsidence is that these planes of weakness are found in all the rocks affected by subsidence. Deformation by faulting has occurred throughout the geologic history of the mine area. Each succeeding faulting event has substantially weakened the host rocks so that the effects of ore removal from below the surface are easily transmitted to the surface along these and other planes of weakness.

The San Manuel fault has resulted in the placement of a thrust sheet of conglomerates against the older igneous rocks of the basement. This single feature (Figs. 3 and 4) transcends all other structural elements in the area in its influence on the character of subsidence. There is a greater abundance of parallel faults in the portion of the thrust sheet surrounding the subsidence pits than to the north of the San Manuel fault.

Northward expansion of the South Ore Body pit has been checked by the presence of the Cholla fault. This steeply dipping fault can be easily traced on aerial photos (Fig. 1) due to the marked compositional and tonal differences shown by the San Manuel and Cloudburst formations (both conglomerates) on the south and the north sides of the feature, respectively. Immediately within the northwest corner of the South Ore Body pit, a series of genetically related en echelon faults have caused subsidence movement quite close to the Cholla fault which has opened in the manner of a peripheral tension fracture. The Cholla and its related faults have deterred the advance of the pit for more than 4.5 years in a northeasterly direction. The faults have relieved a great deal of tensile stress by releasing large masses of conglomerate from the surrounding stable ground.

PHYSICAL CHARACTER OF THE ROCKS

The rocks involved with subsidence at the mine are not strong when considered as gross units. As small blocks, free of structural discontinuities, the sedimentary rocks of the overlying caprock have a great deal of strength. However, the discontinuities in both the caprock and the ore body, and chemical and hydrothermal alteration in the ore body, have destroyed any appreciable strength when the units are considered as a whole. Rock strengths have been referred to in general descriptive terms due to an inability to make meaningful bulk strength measurements. In the future, it is hoped that worthwhile strength measurements may be made that will take into consideration the large number of discontinuities present.

The Ore Body (Quartz Monzonite and Granodiorite Porphyry)

An important requisite and key to success in mining the ore body is the relative ease with which the rocks will break into relatively small fragments for block caving. Blasting is routinely used in developing the draw raises, but the flow of rock is largely unhampered once it has begun. The structural weakness of the ore is entirely due to the widespread fracturing and completeness of alteration that Creasy (1965, p. 11) noted as being common throughout the ore body. Creasy found that this rather intensive alteration consisted of varying amounts of unidentified clay minerals, as well as sericite, chlorite, calcite, and epidote. Utilizing the rock classification for engineering purposes proposed by Coates (1964; as modified by Burton, 1965), the rocks of the ore body show "blocky continuity" with fracture spacing ranging from about 3 inches to 6 feet. Throughout those portions of the deposit that have been mined, the rocks have been highly fractured, jointed, and sheared. The ore body contains a variety of aplite, diabase, and rhyolite dikes that have occupied pre-existing fractures. These dikes have the general effect of weakening the rocks, and they have also facilitated block caving.

The Conglomerates

Cementation and induration of the coarse and angular fragments provide the caprock with considerable strength. Lensing is prevalent throughout, and nodular tuff beds, a few inches thick, are present in the caprock. The structural integrity of the unit has been weakened with the addition of poorly bonded contacts at the bedding planes separating the tuff beds and the surrounding rock. Shearing strength of the individual blocks of the conglomerate is relatively high; the critical (maximum possible) height of the boundary escarpments of the pits approaches 50 feet. Natural breaks along these walls will occur at approximately 90 degrees and the faces will stand for months. Collapse will occur only when weathering and saturation by rain have combined with removal of support at the toe of the face.

Thickness of the conglomerate caprock affects the process of subsidence in three ways: (1) it increases the lithostatic pressure (or vertical applied stress) on any given portion of the caprock that is bounded by faults; (2) it increases the thickness of the slab of conglomerate that must be broken, as support is lost through caving activities going on at depth; and (3) it aids in breaking underlying ore after the initial stages of subsidence have occurred.

GROWTH AND DEVELOPMENT OF THE SUBSIDENCE PITS

The first effects of subsidence to occur on the surface of a mining property are naturally time dependent upon the rate and amount of ore withdrawn from the mine, and upon the geologic conditions of the ore body and the caprock (if any caprock exists). In addition to these factors, climatic conditions will probably play an active part in the behavior of the surficial materials in subsidence pits. It is difficult to assign a definite time span to the formation of the features that occur during the process. However, the character of subsidence at San Manuel Mine has been marked by two general stages: a preliminary period of tensional fracturing, together with gentle settlement of the ground surface, and a final period of slowly developing, pit-like sinking of the ground surface.

The first noticeable effects at the ground surface came after approximately 100 days of draw, and are referred to in mining terminology as "pipes". These features are not to be associated with those of "piping" in soils caused by removal of the fine fraction by flowing water. Pipes are upward extensions of caved areas that provide a path for the transport of broken fragments of conglomerate into the mined area below. Pipes are illustrated diagrammatically in Figure 13 and can be seen in Figure 1 as crater-like depressions resembling those formed by chemical explosives in soil. Formation of the pipes is partly due to the gravitational effect of vertical loading induced by masses of broken conglomerate resting above, and also due to stresses activated by mining about their lower extents.

Peripheral Tension Fractures

Contemporaneous with the formation of pipes at the surface is tension fracturing of the ground surface. Initially, these fractures are produced by a gentle settling of the ground surface accompanying the first withdrawal of ore. With time, these fractures widen and become as deep as 300 to 500 feet. Continued widening of the fractures results in varying amounts of vertical and horizontal movement, together with a general shifting of individual blocks of conglomerate toward the center of each subsiding area.

The South Ore Body was the site of the first ore withdrawal by block caving at the mine, and the fractures first formed over this locality. After two years of active mining, the first well-developed boundary escarpment began to form along the former fractures, and a nearly equidimensional subsidence area was outlined (Fig. 7). As each of the ore bodies was caved and drawn, tension fractures developed and ringed each pit with a series of both straight and arcuate fractures. In all cases, these fractures have remained remarkably parallel to existing escarpments and have reflected the future expansion of each pit. Growth of the tension fractures has followed a definite pattern. This begins with an initial parting along a pre-existing fracture or fault. Initial vertical movement in the freed block is followed by horizontal movement. Rotational movement occurs as the toe of the downhill mass is removed through transport of material into one of the pipes that exists at the pit bottom. Stable ground surrounding the fractured areas remains intact until tensile stresses are concentrated against it to form new peripheral fractures.

Orientation of the straight fractures follows two modes: (1) parallel to (and including) existing normal faults in the area; and (2) parallel to both the long dimension of the caved area below and to the boundary escarpment that marks the longest dimension of each pit (Figs. 1 and 3).

The tension fractures are formed as a result of stresses as the caprock attempts to form an arch to resist widespread caving. The fracturing arises in response to vertically applied loading imposed by the weight of the caprock itself. The loading also induces further breakage in the caprock as underlying support is removed through the drawing of ore in the mine workings below. The caprock has been faulted and fractured into a large, but unknown, number of individual blocks. Due to this condition, the effects of vertical loading and the resulting tensional failure cannot act laterally for appreciable distances beyond the boundaries of the caved mass of ore lying directly below the conglomerate. Along with this line of reasoning, there should be a fairly close correspondence between the outline of the caved area of the ore body and the location of the newest tension fractures at the surface. At the San Manuel Mine this is not the case, because the outer limits of the subsidence pits far surpass those of the underground caved areas (Fig. 2). The reason for this condition probably lies in the fact that the vertical loading of the underlying structurally weak rocks of the quartz monzonite and granodiorite porphyry permits a secondary phase of breakage under the weight of displaced blocks of conglomerate (discussed later).

The general state of stress surrounding the formation of peripheral tension fractures is shown on Figure 5. The underlying igneous rocks are structurally weaker than the overlying conglomerate and a cantilever effect, combined with the varying thickness of conglomerate, makes vertical stress S_{v1} greater in effect than S_{v2}. Also, tension is concentrated at the pit escarpments where a lack of constraint toward the pit will cause failure of the slope in tension. Differences in magnitudes of S_v arise from the differences in thickness of caprock in the overlying thrust sheet (Fig. 5). If uniform removal of support occurs during mining, that portion of the area under the greatest vertical loading will certainly rupture first. This assumes that all other factors are equal, and there is virtually no lateral constraint. A lack of lateral constraint occurs wherever the conglomerate has been removed along a tension fracture through rotational or translational movement.

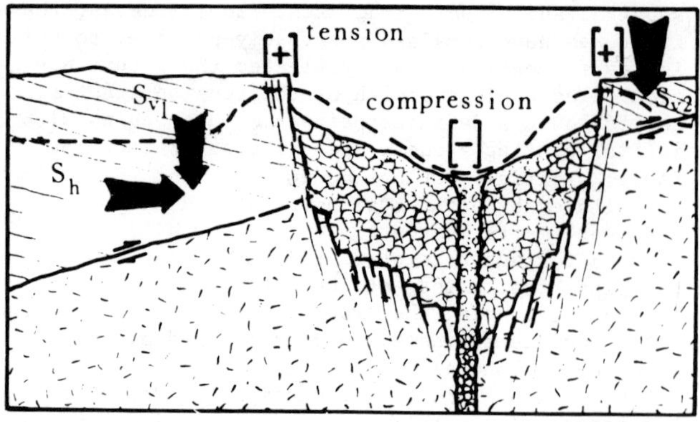

Figure 5. Diagram showing the effect of attitude of bedding and removal of lateral constraint on the formation of tension fractures.

General Trend of Development

Growth rates for the pits have been plotted (Fig. 6) showing the total footage of the periphery of each pit measured at regular intervals. The plot is semilogarithmic and shows that the growth rate is quite uniform throughout the process. A flattening of the curve indicates a trend toward eventual stabilization of the process. This stabilization trend began to appear after some seven years of continual mining of the South Ore Body. The growth rates of the two remaining pits had begun to flatten before they were inactivated in 1964. The initial points of each of the three curves represent the first appearance of a well-defined, continuous boundary escarpment at a particular pit (see Fig. 7, growth lines for July 1956).

South Ore Body Pit

In the first stage of pit development, fracturing was noticed on the surface about three months after draw had been initiated. These fractures were followed, at seven months, by the breakthrough of a pipe. At this time, Storms and Hardwick of the U.S. Bureau of Mines (1957, Structural analysis of block-caving operation, San Manuel Copper Mine, Pinal County, Arizona: U. S. Bur. Mines, Tucson, Arizona, progress report, June 1953-January 1957 (unpublished p. 13) estimated the pipe to be about 100 feet in diameter. The same observers also noted that nearly two years after mining had begun, subsidence had attained a rate of about 0.26 feet of vertical movement per day over the center of mining activity.

If the caprock is considered as a beam overlying the mined area, the fracturing resulted from deflection of the mass of the beam of conglomerate (Fig. 8). Deflection of the original beam resulted in a concentration of compressive stresses above the upward-advancing pipe and tensile stresses along the margins of the pipe. The beam yielded continuously, in an elastic fashion, with some localized vertical slippage along faults and other planes of weakness. The beam then ruptured at the time the advancing pipe penetrated the full thickness of the conglomerate, and appeared at the surface. This action must be considered to be localized if the penetration is taken to be merely a hole in a plate. However, with the failure of the beam as a unit, tensile stresses were redistributed along the margins of the pipe and there caused breakage of the rock along planes of weakness and at areas of critical stress concentration (Fig. 9). Areas of critical stress concentrations occur around points at which the applied tensile stress exceeds the bulk tensile strength of the conglomerate making up the surrounding stable ground. As is noted in Figure 8, there is a degree of tensile stress surrounding the area undergoing application of compressive stresses. The upward direction of these tensile stresses tends to elevate the ground surface. Steel survey pins placed in concrete as a project of the U.S. Bureau of Mines actually reflected this movement.

The subsidence area over the South Ore Body began its growth at approximately the center of the present pit and expanded rather unevenly with respect to time. Initial development of the escarpments favored growth in a northwesterly direction (Fig. 7). A further consideration of pit growth has shown that the effects of subsidence extended first into the area of the thinnest caprock (to the northwest), and later into the thicker portions of the caprock (southeast). This initial growth was from a central area not coincidental with the center line of the pit. Recent mining activity at depth has brought the center line of the pit into coincidence with the center line of the caved blocks below (Fig. 2). There is no indication that there is any normal linear relationship existing between the depth of the mined area and the horizontal influence of subsidence at the surface.

The effect of faulting is such that there has been some parallelism between known fault locations and the positions of the boundary escarpments. However, this has not always been the case as is noted in Figure 7. The anomalies are due to the presence of unrecorded faults and fractures, and by local modifications of the escarpments caused by spalling from their free faces. Faults were influential at the surface in sculpturing the escarpments and the presence of faults acted to slow the process of growth between succeeding pit boundaries. After the previously noted pit boundary stabilization, which occured after about seven years, subsequent growth of the pit has been localized and has been

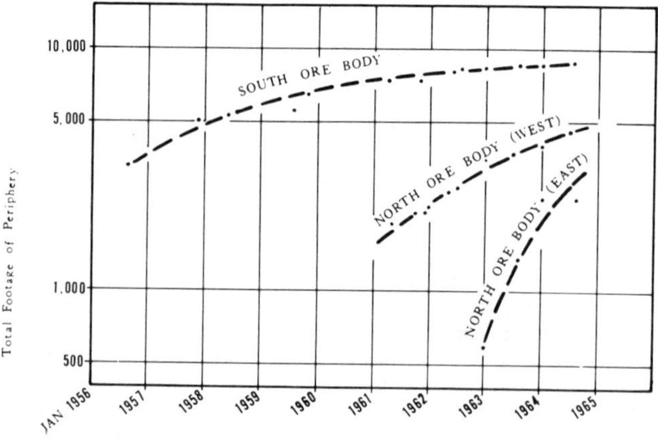

Figure 6. Peripheral growth as a function of time.

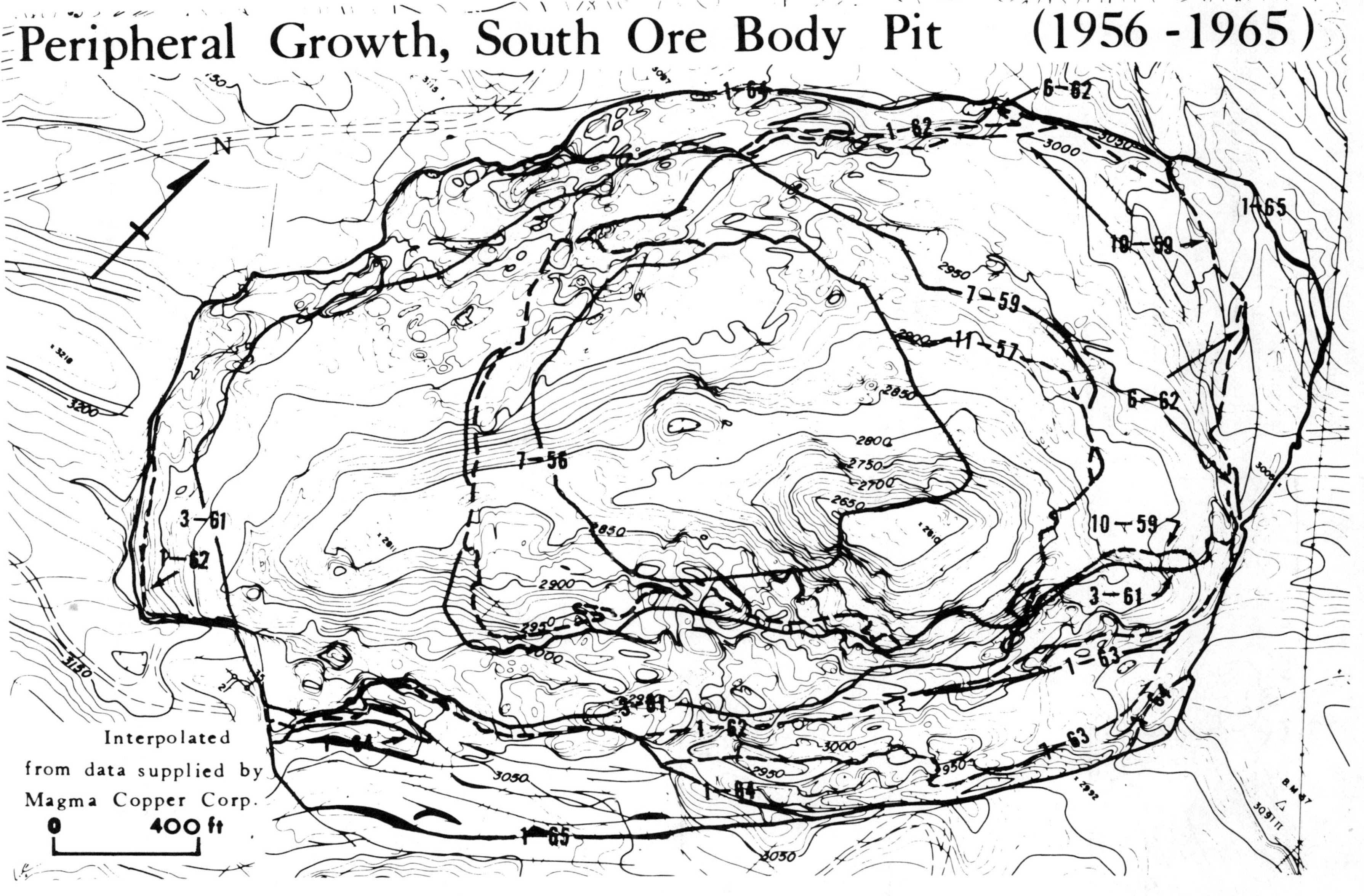

Figure 7. Peripheral growth, South Ore Body pit, 1956-1965.

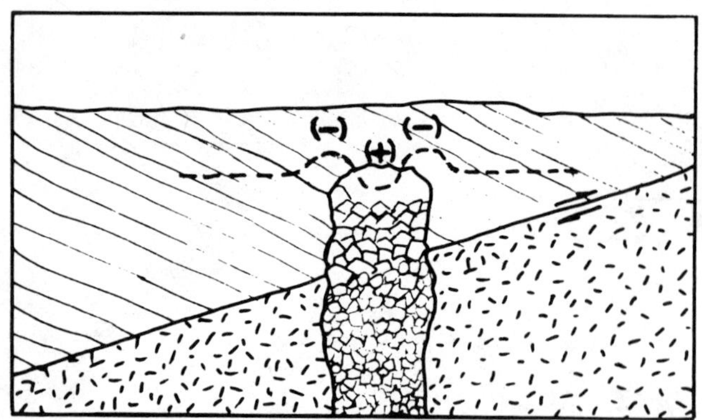

Figure 8. Diagrammatic sketch showing concentrations of tensile stresses that cause breakage around the margin of the growing pipe, while compressive stresses aid in breakage overhead.

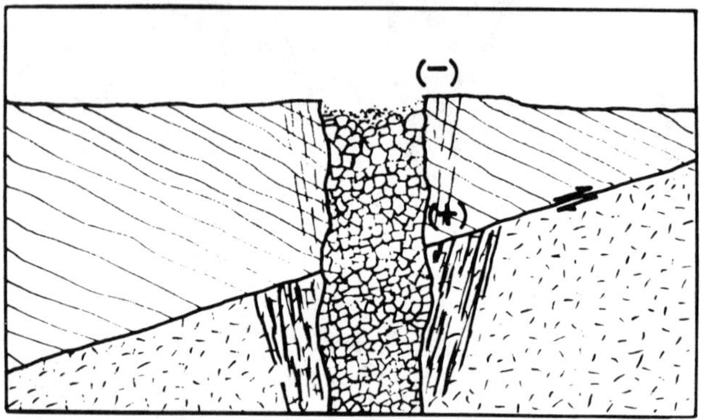

Figure 9. State of stress in the conglomerate caprock at the time of penetration of the unit by a pipe of broken rock. Presence of the weaker porphyry below the caprock induces flexural stresses in the surrounding conglomerate, much like those found in flexure of a cantilevered beam.

caused by lowering of the pit bottom. There has also been ensuing removal of material from the bases of the escarpments through mass wastage.

North Ore Body West Pit

Conditions in this intermediate-sized pit have been substantially different from those of the other two pits. A total time lapse of 560 days occurred between initiation of draw and measurable ground settlement (C. L. Pillar, 1965, personal commun., former mine superintendent). During this period, it is most likely that the caprock was arching over the caved blocks. When settlement reached a critical point, the ensuing subsidence was rapid, and large amounts of conglomerate broke and caved into the forming pit. Thickness of the caprock in the vicinity of this pit ranges from 350 to nearly 1100 feet, due to thinning on the nappe of the thrust sheet. Density of faulting is greater than at any other location over the mined area (Fig. 3).

Pit growth has been much the same as that of the South Ore Body pit, with the exception that growth lines have exhibited a greater degree of regularity with a more even progression from one line to another. Northeast and southwest escarpments have apparently followed the contacts of pre-existing faults. The usual peripheral tension fractures occur, but these are of a greater number and density than in the South Ore Body area.

North Ore Body East Pit

Curtailment of draw under this pit has limited its growth. Conditions here may be taken to represent a subsidence pit in its early stages of growth. Thickness of conglomerate over this pit is minimal, and growth is centered about a tongue of quartz monzonite exposed at the surface. Because the pit did not have sufficient time to grow and become stabilized, the area covered by pipes is abnormally large. Peripheral tension fractures have extended laterally to a greater distance than in the other two pit areas. This is partially due to the slight thickness of the conglomerate caprock causing greater fracture development and in part due to timing of mining. The pit began to form in 1962, which means that some degree of fracturing by mining stresses probably originated, over the previous six years of mining, under the areas immediately adjacent to the pit.

MECHANICS OF SUBSIDENCE

There are a number of parameters acting in the formation of these large pits. These factors are both geologic and man-initiated, and must be considered in relation to each other and in terms of time and of relative magnitude. At the San Manuel Mine, it is quite evident that the surface pits have been created by the removal of support at depth and by a general collapse of the igneous rocks and overlying conglomerate. What is not nearly as evident is the order of importance of the factors that have influenced the formation of each pit. The initial action began in the porphyry copper ores at depth, and then was transmitted upward into the conglomerate. Once broken, the caprock added the effect of a displaced mass of rock acting downward upon the underlying porphyry. In both the undisturbed conglomerate and the porphyry, there had been a balance of two general stress fields: (1) vertical force represented by the weight of the igneous rocks and the conglomerate; and (2) horizontal forces related to both the vertical force and possible tectonic stresses.[1] The first result of the imbalance between the two stress fields was initial sagging of the caprock followed by rupture of the conglomerate in response to the removal of support from depth. This, in turn, caused conditions of localized loading, and also spread breakage within the body of quartz monzonite and granodiorite porphyry lying below.

---

[1] In a general sense, the relationship between vertical stress (S_v) and horizontal stress (S_h) is dependent upon the bulk Poisson's ratio for the conglomerate (μ): $$(Sh) = \frac{\mu}{1 - \mu} (Sv).$$

Behavior of the Igneous Rocks

For any given area of the mine which is undergoing an initial draw, it is believed that breakage in the prophyry progresses upward toward the ground surface with little lateral deviation. Because the ore is heterogeneous with respect to strength characteristics and the fracture systems are not at all uniform, there may be some local variation in the angle of break (angle of draw) in the rock and in the upward direction of growth of the advancing break line in the porphyry. These changes are caused by anomalous stress concentrations and by the presence of areas of varying rock strength. Residual stresses within the mine workings are accumulated as mining shifts loads from one area to another. Concentrations of these stresses damage drifts in the mine from time to time. These stresses have required placement of a great deal of steel and concrete supporting material in crosscuts and drifts. Rock bursts are not known, but slabbing of the back and slow closure of drifts have been common in localized areas at varying times.

The term angle of subsidence is often used to record data concerning the inclination of a line drawn between the outer edge of the underground workings and the boundary escarpment of the overlying subsidence pit (Fig. 11). Earliest theories held that the failure surface between the underground workings and the surficial extent of subsidence was curvilinear. However, it is more reasonable to think of the breakage boundary as a series of parallel, planar surfaces, such as those seen in the left portion of Figure 11. These surfaces have been formed in shear through vertical loading imposed by the dislocated masses of conglomerate. This condition is brought about by removal of uniform support for the mass of the caprock after the beam or arch of overlying conglomerate has been destroyed by the pipes of broken rock. The individual fragments of conglomerate lying broken within the pit exert the vertical loading. The process is demonstrated by the sequence of events (item 4) associated with subsidence (Fig. 11). Angles of subsidence at San Manuel Mine range from 78 to 90 degrees for the three pits.

Rock Breakage along the Pipes

Growth of the pipes is directly upward, without appreciable widening of the diameter of each pipe. There is probably a slowing of the process upon reaching the interface between the granitic rocks and the overlying conglomerate at the San Manuel fault plane. Direct transmission of stresses across the interface should occur after the short period of time involved in the initiation of failure in the dipping beds of the conglomerate. After penetration to the surface by a pipe has been completed, and large masses of conglomerate have been displaced into the forming subsidence pit, localized breakage occurs in the quartz monzonite and granodiorite porphyry making up the walls of the pipes. Pressure distribution from the moving masses of broken conglomerate cause renewed breakage from the upper levels of the pipe walls downward (item 5, Fig. 11). Shear failure, then, is responsible for a reversal of breakage direction and a lateral expansion of the subsidence area is achieved.

Behavior of the Conglomerates

Initially, the state of stress in the conglomerate may be likened to a beam of heterogeneous and anisotropic material spanning an unsupported gap. Arching is represented by a condition of spanning without deflection (or sagging) of the beam. The ultimate degree of arching short of failure is directly dependent upon the structural defects located in the beam of caprock. While the unit is intact and benefits from total support at its interface with the underlying igneous rocks, there is no deflection, and there are essentially two primary stresses in the mass of conglomerate. These are vertical applied stress from the lithostatic pressure of the conglomerate and horizontal applied stress in accord with the relation stated previously. Horizontal stress is relatively unimportant in regard to deflection because, even in homogeneous, isotropic material, it has been found to constitute less than 10 percent of the total bending moment of a simply supported beam.

To an extent, stresses are accommodated within the mass by readjustment along the many discontinuities that are present in the rock. However, whatever degree of arching (i.e., lack of appreciable deflection) that is present is eventually destroyed by increasing amounts of tensile failure and by parting of bedding planes in the caprock. Localized tensile failure also destroys the effective clamping or holding of beams over given voids (pipes short of penetration). Removal of support leaves gravity-loaded slabs of rock partially unsupported over these voids. With movement of blocks of conglomerate into the subsidence pits, there is a removal of lateral constraint at the escarpments, and rock is free to spall from the faces bordering the pit at the surface.

Absence of "Doming" as a Failure Mode

The formation of a "dome" is often associated with subsidence in which a convex surface breaks upward in the mined area (Fig. 10). There is no indication that this phenomenon has been present at San Manuel mine. The factors limiting the development of such a feature are the presence of discontinuities and dipping beds in the conglomerate,

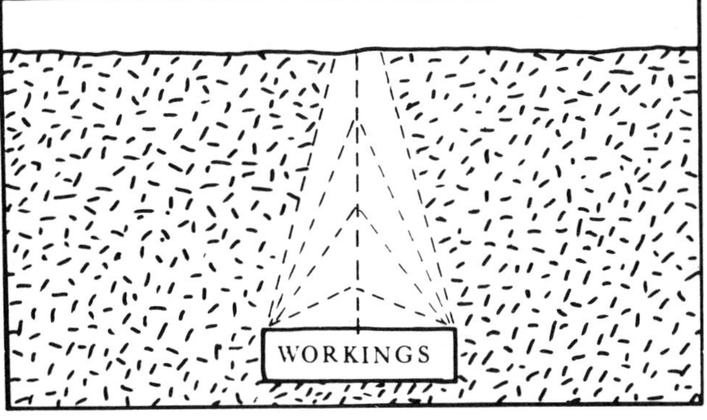

Figure 10. Modification after Crane (1931) of diagram showing successive stages of development of an idealized subsidence pit with initial "domes" formed over the underground draw areas.

together with localized areas of draw and varying rates of draw. These factors preclude the keying or reinforcing of the flanks of any developing arch. Also, the perpendicular orientation relating strike of conglomerate beds to the long axis of the caved areas and pits (Fig. 2) has created a condition allowing for comparative ease of breakage of the overlying conglomerate through the exposure of many edges of bedding in any given area of a pipe or coalescence of pipes.

Rock Failure

Failure in tension will occur preferentially along discontinuities that previously exist in both the ore body and the caprock. After sufficient downward movement of material has created free faces (escarpments) at the surface, a gradual undermining of the support of any face will lead to a failure of the block of conglomerate behind the face. Failure will occur along a critical plane that depends upon the height and strength of the rock (many times occurring along a convenient plane of weakness). After breakthrough to the surface is complete, shear will probably appear along pipe walls. A coalescence of pipes could then occur in the mined area. There is also some amount of spalling through parting along bedding planes.

Effect of Faults

In addition to the over-all effect of faults in creating structural weakness in the caprock, these features have also affected the angle of subsidence and general surface growth at the property. At the stage of development in 1965, the short-dimension escarpments of the pits had formed directly on fault planes. The presence of faults and fractures in the caprock is such that the conglomerate is more likely to break into smaller fragments in the areas paralleling the long dimensions of the pits. The angle of subsidence is largest when measured in a plane parallel to faulting at the mine. The greater ease of breakage has permitted the pits to grow faster in their long dimensions and, hence, has preferentially enlarged the angle of subsidence.

Attitude and Thickness of the Caprock

Attitude and thickness of the conglomerate has a gross influence on subsidence. The relationship is probably such that an excessive thickness of conglomerate occurring in conjunction with break lines in the porphyry combine to cause the surface breakage to occur nearer the caved area. In both the South Ore Body and North Ore Body West pits, this asymmetry occurs, placing the closest escarpment in the area of greatest thickness (thickness increases to the southwest paralleling the lip of the San Manuel thrust fault).

Formation of Escarpments

The conglomerate always tends to break along pre existing planes of weakness. Where an escarpment forms parallel to the outlines of the underground caved area, failure comes in response to stress concentrations along the edges of the broken and caved ground beneath the surface. As shown by Figure 7, a period of time is required to stabilize the process and to allow for caving to advance so that it acts along a front. The ultimate result is a nearly straight line of breakage that forms the escarpments of the long dimensions of the pits. The escarpments also advance laterally due to mechanical weathering and sloughing of fragments into the pits. This action is aided: (1) by loss of constraint at the slope face and removal of support at the toe of the escarpments, and (2) by mechanical weathering and desiccation of the conglomerate forming the faces. Escarpments always fail along nearly vertical faces.

Summary of the Subsidence Process

A chronological summary can be developed for the subsidence process at San Manuel Mine. The following stages are keyed to the two hypothetical cross sections found in Figure 11:

(1) Static loading begins with removal of ore and disturbance of the material supporting the overlying caprock. Upward, near-vertical caving occurs mainly along stopes in the form of pipes. Voids formed by the pipes occur directly over the areas of draw with little or no lateral deviation at the surface. Pipes have always occurred within the boundaries of the caved area. These features form in the granitic rocks largely by tension, induced by lithostatic pressure, and in the conglomerate by both tension and spalling along bedding planes. Spalling and uneven breakage of the granitic rocks occur at the free faces bordering the pipes in response to relief of residual and active stresses in the immediate vicinity.

(2) A pipe is shown extending to the interface between granitic rocks and the conglomerate at the plane of the San Manuel fault. A small amount of lateral breakage occurs, followed by readjustment and continued vertical expansion of the pipe into the conglomerate.

(3) As a pipe reaches the surface, downward pressure begins as masses of conglomerate begin to weaken and fail along fractures and faults within the mass. Breakage is the result of tangential stresses surrounding the surface of the pipe. Because altered rocks are weakest under shear stress, the pipes are probably enlarged laterally in this fashion.

(4) Tension fractures open along planes of weakness in a downward direction from the surface. Initially, the fractures do not transect the entire thickness of the caprock. Tensile breaks occurring in the sound ground have a characteristic break of 85 to 90 degrees as measured from the horizontal. Tensile stresses are predominant, and if contemporaneous compressional stresses exist, they would appear at the lower edge of the mass of intact conglomerate. This condition is analogous to the flexural stresses developed through normal loading of a cantilevered beam.

(5) Displacement of independent masses of conglomerate increases as a pit develops at the surface. This forces an increased area of breakage along the walls of the pipes (stopes). There is some breakage of rock along the throats of the pipes due to stress relief at the free face of the pipe boundary. Increased widening and coalescence of pipes

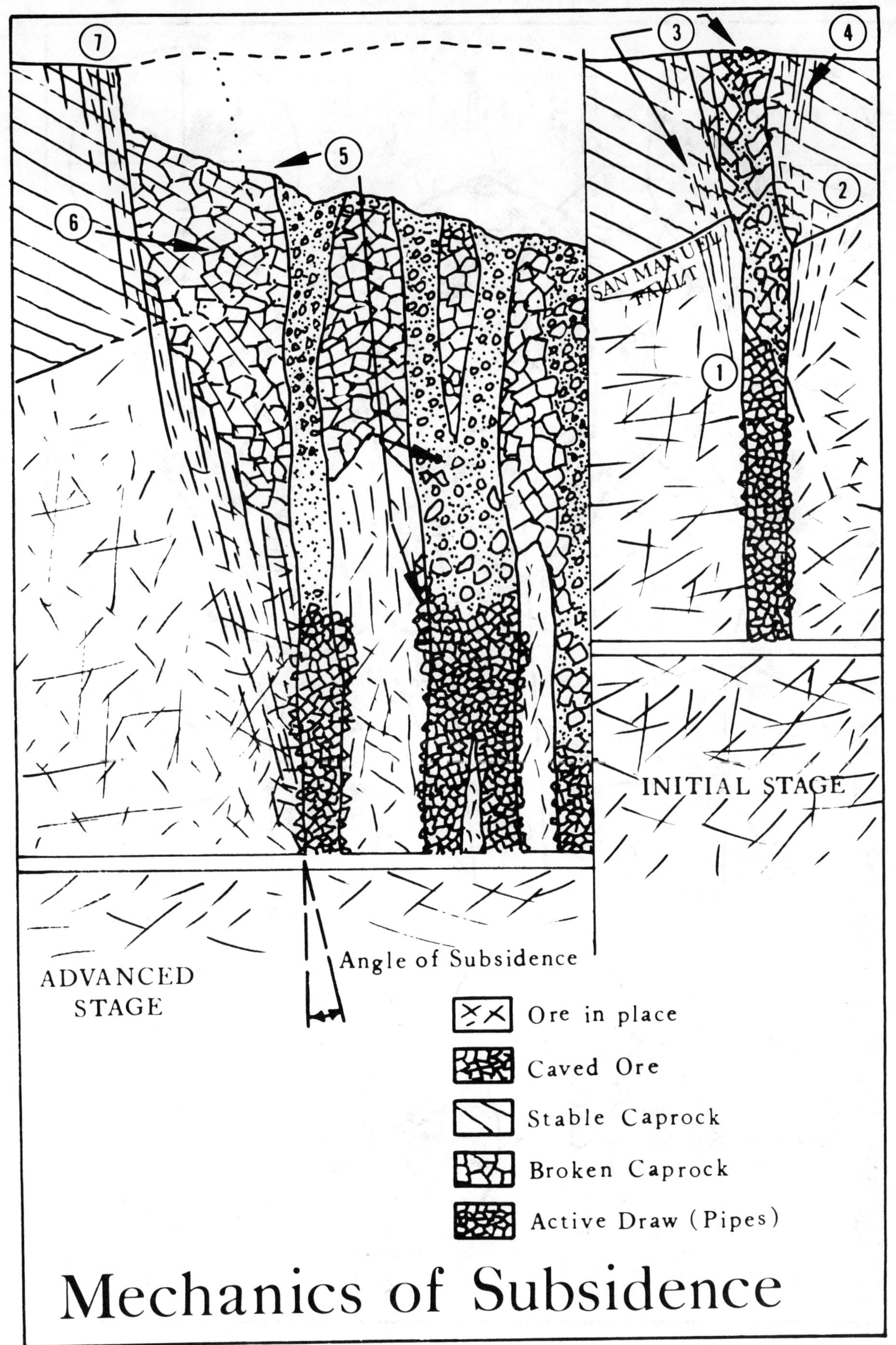

Figure 11. Mechanics of subsidence.

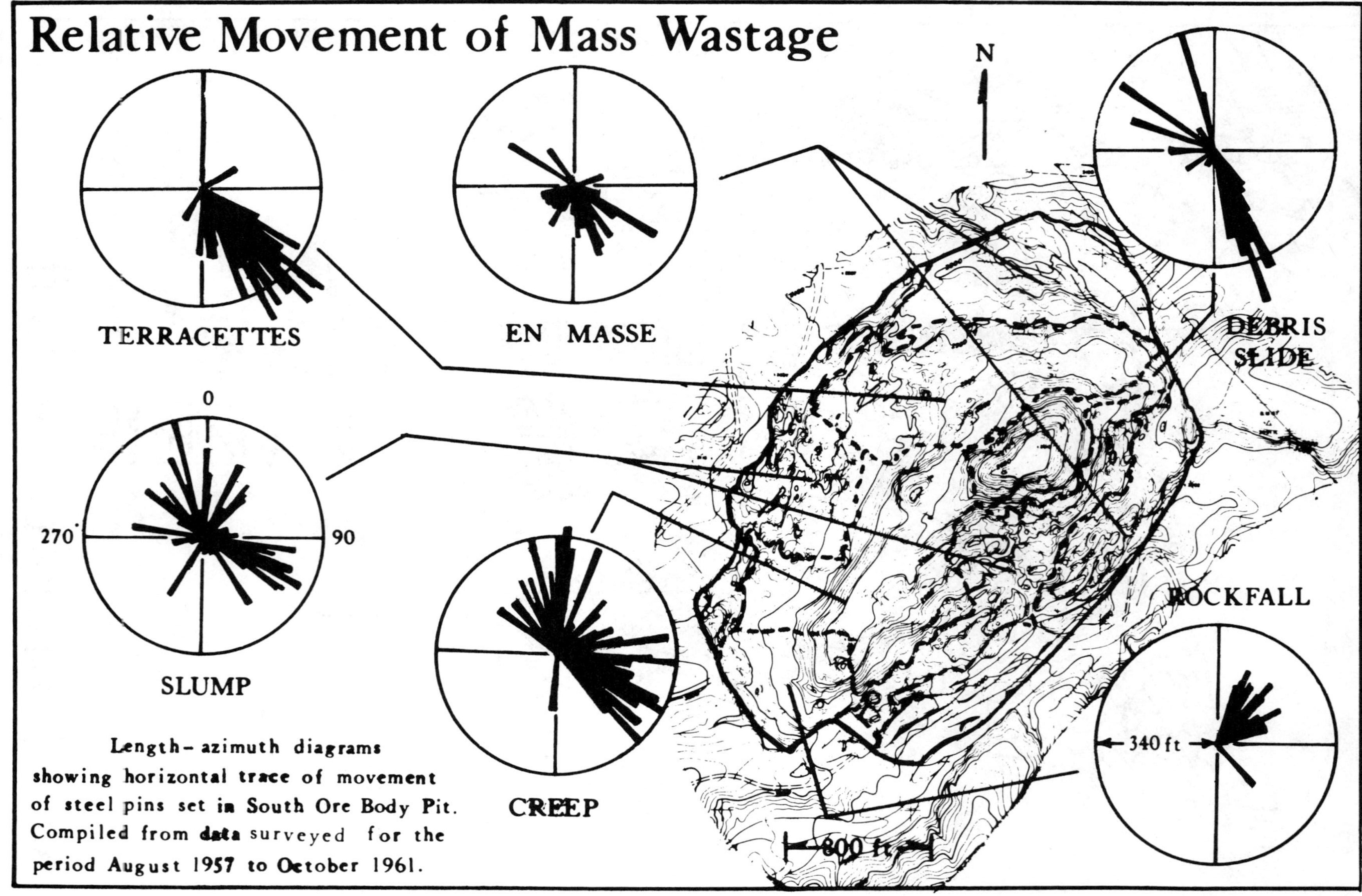

Figure 12. Relative movement of mass wastage.

occur with contemporaneous weakening of the caprock. The pipes are probably eventually filled in their upper reaches by conglomerate. There will be some filling in at lower levels by lateral movement of broken porphyry. An uneven blanket of broken conglomerate is formed over the entire caved area.

(6) A lateral component of movement is introduced. This is produced by the shifting of broken conglomerate toward the center of the pit as sufficient amounts of rock are removed through the pipes. Conglomerate is crushed on its way to the pipes, and the material is further reduced in size by the action of weathering. Continued breakage of the caprock into unsupported and unattached masses results in shearing of porphyry at depth. Boundaries of the pits expand in this fashion.

(7) Faults occasionally limit the lateral growth of the pits by releasing large masses of caprock. Eventually, incipient tension fractures develop into escarpments as failure occurs along existing faces.

MASS WASTAGE

As soon as subsidence has developed sufficiently to allow for vertical movement of broken rock at the surface, there is an introduction of downslope movement of broken conglomerate toward the pit bottom. Distinct types of movement and units of movement appear in time and are easily recognized on aerial photographs. These units may be seen in Figure 1 when correlated with labeling in Figures 12 and 13. There is a continual modification of the surface of each unit through its downslope movement over a period of time. All units of mass wastage represent landslide movement.

The essential factor involved in mass wastage is the withdrawal of ore in the mine, which has resulted in the formation of a center line along the lowest level of each subsidence pit. Appreciable amounts of material at the surface are fed into the subsurface through the ever-present pipes, and there is a constant removal of material from the lower levels of the pits. Stimuli for movement changes slightly with time, and in response to the changing areas of draw in the underground workings. Because rock breakage continues during downslope travel, there is a change in the physical character of each unit of mass wastage during the transport from one point to another. There are currently eight distinct types of mass wastage active in the South Ore Body pit (Fig. 12).

The greatest difference between mass wastage in the subsidence area and that associated with the classical type of landslide movement is a lack of a well-defined failure surface. Because block caving has displaced the supporting material from beneath the subsiding ground, the fractured and broken ground continues beneath the sliding mass. An evolution of one distinct type of mass wastage occurs only as rotational and translational movement continues downslope. Continued sloughing from the escarpments of the enlarging pits furnishes a regular supply of rock at the unstable heads of the masses. Rain probably affects the process of wastage by aiding, to limited amounts, the consolidation of finer particles, flowage with increased weight caused by saturation at the surface and increased pore water pressure, and surficial transport of fine particles by runoff.

Classification

Five types of mass wastage surround the centers of the subsidence pits. These give rise to three additional types of movement in the lower areas of the pits. The three latter types of movement are made up of finer particles and are labeled debris slide, creep, and pipes. These three types are noticeable in the finer-textured portions of the aerial photographs (Fig. 1). Debris slides and creep occur only in the South Ore Body pit due to a lack of sufficient time for development or a restriction of space in the two remaining pits. The evolution is presented on Figure 14.

The classification presented here conforms as closely as possible in terminology to that presented in the Classification of Landlides for Engineering Practice, (Eckel, E. B., Editor, 1958).

En masse movement. A pronounced, but gentle, settlement occurred early in the history of subsidence. En masse subsidence consists of initial vertical settlement of the area resulting in the lowering of large masses of conglomerate. Interior breaks form, or open, in tension along discontinuities and parallel to areas of maximum vertical subsidence. Separation along the breaks is minimal, and sufficient lateral restraint is present to limit rotational and translational movement. En masse units initially develop, in all areas of the pits, and are not dependent upon the thickness of physical character of the conglomerate caprock. Areas now supporting en masse units range in thickness from less than 90 feet to much more than 800 feet. As the increased subsidence process develops, and continued vertical movement and increased lateral movement occurs, there is a transformation of en masse wastage into other types of movement. En masse subsidence is the parent phenomenon for other types of mass wastage, with the exception of pipes.

Slump. Increased lateral movement in the pits is brought about by relief of lateral constraint on the blocks of conglomerate in the pits. En masse units deform by breaking into smaller independent blocks, and there is a general shifting and readjustment of the unstable mass to allow for some rotational movement. Blocks of conglomerate then rotate slowly, over a period of time, both backward and forward. At times the blocks tumble very short distances downslope, but movement is generally slow until an appreciable change in slope is achieved. Continued transport and weathering erodes the blocks until they are no longer found in these areas. Blocks moving into the pit bottoms remain there only for short periods of time before breaking up, as implied by their absence in these areas (Fig. 1).

Slump blocks are also derived directly from the escarpments, where faulting and fracturing break blocks into units measuring 3 to 10 feet in diameter. Slumping is perhaps the most typical of all movement in the pits, and is intermediate in the translational history of the conglomerate. It is furnished directly from en masse subsidence and from the pit escarpments. In turn, the slump blocks furnish material for further transport by debris slide and

Figure 13. Units of mass wastage shown in the northeastern portion of the South Ore Body pit. Symbols depict the type of wastage involved: D = Debris slide; made up of material spalling directly from the escarpments and of small particles and fragments weathering from the surrounding masses. Note the fine detrital cones developing under the slump blocks. E = En mass; large unit of conglomerate separated from the southeast escarpment by a wide and deep tension fracture. Slump blocks are furnished directly from this unit. S = Slump blocks; showing rotational effect and also angularity coming from breaks forming along discontinuities. T = Terracettes.

creep. They are also removed through pipes.

Terracettes. Terracettes are a result of en masse subsidence and form terraces of broken conglomerate in the subsidence pits. The same general form also occurs wholly within the exposed quartz monzonite of the North Ore Body pit, where the unit is bordered by two major faults (Figs. 1 and 3).

In the South Ore Body pit, terracettes follow the outlines of former peripheral tension fractures that were formed early in the subsidence history. Fragmentation of the conglomerate is greater in this unit than in the corresponding slump blocks. The finer resulting material is carried in waves of irregular steplike character toward the edge of the debris slide unit, and thence into the pipes below. Terracettes are the result of freedom of lateral movement imposed upon previously parted blocks of rock. They do not occur over drawn blocks.

Rockfall. Sloughing directly from the steep southwestern face of the South Ore Body pit accounts for a large area of rockfall and its associated talus. Rockfall has occupied this portion of the pit only after en masse subsidence had lowered the surface of the area enough to have formed dominant scarps along its border. Fragments that slough from the free face of the escarpments are generally small in size, and move rapidly downslope for as much as 650 feet. Travel beyond, into the creep area, is slower.

Rockslide. A single exposed bedding plane in the extreme southeast corner of the South Ore Body pit is the only current site of a rockslide. Fragments of conglomerate slide down a dip slope, moving directly into the opposing slump area (Fig. 12). Probably rockslides have been active in several areas along the southwest escarpment of the pit where bedding planes are prominent.

Creep. The largest area of movement, entailing sand and gravel-sized particles of broken conglomerate, is due to creep. This unit also shows the most profound relief changes of all units undergoing mass wastage. The slightly hummocky surface is covered with grass and fragments of conglomerate that move downslope slowly. The movement is primarily along the surface and at the toe, where material is fed into the pipe areas at the pit bottom. Finer materials are also carried directly to the pit bottom by runoff. This type of movement will cover larger areas of the pit as mining continues. Creep has already touched the escarpment of the South Ore Body in one location, and its lateral margins are spreading into adjoining units as large blocks of conglomerate disintegrate.

Debris slide. The character of debris slide movement is quite analoguous to creep in its appearance. Movement is rapid, however, over steeper slopes than those undergoing creep. Particles sometimes travel continuously along the entire slope of the pit from top to bottom (Fig. 12). Because movement here is constant and rapid over the entire slide area, grass does not grow on its surface. There is a sharp contact between debris slide and areas in which other types of movement occur. Debris slides often undermine adjacent areas and result in their erosion. The slide touches the east escarpment of the pit where it has exposed a body of quartz monzonite. Individual detrital cones have formed along the upper edge of the slide area where slump blocks have been disintegrating.

Pipes. Pipes constitute the ubiquitous type of movement that takes place at the onset of subsidence and which occurs throughout its development. Pipes are centered directly over the active blocks in the mine. Eight of these features were active in the South Ore Body in 1965, and they have been prevalent in the other two pits on the property. It is important to restate that all movement in the pits tends to be channeled toward the pipe areas. At one place or another, all types of mass wastage furnish material to pipes for transport to sublevels of the mine.

EVOLUTION OF THE MASS WASTAGE UNITS

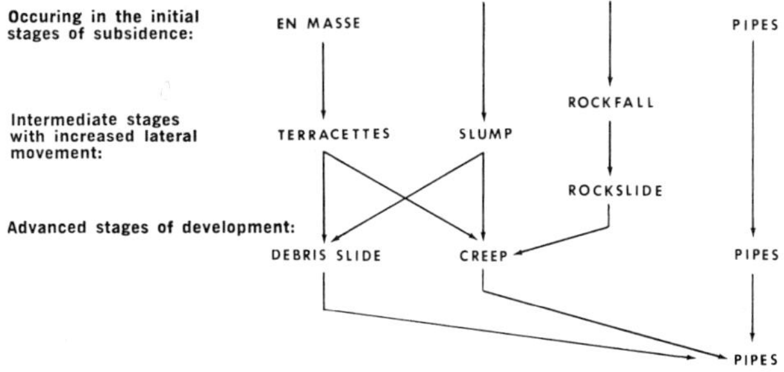

Figure 14. Evolution of the mass wastage units.

REFERENCES CITED

Burton, A. N., 1965, Classification of rocks for rock mechanics: Internat. Jour. Rock Mechanics and Mining Sci., v. 2, p. 105.

Coates, D. F., 1964, Classification of rocks for rock mechanics: Internat. Jour. Rock Mechanics and Mining Sci., v. 1, pp. 421-429.

Crane, W. R., 1931, Essential factors influencing subsidence and ground movement: U. S. Bur. Mines Inf. Circ. 6501, 14 pp.

Creasy, S. C., 1965, Geology of the San Manuel area, Pinal County, Arizona (with a section on ore deposits by J. D. Pelletier and S. C. Creasy): U. S. Geol. Survey Prof. Paper 471, 64 pp.

Eckel, E. B., Editor, 1958, Landslides and engineering practice: Highway Research Board Special Report 29, 232 pp.

Hatheway, A. W., 1966, Engineering geology of subsidence at San Manuel Mine, Pinal County, Arizona: M. S. thesis, Univ. Arizona, 110 pp., 45 figs. (unpubl.)

SWELLING OF ROCKS IN FAULTS IN THE ROBERTS TUNNEL, COLORADO

E. E. Wahlstrom (Professor of Geology, University of Colorado, Boulder, Colo.), and
C. S. Robinson and T. C. Nichols (Engineering Geologist, U. S. Geological Survey, Denver, Colo.)

Abstract

Samples of crushed and altered rock associated with faults were collected during the construction of the Roberts tunnel, Colorado, and were tested during a postconstruction study of the geology of the tunnel. The swelling properties and the mineralogy were determined and correlated with rock type and rock alteration. It was concluded that use of such tests at the time of construction could have aided in identifying potentially troublesome ground. Also, such tests, used in conjunction with surface examinations prior to the construction of a tunnel, would have indicated if swelling ground would be encountered during construction, and the order of magnitude of the swelling.

CONTENTS

ILLUSTRATIONS

INTRODUCTION

During the construction of the Harold D. Roberts Tunnel in Colorado, undesirable swelling and squeezing ground was encountered in several sections. It required initial installation of steel and timber supports, and later extensive remedial measures. As part of postconstruction analysis of the geologic and engineering features by the U. S. Geological Survey, the potential volume change (PVC) swell index, swell capacity, and mineralogy were determined for several samples from within and near faults that localized the swelling ground. The determinations permit comparison of the interrelationships of swelling properties, rock type, and rock alteration. Had such relatively simple and rapid tests been made on samples at the time of construction, the results would have indicated areas of potential trouble. They would also have indicated the need for the installation of additional support as a part of normal construction procedures.

The Roberts Tunnel is a 23.3-mile-long dogleg tunnel that extends from near Dillon, in Summit County, to near Grant, in Park County, Colorado (Fig. 1). The tunnel was constructed by the Denver Board of Water Commissioners to bring water for domestic use from the headwaters of the Blue River on the western side of the Continental Divide to the North Fork of the South Platte River on the eastern side of the Divide.

The tunnel intersects many varied rock types and geologic structures (Fig. 2). The sequence from the west portal to the east portal includes: Mesozoic sedimentary rocks; thoroughly shattered and altered Precambrian igneous and metamorphic rocks above a major thrust fault; hard, brittle, baked Mesozoic shales; quartz monzonite in an intrusive stock of Tertiary age; and a thick, structurally complex succession of Precambrian schists, gneisses, and granite. All rock types locally have been subjected to folding, faulting, and hydrothermal alteration. Also, near Montezuma, the tunnel intersects the Front Range mineral belt, a wide zone of faulting, alteration, and economic mineralization.

The Roberts Tunnel was driven from four headings: a heading from each portal, and two headings from an access shaft. Except for shales near the west portal, the kind of rock that was present was of small consequence, and, for engineering purposes, the petrographic classification of the rock was not very significant. It soon became apparent, however, that the progress of construction and the need for installation of steel and timber supports were directly related to the intensity of faulting and alteration and, to a lesser extent, to the presence and concentration of joints. Special difficulty was experienced in driving the tunnel through several fault zones containing extensive gouge, some accompanied by hydrothermal argillic wall-rock alteration and by ground water under high pressure. Also, in the Montezuma stock, several months after the rocks were exposed, many sections of the tunnel showed the effects of squeezing and dislocation of the supports, and additional supports were required to maintain the tunnel opening. In these sections considerable time, money, and delay could have been avoided if the subsequent behavior of the rocks had been anticipated at the time they were exposed in the tunnel heading. In retrospect, if field tests of the type outlined below had been made, sufficient support could have been installed as a part of the routine tunneling operation to maintain the tunnel opening safely until the lining was placed.

TEST PROCEDURES AND RESULTS

Samples were systematically collected from the faults and associated rocks that required support, as these were exposed by the tunnel heading. Samples from the Montezuma stock were collected later when it became apparent that additional supports would be required for that section.

Table 1 lists the samples that were selected for testing, and gives a brief description of the sample and associated wall rock.

The PVC (potential volume change) swell index, the swell capacity, and mineral composition of samples of swelling ground, from within and near faults in the Roberts Tunnel, are summarized in order of increasing PVC swell index in Table 2. Included also are PVC rating and classification, final heave, and final swell pressure, as determined by calculation.

The PVC test and instrument were initially devised to determine swelling characteristics of soils in foundations. Procedures and instrumentation are described by Lambe (1960), who developed the technique under the auspices of the Federal Housing Administration. Lambe (1960, p. 2) defines the swell index ". . . as the pressure exerted by a compacted sample that has been immersed in water for two hours."

Lambe's procedure was somewhat modified in the present investigation. Because samples from the Roberts Tunnel generally contained numerous hard rock fragments, it was necessary to crush the samples so that they would pass a 20-mesh screen before placing them in the container in the PVC meter. The samples were moistened (equivalent to Lambe, 1960, p. 20, "Moist classification"), and then compacted in the consolidation ring of the test device by using seven blows of the standard compacting hammer for each of three layers. Otherwise, the procedure was the same as that described by Lambe. The compacted samples were placed between two porous stones, a cover added, and adjustments were made in the proving ring and dial. Water was added to cover the samples and stones, and the pressure developed after 2 hours was recorded as the swell index pressure.

The potential volume change (PVC) was read from a chart prepared by Lambe (1960, Fig. 20), who defines PVC (p. 4) as ". . . the maximum possible volume change that the soil could undergo from water content changes." The potential volume change, as determined from the swell index pressure, provides an arbitrary rating based on correlation of heave relative to an air-dried condition, water content at 85 percent relative humidity, plasticity index, activity, and linear shrinkage.

The values of heave listed in Table 2, which were obtained from a diagram prepared by Lambe (1960, Fig. 16), give an estimate, accurate to about 50 percent, of the percent volume increase, or heave, after the sample comes into contact with water. Final swell pressure obtained when constant volume in the wetted sample is maintained also may be estimated from the swell index pressure. Values above 3000 psf are only of limited value, as discussed by Lambe (1960, p. 36), but do indicate that the pressure increases in confined samples.

Swell capacity was determined by following the procedure outlined on Data sheet No. 251 of the American Colloid Company (Knechtel and Patterson, 1952, p. 5). A 2-gram sample of air-dried minus 40-mesh material was added slowly to 100 ml of distilled water. After 1 hour the resulting volume increase, in milliliters, was recorded as the swell capacity.

The mineralogy of most of the samples was determined by means of an X-ray diffractometer by A. J. Gude, III. Relative order of abundance of the minerals is indicated by numbers for each sample (Table 2). Figure 1 indicates the most abundant mineral, and if that number is underlined, it indicates that the mineral is by far a predominant constituent. The numbers do not provide a basis of comparison of abundance among samples.

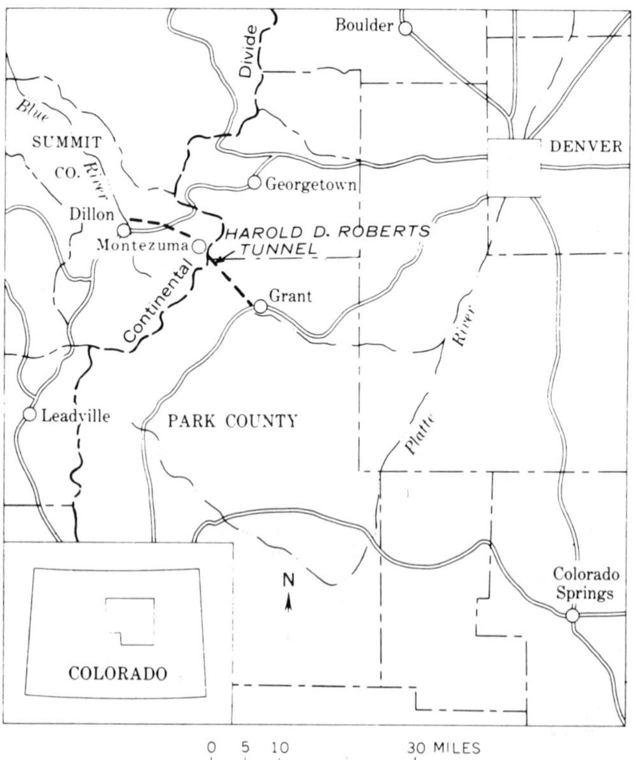

Figure 1. Index map showing location of Harold D. Roberts Tunnel, Colorado.

Table 1. Description of samples and associated wall rock (See Fig. 2 for location of sample.)

Sample number (station, or drill hole and footage)	Sample description	Wall rock
216+52	Crushed, gougy material from 20- to 30-ft wide fault zone. Heavy squeezing ground.	Brittle, silicified, mylonitized Precambrian gneiss.
230+00	Gouge from 1-ft wide fault-----------	Do.
231+44	----do-------------------------------	Do.
235+00	Gouge from closely faulted interval.	Do.
273+00	Crushed, gougy material from 100-ft wide intensely sheared zone. Heavy squeezing, water bearing.	Do.
298+50	Gouge, crushed material from 200-ft wide fractured zone. Heavy squeezing.	Closely fractured baked Mesozoic shale (hornfels).
337+12	Gouge in narrow fault----------------	Closely jointed baked Mesozoic shale (hornfels).
399+20	Gouge and bleached rock from 10-ft wide shear zone.	Hydrothermally altered Montezuma quartz monzonite.
459+43	Bleached, soft rock from wide fractured zone. Heavy squeezing ground.	Do.
473+75	Bleached rock in fracture zone between gouge slips.	Do.
510+00	Gouge and crushed rock in 6-inch fault.	Slightly altered Montezuma quartz monzonite.
530+90	Bleached rock in closely fractured zone.	Hydrothermally altered Montezuma quartz monzonite.
545+50	Gouge and crushed rock in strong flat fault.	Slightly altered quartz Montezuma monzonite.
569+80	Gouge and bleached rock in fracture zone between faults.	Hydrothermally altered Montezuma quartz.
594+15	Crushed rock and gouge in closely fractured zone.	Montezuma quartz monzonite.
757+45	Gouge and altered rock in shear zone. Heavy squeezing ground.	Altered Precambrian biotite gneiss.
770+07	Gouge in 6-inch to 1-ft fault--------	Precambrian biotite gneiss.
834+10	Gouge and crushed rock from closely fractured altered zone.	Do.

Table 1. (Continued)

1207+32	Gouge in altered closely fractured rock.	Precambrian biotite gneiss and schist.
C-188 (276+00)	Gouge and bleached rock from drill core, about station 276+00.	Silicified, mylonitized Precambrian biotite gneiss.
D-925 (Access shaft)	Bleached soft rock from drill core.	Montezuma quartz monzonite.

Table 2. Swelling properties and mineralogy of samples of crushed and altered rocks associated with faults

Sample number (location by stations or drill hole and footage)	PVC* swell index (psf)	Swell* capacity (ml)	Final heave (percent volume change)	Final swell pressure (psf)	PVC rating
PVC CLASSIFICATION: NONCRITICAL					
216+52	330	2.5	0.5	300	0.015
594+15	520	2.5	1.5	600	.04
598+65	1,050	2.5	3.5	1,400	1.2
545+50	1,070	2.5	3.5	1,400	1.2
770+07	1,150	2.4	3.75	1,600	1.3
273+00	1,140	2.7	4.5	1,900	1.6
510+00	1,500	3.0	5.0	2,000	1.7
PVC CLASSIFICATION: MARGINAL					
298+50	2,000	2.0	6.75	2,700	2.4
834+10	2,640	3.0	8.75	3,800	3.2
C-188	2,700	2.8	9.0	3,900	2.9
1207+32	2,700	2.9	9.0	3,900	2.9
D-925	2,700	3.0	9.0	3,900	2.9
473+75	3,000	2.5	10.0	4,250	3.7
337+12	3,070	2.5	10.25	4,500	3.8
PVC CLASSIFICATION: CRITICAL					
235+00	3,220	2.0	10.75	4,900	4.0
569+80	4,200	2.7	13.75	8,000	5.3
PVC CLASSIFICATION: VERY CRITICAL					
230+00	4,920	2.6	16.0	9,500	6.3
231+44	5,830	2.5	18.5	11,800	7.5
459+43	6,730	3.8	21.5	12,000	8.1
399+20	7,470	4.0	23.5	12,000	8.6
757+45	9,650	3.5	25	12,000	10.4
530+90	9,890	3.5	25	12,000	10.8

* Analyst: J. C. Thomas

Table 2 (Cont.). Swelling properties and mineralogy of samples of crushed and altered rocks associated with faults

Mineralogy **							
Montmoril-lonite	Kaolinite	Mica	Potash feldspar	Plagioclase feldspar	Quartz	Carbonate	Pyrite
PVC CLASSIFICATION: NONCRITICAL							
2	4	3	--	5	1	6	--
--	--	2	--	--	1	--	--
nd	nd	nd	nd	nd	nd	nd	nd
4	3	2	6	--	1	5	--
--	--	1	--	--	2	--	3
3	2	5	--	4	1	6	--
2	5	4	1†		3	--	--
PVC CLASSIFICATION: MARGINAL							
--	2	3	--	5	1	4	--
5	--	2	4	3	1	--	--
2	5	6	4	3	1	--	--
--	3	2	--	4	1	--	--
nd	nd	nd	nd	nd	nd	nd	nd
3	5	2	1†		4	--	--
1	4	3	--	--	2	--	--
PVC CLASSIFICATION: CRITICAL							
1	3	6	--	2	4	5	--
--	--	1	--	--	3	--	2
PVC CLASSIFICATION: VERY CRITICAL							
1	4	5	--	3	2	6	--
1	5	7	3	4	2	6	--
1	6	2	5	--	3	4	--
1	6	4	5	--	2	3	--
1	7	2	5	6	3	4	--
1	6	4	5	--	3	2	--

Analyst: A. J. Gude, III

** Minerals numbered in order of decreasing estimated relative abundance;

† 1 indicates mineral is predominant.

(nd, not determined; --, not present)

DISCUSSION

The use of modified soil PVC measurements in the study of altered rocks in faults appears to be justified by the data presented in this paper. The instrument quickly measures a partially confined pressure exerted by a water-saturated sample and indicates, at least semiquantitatively, the expected reaction of such rocks under several conditions encountered in rock excavation, especially where water is present. The meter measures pressure in only one direction. Similarly, in any opening in incompetent rocks, pressure will be relieved by movement into the opening, and at any point the force exerted as a result of the movement can be regarded as unidirectional.

Steel ribs and struts, together with the timber lagging that is placed around them, are designed to prevent rockfalls and to withstand pressures of surrounding masses that might otherwise move into the tunnel opening. Pressures can be completely resisted only by installation of steel liner plates or gunite rings, and finally, the concrete tunnel lining. Pressures exerted on steel and timber

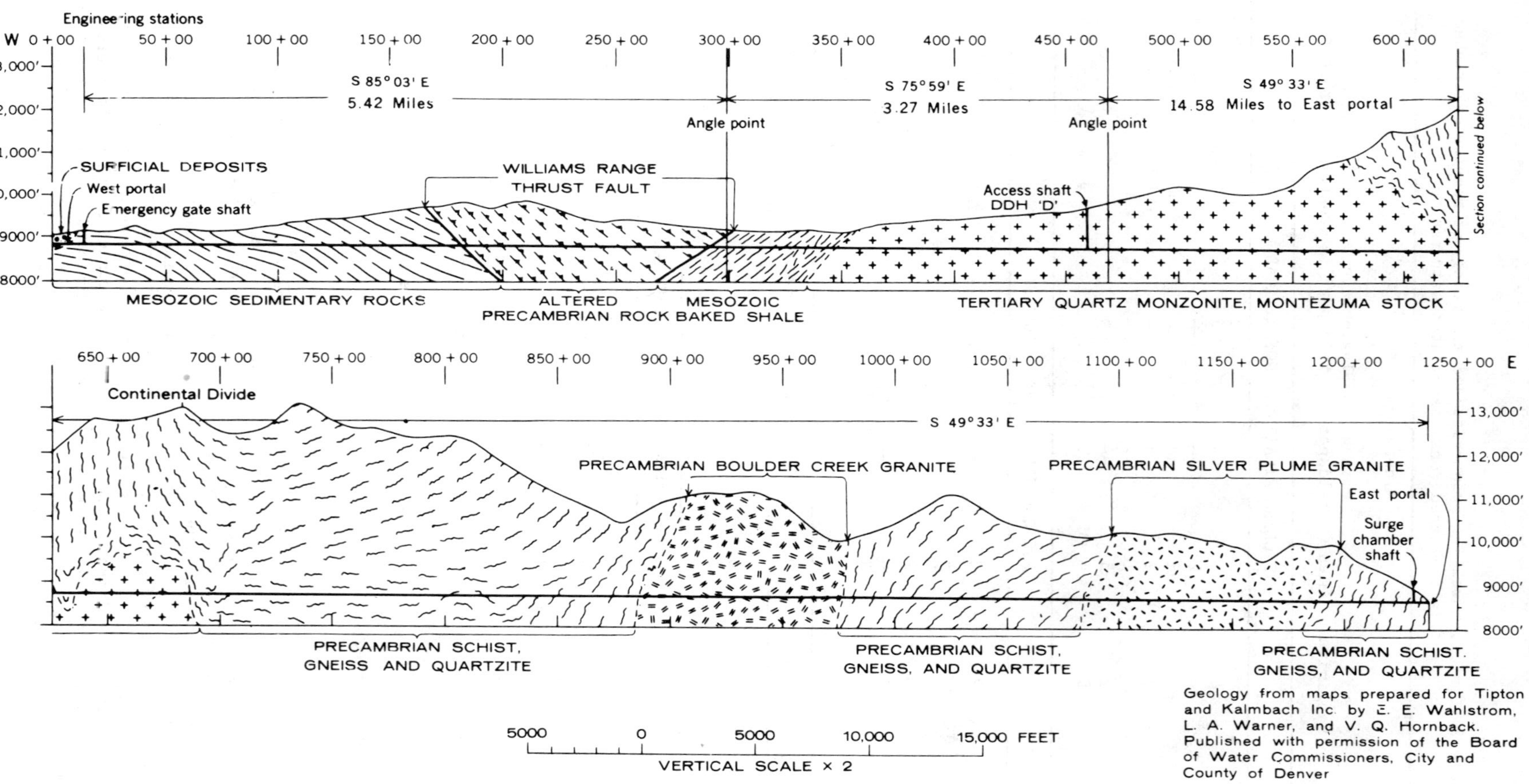

Figure 2. Generalized section through the Harold D. Roberts Tunnel, Colorado.

sets, because of irregularities of contact with the rock they are supporting, are unequally applied. The response of the supports to the applied pressures is complex, and, commonly, cannot be analyzed by any simple procedures of stress analysis. Also, the pressures of swelling ground may be in part relieved by plastic or semiplastic flow into the tunnel opening through openings between steel sets and timber lagging. Nevertheless, if some idea of the immediate or long-term pressures that will be applied to the supports is obtained by use of the PVC, or a comparable method, sufficient support can be installed to provide an adequate margin of safety until completion of the lining of the tunnel.

The initial pressure exerted by a swelling material will be roughly the swell index pressure plus pressures that may be dependent on other geologic factors. After supports are emplaced, the pressure will gradually increase to about the final swell pressure plus pressures of other kinds, except insofar as the pressure is reduced by flow through openings in the supports. The final heave, which is the percent volume change on hydration, indicates the potential amount of swelling. Terzaghi (1946, p. 82) suggests that swelling pressures on supports may be relieved by removing material for some distance beyond the supports and by allowing the swelling material to fill the void thus created. Experience might enable an estimate to be made from the final heave, as determined by testing, as to the amount of such excavation that would be required.

The swelling properties of fault gouge and rock alteration products are dependent upon the amount of water that is adsorbed, which in turn depends upon the amount and kind of clay present. Grim (1962, pp. 247-251) discusses in detail the many factors that influence the adsorption of water and the resultant swelling, most important of which is the type of clay. Aside from one exception, those samples from the Roberts Tunnel in which montmorillonite is the predominant mineral have the highest swell index (Table 2), and these results agree with the findings of Grim (1962, pp. 247-251). The montmorillonite is associated with faults showing effects of hydrothermal alteration.

Most tunnels are driven below the water table, as was the Roberts Tunnel, and the pore space and fractures within the rock are filled with water. As mining removes support, the clayey material will start to swell and squeeze into the tunnel opening. At the same time, the water present in the fault and surrounding rock will be adsorbed and the clay will swell further. Terzaghi (1946, pp. 73-74) contends that the swelling is not the result of adsorption of water from the moist air of the tunnel, as is commonly believed, but results from the adsorption of water from the fractured rock in the wall of the tunnel. This generalization probably is valid for most sections of the Roberts Tunnel; but, in part of the Montezuma stock, sections contained swelling ground where no visible water flowed from cracks. However, surfaces promptly protected from the tunnel air by a tar coating did not swell. At one time, consideration was given to spraying a tar coating on the tunnel walls to reduce or stop the swelling. Tunnel temperatures were more than 80°F; the relative humidity was high; and, at least in some locations, swelling could be attributed only to adsorption of moisture from the tunnel air.

A direct correlation between swelling indices and support requirements was not possible. Equally, or more, important in determining the support requirements were the width of the fault zone and the presence or absence of water under high pressure. Narrow faults and altered zones, although they included clays of high swelling indices, did not require abnormally heavy support because the clay was contained between masses of more competent rock. Maximum difficulty in support was encountered in wide complex fault zones that contained gouge, altered rock, and water under high pressure. A knowledge of the swell indices of altered rock would have been of maximum benefit in those sections of the tunnel where the altered rock hydrated and swelled slowly over a long period of time. These sections required the installation of additional support after the heading had advanced beyond the altered rock, and this work impeded the normal progress of the tunneling operation.

SUMMARY AND CONCLUSIONS

As a routine operation, measurement of swell index pressures and similar properties of altered or crushed rocks would appear to be very useful in construction involving rock excavation. Such tests, modified from the procedures described by Lambe (1960) so as to adapt them for use with crushed and screened rock samples, can yield information on the probable swelling behavior of a rock at a time when such information is of maximum benefit. Such tests would have warned of potential trouble in the Roberts Tunnel, and adequate support could have been installed as a part of normal construction procedures. Also, it seems probable that the measurement of swell index pressures on samples collected from the surface prior to the construction of a tunnel will indicate the order of magnitude of the final swelling pressures that will be developed from crushed and altered rock encountered during construction of a tunnel. This conclusion is based on the reasonable assumption that the origin, and therefore the composition and physical properties, of the crushed and altered rock at the surface is the same as that at tunnel level. It should be kept in mind, however, that in some environments the composition and the physical properties of the clays of the crushed and altered rock in outcrop may have been changed by weathering and ground water. Such factors are discussed by Grim (1962, pp. 40-42).

REFERENCES CITED

Grim, R. E., 1962, Applied clay mineralogy: New York, McGraw-Hill, 422 pp.

Knechtel, M. M., and Patterson, S. H., 1952, Bentonite deposits of the Yellowtail district, Montana and Wyoming: U. S. Geol. Survey Circ. 150, 7 pp.

Lambe, T. W., 1960, The character and identification of expansive soils: Federal Housing Adm. Tech. Studies Rept., FHA 701, 46 pp.

Terzaghi, Karl, 1946, Rock defects and loads on tunnel supports, in Proctor, R. V., and White, T. L., Rock tunneling with steel supports: Commercial Shearing and Stamping Co., 99 pp.

Publication authorized by the Director, U. S. Geological Survey

Engineering Geology Case Histories

Number 7

Legal Aspects of Geology in Engineering Practice

GEORGE A. KIERSCH
and
ARTHUR B. CLEAVES
Editors

Prepared for the Division on Engineering Geology
of
The Geological Society of America

1969

Reprinted 1973

Published by

THE GEOLOGICAL SOCIETY OF AMERICA, INC.

3300 Penrose Place

Boulder, Colorado 80301

Printed in the United States of America

CONTENTS

Some Legal Aspects of Geology in Engineering Practice

PREFACE

This is the seventh volume in the Case History series of the Division on Engineering Geology of the Geological Society of America, initiated in 1957. This volume is composed of case histories involved with Legal Aspects of Geology in Engineering Practice.

Because of growing needs, the engineer is required to utilize many sites with very adverse geologic features. Due to this and the growing emphasis on contract specifications, the number of engineering projects that ultimately become involved in a legal controversy is on the increase.

The manuscripts for Volume 7 were submitted in 1965 and 1966. Acceptance of modifications by the authors are appreciated in order to conform with the publication format and limitations for the sake of cost. Because case histories dealing with the legal aspects are certainly controversial by their very nature, the Geological Society of America does not necessarily concur with the views expressed herein by the authors.

Attorney Robert B. Jarvis, an engineer and specialist in contract law, recently expressed the situation when he posed the question, "Are engineers the largest group practicing law without a license?" (Civil Engineering, June 1963). Attorney Jarvis has raised the often repeated question, "Who pays for the unexpected in construction?" The case histories in this volume demonstrate why this question arises for a number of geological circumstances common to engineering works.

George A. Kiersch and
Arthur B. Cleaves
Editors

McGraw Hall
Cornell University
Ithaca, New York

THE GEOLOGIST AND LEGAL CASES--RESPONSIBILITY, PREPARATION, AND THE EXPERT WITNESS[1]

George A. Kiersch (Department of Geological Sciences, Cornell University, Ithaca, New York)

Abstract

The geologist is becoming increasingly involved with court cases and the legal solution to problems and disputes in engineering practice; contract adjustments, construction claims, water rights, and safety violations are typical subjects. The most frequent circumstances that give rise to litigation involving geology result from a few major categories such as: surface water, ground water, sedimentation-channels, mineral resources and land, public health and safety, foundations-open cuts, and underground openings.

The responsibilities of technical personnel on engineering projects to the owner, contractor, public-at-large, and others concerned are complex and diverse. Consequently, interrelated responsibility is a common basis for litigation in engineering work. For example, who is responsible for problems which arise because of the geologic conditions, both known and unforseen.

Major guidelines are innumerated for the expert witness who is preparing his testimony or appearing in court to present facts or opinion.

CONTENTS

---

[1] Paper presented at the "Eugene Stebinger Memorial Symposium on Foundation Problems," Northwestern and University of Illinois at Chicago, December 3, 1960.

INTRODUCTION

The appearance of professional men in court to express opinions has been very common for several decades in medical cases and by professional appraisers for condemnation suits. Other professional groups rarely appeared in court, until recently when public accountants were called for tax cases and architects, engineers, and geologists began to appear in connection with construction programs to testify about facts or to express their opinions. The trend for a greater reliance on a legal solution to professional problems has been paralleled by another change which is interrelated--the requirement for specialization in all professions. As an outgrowth, today the specialized trial attorney and the specialized engineer or geologist find themselves working together frequently on a case in the court.

The trends of specialization and legal involvement require that the geologist be acquainted with certain laws, statutes, ordinances, and working agreements. Furthermore, if he is to perform his work in a safe and responsible manner, the geologist must be advised or aware of many forms of contract agreements and related documents, and how they pertain to the geologic conditions and events that affect engineering works. Some common geologic phenomena and engineering techniques that fall in this category and form the basis for much litigation are the following.

Surface Water

Water-rights; responsibility for water-logging or damage to land, correction measures and consequences; water retention and recharge, supply and development; "stream" piracy.

Ground Water

Water-rights, codes regulating natural underground flow, reservoirs; springs, loss of source, damage from infiltration, changes caused by flooding or decreased water level; riparian rights; leasing terms; depletion allowance; geothermal resources, leasing, replenishment, and depletion.

Sedimentation-Channels

Accretion or avulsion of sediment, riparian land rights lost or gained; natural sediment process, artificial changes and damage caused.

Mineral Resources and Land

Economic mineral values, mineral claim validation, ownership, royalty, rental, freight rates, evaluation and fair price by condemnation; public law No. 167 (1955) on unique and distinct value of selected mineral materials, intent, scope and acquisition; tax depletion and evaluation; disposal and waste areas, access.

Public Health-Safety

Surface or ground-water pollution, or both; disposal of atomic and industrial wastes; vibrations from blasting, damage; sliding or cave-ins and workers safety.

Foundations-Open Cuts

Excavation classification; failure of cut slope; responsibility for changed conditions and the unforseen; blasting, damage to nearby property; surface runoff; ground-water inflow, dewatering.

Underground Openings

Tunneling, over excavation, safety and support needs; blasting, damage to nearby property; ground-water inflow, effect on surface water rights; gas encounters; health hazards.

Participation in a court case requires that the geologist be aware of many critical aspects such as: responsibility of the diverse parties; preparation of the supporting data; and the actions of an expert witness in and about court in presenting his testimony. Major guidelines for the witness are innumerated. These have been assembled while participating as a consultant and witness on a large number of cases; they may help to avoid the pitfalls that can arise.

RESPONSIBILITY

Typical Circumstances

The responsibility of technical personnel to the owner, contractors, designer, the public-at-large, and others concerned with engineering practice and works are complex and conflicting. Today, workable solutions to our mixed problems require an increasing co-operation among those responsible and the professional team. The following tabulation of typical circumstances demonstrates a common basis for litigation in engineering practice, that of responsibility for the legal problems which arise from the geologic conditions.

The Responsibility of: project owner to the public; designer to owner (or engineer); owner to contractor; and contractor to owner.
The Responsibility of consultant to employer: owner,-designer, or contractor.
The Responsibility of: insurer to insured.
The Responsibility of designer versus owner for: limits of accuracy of the project design; and, factors of safety--and accepted standards.

The Responsibility of geologist and engineer to owner for: Subsurface interpretation--assume rightful calculated risk; accuracy of data; familiarity with construction methods, equipment, and uses. Legal liability--negligence, fails to exercise ordinary care in report data, erroneous opinions, fails to comply with ordinances and statutes. Fraud--fails to disclose adverse geologic facts affecting property, project and plans.
Presentation of data by owner: availability known geologic facts to contractor, ambiguities; disclaimer clauses (largely invalid); and interpretation owner's specifications by contractor.
The Responsibility for changed conditions: owner to designer; owner to contractor; designer to owner; and contractor to owner.
Unwarranted delay or suspension of work: owner (engineer) to contractor.

Depot Construction Corp. versus State, changed conditions, disclaimers

Construction claims today comprise a high percentage of the litigation which is based on the influence of geologic circumstances. The number of such court cases based on a plea of changed conditions is on the steady increase. A 1964 ruling by a New York court Depot Construction Corporation versus State, NYS. 246, 2nd 527 (1964) illustrates one type of construction claim frequently sought that is based on a plea of insufficient knowledge of the subsurface or changed conditions and disclaimer clauses.

Depot Construction Corporation was awarded a contract by New York State for construction of a hospital building. On completion of the project, Depot sought extra compensation for the contract item of rock excavation--contending there was a much greater quantity required than indicated by the test boring data supplied by the state. The contractor was required to bid the project on the basis of specifications and test hole information that contained the following disclaimer:

> Holes drilled on the site are shown on the plan drawing. Test hole data are not guaranteed by the state in any respect, nor represented by it as being worthy of reliance. The state makes them available as information in its possession without intent or attempt to induce the bidders to rely thereon; they should make their own independent evaluation.

The court awarded a judgment to Depot and said the contractor was not bound by the unit price in the contract and could recover the fair and reasonable value of extra excavation work. The basis for the decision was the court's following statements.

(1) The disclaimer clause on test hole data is considered confusing, illogical, contradictory, and even deceptive.

(2) The disclaimer clause was conceived to permit easy escape on the part of the state from its prime responsibility of presenting to the prospective bidders as complete and efficient data on the subsurface as modern technology and equipment allows.

(3) The state was haphazard in the manner they took the test borings.

(4) Unacceptance of the state's rationale, whereby, for such purposes as bidding the borings are described as unworthy of reliance, while otherwise the state prepared its estimate of the amount of rock excavation and foundation design on the same 'unworthy borings.'

(5) The contractor is required to base his bid upon the 'unworthy borings' as there is neither time nor opportunity to permit, nor a moral right on the part of the state to demand, independent borings by the contractor prior to acceptance of its bid.

(6) By the very nature of public bidding methods, the state makes its borings a necessity of the bid. Furthermore, there is no provision in the bid notice that the bidder would have to make his own borings in preparation for his bid.

PREPARATION--COURT CASES

General Guides

The following suggestions may assist the inexperienced witness in preparing his testimony and avoiding some difficulties common to legal suits.

(1) Do not accept participation in a case against your best judgment. If circumstances and facts do not appear consistent, and you cannot willingly and ethically support them, withdraw.

(2) Investigate the client before agreeing to work for him, even though the case is judged worthy. He may be objectionable as an unethical operator.

(3) Do not risk your reputation with a careless lawyer. An ignorant lawyer is bad, a careless lawyer is a menace to the profession--avoid employment with latter.

(4) Prepare yourself adequately for the case with the necessary field or office investigations, or both. Lawyers frequently want to restrict severely the amount of time the expert spends on preparation, in order to reduce costs. Prior to accepting participation, have a definite understanding regarding the need for adequate geologic investigations. The lawyer may mistakenly feel that geologic facts are less than critical to the outcome of the case and he can simply "out-argue" the opposition.

(5) Profusely document conditions bearing on the case with drawings, maps and photographs that clearly demonstrate the facts, your interpretations and conclusions; these can be used to demonstrate geologic principles, origin of features, changes induced to natural (in situ) conditions by works of

man, changes incurred with time, and the like. Where possible use three-dimensional drawings to document subsurface geologic conditions and demonstrate their legal relevance to the surface features, processes, and events.

(6) Review your findings and interpretation-conclusions with the lawyer well in advance of court date. Do not forget--it is part of your job to educate the lawyer regarding the way in which geologic circumstances and processes are decisive to the case.

(7) Do not go into court until sure your lawyer knows all of your findings and conclusions. Refuse to go if lawyer is not fully aware of them.

(8) Plan your presentation of testimony in a general way with the lawyer prior to court date. It may prove advantageous for the lawyer to prepare an outline (with your help) and questions, so that the relevant facts and your conclusions can be brought forth in logical sequence. This will insure that your testimony is complete (as is warranted). Remember, however, not to load testimony with "filler" and unnecessary information. Remain strictly within the bounds of the case and present data only to elucidate.

(9) Do not take the stand unless you have carefully prepared your testimony. At that time be ready to face skilled, and sometimes unfair, cross-examination of your testimony. Consequently, to assist your preparation, review all work and reports on the project (yourself and others), as well as your professional writings that may have a bearing. This may constitute 5 to 10 hours of preparation for each hour on the stand.

(10) Remember that the lawyer is the captain of the professional team for the case; be directed by his requests and lend him your full support and loyalty.

(11) Co-ordinate your testimony with the conclusions of the supporting experts of your team through discussions with the lawyer, and if directed, the other experts.

(12) Where appropriate, study the pre-trial deposition of any witnesses or experts concerned with your area of testimony.

(13) If the opposing lawyer takes your deposition (pre-trial "testimony") as an expert, be aware that he can be authorized to request copies of your notes, correspondence, reports, and so on, which you may have in your possession at the time of the taking of the deposition.

(14) For background on a lengthy case, study the transcript of relevant court proceedings given prior to your appearance.

(15) Use notes to refresh your memory on a point or series of data; these are permissible if reference to them is prefaced by appropriate remarks. (An outline of testimony is not permissible however). The court may ask to inspect such notes, so be prepared.

THE EXPERT WITNESS

Court Appearance--Testimony[2]

(1) On answering questions--keep your eyes on lawyer, listen carefully throughout his question; then direct your attention and answer straight to jury (trial) or judge (hearing), not to the lawyer. (He should know what you are going to answer.) If you lose the attention of the jury (or the judge) you can easily lose the case.

(2) State your professional qualifications clearly and completely (as appropriate); present professional qualifications that have bearing on the case. Include education, practical experience, and whether you are professionally registered (licensed) as a geologist or engineer.

(3) Display total impartiality. Remember that you are testifying for the party because you are convinced that the facts support him and that his contentions are justified.

(4) One of two methods for the presentation of testimony is generally followed by an expert witness: (a) set forth all evidence bearing on case during direct examination by your lawyer; (b) withhold some of the evidence which is damaging to the opposing side (during direct examination by your lawyer) in anticipation that the opposing side will "rise-to-the-bait" and request it during cross-examination, in the belief that you omitted a statement on the subject because it contradicts your conclusions. This delayed introduction of evidence has the effect of both strengthening your case and reducing the opposition's enthusiasm for further questioning. If, however, the opposition does not ask questions on the omitted subjects, your lawyer should then question you on these points following the cross-examination; he may do this in re-direct examination.

(5) Do not speak in low tones. As you have something worth hearing, raise your voice and speak to the jury or judge in an authoritative manner.

(6) Omit uncommon words and wherever possible technical terms in your testimony. Use the plainest language possible to convey your ideas. To explain the technical points essential to the case, employ well-thought-out illustrations as exhibits.

(7) Answer the question as asked (if you can). Do not enlarge upon the requested answer with "BUT."

(8) Do not, however, be content with a cursory answer to a complex question; present the reasons for reaching your conclusions if those reasons clarify the basic points in question.

---

[2] In part from "Don'ts for Expert Witness" by the late Prof. Lyman P. Wilson, Cornell University Law School, and Hon. Alva W. Burlingame, Jr., Judge of the Court of Special Sessions, New York City.

(9) Do not fail to help the lawyer who is examining you by rephrasing a question, the meaning of which is not precisely stated. You can say: "Do you mean . . . ?" You may find a cross-examiner asking a misleading question or one who is ignorant of the subject matter.

(10) Do not exaggerate in your response; it is a potential trap during cross-examination.

(11) Appear modest on the stand; a jury or judge is inclined to be suspicious of undue assertiveness. Rely on your professional credentials to reveal your authority in the field.

(12) Do not guess; if you do not know the answer, say so.

(13) Do not be afraid to admit a mistake or qualify an answer; an impression of and reputation for honesty and sincerity is valuable.

(14) Do not forget that you may compel a lawyer to delve more deeply into the subject. If you answer "sometimes," "usually not," or "under certain circumstances," he is almost forced to ask you to explain and the door is opened for a complete statement.

(15) Keep notes to refresh your memory about your own testimony and that of others from day to day. If possible, read the transcript of the preceding day's trial and study your testimony before returning to the witness stand the next day.

Court Appearance--Cross-Examination[3]

(1) The opposing lawyer will treat an expert witness in one of three ways during cross-examination: (a) as if the expert does not know his subject or the facts of case, and being ignorant is not an expert; thereby he discredits the witness; (b) as if the expert is unsure about important facts or aspects of the case; thereby he discredits the testimony and obtains conflicting statements for the record; or, (c) as if the expert is well prepared and truly an expert; in this event he asks few questions because he fears damaging answers. It is this attitude an expert should try to elicit from the opposing attorney.

(2) Do not rush your answers; never allow a lawyer to force you to hasty conclusions. Theoretically you have unlimited time to answer a question. For example, a correct answer may require several hours of calculations; if so, state and await court's instructions.

(3) Do not accept confusing rapid-fire questions; be deliberate and selective. The lawyer gains nothing by asking questions which are not answered.

(4) Do not attempt to answer several questions at once; have the lawyer choose the one you are to answer.

(5) Do not hesitate if an answer is obvious. It is highly effective to answer promptly as this leaves the jury or judge waiting for the lawyer.

(6) Beware of trick questions by opposing lawyer. You may properly be compelled to answer "yes" or "no" as the trick question indicates, but meet it this way. "Yes I can explain that." If the opposing lawyer avoids the explanation, the jury is immediately awakened to the attempted trick. When your own lawyer hears you say "I can explain that" he will take note; rely on him to call forth the explanation on re-direct examination. Another approach to trick questions requiring a "yes" or "no" is to reply: "I will be happy to answer if the court will allow me to qualify my answer."

(7) Do not allow the opposing lawyer to disturb your composure; just grin and be courteous to him and the court. (The jury or judge likes to see a badgering lawyer fail.)

(8) Do not try to be "clever;" the lawyer is playing on home ground and has the advantage. If you forget this, he'll show you a few tricks that you might not have heard about.

(9) Do not forget that the opposing lawyer may be better informed on some point than you are. However uninformed he may seem, do not underestimate his grasp of the facts; some lawyers act uninformed with a witness in order to catch him on a technical point and discredit him.

(10) Do not be too eager to agree with authorities in your field or accept a book as authoritative unless you know its contents. (The laywer may have a copy under the table.) (a) Do not hesitate to question statements in text books by alleged authorities, if you disagree. If you know it is an old book and out of date, ask "When was that book written?" or "What edition do you find that statement in?" or "I'm afraid the author had no practical experience when he wrote that," or "You can find authority for almost anything in books that won't stand up upon close examination," or fall back on "I can explain that." If called upon to explain, name the books which are most authoritative and which support your "expert opinion."

(11) Do not be surprised if certain stock questions are asked, such as: (a) "Have you talked with anybody about this case?" Answer: "Certainly. I talked it over at length with the lawyer who called me here." (b) "How much are you being paid to testify?" The judge is now quite likely to intervene to protect you. If he does not, state the amount frankly and matter of factly and add: "That is what my time is worth." (c) "You knew what you were going to say before you took the stand, did you not?" Answer: "If the evidence disclosed certain facts, yes, for then there would be only one reasonable conclusion--the one I have stated." (d) "Have you

---

[3] In part from "Don'ts for Expert Witness" by the late Prof. Lyman P. Wilson, Cornell University Law School, and Hon. Alva W. Burlingame, Jr., Judge of the Court of Special Sessions, New York City.

not frequently differed from other experts?" Answer: "Perhaps, but in the present instance I see no basis for any difference of opinion" or "Perhaps, but I was then and still am convinced that my opinion was correct." In close cases there may be room for such differences.

(12) Generally, limit yourself to questions asked and offer only your own opinions. When attorneys object to questions, remain silent. Wait until the judge decides whether you may answer or not.

(13) Do not risk confusion with a long, involved, hypothetical question. If in any doubt, ask that the court stenographer read the question and then be ready to answer promptly and deliberately. It will not hurt the jury (or judge) to hear the question twice; it will emphasize the importance of the forthcoming answer. Impress the jury with your desire to be accurate and careful in answering questions.

(14) Do not be misled by compound questions. If a lawyer asks you two questions in one, say "as to the first question 'yes' and the second question 'no,'" or just say "yes and no." You may be asked to explain; the lawyer is not likely to ask many more such questions.

(15) Do not hesitate to admit (if it is a fact) that you have been called upon to testify as an expert many times. The fact that your opinion is much sought after is proof of your knowledge of a specialized subject. Add, if true, "And I am called in consultation very often." The inexperienced lawyer will ask, "How often?"; give him the details.

Actions In and About Court[4]

(1) From the time you enter the courtroom until you leave the witness stand, attention is focused on you. Your testimony may be decisive as to whether your team succeeds or fails in the case.

(2) Do not appear in court until instructed to do so. On the witness stand, do not slump in the chair, cross your legs, or appear tired. Sit straight up and be alert at all times. Remember, you are being closely observed and evaluated by all those who are present in the courtroom.

(3) The jury will be critical of your appearance. Experts who are leaders in the profession command high fees; look accordingly.

(4) While the rules of evidence and court procedure may seem restraining or inexplicable, follow them closely. Each has an important relation to the just determination of the controversy.

(5) Do not discuss any aspect of the case whatever in the corridor or courtroom. Remember, you are being paid to give your expert opinions to the jury (or judge) from the witness stand. Some lawyers post clerks near opposing witnesses in the hall, and at recesses, to use what is overhead to the detriment of witnesses on cross-examination.

(6) Do not attract attention by effusively greeting the opposing expert, even if he is an old college friend; by so doing you are enhancing his image. You are being paid, rather, to make little of him and to destroy his opinion (rightfully) by your superior opinion.

(7) Do not consult with the opposing expert or experts. If you know your subject why consent to prime the opposition?

(8) Unless otherwise requested by your lawyer, leave the court when you have completed your testimony.

(9) Unless asked to do so, do not sit at the counsel's table within the railing.

(10) If your lawyer asked you to remain during the trial and advise him on the testimony given by the opposing experts, do not engage in lengthy note passing with your lawyer, unless he invites such. Provided the matter is not urgent, wait until the next recess--then be aware of who may be listening when you talk with him.

---

[4] In part from "Don'ts for Expert Witness" by the late Prof. Lyman P. Wilson, Cornell University Law School, and Hon. Alva W. Burlingame, Jr., Judge of the Court of Special Sessions, New York City.

JIM WOODRUFF DAM—CHANGED CONDITIONS CLAIM

John W. Gaskins (Partner, King & King, Washington, D. C.) and Arthur B. Cleaves (Dept. Earth Sciences, Washington University, St. Louis, Missouri)

Abstract

Government construction contracts possess a changed condition clause which empowers the contracting officer to make an equitable adjustment in the contract price if subsurface conditions actually encountered were unknown and differed materially from those indicated to exist by the contract. The invitation for bids for a powerhouse located in a river bed had estimated the quantity of cofferdam pumping at 625,000 units of 1000 gallons each, which the contractor, based on prior experience at the site and other information equally available to the Government, believed to be grossly inadequate. The contractor accordingly based his bid on the premise that a much larger quantity of pumping would be required. Where actual performance proved the contractor right, the quantity of water pumped exceeding the Government's estimate by 10 times, was the Government justified in unilaterally declaring the existence of a changed condition and in reducing the contract unit price for pumping from $.40 per thousand gallons to $.08?

CONTENTS

ILLUSTRATIONS

INTRODUCTION

The Jim Woodruff Dam is located in northwestern Florida, on the Apalachicola River where it is formed by the juncture of the Flint and Chattahoochee Rivers. This project, conceived by the U.S. Corps of Engineers, was built to provide flood control, navigation, and power for the area. The contract involved in this case called for the construction of the Gated Spillway and Powerhouse portion of the project. The contractors, in a joint-venture, were highly experienced in this type of work. [1]

The information available to bidders included the logs of numerous borings made by the Government in advance of the invitation for bids, which showed the presence of solution pits, channels, cavities, and other openings in the rock. These disclosures indicated that a serious water condition would be experienced during construction. The joint-venture had previously performed an earlier contract at the site for the construction of the Lock and Fixed Crest Spillway, which was a part of the same project, in the course of which it had pumped approximately 6,000,000 units of water. The joint-venture did not, therefore, regard the 625,000 units of water estimated by the Government for the present contract to be reasonable or realistic, and its bid was based upon the assumption that the quantities of water that would have to be pumped would be at least as great as those which had been handled under the earlier contract.

During the performance of the work, the subsurface conditions actually encountered did not depart in any material respect from the conditions that had been indicated to exist by the subsurface boring program made in advance of the work, and true to the joint-venture's expectation, the quantity of water that had to be pumped far exceeded the 625,000 units estimated by the Government for the two stages, the same approximating 825,500 units for the first stage and 7,564,610 units for the second stage.

The contracting officer, over the objection of the joint-venture, claimed the right to invoke the changed conditions clause on behalf of the Government so as to reduce the agreed unit price of $.40 per unit of 1000 gallons of water pumped to $.08 per unit. This decision was reached three years after the contract had been entered into, and two years after pumping had been in progress on the second stage. An appeal was taken by the joint-venture to the Corps of Engineers Board of Contract Appeals, where the action of the contracting officer, in modified form, was upheld. Litigation is presently pending in the United States Court of Claims.

PURPOSE TO EVALUATE ADMINISTRATIVE DECISION

The purpose of this paper is to evaluate the merits of the decision of the Corps of Engineers Board of Contract Appeals, which was rendered in this case on December 13, 1957 under the docket number C&A No. 865. Extensive hearings were conducted by the Board in advance of its decision, the stenographic transcript of testimony being 1656 pages in length. Much of the record was taken up with expert testimony pertaining to the geology of the site.

The joint-venture contended before the Board that no changed condition had been encountered which would justify reducing the contract price because the water condition actually experienced had been accurately foretold not only by the boring data made available to bidders in advance of bidding, but by a plethora of other information that was equally available to both parties in advance of bidding.

The work was performed in two separate cofferdam stages in order to avoid interrupting the river flow and to permit navigation during construction. Each cofferdam consisted of a series of interconnecting cells formed by steel sheet piling, driving to practical refusal, and filling with sand.

The contract documents had advised prospective bidders that 323,000 units of 1000 gallons each of pumping had been estimated from the first stage cofferdam, and 302,000 units from the second stage cofferdam, but bidders were informed by the contract that they were expected to perform all the work required for the completion of the contract for the unit prices bid, without regard to possible variation from quantities so estimated. The contract also contained the customary changed conditions clause in which the Government promised to make an equitable adjustment in the contract price in the event that the subsurface conditions differed materially from those which had been indicated to exist by the contract documents, or in the event that the conditions so encountered were unknown and of an unusual nature so as not normally to inhere in work of the character provided for by the contract. [2] The joint-venture established: that the site of the work was located in a well-known part of the Florida aquifer system, a system which was generally recognized as being one of the most prolific in the world; that the specifications had warned of fissures, channels, cracks, and joints in the rock which would

---

[1] Perini Corporation, Walsh Construction Company, Ralph E. Mills Company, and Blythe Brothers Company, Inc., operating under the name of Perini, Walsh, Mills, & Blythe Brothers Construction Companies.

[2] Exact context quoted at p. 7.

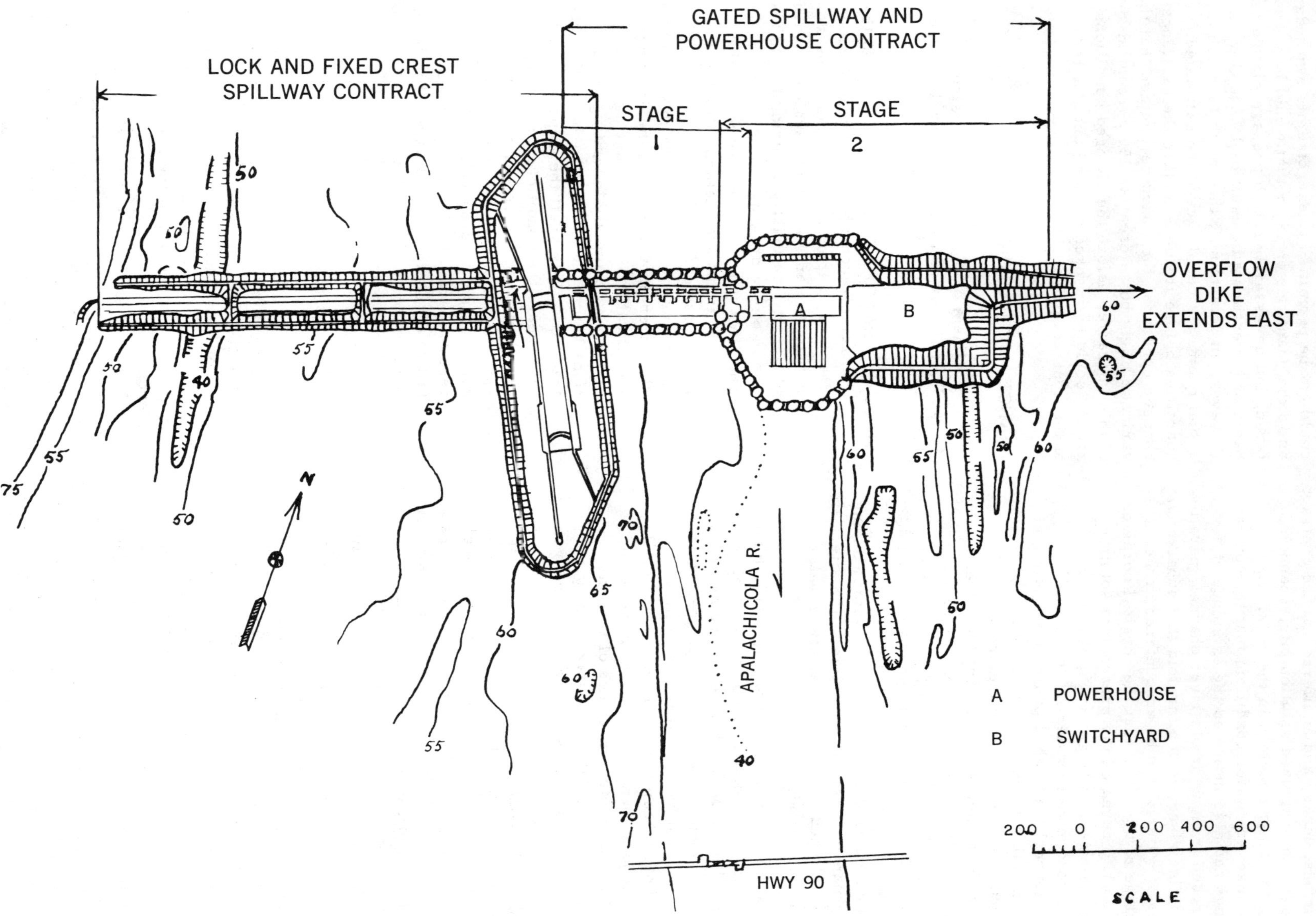

Figure 1. Plan of the Jim Woodruff Dam appurtenant structures.

cause springs, leaks, and aquifers; that the contract-boring logs had described solution pits, channels, cavities, and other openings in the rock, which indicated a severe water condition; that the contract-boring logs had also indicated the existence of a deeply eroded trough, or old river channel, parallel to the existing river, which foretold that substantial quantities of water would be encountered in the second stage of the work; that the physical cores of the borings had revealed the presence of solution channels, cavities, pittings, vugs, and other conditions conducive to water infiltration; that scientific journals published prior to the letting of the work in both the fields of engineering and geology had drawn attention to the presence of artesian water at this site; and that the Government's geologists had warned prospective bidders, including the joint-venture, at the time of investigating the site, that the old river channel would present a difficult problem of dewatering. This information, coupled with information gained during the performance of earlier contracts at the site, clearly established that a vast quantity of water would be encountered.

The joint-venture, before the Board, further contended, and proved, that the estimate which the Government had made in advance of bidding, which had fixed the amount of water to be pumped at only 625,000 units, had been arrived at in a manner which was both irresponsible and unscientific. In addition, the joint-venture relied upon the legal premise that only mutual mistakes of fact, as distinguished from unilateral mistakes, may be relied upon as a basis for setting aside or reforming a contract which condition, it established, did not prevail under the contract. It was also the joint-venture's contention that its obligation, according to the language of the specifications, was to pump all the water that might be required at the unit prices agreed upon, regardless of whether such water exceeded the quantities that had been estimated, and that the Government possessed no authority to revise or change this provision in the specifications.

The Board, in its decision, found all of the facts in favor of the joint-venture, and concluded that the variation in pumping between the estimated quantities and the actual quantities could not be explained by the discovery of subsurface physical conditions at the site which differed materially from those described in the contract plans and specifications. Quite to the contrary, the Board found that the various data, descriptions, prior experiences, and other warnings of subsurface physical conditions to be encountered at the site accurately foretold the conditions that were actually experienced. The Board, however, still concluded that a changed condition had been encountered solely as a result of the estimated quantities having been exceeded by the actual quantities. The Board stated that a bidder could not be permitted to exercise his own judgment as to the quantities of water that would be encountered, but necessarily had to prepare his bid on the basis of the estimated quantities by the Government. However, in determining the extent to which the actual quantities could exceed the estimated quantities before it would declare the existence of a changed condition, the Board recognized that the variance might be as great as 150 percent in a case of this character before the changed conditions article could be invoked to bring about a reduction of the contract price.

CHANGED CONDITIONS

Article 4 of the contract contained the changed conditions clause, and as its language is pertinent, it is hereinafter quoted:

Article 4. CHANGED CONDITIONS--Should the contractor encounter, or the Government discover, during the progress of the work subsurface and/or latent physical conditions at the site materially differing from those shown on the drawings or indicated in the specifications, or unknown physical conditions of an unusual nature differing materially from those ordinarily encountered and generally recognized as inhering in work of the character provided for in the drawings and specifications, the contracting officer shall be notified promptly in writing of such conditions before they are disturbed. The contracting officer shall thereupon promptly investigate the conditions, and if he finds that they do so materially differ the contract shall be modified to provide for any increase or decrease of cost and/or difference in time resulting from such conditions. If the parties fail to agree upon the adjustment to be made, the dispute shall be determined as provided in Clause 6 hereof.

GEOLOGY

Inasmuch as the case involved the extent to which geological data, available both to the Government and to the joint-venture in advance of the invitation for bids, had revealed the probability that a major water condition would be encountered, it is important that the geology of the area be carefully considered.

The rock formations at the site, upon which the foundations of the structure were to rest, were limestone and dolomite, deposited by the action of ancient seas. At the time of bidding, the physical characteristics of the rocks were visible from the exposed faces of the east and west banks of the existing river at the site; from the excavation performed, not only by the joint-venture for the earlier contract for the Lock and Fixed Crest Spillway, but also accomplished by another contractor who had previously constructed the Overflow Dike; and from specimens of rock which the Government had obtained by its exploratory drilling for such earlier contracts, all of which information had been made available to bidders.

The first layer of material at the site consisted of bedded, sandy clays, silts, sands, and gravels which lay unconformably upon the underlying rock.

The second layer consisted of a stratum of sandy and earthy limestones and dolomites, interbedded with light-green clay beds, with a brecciated condition at the lower boundary. This stratum was known as the Tampa Formation of the Miocene epoch, and ranged from 0 to 20 feet in thickness. It was highly porous and permeable, possessing a large number of interconnected cavities, solution pits, open solution channels, and vugs, which had been created by the erosive solution effects of artesian water passing through the rock.

The third layer was a stratum of sandstone and dolomites estimated to range from 150 to 200 feet in thickness, which was part of the Suwannee Formation of the Oligocene epoch. The contact between the Suwannee and the overlying Tampa was an irregular one marked by the presence of light-gray clay beds and occasional sand lenses. The Suwannee Formation also presented numerous interconnected channels and cavities, and possessed even greater permeability and porosity than the Tampa Formation. The Suwannee Formation also contained a very permeable coquina limestone, which possessed the capacity to deliver large quantities of water.

The fourth layer consisted of a stratum of very porous and permeable limestone, sometimes dolomitized and estimated to approach 300 feet in thickness, belonging to the Ocala Formation of the Upper Eocene epoch. The Ocala possessed a more highly developed system of interconnected solution channels, cavities, pits, and vugs than either the Tampa or Suwannee Formations due to a greater volume of artesian water which flowed through the formation. The Ocala Formation also contained some of the very permeable coquina limestone.

The Tampa, Suwannee, and Ocala Limestone Formations at the site were each a part of the well-known and much publicized aquifer system that underlies almost all of Florida. These formations are not horizontal, but instead are inclined upward in an upstream direction, with the result that each formation intersects the river upstream of the site where they are progressively exposed in shingle-like fashion, and are recharged with river water which passes through the numerous solution passages and channels in each formation to gain access to the excavated area at the site. This recharging produced an artesian water condition. Whereas all three formations contain artesian water under considerable pressure, the artesian pressure in the Suwannee Formation proved to be greater than in the Tampa, and the Ocala was even more prolific than the Suwannee.

The Government's Definite Project Report which it made in 1948, prior to its decision to go forward with the project, described the existence of a deep trough parallel to the present river which ran through the site, the trough being filled with alluvial sand and gravel. The Definite Project Report postulated that the trough had resulted from solution and scour of the soft Tampa Limestone Formation along a former river channel, and also observed that the water table at the site fluctuated with the water level in the river. The Report indicated that the problem of unwatering the site would be "quite a formidable task" and that "all indications point to a difficult cofferdam problem which is likely to entail substantial expenditures to overcome." The trough had also been indicated by the boring data made available to bidders.

The specifications for the present work likewise directed attention to springs, leaks, and/or aquifers at the site of the work, and the logs of the borings showed that the rock possessed a high degree of cavitation. In some instances, the logs reported low core recovery, as well as loss of drill water.

Quite apart from the above, a large amount of published data in existence prior to the invitation for bids had indicated the water condition that existed in the area of the work. Florida Geological Survey Bulletin No. 31, published in 1947, had called attention to Blue Springs, located approximately 5 miles from the project, where flowage over a 20-year period had been measured to average 105,000,000 gallons per day. The general water conditions in Florida were also described in Information Circular No. 3, published in June 1950 by H. H. Cooper, Jr., and V. T. Springfield. This report emphasized that artesian water occurred in the extensive limestone aquifers of the Eocene, Oligocene, and Miocene epochs. A geological map was also in existence which indicated that in the area of the work, limestone aquifers would rise in tightly cased wells to elevation +70, or approximately 70 feet above sea level. As the top of rock at the site is approximately elevation +15, and as sea level is 0.0, such artesian water would have risen for a distance of 55 feet above the top of rock, indicating very substantial artesian flow.

Much other published material of a similar nature was available to the parties, and, as determined by the Board, quite accurately foretold the conditions which could be expected at the site.

PROBLEMS AND DIFFICULTIES POSED BY THE BOARD'S DECISION

Whether the Board's decision will be modified by pending litigation may not be determined for months to come, and as it presently stands, the decision of the Corps of Engineers Board of Contract Appeals reflects the views of the Corps of Engineers in a changed condition case involving a variation between estimated and actual quantities. The principal problem which the decision poses is the unwillingness of the Government to permit a bidder, in the preparation of his bid, to use his own intelligence and experience in attempting to prophesy the conditions which will prevail during the performance of the work.

In this case, the joint-venture, on the basis of its own experience, believed that the Government had committed gross error in estimating that only 625,000 units of water would be pumped, and therefore concluded that the Government's estimate upon which bids had been requested had not been prepared in either a scientific or responsible manner. That the joint-venture ultimately succeeded in proving this to the Board, not only through an analysis of the methods pursued by the Government in estimating the quantities of water to be pumped, but also on the basis of the quantity of water that was actually experienced, has been swept aside as unimportant upon the highly disputable premise that bidders should not entertain secret views as to the quantity of work to be performed, but instead must rely on the estimates provided them by the Government. Taking into account that the Government may reduce the contract price by invoking the changed conditions clause, and may do so apparently without regard to whether it has arrived at its estimate in a responsible manner, the Board's decision would appear to place a premium upon the adoption of procedures which, through carelessness or for any other cause, might bring about an underestimation of the quantities of work to be performed, with the expectation that the contract price could be further reduced in the light of the necessity of performing the quantity of work that was reasonably foreseeable at the time bids were taken. This type of reasoning, it is believed, strikes at the heart of the competitive bidding system by depriving the contractor of the use of his own initiative in anticipating what the actual conditions may be, and requires him instead to rely upon a palpably incorrect premise in submitting a price for the work.

The decision has also disturbed another aspect of the law in relation to contracts. Generally, it is recognized that contracts may be rescinded, or in some cases reformed, only when both parties to the contract labored under a mistake of fact concerning the conditions under which the work was to be performed. Where the mistake is not mutual, and is unilateral only, the other party having been correctly advised of the conditions surrounding the performance of the work (as was the joint-venture here with respect to the quantity of water to be pumped), the courts have uniformly refused to intervene. The Board's decision has disregarded this concept of the law, for the proof received overwhelmingly established that no such misconception existed on the part of the joint-venture as to the quantity of water to be pumped.

The purpose of the changed conditions clause has been frequently defined by the courts as being to induce bidders to eliminate from their proposals money allowances to cover unknown difficulties in the performance of subsurface work which conceivably increase a contractor's cost, and to rely instead upon the promise of the Government to make an equitable adjustment of the contract price if such difficulties should in fact be encountered. In this way, the cost to the Government of work is ultimately reduced. The present decision of the Corps of Engineers Board of Contract Appeals, which reduces the contract price to eliminate the anticipated cost of an event which actually occurred, is far removed from the announced purpose of the changed conditions clause, and can serve only to encourage irresponsibility in the preparation of engineering estimates.

VIBRATION ANALYSIS AND CONSTRUCTION BLASTING

Marvin Ehrlich (Research Assistant, Department of Earth Sciences, Washington University, St. Louis, Missouri), H. LeRoy Scharon (Professor of Geophysics, Washington University, St. Louis, Missouri), and Emil J. Mateker, Jr. (Assistant Professor of Geophysics, Washington University, St. Louis, Missouri)

Abstract

Vibration analysis was employed to control the amount of explosives used in highway construction through residential areas. Measurements of the induced ground vibrations were made by portable blast seismographs. The recorded vibrations were analyzed in terms of Crandell's (1949) energy ratio. Crandell's empirical formula relating the amount of vibration transmitted to the weight of explosives detonated at any instant was used to determine the amount of explosives per delay which could be used at a given distance without causing structural damage.

Examples are cited which illustrate how vibration analysis was used to disclaim structural damage to private residences by blasting.

CONTENTS

ILLUSTRATIONS

INTRODUCTION

Any explosion in the earth, at its surface, or in the air near its surface produces vibrations which are transmitted outward in all directions from the source. (It should be pointed out that the impact of "headache balls," pile drivers, and even the movement of heavy equipment also produce vibrations.) If the explosive source is sufficiently large and the earth material near the source does not strongly attenuate the induced vibrations, measurable elastic energy will propagate to large distances, and could produce structural damage several thousand feet from the source.

The science of seismology is concerned with the elastic vibrations from earthquakes and explosions. The principles of measurements applied in earthquake studies have been extended to monitor vibrations from construction and quarry blasting. A seismologist can utilize recorded ground motion from construction explosions to determine whether the induced vibrations at a given location are potentially damaging to a building. More important, however, the seismologist can use these same data to set "safe" shooting criteria that will allow maximum efficiency in excavation or rock removal procedures while maintaining levels of vibration that will not produce damage to nearby structures. In this paper, the nature of vibration analysis is briefly reviewed and the development of a safe shooting procedure on a major construction project discussed.

INSTRUMENTATION

Elastic vibrations are measured at the surface of the earth by seismometers. When the instrument is equipped with its own recording device, it is a seismograph. The principles of seismometer design and operation are discussed in all textbooks on seismology and in summary form by Hautly and Kisslinger (1959).

In explosion vibration studies, special portable seismographs are used. These consist of three individual seismometers, two to measure orthogonal horizontal motion and one to measure vertical motion, mounted in one case with the recording system. The natural period of the seismometers can be designed to make the instrument respond to ground acceleration, velocity, or displacement.

In the study reported herein, Sprengnether portable blast seismographs were used. This instrument has hinged mechanical pendulums for the seismometer and an optical-photographic recording system. The instruments used have a dual static magnification on all components so that the ground motion is magnified either 50 or 100 times, or both.

VIBRATION ANALYSIS

The nature of the induced ground motion at a given location depends upon many factors: some are geologic, others are related to the source location and geometry, and others are the azimuthal location and distance of the site from the source. Figure 1 shows a tracing of the ground motion recorded

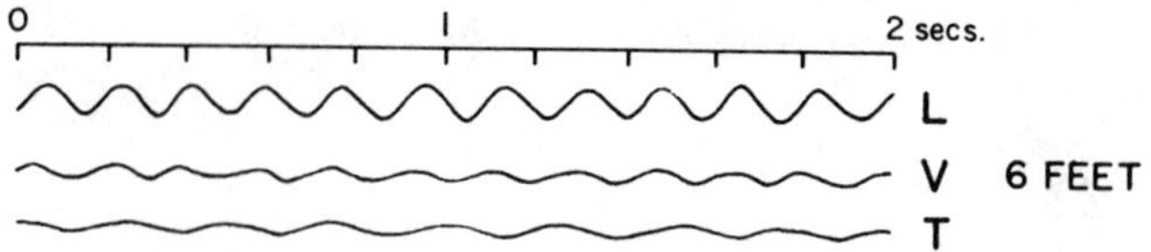

Figure 1. Tracing of recorded ground motion from compressor.

approximately 6 feet from a compressor room in which three large compressors were in operation simultaneously. The motion on all components is approximately constant frequency and is continuous because the source, the compressors, is a continuous simple harmonic oscillator. Figure 2 is a tracing of the recorded ground motion at a given site from two explosive sources, which were different distances from the recording site and different

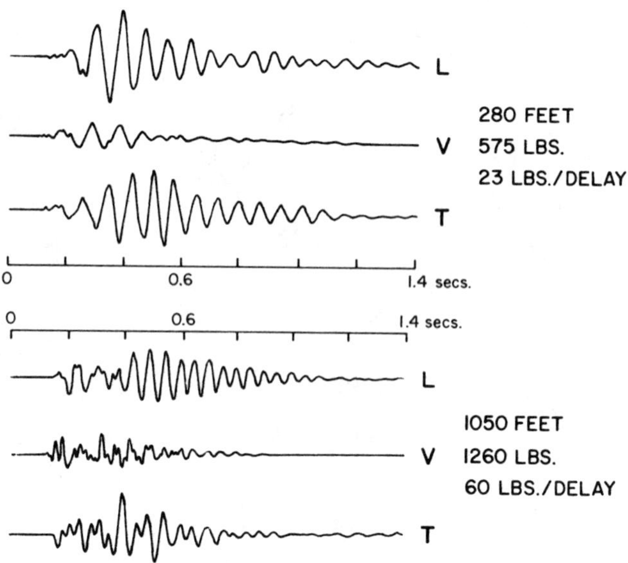

Figure 2. Comparison of recorded ground motion at one location from two different shots.

charge sizes. Note that the motion is not simple harmonic, has an abrupt beginning, which is not the maximum amplitude, and differs on each component. An explosive source is not continuous, and in this case was not located at a point nor fired instantaneously. However, the details of the shot geometry and delays are not important to this discussion; what is important is that the latter record is more typical of the type of ground motion induced by explosions than is that in Figure 1, and the motion is, in general, aperiodic. It should also be noted that the ground motion from the explosion decays rather rapidly.

If one is to use the recorded ground motion to deduce the possible effects on dwellings, a criterion must be used to distinguish damaging or potentially damaging ground motion from safe ground motion. That a criterion can be established is fundamental to the use of the analysis of recorded ground motion in court as evidence of the effect of quarry or construction blasting on dwellings. Criteria which are accepted by law in some states have been developed and form the guidelines for developing a safe-shooting procedure in the study reported herein.

Although the recorded ground motion is not simple harmonic, it can be thought of as periodic over short intervals. Thus, it is possible to measure the frequency of the motion as well as the amplitude over a given short interval of time, and use these quantities in the analysis of the data. It is not pertinent to this paper to discuss the techniques of converting the recorded motion to actual ground motion, although it is important to realize that the recorded motion represents the response of the seismometer to the ground motion. This response is magnified, optically, by the instruments used in this study. Thus, to convert the recorded motion to actual ground motion, one must know the transfer function of the instrument.

Once the data have been reduced to proper form, several quantities can be calculated which can be used for comparison with established criteria. One such criterion is acceleration. Thoenen and Windes (1942), using data largely from damage inflicted on old structures by shaking machines which were placed inside them, deduced the following equation as a measure of damage:

$$a = 0.1 f^2 A,$$

in which, $\underline{a}$ is the acceleration in terms of gravity, $\underline{f}$ is the frequency of the ground motion (cps), and $\underline{A}$ is the amplitude of the ground motion in inches. They concluded that an acceleration of 1.0 $\underline{g}$ would be damaging.

Crandell (1949) arrived at a criteria based on an energy ratio (ER) for explosions close to dwellings in urban areas. The energy ratio is determined from

$$ER = \frac{a^2}{f^2},$$

in which $\underline{a}$ is peak acceleration in units of ft/sec^2, and $\underline{f}$ is the frequency in cycles/record. Crandell states that an energy ratio below 3 indicates that the ground vibrations are safe for buildings of average workmanship and good materials; an energy ratio between 3 and 6 is indicated as a caution range and should be avoided in repeated explosions; above 6, the ground motion is potentially damaging.

Some European investigators prefer to use Zeller's "vibration intensity," Δ, (Alpan and Meidav, 1963) as a criterion for damage. This quantity is proportional to the power transmitted to a unit mass during a quarter cycle and is given by the equation

$$\Delta = a^2/f,$$

or in terms of the energy ratio,

$$\Delta = (ER)f.$$

As the result of a recent statistical study of a large amount of vibration measurements under a variety of conditions, Duvall and Fogelson (1962) determined that wave motions having a peak particle velocity of less than 2 in/sec have a very low probability of causing any damage to residential structures. This is comparable to the results presented by Edwards and Northwood (1960) from a series of controlled blasting tests along the St. Lawrence Seaway.

It is not the purpose of this paper to give a complete treatment of the techniques of vibration analysis, nor to compare criteria. However, it is important to note that the criteria listed above are empirical results based upon many experiments, and each has proven useful in certain types of problems. In the study presented in this paper, Crandell's energy ratio was used, but at a lower "safe" level.

TYPE OF CONSTRUCTION PROBLEM

A contractor was to construct a series of interstate highway sections, with "clover-leafs," through a metropolitan area in which the bedrock consists of horizontal beds of massive limestone and some shale; the topography is one of gently rolling wooded hills with a relief of a few hundred feet. In the initial stages of the construction, after the right-of-way had been cleared, small blasting operations were conducted twice daily approximately one-third of a mile from the nearest structure. However, plans called for daily blasting at several locations in which the limestone ledges to be blasted were only 100 feet from a residential building.

The structures in the area ranged from older one and two story brick, stone, or frame houses to newly constructed ranch type and split level dwellings. Some of the houses were founded on limestone bedrock, but most on clay-fill. Most of the homes were built on the hilltops, although some were on steep slopes and a few in the valleys.

As a preliminary caution, the contractor's insurance company had investigated all structures within a one mile radius of the designated blast zones and informed the residents of the blasting schedule. The insurance company recorded some blasts while the blasting area was a considerable distance from the residential sections. Although the recordings indicated that the induced ground motion was not potentially damaging beyond a few hundred feet from the source region, complaints from residents as far as one-half mile from the shot region were being made.

In this initial stage, the blasting procedure being followed had been devised by the chief blaster, a man with over 25 years experience with explosives. Because the plans called for blasting much closer to residential areas, the contractor and insurance company requested that a safe-shooting procedure be established to reduce ground motion to a safe level.

DEVELOPMENT OF BLASTING CRITERION

A safe blasting criterion is one in which the maximum amount of explosives can be detonated at a given distance from a structure without damaging the structure, yet provide enough broken rock for an efficient excavation operation. Since each shot contains several holes to fracture the maximum amount of material, the explosive in adjacent holes can be fired after a slight delay to reduce the induced ground vibration. Thus, the critical value to be determined is the amount of explosive used per delay rather than the total amount to be detonated. However, since every attempt should be made to be as safe as possible, a maximum value of total pounds of explosives per blast should also be established.

To develop a safe criterion for the construction project under discussion, it was decided that a criterion devised by Crandell (1949, 1960) would be utilized, and a series of experiments designed to yield the necessary data were carried out.

The ground motion from six blasts in the same region were recorded with two Sprengnether instruments at distances varying from 150 feet to 1 mile from the source. The experiments were designed to obtain data from a variety of blasting techniques which included changes in: (1) total pounds of explosives per blast; (2) type of explosive; (3) pounds per delay; (4) number of holes per blast; and (5) the depth and pattern of the shot holes. The blasts included shallow shots which fractured the surface materials to well-contained deeper shots for displacing large blocks. Because the contractor believed that up to 2000 pounds of explosives per blast might be necessary in some locations, it was set as an upper limit with seventeen delays at 25 milliseconds. Later experiments with 21 to 36 delays allowed for an increase in total charge size by increasing the number of delays.

None of the blasts in the experiments were damaging to any structures in the area. Since future blasting procedures were to be similar to those from which data had been recorded, it was believed that sufficient data were available to establish a safe blasting criterion for that area.

Crandell (1949, 1960) deduced the following relationship between the recorded parameters of the ground motion, amplitude, and frequency, and the source and distance parameters:

$$\left(\frac{50}{D}\right)^2 C^2 K = 16\pi^4 f^2 A^2,$$

where: D = distance in feet from the center of the blast zone to the recording instrument;
C = average pounds of explosives per delay;
K = the transmission coefficient, a constant related to the velocity of propagation of a wave in the near surface material; therefore, dependent upon local ground conditions;
f = frequency of ground motion;
A = amplitude of the ground displacement in feet.

From the experimental data it is possible to determine the transmission coefficient, K, for the region, because all other quantities are known. Once K is determined a safe criterion can be set from the following relation

$$ER = \left(\frac{50}{D}\right)^2 C^2 K.$$

Thus, once one decides upon the maximum energy ratio (ER) desired, it is possible to calculate the pounds of explosives per delay, C, for a series of distances, D, and make a chart similar to that in Figure 3. This chart is one that was used in one of the areas in which the bedrock was limestone.

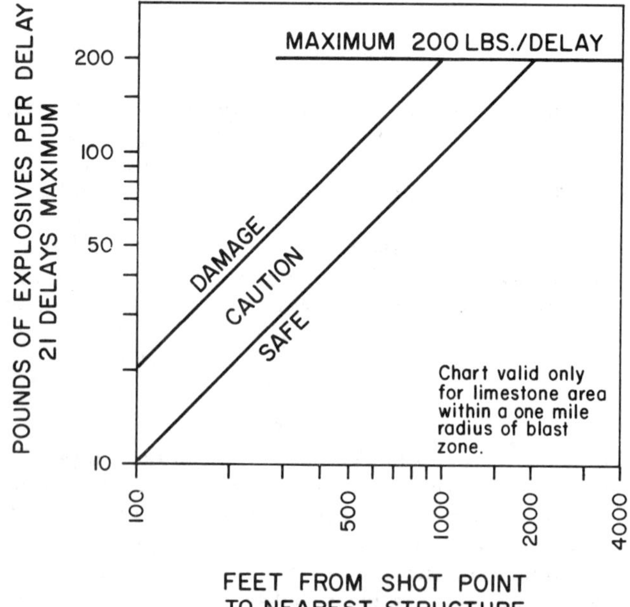

Figure 3. Blasting criterion chart.

In the project under discussion, a maximum value of energy ratio of 0.2 was selected. Even though Crandell's experiments showed that energy ratios less than three are not damaging to buildings, one must consider the effects of energy ratios of that magnitude on the residents of the area. This is particularly true when the blasting program is to continue over an extended period. Figure 4 gives a comparison of various blasting criteria in terms of ground amplitude versus frequency. The curve labeled "a" shows the level of ground motion easily detected by people. A comparison of this curve with curve "g" shows that human perceptibility is far below Crandell's safe criterion for structures. Curves "b" and "c" show the range in which the vibrations become particularly annoying to people.

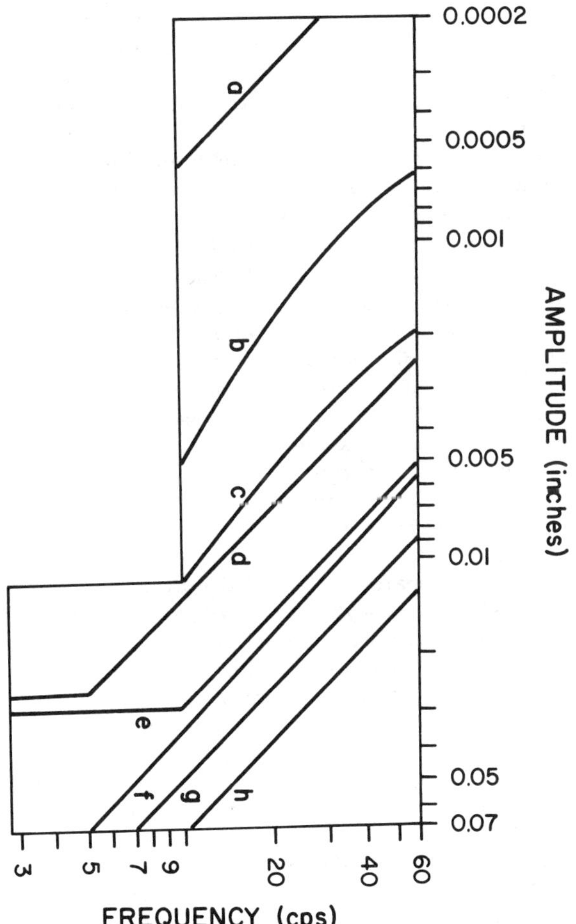

Figure 4. Vibration criteria curves. (a) Easily noticeable to people. (b) Troublesome to people. (c) Severe to people. (d) ER = 0.2, limit used for this investigation. (e) Pennsylvania, Massachusetts, and New Jersey allowable. (f) Edwards & Northwood--Bureau of Mines, V = 2.0 in/sec. (g) ER=3, Crandell's safe limit. (h) ER = 6, Crandell's maximum caution limit, above is damage.

Consequently, the limit chosen, shown by curve "d," although annoying to people, is far below the annoyance that would have been created had Crandell's criteria been chosen. It should also be noted that the level chosen is below that allowable by state law, curve "e," in some eastern states.

The blasting criterion chosen was satisfactory to the contractor when structures were beyond 1000 feet from the nearest structure. However, he felt that the total charge size had to be reduced too much for efficient removal of rock at locations closer than 1000 feet from structures (Fig. 3). In order to supply enough rock to keep his power shovels in operation, and still maintain the small vibration level required by the criterion established, it was necessary to blast three to five times a day in these regions rather than twice. Although this procedure increased the contractors operating costs, the induced ground vibrations did not cause damage to structures.

VIBRATION ANALYSIS AND DAMAGE CLAIMS

After a damage complaint has been made, most insurance companies will record the next two or three blasts in or near the structure concerned. This method was followed during the period following the development of the safe-shooting criterion. It was believed that the level of ground motion could have been increased approximately five times without causing damage to normal structures. However, some complaints continued to be made after the contractor had shifted his blasting procedure to that recommended.

Most of the complaints were from areas several thousand feet from the blast area, whereas only a few were received from residents within several hundred feet of the blasts. The locations from which the complaints were received were plotted on a geologic map of the area, and it was found that most were coming from residents whose homes were situated on a shale bed overlying the thick limestone formation in which the blasting was done. No energy focusing effect was observed from the data recorded in these areas. However, the velocity of compressional seismic waves in limestone is about 10,000 to 15,000 ft/sec, whereas in loosely compacted shale and overburden it can be as low as 1000 ft/sec. As the energy propagates from the limestone into the shale the frequency decreases, but the amplitude increases. However, the total energy decreases with distance. This change to lower frequency and larger amplitude produces a noticeable difference in the vibration level for people in the shale area over that felt in the limestone area. Consequently, the residents in the shale area were aroused to complain.

All the vibration data recorded from the shale areas yielded very low energy ratios which were well within the safe limit set. Thus, even though the amplitudes were larger, frequencies lower, and human agitation greater, no damage was done.

In some homes in the shale area, with the owner's consent, recordings were made of vibrations caused by normal walking on floors and stairs, and from vibrating machinery, such as washing machines. The homeowners were surprised to find that these normal household vibrations were higher in intensity than those from the delayed detonation of 2000 pounds of explosives one-half mile away.

The results of this investigation paralleled those of other investigators in similar situations, for example, Crandell (1949) who reported:

> Our experiments showed that it is possible to predetermine the amount of explosives that can safely be used in any blast without injuring the adjacent structure, but claims as to damage were still being made by the owners. Investigation of these claims showed that no structural damage had occurred. From this we can conclude that because the vibration was felt by the people living in the homes adjacent to the blasting, that they assumed damage occurred. This in all probability is because of the ability of people to feel vibration long before any structural damage could occur.

During the course of the construction, most of the complainants proved to be persons who were either agitated by the vibrations or startled by blast noise. One complainant charged that an air shock wave from a blast 1200 feet away had damaged the wall structure. A preliminary investigation gave the following information: (1) the complainant's house, constructed from limestone blocks, was situated on a wooded hill which shielded the house from a direct line to the blast area; (2) no windows were broken or cracked, and no old settling cracks in the wall were reopened; (3) the building embodied many settling cracks; and, (4) 500 feet downhill toward the blast site, no damage of any type was done to a home which was in a direct line-sight with the blast area 700 feet away.

Two blasts (in which the total pounds of explosives detonated in each surpassed that of the blast in question), were recorded at this complainant's home. The maximum peak displacement measured was only five ten thousandths of an inch. For a frequency as high as 60 cps, this would yield an energy ratio of only 0.00973. A sound-level meter was placed outside the house facing the blast site and set to record up to 120 decibels, which is the threshold of pain to humans. Since wind direction was from the blast area, the maximum noise conditions (Leet, 1951), were present. However, the explosions failed to produce a response on the instrument. The charge of damage by the complainant was dropped.

Some of the dwellings which had been moved from the right-of-way had been relocated on filled ground, and were within 100 feet of some of the blast areas. When the blasting site was 700 feet from one of these homes the owner complained of ground damage, stating that his yard was sinking and his newly constructed asphalt driveway was breaking up. Vibration recordings in the yard indicated that the induced ground motion was not damaging. However, due to the nature of the fill, the recorded energy ratios, although still in the "safe range," were approaching the "caution zone." As a precaution the pounds of explosives per delay and the total poundage were reduced for this area.

An investigation of the damage to this complainant's property disclosed the following: (1) the fill, built up on three sides, was inadequately compacted along the slopes and, because of heavy rain, was consolidating more from moisture increase than from blasting vibrations; (2) the unsodded slopes of the fill were too steep, and exhibited earthflow phenomena and other erosional features such as rills, furrows, and small slumps; (3) the asphalt driveway, adjacent to one slope, was undercut by surface runoff and sheetwash, and had slumped and flowed into the erosional voids; in fact, at the end of the driveway a miniature sinkhole about 1 foot in diameter developed where the asphalt had collapsed into a hole, which visibly emerged about 4 feet down the slope; and,(4) the soil adjacent to the building foundation was undisturbed, and showed no cracks or separation from the walls, whereas portions of the yard adjacent to the steep slopes exhibited slips and glide planes parallel to the slopes.

The contractor had not been involved in the relocation of these homes and was not responsible for the way it was done. He and the homeowner were informed that the damage up to the time of the investigation was the result of erosion, and if heavy precipitation were to reoccur shortly, the slopes of the fill might fail. The contractor was further informed that, unfortunately, if the fill material were to become saturated with water, vibrations from a small blast, normally considered very safe, might cause the slopes to fail completely. Fortunately the rain subsided, blasting was continued with reduced charge sizes, and major damage was avoided. It is important to note that the ground conditions at this site were not normal so that the "safe" criterion had to be adjusted for shooting in this location.

Another unusual complaint was received from the owner of a 6-month-old ranch-type house, in the limestone area, but approximately 2500 feet from the blast area. The owner complained that a center wall, two windows on the adjacent wall, and one of the stone window sills had all cracked as a result of the blasting. Vibrations recorded on the premises from four blasts were well within the safe limit, and the intensity of vibration recorded from normal household activities in this home were much higher than those produced by the blasts. In this same region there were other structures much closer to the blast area which were not damaged.

The house was located on the side of a hill, half on sound bedrock, the other half over fill. As a relatively new house it was in the process of settling differentially. The center wall, in a position parallel to the direction of wave propagation, had many small fractures close to the floor and along the adjacent outer wall. Neither the upper portion of the wall nor the ceiling was cracked. The windows of the adjacent wall, one to each side of the center wall, were cracked in the upper corner closest to the center wall. Both windows were tight in their frames and difficult to raise. The outer brick wall was cracked along one of these windows and the fracture extended vertically downward through the cement blocks of the foundation.

All the damage appeared to be from differential settling which caused abnormal stresses. Any structure under such abnormal stresses may be subject to damage from minor vibrations. Even though the vibration analyses clearly indicated that a normal structure would be very safe from the induced ground vibrations at this site, the peculiar subsurface conditions upon which this house was founded made it vulnerable to the normally safe vibration level. Through the efforts of a carefully controlled blasting procedure, the potential damage was minimized, and the use of vibration analysis aided in an out-of-court settlement.

The examples discussed are typical of the types of complaints that were encountered in this construction project. It is believed that vibration criterion developed and rigidly followed reduced complaints to a minimum. Every case in which damage may have been produced proved to be an abnormal situation, usually in the founding of the house.

PRE-BLAST RECOMMENDATIONS FOR CONTRACTORS AND INSURANCE COMPANIES

In addition to the usual examination and description of all structures within a reasonable distance of the blast area in advance of blasting, an effort should be made to inform the public about such things as the characteristics of blasting vibrations, the differences in structural damage caused by blasting from that produced by settling phenomena, the response of loose objects, such as wall pictures or mirrors and shelf knick-knacks to vibration, and the human response to vibration. This can be achieved by distributing literature, showing films, conducting lectures at civil meetings, and broadcasting on radio and television.

An extensive investigation should be conducted to determine the depths and condition of the overburden. This type of information may be obtained by employing seismic refraction or electrical resistivity techniques, or both. This information is useful in the prediction and prevention of excessive vibrations in areas of thick and unconsolidated overburden.

The contractor should detonate only at predetermined times and alert the surrounding public by blowing horns or whistles. Where this was not done, some startled people were known to have run hysterically from their homes thinking an earthquake was in progress, although the vibrations were very slight.

The blasting operation cited in this article lasted for a period of eight months, certainly time enough to aggravate anyone living in that area. Had all the above preventive measures been carried out, many of the complaints probably would not have been made.

Structural damage can be avoided by establishing a safe blasting criteria through vibration analysis. Analysis of heart failure, mounting tension, nervous disorders, anxiety, and hysteria cannot be made with vibration instruments. The value of a well informed public and fulfillment of proper pre-blast procedures is obvious.

CONCLUSION

The examples presented in this article are typical of the types of claims and complaints with which contractors and insurance companies must contend. Ground-vibration recordings and analysis proved extremely useful in settling most complaints without being brought to court. More important, however, vibration analysis was used to establish a criterion for safe blasting and to reduce the number of claims and structural damage. When the contractor rigidly followed the blasting criterion and damage still occurred, it was traced to abnormal conditions in the structure or subsurface. In these cases, the contractor may or may not be relieved of total responsibility for the damage and the likely claims dropped, depending upon the type and amount of subsurface geologic knowledge available to him and stipulations of responsibility set forth in the contract specifications.

REFERENCES CITED

Alpan, I., and Meidav, T., 1963, The effect of pile driving on adjacent buildings, a case history: Israel Institute of Technology, Haifa, Israel, no. 27.

Crandell, F. J., 1949, Ground vibration due to blasting and its effect upon structures: Journal, Boston Society of Civil Engineers, v. 38, n. 2, p. 222-245.

-----1960, Transmission coefficient for ground vibrations due to blasting: Journal, Boston Society of Civil Engineers, v. 47, n. 2, p. 152-168.

Duvall, W. I., and Fogelson, D. E., 1962, Review of criteria for estimating damage to residences from blasting vibrations: U.S. Dept. of Interior, Bureau of Mines Report of Investigations 5968.

Edwards, A. T., and Northwood, T. D., 1960, Experimental studies of the effects of blasting on structures: The Engineer, Sept. 30.

Hautly, R. F. and Kisslinger, C., 1959, Seismograph structures, instruments, and control systems: v. 32, n. 10, October.

Leet, D. L., 1951, Blasting vibrations' effects: Hercules Power Co., Wilmington 99, Delaware.

Thoenen, J. R., and Windes, S. L., 1942, Seismic effects of quarry blasting: U.S. Dept. of the Interior, Bureau of Mines, Bull. 442, U.S. Gov. Print. Off., Washington D.C.

GEOLOGICAL CONDITIONS AND ARBITRATION OF DREDGED CHANNEL AND TURNING BASIN, BLUFF, NEW ZEALAND[1]

Keith R. Miles (Department of Mines, Adelaide, South Australia)

Abstract

A dispute leading to arbitration developed during the dredging and construction of an island wharf in Bluff Harbor, New Zealand, when the contractors claimed that they had been supplied with misleading information as to geological conditions within the areas to be dredged.

The allegedly unexpected geological conditions were the presence of an old land surface of partially weathered basement rock at, and in places above, the level to be dredged. Heavy clay and boulders from this basement rock could not be handled by the suction dredge used by the contractors.

The defendant, the Southland Harbor Board, claimed that the true geological conditions at the site could have been foreseen from pre-tender information available to intending bidders.

It is suggested that the claim for "changed conditions" was the result of inadequate pre-tender site investigation by both the owner and contractor resulting from the failure of both parties to employ qualified engineering geologists to interpret the geological conditions.

CONTENTS

ILLUSTRATIONS

---

[1] This paper is published with permission of the Southland Harbor Board, and by approval of the Director of Mines, South Australia.

INTRODUCTION

The port of Bluff, situated near the center of the southern coast of South Island, New Zealand (Fig. 1), is the principal port and main export outlet for the products of the rich rural province of Southland. Bluff is also the gateway for the provincial capital, Invercargill (population 44,000), which lies 17 miles to the north.

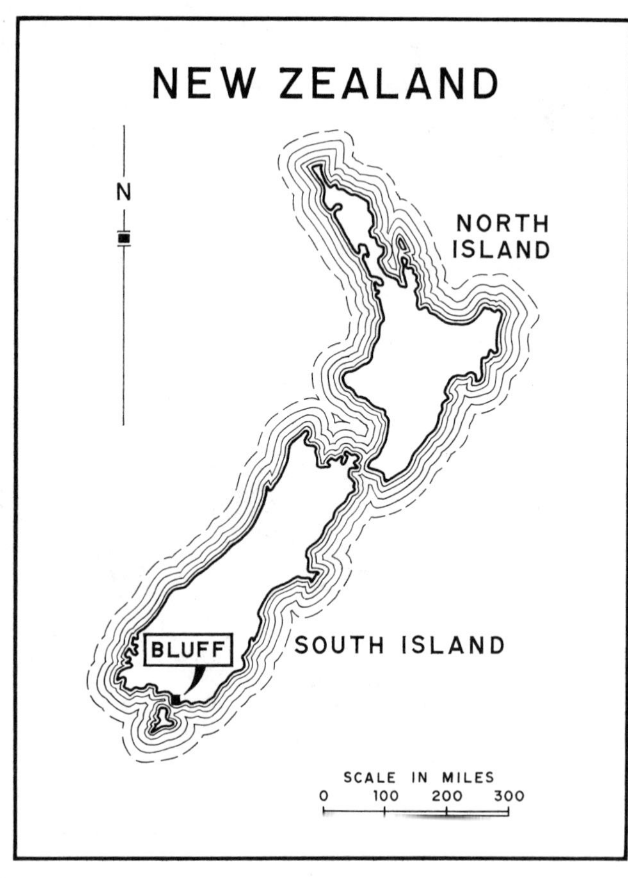

Figure 1. Locality plan, Bluff, New Zealand.

Bluff Harbor comprises an almost circular-shaped basin of mainly shallow water connected to the open sea by a narrow channel opening to the southeast, and linked by another channel which extends easterly to a long arm of shallow water called Awarua Bay.

The harbor is flanked on the south and southwest to northwest by a belt of low rocky hills forming a narrow isthmus or peninsula which provides some measure of shelter from the predominantly strong westerly winds of winter in this region. Tidal flow in and out of the harbor is considerable so that movement and navigation of shipping through the harbor entrance must be closely controlled.

To cope with increased export trade and a growing volume of shipping, both overseas and local, new harbor and wharf facilities were urgently required at Bluff by the 1950's. Consequently in August 1955, tenders for the construction of these facilities were called by the Southland Harbor Board, hereafter referred to as "the Board."

The project called for the construction of an "island" wharf area by reclamation of a continuously sheet-piled area of sandspit within the harbor to the west of the then existing wharf facilities. Additional items were: reclamation of the foreshore immediately opposite the "island;" construction of a road and rail bridge or causeway linking the "island" with the foreshore; and the dredging of berths at the new wharf, and a swinging or turning basin, connecting the berths with the main shipping channel at the mouth of the harbor. The project involved the close integration of the construction and the dredging activities as most of the material used in the reclamation of both "island" and foreshore was to come from the dredged areas. For this reason the successful, almost continuous operation of suitable dredging equipment under variable and often difficult weather conditions was an essential requirement of the whole program.

The contract was awarded for approximately $8,925,000 (U.S.) on May 1, 1956, to a joint-venture of four European dredging and civil construction companies, hereafter called "the Contractor." They proposed to carry out the dredging using a single large-suction dredge. The contract was completed on June 24, 1960.

By mid 1957 the contract was somewhat behind schedule due to delays in the dredging and the necessity of making certain mechanical repairs, and to some apparently unexpected dredging conditions. In March 1958, a second smaller suction dredge was introduced to speed up the dredging program, and in June 1958, the original big dredge was taken away for major repairs, modifications, and fitting with new and more powerful pumps. Despite these improvements, the Contractor continued to have difficulties in dredging parts of the swinging basin due to the claimed presence of material that was markedly different from that anticipated from pre-bidding information supplied by the client and accompanying the tender documents.

The first formal claim for $840,000 (U.S.) compensation on account of unexpected adverse physical conditions in dredging was lodged by the Contractor in February 1958. This was followed subsequently by several supplementary claims, based on the same grounds, before the completion of the contract. These claims were rejected by the Board, and it was finally agreed to submit the matter to arbitration, in accordance with the terms of the contract. In accordance with these terms, the temporary president of the New Zealand Institution of Engineers was invited to select an arbitrator. E. R. McKillop, the recently retired Commissioner of Works in New Zealand, was appointed arbitrator, and the case was heard at Bluff during the period May 28 to June 16, 1961. Judgment was delivered on September 27, 1961.

The final log of claims submitted by the Contractor at the commencement of arbitration amounted to approximately $2,142,000 (U.S.), comprising $2,065,000 (U.S.) compensation in respect of dredging operations, and $76,750 (U.S.) in respect of particular difficulties encountered in the engineering construction work.

GEOLOGICAL ENVIRONMENT

Bluff Harbor is dominated to the south by Bluff Hill, a single hillock of igneous rock (norite) which rises to 869 feet above sea level, and is flanked to the west and northwest by ridges of similar rock forming the Greenhills Range (Fig. 2). To the north, northeast, and east, the country is mainly flat, low-lying, and swampy, comprising peat swamps and overlying sand and gravel deposits. To the southeast, at Tiwai Point, the harbor is flanked by two rock hillocks joined to a long flat peninsula which stands at about 25 feet above sea level and stretches away easterly for many miles.

Bluff Harbor in its present form was developed in geologically recent times. The harbor basin itself consists primarily of basement rock comprising a belt of metamorphosed volcanics--basic lava flows, volcanic ash and sediments, and intrusive sills, the whole belt being intruded to the west by the large body of norite forming the Bluff Hill--Greenhills Range. This basement rock in the harbor is overlain unconformably by some thin discontinuous layers of fresh water or terrestrial sediments which are themselves locally covered by estuarine or main sediments--sands and gravels which are spread continuously over the present harbor floor.

The basement rock is exposed in outcrop at a number of places along the western and southern fringes of Bluff Harbor, that is, at Green Point, between the old Main Wharf and Stirling Point, at Tiwai Point, and also on a number of islands on the northwestern side of the harbor, such as Tikore or Spencer Island, Rabbit Island, Seal Rocks, and Colyer Island (Fig. 2).

Petrographic descriptions of the basement rocks were given by H. S. Service (1937). The individual bands of rock appear to trend roughly NW. to SE. and dip almost vertically. The unweathered rock of all types is very hard and mostly rather brittle. Jointing in the formations is common, with bands of some rock types being very closely jointed and broken. Others may contain only widely spaced joints and can be classified as "massive."

Particularly good examples of close jointing occurs at Tiwai Point where outcrops along the shore line appear to have been slashed by two or three sets of parallel knife cuts. Each set is roughly at right angles to the other and from 6 inches to 3 feet apart, thereby producing a very blocky appearance in the outcrops and shoreline with scattered loose boulders (Pls. 1 and 2).

The age of this belt of volcanic rocks and the accompanying norite intrusion is generally accepted to be late Paleozoic. Since that time, the country was uplifted, very extensively weathered, and eroded down to an irregular land surface which was then locally buried under relatively recent sedimentary deposits. In late Tertiary times the harbor was probably an isolated rock basin, a lagoon flanked to the east by deltaic sand and gravel deposits, and to the south and west by the Bluff Peninsula, a norite ridge. This basin was partly filled with fresh water clays, silts, and fine sands lying on an irregularly weathered rock floor. Subsequently, in recent times, the basin was breached and inundated by the sea, which scoured away the fresh water deposits in some places and distributed sand and fine gravel over the harbor floor.

Borings throughout the harbor in the vicinity of the proposed island wharf revealed a sequence of young sediments overlying the bedrock, namely, sands and some fine gravel at the top over clays,

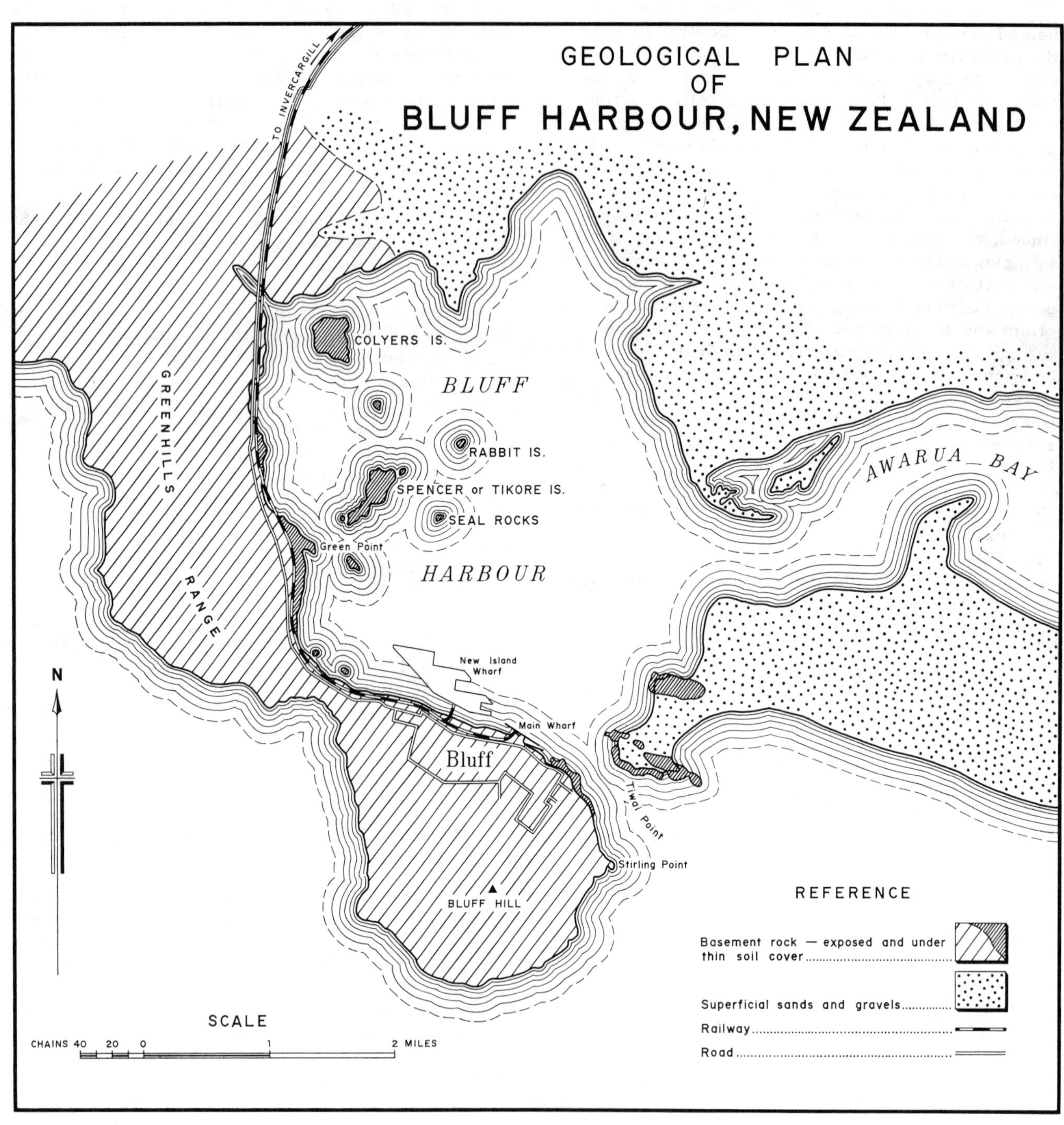

Figure 2. Geological plan of Bluff Harbor, New Zealand.

sandy clays, silts, and sandstone. The core borings also indicated that the first recognizable traces of bedrock marking the old land surface were usually of highly decomposed, or kaolinized, rock with the consistency of stiff clay (Wood, 1958).

Plate 1. Strongly jointed basement rock exposed at Tiwai Point, Bluff Harbor.

Plate 2. Strongly jointed basement rock, Tiwai Point, showing close jointing in a rock floor.

PRE-TENDER INVESTIGATIONS

Test drilling of the proposed island wharf area and causeway site was carried out during 1953 by the Graham Drilling Company in collaboration with engineers of the Bluff Harbor Board (Wood, 1958). Drilling equipment consisted of a portable percussion rig with sand pump sampler for the unconsolidated material, a tube sampler for undisturbed samples of consolidated sediments, and a rotary diamond drill for coring in hard rock. During this program a total of 57 holes were drilled for an aggregate of 2368 feet. All drill samples and cores were retained.

A number of reports were prepared based on logs of the drill holes, an interpretation of the geological conditions, and details of soil tests on undisturbed samples (Graham and Sparrow, 1952; Gedye, 1953). In addition, plans were drawn showing the interpreted contours on both the top of the weathered and unweathered basement rock, and geological sections at the site of the proposed island area. These reports and the samples were made available for inspection by all tenderers.

Some previous drilling in 1913 was done around the foreshore in the vicinity of the proposed causeway area and the eastern end of the proposed island wharf area. Although samples of these early borings were not available, logs of the bore holes with location maps were available to tenderers.

In addition, some jetting was undertaken within Dredging Area B (representing more than 83 percent of the area to be dredged). This comprised two lines, A and B (Fig. 3), with jet sites on 300-foot centers. The jetting consisted of forcing a rigid pipe down through the unconsolidated or loosely consolidated material by pumping water through the pipe until refusal occurred. Depth to refusal was recorded, but no samples were collected.

A total of 27 sites were tested by jet probes in Area B. The locations of these, together with sites of the 1913 and 1951-1952 drilling in the proposed reclamation areas are given on Figure 3. The final location of the shipping berths in relation to the boundaries of the island reclamation do not coincide with the original outline. This is because the outline was amended in 1954 to avoid areas of shallow basement rock revealed by the 1951-1952 site investigations (Wood, 1958, Fig. 1).

Incidently, the jetting procedure adopted in this investigation produced no samples that provided factual evidence of the nature of the material through which the jet probe was forced. The nature of the obstruction which prevented further penetration (refusal) could only be estimated, with the accuracy proportional to the experience of the operators. At several locations it was recorded that small fragments of puggy clay were found adhering to the nozzle of the probe on withdrawal from the point of refusal. This led to the interpretation by the Board that in most cases at refusal the jet probe had reached a surface of "clay pug" which was believed to represent the old land surface of weathered basement rock, elsewhere penetrated by drilling in the proposed island wharf area.

This interpretation was graphically depicted in plan and profile on drawings accompanying the contract and tender documents. It is significant that seven jet holes in Line A and four in Line B failed to reach maximum depth to which dredging was to be taken (Fig. 5). When this is considered, and also the fact that no actual drillings or sample dredging from the main dredging area were supplied by the Board or obtained as pre-tender information by the Contractor, it is understandable that the stage

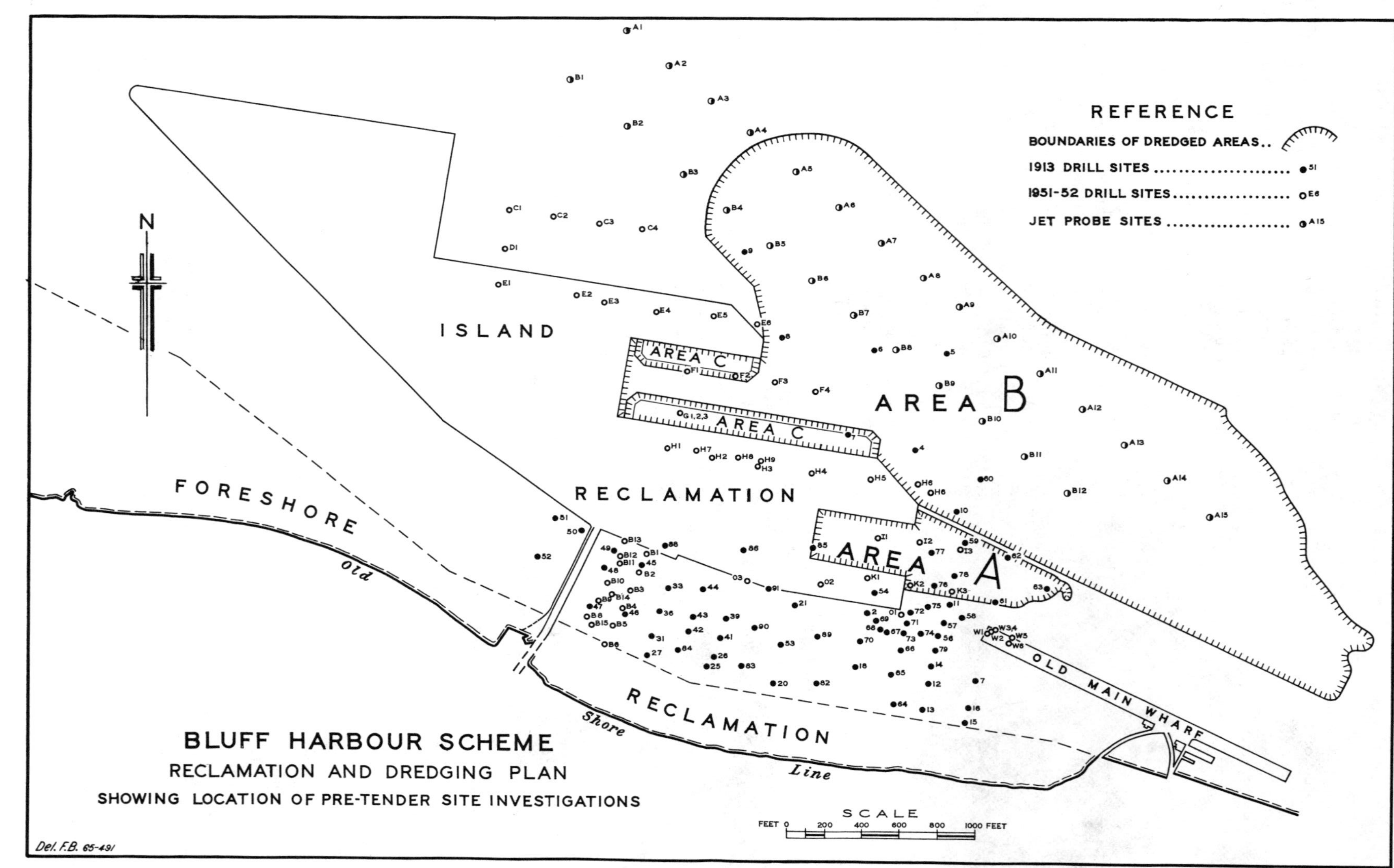

Figure 3. Bluff Harbor scheme--dredging and reclamation areas, showing pre-tender site investigations.

was set for some blind bidding and possible later claims.

THE CONTRACT

The contract required the dredging of some 2,937,000 cubic yards of material from three areas designated areas A, B, and C, and emplacement of this or other materials in specified foreshore and island reclamation areas (Fig. 3). The dredging depth for Area A, berth for coastal vessels, was 26 feet (below datum of Low Water of Standard Tide); for Area B, the main approach channel and swinging basin, 30 feet; and for Area C, the main berthing areas, 35 feet. No specific time schedule for the dredging was laid down in the contract, beyond that inherent in the requirements to complete the entire dredging and construction program.

A general description of the materials expected to be found within the areas to be dredged was contained in a document entitled "General Information for Tenderers" (see below), and under the heading "Conditions of Tender and General Conditions of Contract" it was made clear that all tenderers were expected to satisfy themselves about the geological and other conditions at the site of operations.

Under technical specification No. 6 of the contract, it was made clear that the Contractor in his tender had to submit a complete schedule of the dredging equipment to be used, and it was stressed therein that as full a description of this equipment as possible was required "to enable the Board and the Engineer to completely understand and appreciate" the tenderer's proposals. It would, therefore, appear that approval of the Contractor's proposed equipment is inherent in acceptance by the Board of the Contractor's tender.

An intriguing feature of this case is the basic cause of the arbitration claims--the dredging equipment (suction dredge) selected by the successful tenderer failed to cope with the conditions and materials of the site. It may, therefore, be inferred that neither the Contractor nor the Board fully understood these conditions and materials at the time the contract was let. However, the Board's Engineer, at an earlier stage when the Board was planning to carry out its own dredging on this project, stated his intention to use a bucket dredger which could undoubtedly have handled most of the material that finally caused the Contractor so much trouble.

It is also on record that before the tender was let, the successful bidders were called in to discuss in more detail with the Board the performance of their proposed suction dredging equipment and its capacity to do the required job.

THE CLAIM

Dredging commenced in April 1957, and during the next twelve months many difficulties and delays in dredging were experienced by the Contractor. These delays were due to a number of contributing causes, namely, bad weather; replacement of moorings; repairs and movement of pipeline due to the necessity for frequent change of the dredge location to accomodate construction requirements of the island wharf and foreshore; and mechanical breakdowns, all of which could be attributed to the normal hazards and wear and tear of such operations. In addition to these perhaps normal operational delays, however, the Contractor gradually got into more and more difficulties: (1) in dredging areas of puggy clay, due to mechanical damage to the cutter blades and to clogging of pipelines; and (2) in dredging localized areas containing notable proportions of pebbles, boulders, and fragments of rock which caused the dredge to stop by reason of blocking the ladder filling the stone or trash trap, obstructing the pump, and damaging rubber sleeves in the floating pipelines.

In the general information supplied to tenderers under "Geological Information--Classes of Material to be Dredged," this material was described as mainly "fine sand with a fine gravel fraction in the lower strata;" puggy clay floor material was described elsewhere, but there was no reference to stones or rock boulders. The Contractor admitted his own investigations supported the conclusion that the material to be dredged was 95 percent sand of maximum grain size mm (average 0.2 mm) and 5 percent clay. However, the Contractor took no samples in situ from the area to be dredged, and, therefore, claimed that he had been deceived by information supplied by the Board. More specifically, he claimed that the materials encountered were markedly different from the information supplied to tenderers, and contrary to the geological information contained in the Board's "General Information for Tenderers." Furthermore he claimed these materials were such that he could not be expected to have foreseen them. He, therefore, applied for compensation for: (1) losses of productive time incurred in coping with these unexpectedly different materials, that is, rock boulders; (2) costs of repairs and replacements due to excessive wear of the main dredge; and (3) costs of supplying and operating a subsidiary dredge in an attempt to make up lost time in the dredging program.

The first formal claim was lodged in February 1958, when some 608,000 cubic yards, representing 17.5 percent of the total, had been dredged. Further claims were lodged in November 1958 and June 1959 as more difficulties and delays in dredging were encountered. In the later stages of the contract, it became apparent that the suction dredges were incapable of dealing with portions of the heavy clay which formed the weathered bedrock floor, where this floor was considerably above the base level to be dredged, or with the associated rock boulders that were frequent. A peak of (weathered) rock rising some 19 feet above proposed base level was discovered in one corner of Area B (Fig. 4). This mass contained some 133,000 cubic yards and was later excluded from the contract. The Board also agreed to complete cleaning up of some other difficult bouldery sections in both Areas A and B with a bucket dredge.

GEOLOGICAL ASPECTS OF THE LITIGATION

The basis of the dispute argued before the arbitrator was geological--the matter of correct interpretation of geological conditions in the areas to be dredged and the accurate identification of the materials to be removed by the dredging. Under the terms of the contract, it was the Contractor's responsibility to carry out his own interpretation and assessment of these geological conditions, based not only on factual information supplied by the Board and on such reports, plans, and interpretive data supplied as supplements to the document "General Information to Tenderers," but also augmented by such independent observations and investigations as the Contractor deemed necessary.

As the case progressed, it became apparent that for the materials to be dredged, the Contractor had relied almost entirely upon his reading of the geological information supplied by the Board and some observations of materials taken from the sand spit, site of the island wharf. No evidence was introduced that any field investigations were undertaken by a geologist for the contractor prior to bidding.

Under "General Information to Tenderers," the description of "Classes of Materials to be Dredged" was given in the following terms--"Jet soundings were made on the lines shown on Plan No. 1361 (Figs. 3 and 4). These soundings indicate that the majority of the work consists of dredging fine sand with a fine gravel fraction in the lower strata. On Line A there is an area of clay pug, but it is expected to be similar to the "Green Clay" or weathered rock. . . . "

The "Green Clay" was described as--"This material has resulted from the weathering of the base rock and as its depth increases so does its hardness. The particle size and range varies considerably and on the basis of the U.S. Bureau of Soils Classification ranges from a clay to a loam. The average unconfined compressive strength is 35 lbs per square inch."

THE CASES SUBMITTED

The Claimant

The case for the claimant (the Contractor), with regard to the dredging, rested almost entirely on the unexpected nature of the material which had to be dredged, and the misleading nature of the information about this material supplied by the defendants (the Board). In defense of his own lack of pre-tender site investigations, he drew attention to the "unusually full" geological information supplied with the tender documents, and to the long experience of the Board's engineers in dredging within the harbor. He brought a number of eminent experts in marine engineering as witnesses to support his claim that the dredge selected by them was suitable to deal with the class of material expected, and that, in fact, the dredging was carried out "in accordance with the standards and practices of an experienced contractor and professional engineer," a requirement under the contract. He also produced witnesses to confirm that the adverse physical conditions responsible for the claim were such as an experienced contractor could not reasonably have foreseen.

The Defendant

In refuting the claim, the defendant's case was based on two main points: (1) the actual geological conditions within the areas to be dredged could have been foreseen, and that, in any case, the geological information supplied to tenderers was substantially correct and not misleading; and (2) the dredge was and always had been incapable of the performance claimed by the contractors and required of it to carry out the contract, and that the Board had been deceived in this by the contractors at the time of tendering.

The second point is outside the scope of the present paper, and will not be discussed further.

The first point of the defendant's case was supported by expert geological witnesses, and an exhibit submitted in the form of a submarine contour plan and geological sections depicting the upper surface of the "clay pug" exposed in drilling around the island wharf site and extending out to the line of jettings in Dredging Area B (Figs. 4 and 5). All the information from which the plan and sections were prepared was available at the time of tendering. Furthermore, it was contended that the interpretation that the "clay pug" represented the old buried land surface of weathered basement rock was entirely reasonable. The geological profiles clearly suggested that this old land surface did not continue to deepen to the north of the proposed wharf area, but rather rose to form a ridge, part of which was above the 30-foot depth to be dredged in Area B.

The author was responsible for the preparation of this exhibit and the submission of this geological interpretation at the arbitration hearing. From the analysis of the geological conditions outlined above, it was apparent that there was available no factual information as to the exact whereabouts of the old basement rock surface on the northern side of

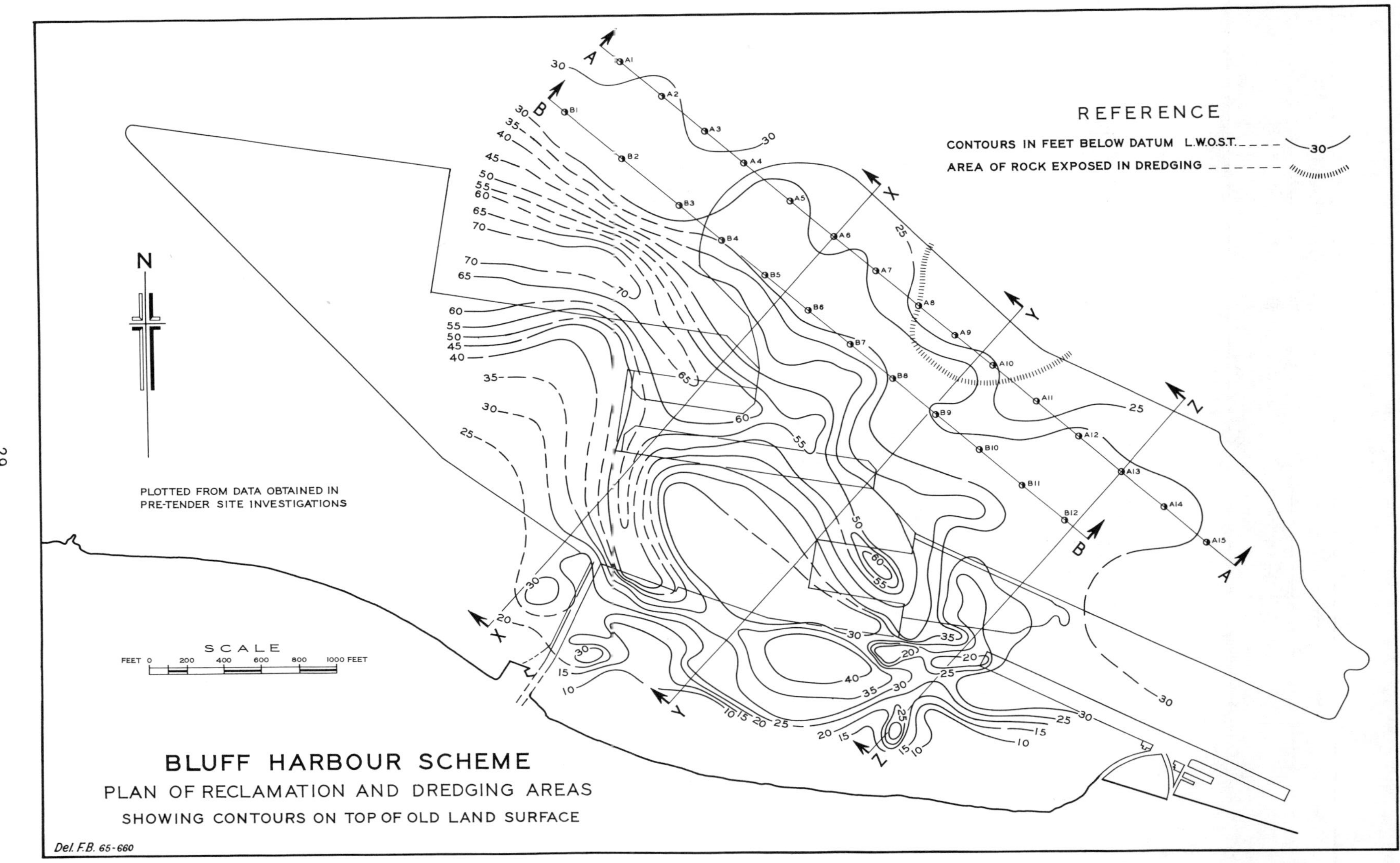

Figure 4. Contours of old basement-land surface below dredging and reclamation areas--from pretender site investigations.

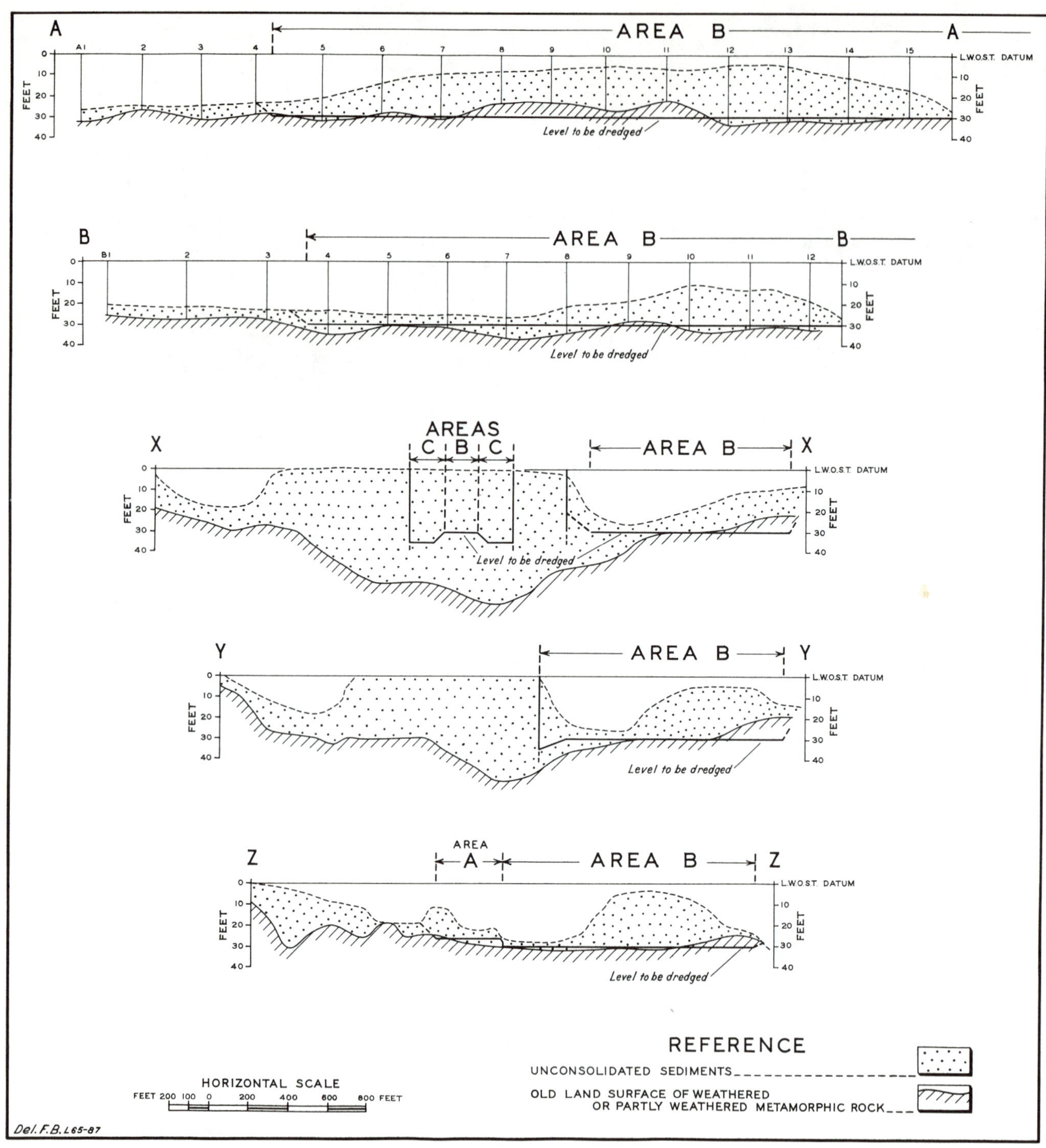

Figure 5. Geological profiles from Figure 4.

the turning basin (Area B) at the time the contract was let. Indeed the later discovery of the peak of weathered and hard rock in this area, subsequently deleted from the contract, came as a surprise equally to the Board and the Contractor. Thus, while sufficient geological facts were available about subsurface conditions at the island wharf area, geological information on the dredging areas was inadequate.

Evidence produced in support of the claim that some boulders and gravel in the material to be dredged should have been foreseen was based on a geological appreciation of the probable nature of the old basement rock surface. The strongly jointed and broken condition of portions at least of the basement rock is clearly visible along the shoreline of the harbor (Pls. 1 and 2). A profile of weathered basement rock was penetrated by drilling at the island wharf site, and a diagrammatic representation of a fairly strongly jointed section of this was exhibited (Fig. 6). Removal of the upper clay surface of this weathered rock either by natural means, such as wave erosion, or erosion by stream channels during the early history of the formation of the Bluff Harbor, or by mechanical means of dredging, particularly suction dredging, would tend to expose or release residual fragments or boulders of unweathered basement rock at the newly formed surface (Fig. 7). It was claimed that this was a valid and reasonable interpretation of the probable geological conditions which could have been made at the time of tendering by any competent and experienced geologist employed by a competent and experienced contractor. Certainly the interpretation adequately explains the conditions encountered in dredging the Areas A and B.

It was further claimed that with the numerous exposures of medium-to-fine gravel beds around

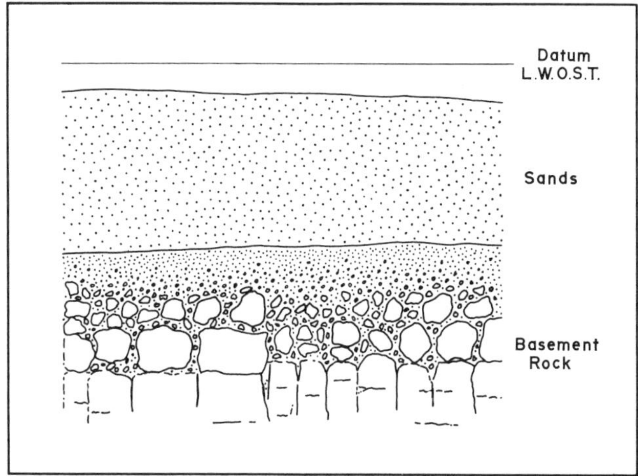

Figure 6. Diagram--partly weathered basement rock surface under submarine unconsolidated sands.

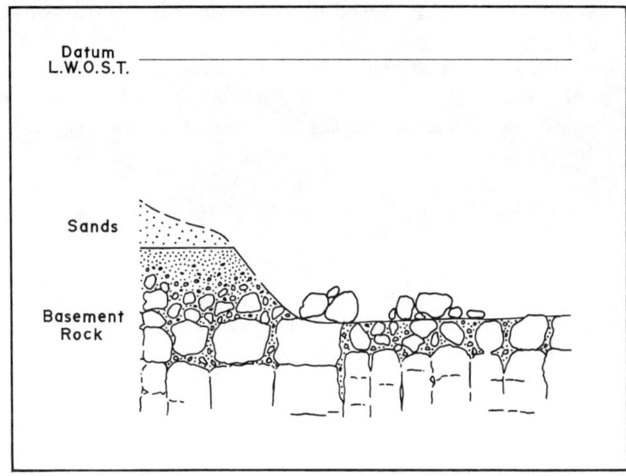

Figure 7. Partly weathered basement rock surface after mechanical removal of unconsolidated material--showing relict boulders.

the eastern and southern sides of the harbor, the presence of some gravel in the areas to be dredged could under no circumstances be considered unforeseeable. Finally, evidence was produced that the bulk of the material actually dredged was 95 percent or more fine sand which was: (1) substantially what had been described in the "General Information for Tenderers;" and (2) what had been expected by the Contractor.

THE JUDGEMENT

The Arbitrator determined that the Board was liable to pay the sum of $1,068,000 (U.S.) to the Contractor in respect of the dredging claims, and that the Contractor was entitled to an extension of four weeks in the time for completion of the contract.

COMMENTS

This case was the direct result of inadequate pre-tender site investigation, and the failure of both the Board and the Contractor to understand fully the geological conditions at the site of operations. There is little doubt that had the Board adequately drilled and sampled the areas to be dredged they would never have accepted any tender involving the use of a suction dredge as the sole means of submarine excavation to the designed depths.

Had the Contractor called for or otherwise obtained representative samples of the material to be dredged before tendering, as is common practice, it is doubtful that he would have risked doing the work with the type of suction dredge used.

It is the author's belief that the case would not have arisen if the Board and the Contractor had each engaged the services of competent engineering geologists to direct the site investigation to assist in preparation of the tender documents (Board), and to interpret the geological conditions and advise on the probable nature of the materials to be dredged (Contractor).

REFERENCES CITED

Gedye, N. T., 1953, Bluff Harbor development scheme: Report on geological investigations of sand bank area and soil mechanics, April 20, 1953 (Unpubl.).

Graham, C. J. and Sparrow, W. R., 1952, Bluff Harbor development scheme: Report on drilling data of sand bank area, November 24, 1952 (Unpubl.).

Service, H. S., 1937, An intrusion of norite and its accompanying contact metamorphism at Bluff, New Zealand: Trans. Royal Soc. N. Z., v. 67, pp. 185-217.

Wood, B. L., 1958, Submarine geology of Bluff Harbor: New Zealand Jour. Geology and Geophysics, v. 1., pp. 461-469.

GEOLOGIC CONDITIONS AND CONSTRUCTION CLAIMS ON EARTH- AND ROCK-FILL DAMS AND RELATED STRUCTURES

E. B. Waggoner, J. L. Sherard, and W. A. Clevenger, Partners (Woodward-Clyde-Sherard and Associates, Consulting Engineers and Geologists, San Francisco, California)

Abstract

The number and magnitude of disputes with contractors over claims for extra compensation in connection with earth dam construction have been increasing in recent years. Frequently, these claims are based on problems within the scope of activity of the geologist and soil engineer. A number of the more common geologic situations which have led to disputes over claims are described. Arguments are given that it would be desirable to give bidding contractors a report which describes the engineer's and geologist's interpretation of the results of the field explorations and of the probable construction problems.

CONTENTS

ILLUSTRATIONS

INTRODUCTION

In recent years there have been an increasing number of dam construction projects which ended with the contractor making large claims for extra compensation because of "changed" conditions, "withheld or incomplete" information, and other reasons. Frequently, these claims are disputed by the engineer and owner, and are finally settled by a board of arbitration or the courts. A large proportion of construction claims arise from problems related to soil engineering and engineering geology.

The authors have been involved as consultants in a considerable number of these disputes, sometimes for the contractor, and sometimes for the owner. In addition, they have studied the geological problems underlying a number of claims in which they were not personally involved. This paper describes geologic conditions which have been at the heart of some major claims. It also presents arguments for the authors' conclusion that most of these problems could be reduced in magnitude or eliminated if the bidding contractors were given a competently prepared report describing the engineer's and geologist's interpretation of the exploratory data and of the probable construction problems.

EMBANKMENT SOILS

Difficulties in excavating and placing soil embankment material have been responsible for an appreciable number of claims. Typically, for an earth dam, the embankment soil is excavated in the borrow area, transported to the dam site, and spread and compacted in a more or less continuous operation, with some interruptions and delays for controlling the water content and blending the material. These delays are not usually great with respect to the cost of the total operation. If conditions arise during construction which result in appreciable delays in the excavation-hauling-placement operations, and if these delays have not been anticipated in the contract documents, the contractor may have a justifiable claim for extra compensation. Such conditions are frequently associated with certain soil types and borrow pit conditions, some of which are discussed in the following sections.

Pervious Material Below the Water Table

It has been a fairly common practice to build pervious zones of dams from sands or gravels excavated with drag lines from below the water table in river bottoms. If the material is very clean, it can be loaded directly into trucks, transported to the dam, and placed according to specifications. However, experience has shown that a large percentage of apparently clean river sands and gravels contain enough silty fines that water does not drain out easily. This kind of material must be worked or stockpiled, or both, in order to make it manageable. A very well-graded, sand-gravel mixture containing a small percent of silt sizes (finer than the No. 200 sieve) can be especially deceiving (Fig. 1). Because these materials compact into a very dense state, a few percent of fines can have an appreciable influence on the permeability. In some cases, the only way the water can be drained out of these soils is to stockpile them for weeks in small piles. If the contractor had not expected this, he may be seriously hurt. Some examples of trouble of this kind are described in Sherard and others (1963, pp. 639-644).

Selective Excavation

Selective excavation in borrow pits is often required by the specifications in such a way that the inspection personnel must judge the suitability of each excavated load of the material being excavated for use in a given zone of the dam. Such situations have led to several hotly disputed claims.

One common example is the borrow pit of thinly stratified or lenticular river alluvium where the design calls for placing the cleaner material,

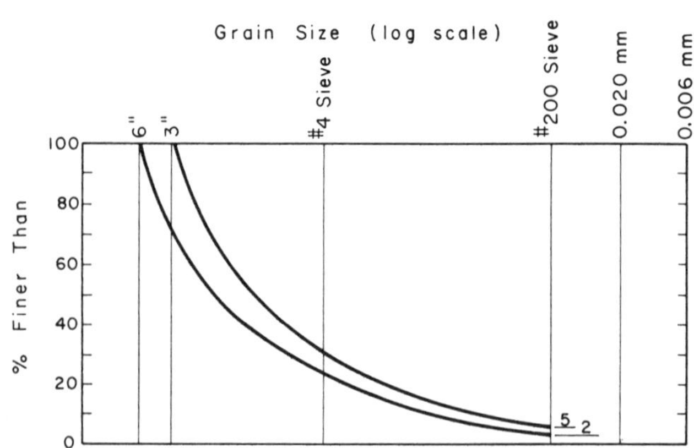

Figure 1. Well-graded mixture of sand and gravel which has been especially difficult to handle when excavated below the water table.

selectively excavated, into special pervious drains or filters and the dirtier material into zones in which the gradation is less important. This design technique, unless the construction problems are carefully thought out and provided for in the contract documents, has a high likelihood of leading to unexpected construction delays, arguments, and claims. The problems arise when the contractor claims that the inspectors have unduly delayed his operation by taking excessive time to make decisions, by forcing him to work his excavating and hauling equipment in an inefficient manner, and by sometimes forcing him to excavate and stockpile material which cannot be placed in the dam immediately.

Difficult Moisture-Density Control

Rigid adherence to water content specifications for the compaction of clays of high plasticity (liquid limit above 50 percent) has caused important construction delays and disputes, especially when the natural water content is above the specified maximum value. Drainage ditches in the borrow pits may be of no value whatsoever in reducing the water content of the material, and, hence, the only available way of lowering the water content is by drying on the surface of the borrow pit to a depth of a few inches and by working the material on the embankment with discs. In the worst case, the maximum practical rate at which this can be done is below the rate at which the dam must be built, causing an impossible situation.

The use of cohesionless silts and fine uniform silty sands has led to expensive construction delays primarily for two reasons. On some jobs, it has been practically impossible to obtain the density requirement specified, regardless of the water content, type of roller, and thickness of layer, and much experimentation and time was required to solve the problem. Also, when wet, some of these soils are extremely difficult to handle. On some jobs, the problems of getting the density and keeping the construction surface stable have been so difficult that both the engineer and contractor wished that they had never been associated with the project. Gradation curves for some materials that proved difficult to handle are shown in Figure 2.

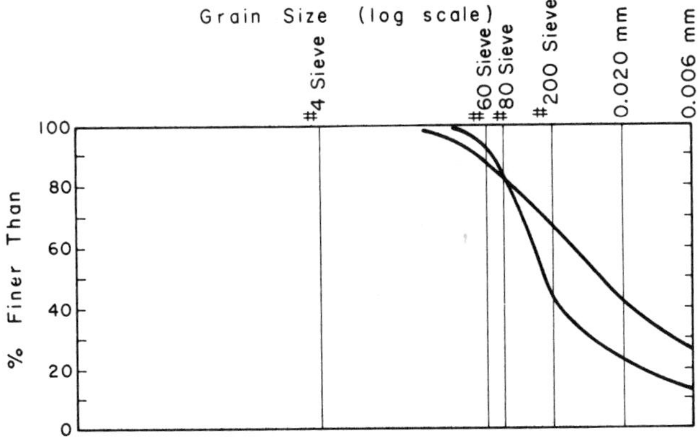

Figure 2. Cohesionless soils which have been very difficult to compact.

For some soil types which may be used for dam cores, the results of laboratory compaction tests are dependent to a large degree on the test procedure. There are several different kinds of these soils: for example the properties of some volcanic clays are permanently changed by changing the water content; or some residual soils with gravels of rotten rock become much finer with handling and compaction. For such materials, unless the water content control procedures are especially devised to suit the peculiar soil properties and specified in advance, the contractor can make a good argument for a "changed conditions" claim simply by demonstrating that the material is not amenable to standard control procedures.

A similar large claim arose over the use of fine silty sand. In this case, the specifications, which were more or less conventional, required that the material be compacted at a water content which was in the range between 2 percent below, and 1 percent above, Standard Proctor optimum water content. The claim was based on the fact that for this soil there was no well-defined "optimum" water content in the ordinary sense; that is, the density obtained in the laboratory compaction test was essentially constant over a considerable range of water content. During construction, the contractor had a considerable amount of trouble because the engineer insisted on adding water to the material to the point where the construction surface was continuously soft, and pushers were required for the hauling equipment and the rollers. The contractor claimed that the engineer forced him to add more water and to work with a much more unstable embankment than was justified under the specifications.

Segregation of Coarse Core Material

One of the best core materials for an earth dam is a well-graded mixture of sand, gravel, and clayey or silty fines. However, when there are too many gravel-sized particles in the mixture, the most careful construction procedures are needed to avoid excessive segregation of the coarser particles. When the engineer finds that he is having trouble with segregation during construction, he may be forced to ask the contractor to use special techniques for excavating, transporting, and spreading the material which, if not specifically foreseen in the specifications, can be a justified basis for a "changed conditions" claim. A rough correlation providing a guide to the probability of trouble of this kind is given in Figure 3.

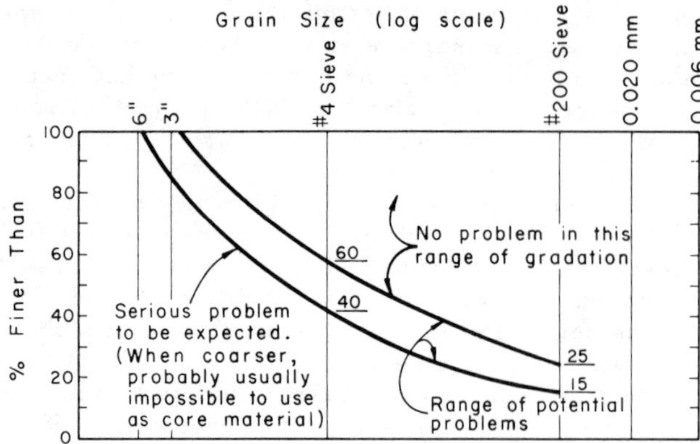

Figure 3. Rough correlation between gradation of coarse, well-graded core material and the problem of segregation during construction.

Halloysite

Core materials which contain an appreciable content of the mineral halloysite have been the source of considerable construction trouble and major claims. When examined in their natural state in the walls of exploratory test pits, these materials may have the appearance of an ordinary soil which can be easily placed and compacted at its natural water content. However, once excavated and spread on an embankment, they may change rather mysteriously. The phenomenon is evidently caused by water which is chemically bound to the clay mineral in the natural state, but which is freed as the soil is manipulated at normal temperatures. In some cases, the more the material is worked on the embankment, by discing and rolling, the more wet appearing and soft it gets, even in dry weather.

In the Pacific Northwest, there are two extreme types of halloysitic soils; one which causes the trouble described above in the very worst manner, and one which is wholly stable and can be used easily without trouble in dam embankments. These extreme materials cannot be differentiated by visual examination (in the walls of exploratory test pits) even by an experienced soil engineer, and there are no well-established petrographic or soil engineering tests to distinguish between them. They can be differentiated, however, relatively easily in the field by spreading out the material in a small test embankment and compacting it with a tractor. The critical material will visually change under a few passes of the tractor from an apparently dry and stable soil to a sloppy mess. The other will compact easily as an ordinary soil at good water content.

At the high Hill's Creek Dam in Oregon, constructed during 1958-1960 the problem was so bad that the designers finally decided to abandon all attempts to compact the core material with a roller and placed it in a very wet, plastic state only with bulldozers (Brown, 1961). Terzaghi (1958) and Dixon and others (1958) give an excellent description of the design and construction problems associated with the use of an halloysitic clay at the Sasamua Dam in Kenya.

Contamination of Pervious Borrow Soils

One of the most common changes in the design of earth dams, necessitated by conditions encountered during construction, results when the material in the pervious soil borrow area is found to be much less pervious than anticipated. This may occur when a deposit of stratified sand and gravel is excavated on a vertical face mixing all the material together with a small quantity of silt or clay, so commonly present and not identified in the exploratory borings and pits. In several cases, this mixing action in apparently very pervious soil deposits has resulted in embankment material with a permeability approaching that of the core. This has necessitated delays for study and changes in the design or the construction specifications.

Shrinkage

Shrinkage of embankment material between the borrow pit condition and the compacted condition in the dam has led to important claims. If the shrinkage is much higher than would be considered normal, or has not been anticipated by the contractor on the basis of the bidding information, he can make a claim for extra compensation. This is provided the extra excavation required caused him a longer average haul distance or necessitated extending his work through another winter.

ROCK QUARRIES

One of the most frequent sources of claims is trouble in the rock quarry. Whether the product is to be used for transition zone, shells, or riprap is usually not too pertinent to the claim. Most rock quarry claims are based upon "changed conditions," usually geologic in origin. Commonly, the claims will be one or more of the following: (1) overburden stripping far in excess of that indicated by specification logs or quantity estimates; (2) rock of poorer quality than was indicated by specification logs or specification requirements resulting in excessive quarry waste; (3) specification requirements for gradation or quality which prove to be infeasible or abnormally difficult to attain in an owner-designated quarry; (4) unreasonable inspection; (5) the unforeseen need to install special processing equipment to produce an acceptable product; (6) excessive handling and rehandling of rock before placing, in order to meet owner requirements.

All of the above listed quarry claims involve disputes concerning both main characteristics of the quarry; that is quantity and quality. In the authors' opinion, too many decisions concerning the quantity and quality of a rock from a quarry are based on a simple visual inspection, perhaps one or two exploratory holes, and sometimes tests of a few hand samples picked up at the site. All too often, in a quarry explored in this way, the rock breaks into undesirable sizes, is too soft, requires special handling, or has too high a waste ratio.

The critical geologic properties of a quarry rock cannot be determined reliably by surface examination and coring alone. Whenever any significant volume of quarry rock is to be produced, it is highly desirable to explore the quarry by opening one or more faces with test blasts. During such tests, records of methods and cost of drilling, blasting, picking up, handling, and placing should be obtained. The sizing and waste content should be observed and recorded, and the properties of the quarry product should be determined by tests similar to those for aggregate soundness. With such information, a designer can know better what kind of rock he will obtain and write better specifications.

One special quarry problem has to do with the use of quarry spalls (generally less than 6 or 8 inches) for transitions between earth and rock-fill embankment zones. Often the designer writes the specifications in such a way that the contractor is given freedom to obtain these spalls in any way he wishes, but with the implication that he can probably get them by selective excavation in the quarry. In several cases of this kind, it has proved to be impossible or excessively difficult to get the desired material in sufficient quantities by selective excavation. For each, it was necessary to go to a grizzly with attendant disputes over who pays for the extra work.

Many other quarry problems which lead to disputes and claims can be eliminated if grizzlying is specified in the contract documents. Because of this, and because separation of quarried rock at the 6 inch or 8 inch size divides it into two materials which are excellent for the design of the safest type of dam (good transition material and strong, pervious rock-fill shells), the authors believe that grizzlying should always be given serious consideration and should be used more widely. It is interesting, and not always well recognized by geologists and engineers, that the cost of grizzlying may be less than the value of the extra volume of embankment material obtained as the result. The rock fines screened out are largely materials which would otherwise fill the voids in the rock-fill sections of the dam. Hence, the total volume which a ton of screened rock occupies in the embankment is likely to be 10 to 15 percent greater than the volume which it would occupy if it were not screened. The value of this extra volume can easily equal the cost of grizzlying and, hence, the rock can be screened and divided into two parts without adding to the cost of the project.

FOUNDATIONS

Unexpected problems with excavations in foundations and abutments have been a major source of claims. Most of these are within the scope of the geologist and soil engineer.

Slides

Slides of the abutments, foundation excavations, cut slopes for access roads, excavations for spillways, and so on, have been a major cause of claims for extras. The most careful study of rock bedding and joints and their relationship to potential slides caused by excavations on abutments and appurtenant structures is always justified.

Only on rare occasions do specifications include geologic data on access roads which must be constructed by contractors before they can reach the site of the project. Yet slides on access roads have led to major claims. Not only can they seriously delay the construction by cutting off the traffic, they can cause loss of life and damage to the completed structure when above the dam.

Slides have frequently occurred where the foundation excavations unexpectedly had to be carried to a greater depth (and hence at steeper average slopes) than had been anticipated. Hence, a careful study and an accurate estimate of the probable depth of excavations has two benefits from the standpoint of claims: (1) minimizing slides; and (2) minimizing unexpected excavation volumes.

Old slides on the abutments, not recognized until construction began, have, on occasion, necessitated the excavation of millions of cubic yards more than had been anticipated. In extreme cases, this has led to the abandonment of the project. Any extra excavation which is greatly larger in volume than the anticipated volume may change the complexion of the job and justify new unit costs.

Rock Defects Requiring Special Treatment

The discovery of unforeseen fault zones, caverns, or similar features in the foundation excavation is a common source of claims. Such conditions always cause delays to the contractor due to: first, the time required to inspect and study the condition; and, second, the time needed for treatment (for example, over-excavation and backfilling, sealing, or some type of dental work requiring special equipment or hand work). Geologists should automatically consider the possibility of such conditions existing at every major construction site and design their explorations accordingly.

Over-excavation

Foundation excavation from other causes (buried valleys, deeper abutment stripping, and so on) which is greatly in excess of the anticipated value can lead to justifiable claims.

In the circumstance where the dam is to be supported directly on the natural soil foundation with only the surface organic material removed, the volume of this surface stripping has sometimes been grossly underestimated. Especially for long dams of low height, this item can be important relative to the cost of the project. Care should be especially taken in estimating the volume of surface stripping in areas where there are deep roots.

Foundations in limestone have been responsible for large unexpected quantities of excavation. In the case where the foundation consists of large blocks of hard massive limestone separated by clay-filled joints and seams, the borings may show nearly 100 percent core recovery of good rock. If, in addition, the surface exposures give every indication of massive bedrock, the designer may conclude that little excavation will be required to give a sound foundation for the dam. During construction, the inspector may direct the contractor to pluck out a few of the larger chunks separated by clay-filled joints. Then he looks at the hole left, directs a few more to be removed, and finally finds he has required the contractor to excavate many times more material than planned.

Excavations exceeding the volumes anticipated by the designer can be particularly troublesome in the circumstance where the spillway (and/or any other large appurtenant structure) is located very near the dam. Over-excavation for the dam foundation or for the spillway can result in a need for change in the design--if the two structures were formerly separated from each other and the over-excavation causes them subsequently to abut. Any such change in the design causes a delay which can lead to a dispute over extra payment unless the problem is foreseen during planning.

Very Irregular Rock Foundation

Where the rock surface against which the core of an earth dam is to be placed is rough with sharp irregularities, the earth material immediately adjacent to the rock cannot be compacted effectively with the heavy roller. Consequently, specifications commonly require that this material be placed in thin layers (3 or 4 inches thick) and compacted by hand labor. This work must be carried to an elevation at which there is enough working space so that the heavy roller can be used effectively.

Usually there is no provision for extra payment for the hand compaction, and the contractor is paid for this work at the same unit price as the core material placed and compacted by machinery. Since it costs considerably more to place and compact a cubic yard of material with hand labor than by machinery, the contractor, in preparing his bid, has to make some assumption in advance about the amount of material to be hand compacted. This estimate will depend, of course, on the shape of the rock surface. If the rock is smooth and gently sloping, probably almost no hand compaction will be required; if it is steep and irregular or jagged, a considerable average thickness of soil in contact with the rock may be required to be hand compacted.

The trouble and potential claims arise in those situations where it is not possible in advance of construction to obtain a good view of the rock surface because it is covered with a soil mantle, and where, when exposed during construction, the surface is found to be very irregular. The contractor will claim that he could not have anticipated this extra work, and, hence, he should be paid extra for it. Also, since the placement of the hand compacted material goes considerably more slowly than the material placed by machine, this problem can have an important influence on the time required to build the dam.

The geologist can minimize the frequency and magnitude of these problems by devoting adequate time and effort to scrutiny of the question "What is the appearance and shape of the final clean rock surface likely to be?" The engineer can help by providing in the contract documents in advance for extra payment for the hand compaction where the item is likely to be of appreciable magnitude.

The problem is likely to be relatively more severe for low dams, because these have a greater area of foundation contact surface per cubic yard of core material than high dams. (The volume of an embankment increases roughly with the square of the height, whereas the foundation contact area increases approximately with the first power of the height.)

One general type of situation which has led to this kind of problem is shown schematically in Figure 4.

Overbreak in Rock Excavations

In rock excavations for spillways, powerhouses, tunnels, and so on, the contact is commonly

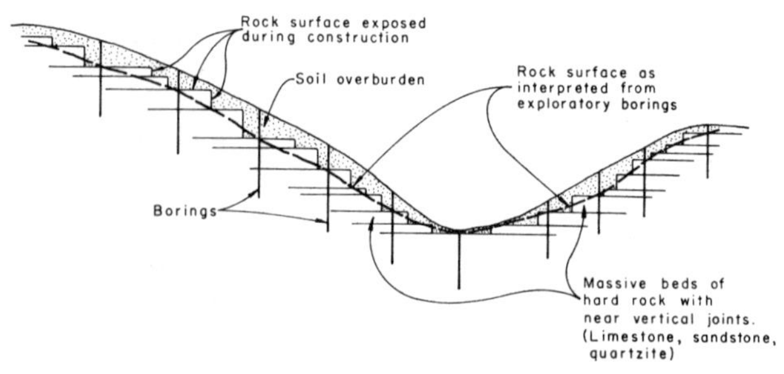

Figure 4. Irregular rock surface masked by soil overburden.

set up to pay for the excavation and concrete up to an arbitrary "pay line." Beyond this pay line, any rock overbreak or necessary concrete backfill is not paid for. The purpose of this method of payment is primarily to induce the contractor to use careful blasting techniques and to avoid any more rock excavation than is absolutely necessary. Therefore, in preparing his bid, the contractor must make an evaluation of the character of the rock and estimate what volume of overbreak beyond the pay line may possibly occur and what volume of unpaid-for concrete backfill may be needed.

One of the most frequent sources of hotly contested claims arises from the situation where there is a large quantity of overbreak and additional excavation beyond the pay line for some reason. Commonly, the contractor claims that the geologic conditions made it literally impossible to excavate the rock to lines which approximate those shown on the drawings; that is, using the very best blasting techniques, the walls simply will not stand and the material falls into the excavation. In reply, the engineer frequently takes the position that the overbreak was due solely to the fact that the contractor used inexpert blasting techniques. The true situation on any given job can vary anywhere between these two extremes.

Since troubles of this kind occur so frequently and involve so much money, the problem justifies a great deal of attention from the geologist and engineer during the investigation and design stage of the project. The geologist's job is to evaluate the rock and the possible excavation problems; the engineer's job is to make a design which minimizes the potential trouble associated with overbreak, and a construction contract which anticipates, as well as possible, the worst contingency which could arise.

Ground Water and Dewatering Problems

Problems dealing with ground water both in foundations and in soil-borrow areas have frequently been the basis of claims. One common source of trouble has been where water levels are inadvertently left off the boring logs for some reason. Even though there are many borings in a given area which all show the water level at a given elevation or in a given condition, if one or two borings of the group do not show water, it can be argued convincingly that the contractor had the right to assume that there was no water at these points.

Another common trouble arises when no water level was measured because the hole caved, drilling mud was used, or some other reason. In such cases, it is imperative to note that no measurement was made of the water level and to state the reason. Those holes which were measured and found dry should be so marked. The presence of artesian flows is extremely important.

When artesian conditions are encountered, all the observations made in the field should be recorded on the boring logs. Too often these are abbreviated into the single note "artesian water" which could indicate anything from a minor condition to a major potential construction problem.

Important disputes have arisen when the ground-water level at the time of construction is higher than observed when the explorations were made. Hence, as a minimum, when borings are made in the dry season of the year, the date of the boring must be noted, and, further, the bidder's attention should be called to the possibility that the water could be higher at other times. If a real problem exists, it would be better to make an estimate of the maximum height to which the water is likely to rise as a seasonal variation. The contractor then does not have to assume that the water will be above this elevation. Some provision for payment should be included in the event that the water rises above the estimated maximum elevation.

Dewatering of excavations, an important and sometimes very critical part of the construction of dams, begins early in the construction, and continues at least until the foundation and cutoff have been excavated and replaced to a level above the river surface. Most dewatering claims are based on what the contractor describes as excessive water or instability of materials to be dewatered, or both. Contractors' estimates and bids are based upon geologic data supplied to them in specifications, and such data should include dependable permeability figures, accurate descriptions of river bed materials, accurate bedrock profiles, and reasonably accurate stream gradients. Permeabilities are best obtained from good pumping tests in sufficient number to allow a good statistical average of the permeability of the alluvial materials. The cost of such tests where the dewatering is going to be a big job is minor in comparison to the potential claim for handling excessive and unforeseen volumes of water for long periods of time.

Claims have arisen out of the fact that "quick sand" developed in the bottom of excavations in sand. In one example, extra compensation was awarded when an excavation in sand below the water table became "quick" and difficult to handle--even though the boring logs clearly showed sand and a high water table. Basis for the award was that the contractor was not alerted to the sand as relatively loose and, therefore, more susceptible to trouble than a denser sand. The owner's engineer argued that any sand can become quick under certain conditions (which, of course, is quite true) and that the contractor should have anticipated this possibility in his bid. However, the board of arbitration concluded essentially that the contractor could not have known that the material was very likely to become quick from the data supplied and consequently, he was entitled to assume, when preparing his bid, that it would not quicken; that is, that he was entitled when bidding the job to take an optimistic view of the situation. In this case, to have avoided paying the claim, it would have been necessary for the engineer to have shown the contractor that the sand was in a loose state and probably to have warned him in writing that a quicksand condition could develop unless special techniques of dewatering were employed.

Slaking Shale

Problems with slaking shale, especially under spillway slabs and in tunnels, have led to claims. Usually, the troubles from this source have come from the necessity of using appreciably thicker slabs and linings (with much greater quantities of concrete) than have been shown on the drawings. The problem is less important with respect to the delay than the increase in volume of material required in the construction of the structure.

Grouting

Surprisingly, the authors do not know of any major claims revolving around the problem of foundation grouting. Probably this is because all concerned with the job fully recognize the difficulty of estimating in advance the amount of grouting ultimately needed. Also, specifications are written with this in mind, and responsibility for advance estimating is not passed to the contractor.

TUNNELS

Most disputed claims in tunnels revolve around problems of estimating: (1) the supports needed; (2) the "overbreak" outside the pay line; and (3) the amount of water entering the tunnel excavation.

Excessive slaking or spalling can result in a need for continuous maintenance during construction, additional backfilling, additional concrete lining, excessive lagging, and resultant delays. Contractors will claim that they were not warned in the specifications of such rock properties and, therefore, had not expected them and were not prepared to handle them. Information on anticipated rock slaking, spalling, caving, squeezing, and overbreak should be provided in every excavation specification by the geologist. As mechanical moles are used more in tunnel excavation, it is increasingly important for the geologist to point out these and similar properties which may influence the choice of type, and affect the performance of, tunnel excavating equipment.

Another source of tunnel claims is the accuracy of the geologic specification data by which the contractor estimates his needs for supports. He must be able to estimate sizes and quantities of steel with reasonable accuracy. He is more concerned with being precise as to the total amount of steel required for a tunnel than knowing the exact details of where each individual support will go. Therefore, the bidders should be given, with the specifications, the best possible opinions by the investigating geologists as to percentages of the various types of rocks to be encountered and how or if they should be supported. Too frequently geologists, to be conservative, overestimate both size and number of supports, and then during construction the inspector, who must approve all supports, will allow much less. As a result, the contractor presents a claim for acquiring, storing, and returning unused supports and for an adjusted higher unit rate for support installation.

The question of how much and what kind of water a contractor has to handle in a tunnel is a common source of claims. The contractor should be given the geologist's estimates of peak volumes and average volumes to be handled, as well as estimated temperatures and pressures, and any information available or opinions concerning chemical constituents which would adversely affect men or equipment.

PRESENTATION OF DATA AND CLASSIFICATION OF EXCAVATION

Ambiguity

The preceding paragraphs contain descriptions of specific geologic conditions which have been associated with claims. Another source of claims common to many geologic specifications are ambiguities or other inadequacies in the presentation of the geologic data. The presentation of these data in the specifications is the geologist's last and sometimes only chance to describe subsurface conditions to the bidding contractor before construction starts. Hence, the presentation must be complete and clear.

Unfortunately, many geologic descriptive terms are ambiguous. For example, frequently words such as "hard," "soft," "weathered," "altered," and so on, when used alone, do not convey an adequate picture of the physical conditions which will be encountered. For instance, a hard granite and a hard shale may be quite different if compared to some constant standard of hardness. Consequently, it is often beneficial to describe the physical properties of the soil and rock in simple, descriptive, easily recognized lay terms which are not subject to arbitrary interpretation.

Completeness

Completeness of geologic descriptions is important, not only to aid in the solution of the main design problem, but to minimize unanticipated construction problems and claims by giving as clear an indication as possible of the stability conditions during construction, and the response to excavation effort with different methods and specific types of construction equipment.

Some important disputes have, quite literally, hinged around the interpretation of one or two words in the specifications, and frequently these are words written by the geologist. The choice of geologic words is of such importance that every geologist responsible for such specifications should keep at hand a glossary of geologic terms, a good dictionary, and reliable reference texts in special fields of geology such as ground water, geophysics, and rock mechanics. He should check himself and the specifications whenever he has the slightest question as to the meaning which a contractor may put upon a word. Commonly, one of the contractor's first acts in looking for support for claims is to start looking at the geologic logs and trying to find meanings in the geologic descriptive terms that will help his claim if they can be shown to be inaccurate, very broad, or ambiguous.

Important errors and omissions, which ultimately have served as support for a contractor's claim, have occurred during the transfer of the geologic field data through the office staff to the specifications. In such cases, a comparison of the bidding documents with the notes of the drilling inspector or field geologist shows that there were condensations, rewording, or mechanical mistakes which changed the original descriptions or meanings, or left out vital information. It is of greatest importance that the final geologic logs put onto the drawings be checked against the original field notes, before the specifications are issued.

One source of trouble leading to claims has been the custom of some geologists or drilling inspectors to describe the material on the boring logs from the drill cuttings rather than to give a description of the material from which the cuttings were derived. In one well-known claim, a hard, dense, tough, glacial boulder till was described only as "sand, gravel, silt, and some clay" in the specifications. In looking back at the original field notes, it was found that the material had been described as a very compact, hard, dense, dark-gray boulder till, containing a matrix of sand, silt, and clay which had to be cored with a diamond bit during the explorations. A major point in the successful support of the claim was the contractor's contention that the description in the specifications misled him.

Another important category of excavation claims has revolved around the word "boulders" on boring logs. Most of the trouble arises from the fact that the contract documents contain only the briefest description of the subsurface material penetrated by the exploratory borings. A typical case occurs when only a few of the logs in a river excavation indicate "boulders up to 4 or 5 feet in diameter." The other logs say only "boulders" or "hard drilling." In this case, if there are many boulders of 10 feet in diameter encountered in the excavation, a claim may be sustained on the basis that the contractor had the right to assume from the study of the boring logs that he would not have had to excavate any boulders larger than 5 feet. If a brief note had been added to the contract documents which expressed the general opinion of the geologist about the probable nature of the deposit, which, for example, might have read "probably a nest of boulders with maximum size estimated in the range between 5 and 15 feet," the support for a claim on this point would have been greatly weakened.

Claims have arisen out of the method of specifying the depth of excavation in the case where the cutoff trench is to be carried into weathered rock. In one case for an earth dam, foundation borings showed weathered rock gradually grading into sound rock, and the specifications required that the cutoff trench be carried down to a level described in the specifications as "suitably sound rock." During construction, the engineer required that the excavation be carried down to fresh, unweathered rock of a quality that would have been satisfactory for a concrete dam--which required the blasting of 5 to 10 feet of slightly weathered rock. The contractor was able to support his claim of changed conditions by demonstrating that the average experienced engineer or engineering geologist would have considered that the rock removed by blasting would have been easily adequate foundation material for an earth dam. Such a problem could occur at sites with almost any kind of weathered rock. This is a case where the claim was based squarely on the wording of the specifications. The design engineer certainly has the right to carry the excavation to any depth he wants; however, if he is to avoid extra payment, he must specify in advance what he is going to require in a more explicit fashion than in this example.

INTERPRETATION OF GEOLOGIC EXPLORATIONS FOR BIDDERS

After studying the details of a considerable number of claims and asking for each the question "What would the geologist and engineer have had to do differently to avoid the trouble which arose?", the authors have reached two main general conclusions. First, even our most conscientious and competent design engineers and geologists frequently are not spending enough time analyzing the problems of construction. Second, in addition to the normal information on the job which is given to a contractor, a report should be prepared describing in detail the designer's and geologist's view of the construction problems which might arise.

For the engineer to be sure that he has a completely safe design, he does not necessarily have to examine in minute detail all of the possible construction problems which could arise. From the standpoint of economics, however, and of minimizing claims, he is well advised to spend a great deal of time studying the probable construction methods which will be used and trying to evaluate and anticipate the problems which could arise. In regard to each problem, he needs to ask the essence of the following three questions: (1) What will probably happen during construction? (2) What is the worst thing that could happen? (3) What would I do if the worst condition occurred? The answer to one question leads to another question and so on, until--with enough effort and imagination--most of the potential problems about which claims could arise can be anticipated in advance, and methods can be devised and included in the contract documents to handle them in a fair and equitable manner.

The report which the authors recommend be given to the bidders should describe, as clearly and frankly as possible, all the thinking of the engineer and geologist concerning the construction problems, and should describe the conditions expected and the risks and uncertainties. Such a prebid report is in direct contrast to current general practice under which the contractor is given only the bare data obtained, without interpretation. The authors believe that there is nothing to be gained by withholding the interpretation. On the contrary, by giving the bidders all the information, the chances of getting a valid low bid and of eliminating unjustified claims for extras are much better. Since the designers usually study the project for a year or more, and the contractor is given only 30 to 60 days to study the job before making a bid, obviously the engineer is in a much better position to have evaluated the construction problems in depth, no matter what the competence and the experience of the contractor and no matter what investigations he may make.

In almost every specific case studied where painful and costly claims or law suits arose, the authors believe that the claim could have been eliminated or made much simpler to resolve if such a prebid report had been written and given to the bidders in advance. In none of the cases studied would such a report, competently prepared, have worsened the problem or have been detrimental to the client's main interest in getting a safe and economical structure. A prebid report would serve three main functions as described in the following paragraphs.

The preparation of a report would provide a strong practical incentive to induce the designer to be sure that he has thought out his problems adequately and the geologist to get enough data to analyze adequately before including it in the specifications. If they know that they have to put their thoughts about the construction problems on paper for the contractor and the client to see, they will certainly make sure that they have spent enough time studying before choosing design details and writing the specifications.

By telling the contractor what the engineer and geologist anticipate, a whole host of problems resulting from possible ambiguities in, or misinterpretation of the intention of, the contract documents can be eliminated. Several examples cited earlier are in this category, for example, if the water table is inadvertently left off one of the boring logs in the specifications, the description of the ground-water problems of the site contained in the report given to the bidders would make it obvious that this was a drafting error and not an indication that there was no water. Another example is given in the paragraph which discusses the problems concerning the word "boulders" on the boring logs.

Finally, the report would have an ameliorating influence on disputes resulting from that large group of problems which arise when the contractor argues that he is forced to do considerably more work on a given aspect of the job than would have been anticipated from a reasonable interpretation of the data available and the specifications. In these cases, if the designer had written down in advance his ideas about the probable construction problems, often there would be no dispute. For this general kind of a problem, if the interpretation of the engineer of the probable difficulties was incorrect on the unconservative side, then the contractor is clearly entitled to some extra compensation.

The authors have discussed the preparation of a prebid report with many engineers, contractors, and attorneys who are interested in various aspects of dam design and construction. Each had some knowledge of the main arguments which are raised in opposition to, and of the hesitations concerning, such a report.

First, attorneys are predominantly opposed. They express general fears that the report may open more avenues for disputes and claims than it would close. However, of the attorneys with whom the authors have discussed the idea, only a few, generally those most experienced with construction claims, have studied the problem in any depth. Most do not realize the extent of the forces which are rising in the construction industry to cause the overall problem, nor have they studied the specific technical details which have accounted for the great bulk of the troubles. Because of this, and because they are naturally conservative in such matters, they generally recommend against any change in the status quo. Also attorneys on the one hand, and engineers and geologists on the other, view the problem from quite different vantage points. While the engineer deplores spending weeks or months of his time in the expensive and unproductive labor of disputing claims for extras, the attorney is usually not inclined to share the view that this is a fundamentally undesirable and distasteful activity. Nevertheless, attorneys are very influential in this matter.

They must be convinced of the essential validity of the recommendations made here that analytical pre-construction reports be written before such a policy can be adopted as a matter of general practice.

Engineers and geologists, themselves, are divided in their initial reaction to the suggestion that reports be prepared and given to the contractor prior to bidding. Some agree completely and immediately, while some are doubtful. The authors believe that this difference of opinion depends to a great degree on the individual's personal experience with claims. The engineer or geologist who has suffered through one or more long and drawn-out arbitration or suit is more sympathetic to these ideas than the man who has not.

One of the most interesting arguments raised against preparing prebid reports is the idea that as a rule the engineer making the design brings to his task years of education and experience in design, but relatively little construction experience. As a result, he is "design-oriented" as opposed to "construction-oriented." For this reason, there may be some question as to whether he is really qualified to write an adequate report. The authors believe that this is an unfounded fear, and whenever true, the problem should be rectified. In fact, in order to design an economic structure, the engineer must be capable of evaluating the main construction problems. Furthermore, there is no reason why he should not call in a good contractor as a consultant to review his thinking concerning the construction problems. This is probably an activity which is justified in many cases and done all too infrequently.

Contractors agree almost universally with the desirability of a prebid report. Most express the long-standing complaint that they are asked to bid on aspects of the construction which are kept needlessly unknown and risky. They say that a report would aid them in understanding the problems and the risks involved, and can only result in lower average bids and fewer disputes. They also point out the great benefit that it would lessen the probability that an inexperienced and hungry contractor will take a job at a figure which is well below his cost--the very last thing the engineer wants on a dam project.

Finally, there is a point of professional ethics involved here which the authors believe to be quite important. Associated with this rash of claims there have been a number of cases where the owner has--after being forced to pay a large claim for extras--turned around and sued his engineer to recoup what he considers to be his losses. Consequently, the engineer himself is becoming liable, and the liability hanging over his head cannot help but have some influence on his technical decisions. When a problem arises during construction of a dam that requires changes of some kind, the engineer should consider, first and foremost, the safety of the structure, and then the economy and fairness to both the owner and the contractor. If he also has to consider the implications of his own liability, he may find, in some circumstances, that his own best interest and that of his client do not coincide. The authors suggest that the possibility that this situation may occur more often as these claims become more frequent is already enough incentive to make the conscientious engineer want to consider a fundamental change in his mode of operation.

REFERENCES CITED

Brown, F. S., 1961, Oral discussion on the problem of the construction of the core of Hill's Creek Dam: Seventh Congress on Large Dams, Rome, IV, p. 50.

Dixon, H. H., and others, 1958, The Chama-Sasamua water supply for Nairobi: Proceedings, The Institution of Civil Engineers (London) April 1958, v. 9, p. 345.

Sherard, J. L., Woodward, R. J., Gizienski, S.F., and Clevenger, W. A., 1963, Earth and earth-rock dams--problems of design and construction: New York, John Wiley, 707 p.

Terzaghi, K., 1958, Design and performance of the Sasamua Dam: Proceedings, Institution of Civil Engineers (London) April 1958, v. 9, p. 369.

MINE DEWATERING AND RECHARGE IN CARBONATE ROCKS NEAR HERSHEY, PENNSYLVANIA

Richard M. Foose (Department of Geology, Amherst College, Amherst, Massachusetts)

Abstract

Mining of high-calcium limestone from the overturned, south-dipping beds of the Annville Limestone resulted in large withdrawals of ground-water from beneath the Hershey Valley, particularly when the mining operation was deepened in 1949. Details of ground-water movement, water levels, and the position and size of cones of depression caused by pumping reveal drastic differences before and after mine deepening.

A program of ground-water recharge to restore ground-water levels by the Hershey Chocolate Corporation resulted in litigation. Continued recharge coupled with a grouting program by the mine resulted in the restoration of normal water levels.

CONTENTS

ILLUSTRATIONS

INTRODUCTION

The Hershey Valley is a small part of the Great Valley, about 9 miles east of Harrisburg and 7 miles west of Lebanon, Pennsylvania (Fig. 1). Its north and south boundaries are low irregular hills that stand about 200 feet above a nearly flat valley floor. The Hershey Valley comprises about 40 square miles, nearly all of which is highly productive farmland. Hershey, the main town in the valley, is well known as the home of the Hershey Chocolate Corporation.

Swatara Creek, the major stream in the valley, flows southwestward to the Susquehanna River 9 miles away. The major tributary to the Swatara in this area is Spring Creek, a small stream with little variation in flow.

This article is modified from an earlier publication dealing with the behavior of ground-water in the Hershey Valley (Foose, 1953).

Louis C. Smith, chief engineer of the Hershey Chocolate Corporation, and many of his staff have been helpful in carrying out the field studies described.

GEOLOGIC SETTING

Ordovician Rocks

Beekmantown Limestone. The Hershey Valley is underlain by Beekmantown Limestone. Rocks of this formation vary widely in physical and chemical characteristics, but generally are dense, laminated, blue-gray magnesian limestones. Locally they are crystalline white limestones or dark-blue dolomites. The Annville Limestone is an upper member, about 500 to 600 feet below the stratigraphic top of the Beekmantown. The Annville member is 250 to 300 feet thick at most places in the area and consists of more than 95 per cent $CaCO_3$.

These valley limestones have been differentially weathered and dissolved by percolating ground-water. Generally, the purer limestones have many more solution openings than the dolomites. Cavities and openings have been developed chiefly along the bedding and cross joints in the limestones, and are naturally larger and more numerous near the surface (Fig. 2A). A 10-foot

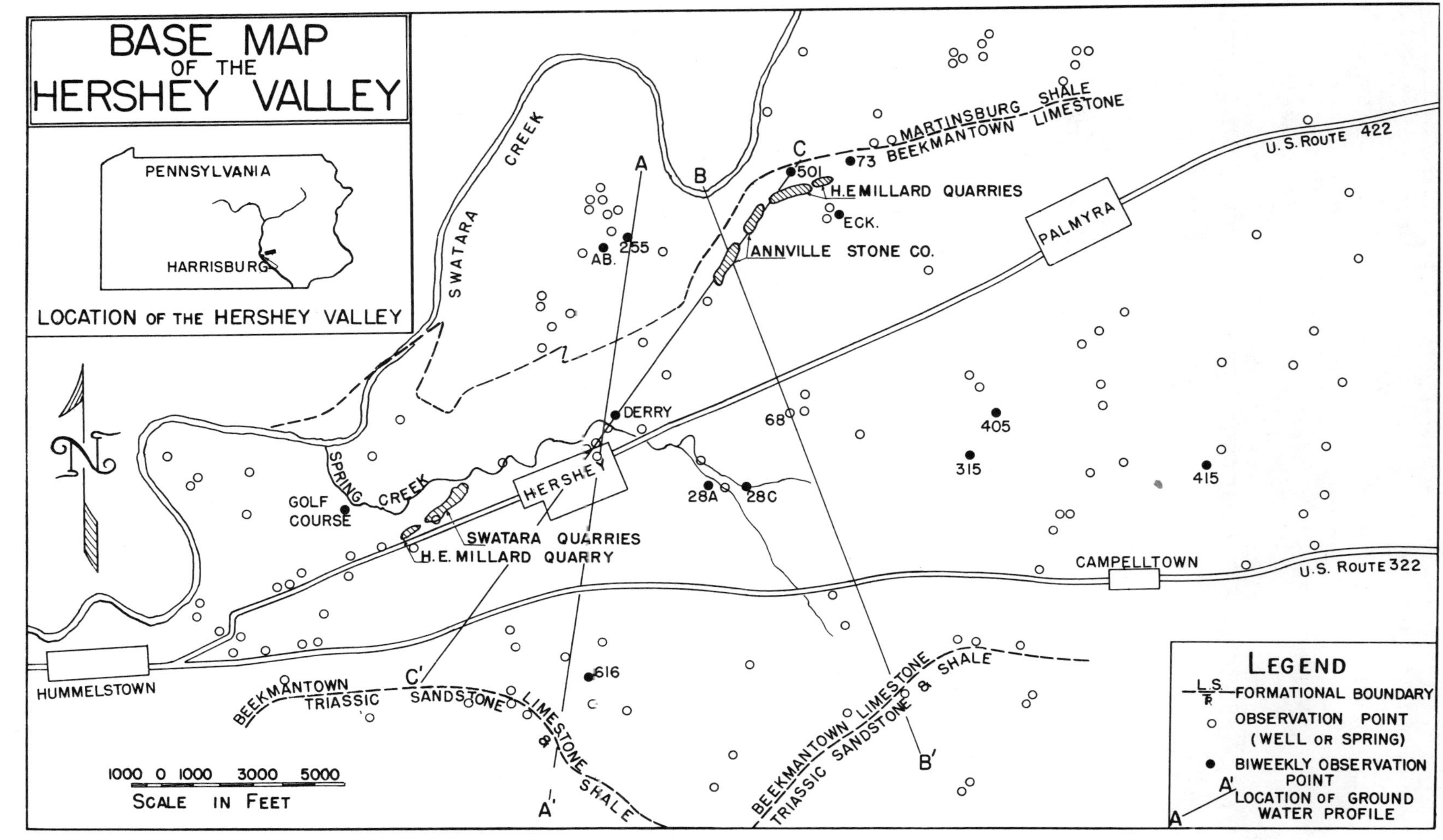

Figure 1. Base map of the Hershey Valley.

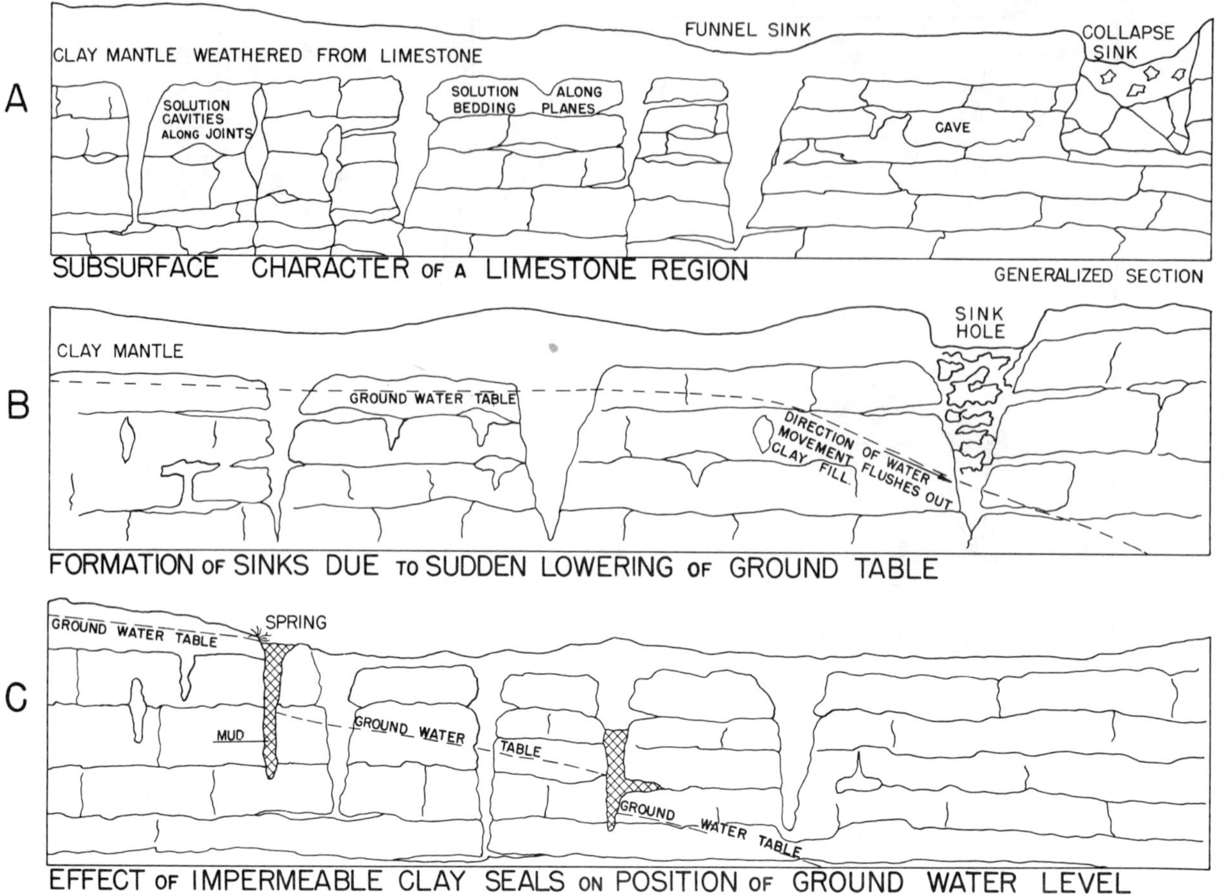

Figure 2. Idealized diagrams of limestone and ground-water conditions.

zone along the contact of the hanging wall of the overturned Annville Limestone and the overlying dolomite is exceedingly permeable (Fig. 3). At many places these openings are tightly plugged with mud (Fig. 2C), most of it residual soil from the weathering of the limestones, that had filtered down into the openings and become packed.

Martinsburg Shale. Conformably overlying the Beekmantown Limestone, the Martinsburg Shale is tan and very thin-bedded with occasional thin beds of sandstone, limestone, and chert. It makes the irregular hills that form the northern boundary of the Hershey Valley and is relatively impermeable.

Triassic Rocks

Gettysburg Shale. Soft red shales and poorly cemented silty sandstones with a few interbedded conglomerates bearing large white and pink quartz pebbles constitute the Gettysburg Shale formation which underlies the hills bounding the Hershey Valley to the south.

Structural Features

The Beekmantown Limestone and Martinsburg Shale strike generally N. 65° to 70° E. and are steeply overturned to the south, having average dips of 45° to 65° S. (Fig. 3). The limestone especially has yielded to severe deforming stresses so that there are many tight folds, some isoclinal and some recumbent in large nappe-like structures. Boudinage (sausage-shaped remnants of beds) is well developed throughout the formation, attesting to the great amount of stretching and flowage.

A prominent set of cross joints and a well-developed set of longitudinal joints are developed in the limestone and to a much lesser extent in the shale. Solution of the limestone has been greatest along the cross joints and the steep south-dipping bedding, and to a lesser extent along the longitudinal joints.

Several faults displace the limestone and shale. Some of these strike northeast and have displaced beds as much as 3000 feet at the surface. Others are cross faults, striking generally north-northwest. Several of the latter cut into the former mining operations of the Annville Stone Company

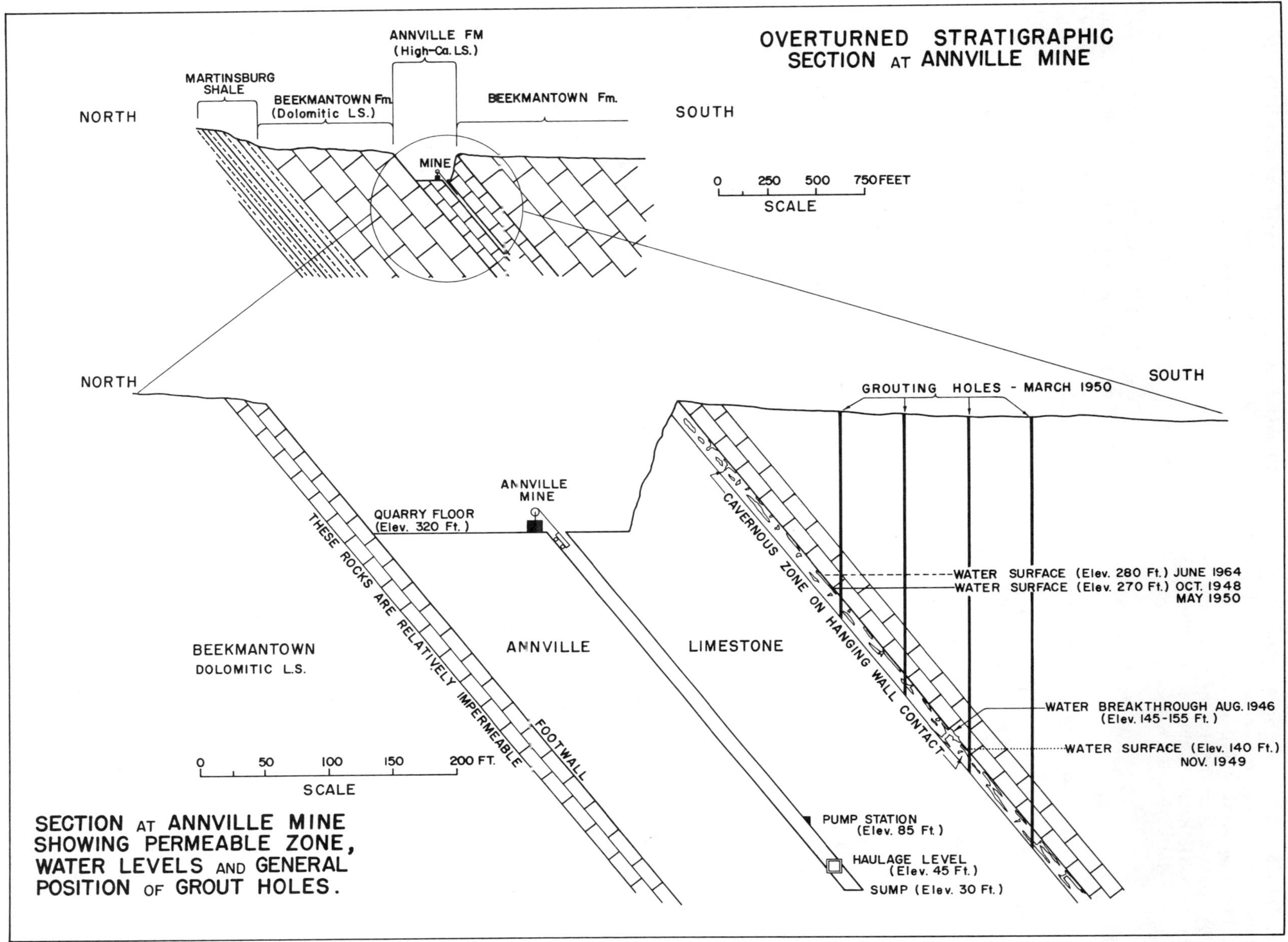

Figure 3. General and detailed section of the Annville Limestone at the Annville Mine, showing permeable zone, water levels, and grout holes.

and the H. A. Millard Company. Locally, along these faults, the rocks have been extensively brecciated. A connecting series of steeply dipping normal faults separate the Beekmantown Limestone along the south side of the Hershey Valley from the younger Triassic rocks to the south. The Triassic rocks have relatively uniform dips of 25° to 35° NW

Mining

The high-calcium Annville Limestone has been quarried and mined for many years at many places within the Great Valley. In the immediate area it was quarried by the H. A. Millard Company at Swatara Station a mile southwest of Hershey until 1954. Their quarry was about 125 feet deep. One and a half miles northeast of Hershey, the Annville Stone Company[1] operated a mine whose workings extended more than 400 feet below the surface. The rock was removed by a combination of room and pillar and shrinkage stope mining methods. The main slope into the mine had its origin on the floor of a large quarry, about 105 feet deep, which was previously operated by the same company (Fig. 3).

GROUND WATER CONDITIONS

Springs

In the vicinity of Hershey and in the valley of Spring Creek, the major tributary to Swatara Creek, there were once many springs, several of them yielding thousands of gallons per minute. One large spring in particular made several large ponds a few hundred feet east of the plant of the Hershey Chocolate Corporation. One reason for the plant's location was this excellent source of water, known first as the Derry Spring. For more than a century these valley springs were used on farms as their only water supply, and in later years to meet the needs of large numbers of orphan boys living on farms maintained by the Hershey Estates. Prior to 1949, none of the large springs ever went dry, even during the longest droughts.

Almost all the springs emanate from solution-widened bedding planes or from the intersection of a prominent cross joint with a bedding plane. The large Derry Spring is situated on the Annville Limestone, which has more and larger solution openings than the rest of the Beekmantown Limestone. The other large springs in the Hershey Valley are either along the outcrop belt of the Annville Limestone or near and connected with it by well-developed cross joints.

Wells

Between 1900 and 1950 many wells were drilled on farms and on properties where new homes were built. Except for very shallow wells, in which the water never stood more than several feet above the bottom of the well, none of these are known to have gone dry. An exhaustive study of 190 wells located throughout the entire valley indicated that lowering of the water table due to drought had never been greater than 10 feet. A series of 12 wells were also drilled by the expanding plant of the Hershey Chocolate Corporation between 1926 and 1948. Yields ranged from 200 to 1200 gallons per minute, and a few had no yield at all, demonstrating how poor an aquifer limestone can be "in between the fractures." The Corporation wells continued to have a constant yield in wet and dry season alike, and they apparently never affected in any way the adjacent springs or wells in the valley.

The drilled wells in the valley range from 50 to 400 feet in depth, and average 150 feet. Average yield is 10 to 30 gallons per minute. Most of the wells with larger yields are situated along the outcrop zone of the Annville Limestone, in or near fractured rock associated with the faults, or relatively near either of these two water-bearing zones and connected to them by joints. This further reveals that many more and larger solution openings were developed in the purer limestones, particularly the Annville. A few such wells are located in other high-calcium beds of the Beekmantown Limestone in the middle and southern parts of the valley.

Several wells that initially had a small yield produced very muddy water when pumped hard and long. These wells ultimately produced clear water and with much greater yield. Obviously the wells had intersected mud-filled solution openings along either bedding or joints. Continuous pumping had eventually flushed out the mud and opened up easier access through which ground-water could move toward the wells. In several such instances, the water level in an adjacent well or wells was lowered, indicating that the mud seal had maintained a perched or elevated water table in the local area (Fig. 2C). In several wells the ground-water level stands many feet above the regional water table; this is attributed to isolation by mud seals or by unusually tight fractures.

Quarries

When it operated, the Millard quarry a mile southwest of Hershey pumped an average of 450 to

---

[1]This company was purchased by the Michigan Limestone and Chemical Company in November 1951, and was operated as a Division of the United States Steel Corporation. In 1953, the Hershey Chocolate Corporation purchased the mine and allowed it to flood. It is now used as a reservoir by the Chocolate Corporation.

500 gallons per minute to maintain their operation under normal conditions.

The Annville Stone Company mine 1.5 miles northeast of Hershey pumped 3000 to 3500 gallons per minute (until 1949) to maintain their operation under normal conditions. When they deepened their operations in 1949 the amount of water pumped increased to as much as 7500 to 8000 gallons per minute. Following grouting operations in the spring of 1950, the average pumping rate was 2500 gallons per minute. Continued flushing of mud seals (that had not been sealed by grout) resulted in an increase to 3200 gallons per minute by March 1952. Further grouting during the summer of 1952 reduced the rate to 2500 gallons per minute (Fig. 3).

Normal Water Table

Examination of the mass of data compiled for the wells and springs of the valley revealed that the water table bore a normal relationship to the topography, the surface drainage, and the rock types. It stood considerably higher in the Martinsburg Shale to the north and in the Triassic rocks to the south than it did in the limestone of the valley. It sloped gently toward the west in the direction of Swatara Creek.

The effect of the many individual wells on the water table was entirely local throughout the valley. Each well had its own cone of depression of varying size depending on the amount of water pumped, the depth at which the pump intake was established, and the potential yield of the well.

The effect of the mine, the quarry operations, and the cluster of wells at the Hershey Chocolate Corporation was greater. The cones of depression surrounding these three places had considerable area and depth. The smallest of the three was that of the Hershey Chocolate Corporation. That cone of depression was the result of the combined cones of nine wells; its area was about 50 acres, and the total depth of the cone was about 15 feet. The cone surrounding the Annville Stone Company mine was elongate, covered an area of at least 255 acres, and was more than 50 feet deep. The cone surrounding the Millard quarry was somewhat more localized, covering about 165 acres, and was about 40 feet deep. In each case the depth of the cone is the difference between the lowest measured water level and those water levels not affected by pumping. The boundary of the cones was

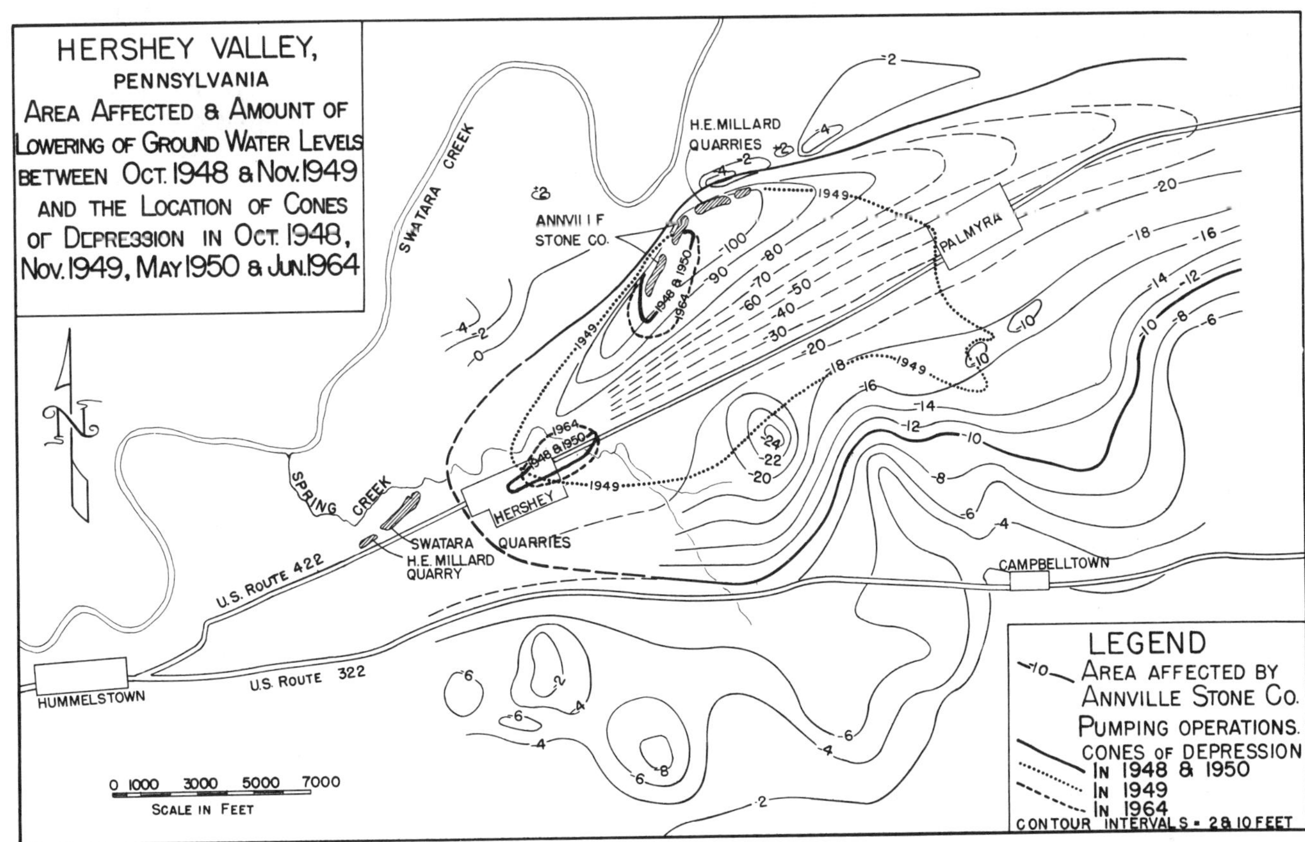

Figure 4. Ground-water levels and cones of depression in the Hershey Valley in 1948, 1949, 1950, and 1964.

likewise determined by water-level measurements that showed no change due to pumping.

Some wells even within these well-established cones of depression were apparently completely dissociated from the others; their static water level was generally higher than the surrounding water table, and some could be pumped down lower than the surrounding water table with no apparent widespread effect. All this indicates the degree of "tightness" of the limestone or the effectiveness of mud seals in the fractures, or both.

When the first water-table map of the Hershey Valley was prepared in October 1948 (Fig. 4), the conditions throughout the valley were essentially the same as they had been for many years. This was true of both mine and quarry operations which had for many years pumped the same amount of water, and also of the Hershey Chocolate Corporation. In addition, it had been neither unduly dry nor wet during the summer and fall months. The 1948 water table could therefore serve as a frame of reference for future observations.

The normal movement of ground-water throughout the valley prior to any wells being drilled or quarries opened must have been toward the southwest in the direction of the gently sloping ground-water table and into the valleys of Swatara Creek, Spring Creek, and their tributaries. As of 1948, the direction of ground-water movement was locally controlled by the many wells and by the larger industrial operations. However, no single well or group of wells, or the quarry and mining operations, adversely affected the supply of any other source of water at that time.

MINE DEWATERING AND GROUND-WATER FLUCTUATIONS

Annville Stone Company Blast of August 1946

During routine mining operations in August 1946, a blast in the hanging wall of the mine exposed a 6-inch-wide solution channel about 275 feet below the surface out of which poured an estimated 8000 to 10,000 gallons of water per minute, flooding the mine in the course of a day (Fig. 3). When this occurred, near-by wells dried up, ground-water seepage into an adjacent quarry ceased, the Derry Spring 1 1/2 miles southwest dried up on the second day, and the water level in two wells of the Hershey Chocolate Corporation rapidly declined. After many months of labor, the opening in the mine was sealed off with a large steel plate. Soon after, the adjacent wells had water in them again, the flow of the Derry Spring was restored, and water levels in the Corporation wells were again normal (September 8, 1948). This incident effectively showed that there were solution openings in the limestone connecting the Derry Spring and some of the Corporation wells with the mining operation.

Annville Stone Company Pumping Test of August 1948

From August 30 to September 4, 1948, the Annville Stone Company pumped an average of 5500 gpm, holding the water at a level of about 200 feet below their quarry floor[2] as a test preliminary to permanent installation of pumps for deeper mining operations. On September 2, the newly drilled Derry Spring well (drilled at the side of the Derry Spring, with a yield of 2100 gpm) had "dried up;" the water level fell from an elevation of 355 feet to 313 feet, which was below the pump intake. In another Corporation well, the water level was lowered more than 30 feet, and it too was "dry." The connection between the wells and the mine was obvious. On September 8, the water level began to rise in the two wells, and within a couple of days normal pumping operations could be resumed.

Both the 1946 incident and the August 1948 pumping test showed that there was easy movement of ground water through the many cavernous passageways of the Annville Limestone, especially in the 10-foot wide zone at the hanging-wall contact with higher magnesian rock. Within 4 days, the normal level of ground-water was drastically altered 1 1/2 miles from the mine.

It was obvious that permanent installation of the pumps at a lower level in the mine would endanger the water supply not only of the Chocolate Corporation but of farms and residents throughout a large portion of the valley.

During the fall of 1948 an exhaustive survey was made of practically every well and spring in the Hershey Valley. For about 200 wells and 30 springs, complete information and a detailed "personal history" were compiled. Answers to the following list of questions were sought from the well owners: (1) Location of residence (on base map of valley); (2) Well present; (3) Do neighbors have wells; (4) Are wells dug or drilled; (5) Name and address of well driller; (6) Depth of well in feet; (7) Diameter of well; (8) Is well cased; (9) Depth of casing in feet; (10) Normal water level (feet below collar); (11) Yield in gpm; (12) Depth of intake; (13) Has level fluctuated; (14) Was well ever dry; (15) Was there a drought when dry; (16) Was the water from well ever muddy; (17) Was water from well ever discolored; (18) Did water ever taste strange; (19) Any unusual behavior; (20) Is the well accessible; and (21) Consent to observe and measure well (yes or no).

From these were selected about 100 wells and springs which, because of their characteristics

[2]Personal communication from Mr. Liles, mine superintendent, in September 1948.

and their location, were ideal observation points (Fig. 1). Elevation of the wells and springs was surveyed, and each was carefully measured to determine: (1) depth to water level; (2) elevation of the water table; and (3) temperature of the water. For the springs, approximate or accurate yield in gallons per minute was determined.

Measurements were made at all observation points in October 1948, November 1949, and May 1950. Subsequent observations have been made many times, the latest in June 1964 (Fig. 4).

The results of water-table measurements made during October 1948 are summarized on Figure 4. These measurements were made when all conditions were similar to those that existed throughout the valley prior to August 1946 and might be termed "normal conditions."

Deeper Mine Development

During May 1949 the mining operation inaugurated its new pumping program designed to permit deep mining at substantially lower levels. About 6500 gallons per minute became the normal discharge from pumps whose intake was about 340 feet (elev. 85 feet) below the surface. The results of this pumping program were phenomenal and critical. Near-by wells were either dried up or their water levels rapidly declined. The Derry Spring well was effectively "dried up." Water levels in two of the Corporation wells immediately declined, and, with time, others were similarly affected. The large springs on the farms dried up. Spring Creek dried up. Irrigation water was unavailable, and crops suffered. Throughout the valley, water levels in wells dropped at an alarming rate, and many wells went dry.

The rate of decline of water levels in the wells and springs during the fall of 1949 far exceeded the rate or amount of decline that might be expected from the lower than average rainfall the area received at that time. (The average lowering of the water table caused by lesser rainfall was shown by those wells located in the Martinsburg Shale, in the Triassic sandstone, and in those parts of the Hershey Valley so situated that the wells could not be affected by the pumping operations of the Annville Stone Company.)

Decline of Springs

Another example of the relatively minor influence of rainfall on the rapidly declining levels of wells and springs was the combined flow of water from spring 28A and the Groy Spring. Early in September 1949 a weir was erected at the mouth of a pond created by these two springs to measure the flow of water (Fig. 5). Between September 12 and 15 about 415 gallons per minute flowed through the weir. Between September 15 and October 24, the flow of water steadily declined to less than 25 gallons per minute. The rate of decline was fairly constant. Only after periods of relatively heavy rainfall did the rate of decline alter, and only after the very heavy rain of October 25 did the flow of water actually increase, and then only for one day, to approximately 40 gallons per minute. On November 2, no more water flowed over the weir,

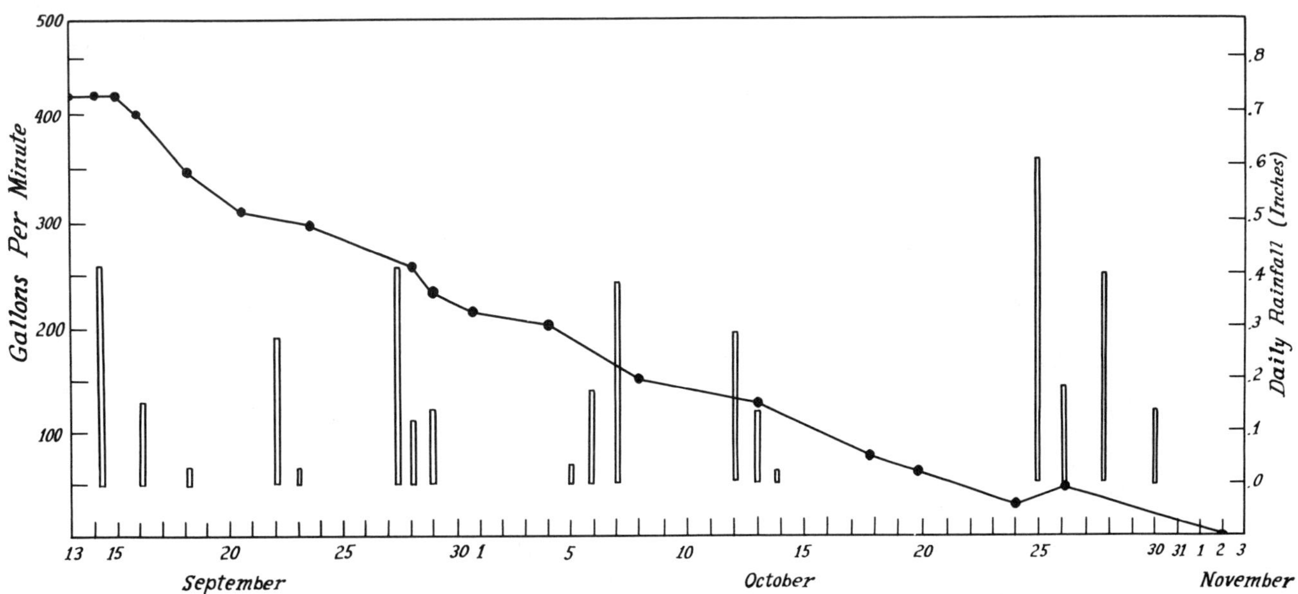

Figure 5. Flow record of Spring 28A during late 1949.

and 2 weeks later the floor of the spring pond was dry (Pl. 1). The phenomenal decline in flow could in no way be attributed to the low rainfall of the period because the springs had never dried up before, even during times of severe drought. Two other large springs, 28C and 28D, located nearby dried before the end of November 1949.

Plate 1. Dried spring, 28A.

Precipitation Data

Precipitation data were assembled for all years between 1930 and 1950 from Federal weather bureau stations at Harrisburg (9 miles west) and Lebanon (14 miles east) (Fig. 6). They show that during the 20-year period there were 5 years that were dried at Harrisburg and 4 years that were drier at Lebanon; in 2 other years rainfall at both places was almost the same. During none of these years did a spring dry up nor did any well in the valley exhibit a drastic lowering of water level.

Development of Sink Holes

During the second month of the mining operation's new pumping program, sink holes began to form in the valley of Spring Creek. At first they were filled, but they continued to form so rapidly it became impossible to keep pace. They ranged from about 1 foot to 20 feet in diameter and from 2 to 10 feet deep. All of them had very steep walls, as though the bottom had suddenly been pulled out and the unsupported earth above had quickly slumped (Pl. 2). Essentially that is what happened. Instead of the ground-water table being near the level of Spring Creek, it had been lowered many feet. Water flowing in the channel of Spring Creek moved downward through whatever openings were available into solution-widened joints and bedding planes of the limestone, carrying along large quantities of silt, sand, and gravel. Unsupported areas in both the channel of the creek and adjacent to it, along its banks and on the flood plain of the creek, slumped and made sink holes (Fig. 2B).

Half a dozen of the nearly 100 sink holes formed in the bed of the creek. In one case an area 2.5 feet in diameter on the surface of the silt under about 2 inches of water caved, leaving a pool 1.5 feet deep. Around the edge of the pool the silt

Plate 2. Sink hole.

was warped down, showing concentric fractures. Water was obviously moving into the incipient sink hole. When probed with a stick it was apparent that no silt remained in the center of the hole. Only gravel was felt; all the silt had been carried downward. The next day this hole was 7 feet in diameter and 6 feet deep. Bedrock was exposed at the bottom of the hole showing a vertical fracture widened by solution. Into the hole poured all the remaining water of Spring Creek (Pls. 3 and 4).

Prior to July 1949 very few sink holes had opened in the Hershey Valley. A careful survey of all the old existing sink holes, some dating back to before 1900, proved that more had opened up in the late summer of 1949 than had existed before. Furthermore, whereas the older sink holes had formed at many places throughout the large Hershey Valley, some of them located along faults and prominent cross joints, the new ones had formed entirely within the area where there was a drastic lowering of the ground-water table.

Many valley residents attested to the fact that the older sinks all had gently sloping sides and a distinct funnel shape even when newly formed, whereas the 1949 sinks had precipitous walls and more of a cylindrical shape, indicating a very rapid lowering of the ground-water table and withdrawal of its support (Pl. 2).

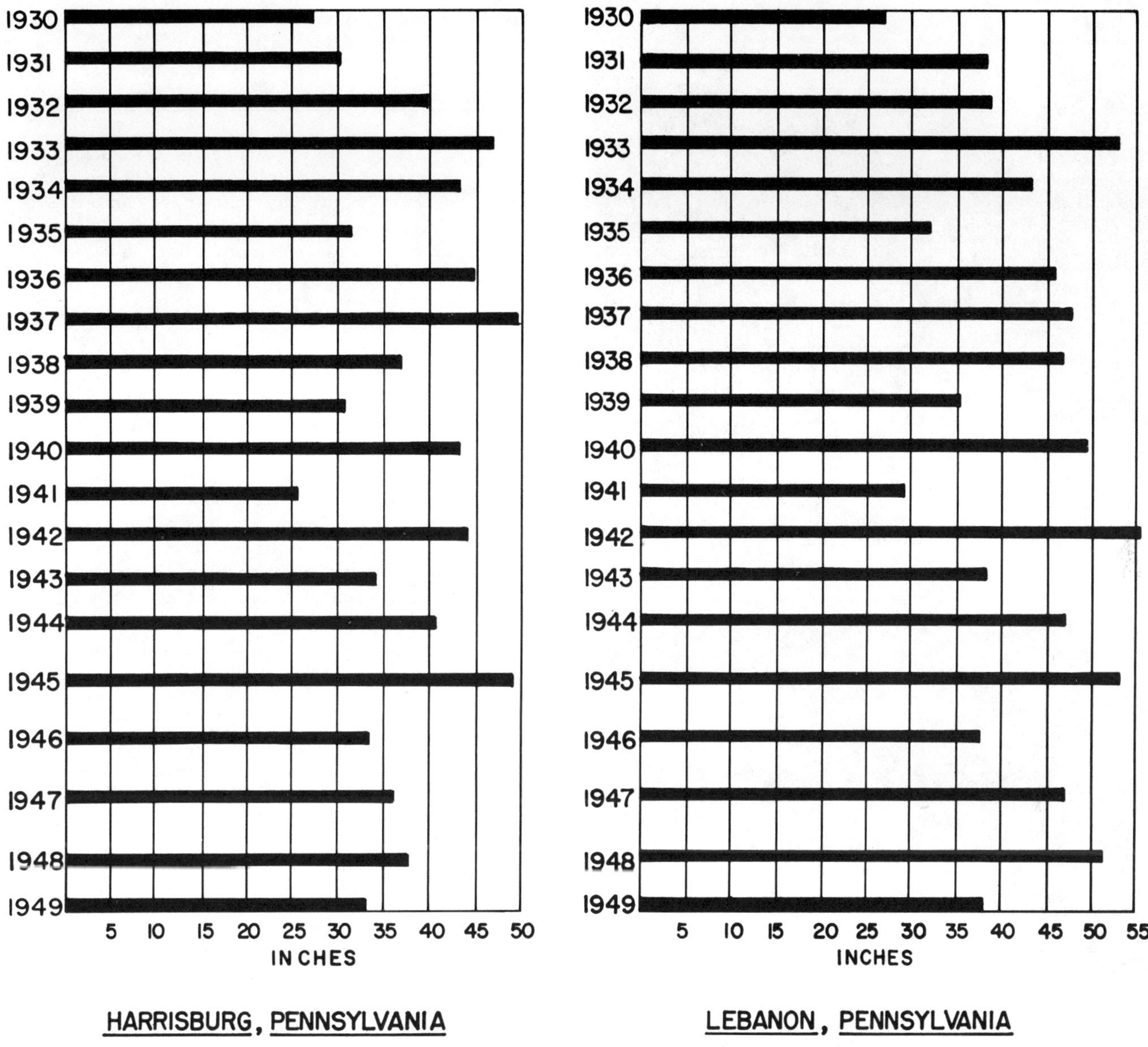

Figure 6. Yearly precipitation data at Harrisburg and Lebanon, Pennsylvania, 1930-1949.

As a direct result of the large number of sink holes, the entire lower course of Spring Creek down to the plant of the Hershey Chocolate Corporation was dry after September 21, 1949 (Pls. 3 and 4). The existence of the sink holes and the rapidity with which they formed constituted a physical danger to life and property. Many areas had to be roped off in the valley of Spring Creek; the abutment of a highway bridge had to be reinforced when a sink hole developed under it; the possibility of serious property damage and failure of foundations was great. The hazard was not eliminated until after the successful raising of ground-water levels by recharging water to the ground.

A large spray pond is maintained just east of the Hershey Chocolate Corporation buildings in the bed of Spring Creek for the purpose of cooling industrial water prior to recirculation. During early November 1949, the water level of the spray pond became very low because the water flowed in a reverse direction (upstream) along the dried-up channel of Spring Creek and sank into small sink holes in the channel. To prevent this, a cofferdam was built at the eastern end of the pond on November 10. A constant danger was the imminent development of one or more sink holes in the bottom of the spray pond or underneath the Corporation buildings. Although the sink holes developed in an increasingly

Plate 3. Spring Creek in 1948.

Plate 4. Spring Creek in September 1949.

larger area week by week, none reached the spray pond or the Corporation buildings.

Direction and Speed of Ground-Water Movement

Coincident with the program of well and spring measurements, detailed studies were conducted to determine the physical characteristics of ground-water movement in the immediate vicinity of the Hershey Chocolate Corporation. The earliest tests, during the fall of 1948, were designed to show the nature of movement within the shallow cone of depression created by the Chocolate Corporation wells. Small charges of 1 and 2 pounds of fluorescein in alkaline solution were flushed into various wells; on arrival at other pumping wells observers noted: (1) time of arrival of dyed water; (2) duration of its observance; and (3) intensity of coloration. Samples revealing the presence of fluorescein were collected at definite time intervals. Ultimately a fairly good picture was gained of the general direction of ground-water movement under various physical conditions, the speed of its movement, and something of the nature of the passages through which it moved. These tests established the fact that most wells and springs are interconnected, though not always by channels through which there is easy movement. Within the Hershey cone of depression, easiest movement is along cross joints, and secondarily along bedding.

Beginning in July 1949, after the Annville Stone Company had begun its new pumping program, some larger fluorescein tests were made to establish the direction and speed of ground-water movement throughout the valley. The results of those tests can be summarized as follows.

(1) Ground water east of the Hershey Chocolate Corporation moved toward the Annville Stone Company mine, covering the full distance in between 40 and 48 hours.

(2) Within an area extending no farther than 1000 feet east of the Hershey Chocolate Corporation, a small portion of the ground-water moved toward the corporation wells that were being pumped, despite the fact that the water table had a regional gradient sloping toward the Annville Stone Company.

(3) Ground-water moving along north-south cross joints toward Chocolate Corporation wells had the least interference from Annville Stone Company pumping operations. Greatest interference was in the general northeast-southwest direction along bedding.

(4) As ground-water levels were lowered throughout much of the Hershey Valley, many of the existing mud "seals" in cavities along joint planes and bedding were flushed out, thereby increasing the ease and speed of movement of ground-water along them (Fig. 2C).

Water Table in November 1949

In November 1949 a new water-table map was drawn from measurements on all the wells and springs. The new ground-water table had a tremendous depressed area which encompassed the Hershey Chocolate Corporation, a large part of the Hershey Valley, and centered about the mining operation (Fig. 4). Total depth of the depression was close to 200 feet. The pumps in the mine were set deeper than the bottom of the cone and were at all times pumping under head. In more than 10 square miles of the Hershey Valley the ground-water surface had been lowered more than 10 feet (Fig. 4). The investigation also proved that maximum lowering of the ground-water surface due to lesser rainfall during the year was only 6 to 8 feet and that average lowering was 2 to 4 feet. These changes were observed in the areas underlain by Martinsburg Shale and the Triassic rocks where no effect of the pumping operations was felt.

The relationship between the water-table levels of October 1948, November 1949, May 1950, and June 1964 is further shown by Figure 7.

Geochemical Tests

Concurrent with the other observations, detailed studies of pH values and dissolved solids concentrations in water samples from all measured wells and springs were made. The objective was to compare these values with ground-water movement and with lithologic changes in the underlying rocks. Both proved to be valuable corroborative techniques. For example, both showed directions of easiest ground-water movement and the position of the Annville Limestone was indicated by higher pH values.

GROUND WATER RECHARGE

The water table throughout the Hershey Valley was falling so rapidly during September and early October 1949, and the development of sink holes was taking place in an area of such rapidly increasing size that it seemed imperative to stabilize the ground-water surface immediately.

All water subsequently recharged was chemically suitable for any use. It was decided to recharge experimentally a small volume of water for a 7 to 8 week period to determine what effect the temperature would have on the Corporation wells. Warm (70° to 76° F) spray pond water at the rate of 1200 gpm was recharged at the Derry Spring well about 1000 feet east of the Corporation buildings beginning October 21. During the recharge period, the wells closest to the point of recharge showed a water temperature increase of 1° to 6° F, and a very small rise in water level (Fig. 8). The Corporation wells farthest west showed neither temperature increase nor water level rise, although their water levels no longer declined.

This experiment proved what had previously been demonstrated by the fluorescein tests--namely, that from the point of recharge water would move outward in all directions subject to the major factors controlling the movement, such as size, shape, and number of openings; the existence of mud "seals;" and of course the regional gradient of the ground-water table. The experiment also corroborated the expectation that ground-water temperatures would be changed by mixing, and that a gradual temperature gradient extended outward from the point of recharge.

Beginning at noon December 12, 1949, about 3500 gpm of low-temperature water was charged to the old pond area adjacent to the dried-up Derry Spring about 1000 feet east of the Corporation buildings. These waters were from various sources. A large percentage had already been used for cooling purposes and returned to the spray pond. Others came from Swatara Creek, from a reservoir in an abandoned quarry one mile southwest, and by pipeline from the Annville mine. This was the beginning of a general program for recharging low-temperature, noncontaminated waters to the ground in amounts up to 8000 to 10,000 gpm in an effort to stabilize or to raise, if possible, ground-water levels in the affected area.

The results were almost immediately beneficial. Figure 8 shows the effect on water temperature and water level in five of the Corporation wells and in Well No. 68, located 6000 feet east of the Corporation buildings and 5000 feet southeast of the Annville Stone Company (located on Fig. 1). Well No. 68 had suffered a serious decline in water level during the fall of 1949. The resulting rise of the ground-water table was shown by many other wells throughout the entire valley. These results demonstrated that recharged waters moved outward in many directions, even though they were under the constant influence of the steep regional gradient on the ground-water table sloping toward the Annville Stone Company. Undoubtedly there was a local "hump" on the ground-water table at the recharge point, which partly explains how the water would move opposite to the regional gradient.

LITIGATION

During January and February 1950, the Annville Stone Company and Hershey Chocolate Corporation were actively engaged in litigation growing out of the abnormal ground-water conditions in the Hershey Valley, and specifically from the program of ground-water recharge practiced by the Hershey Chocolate Corporation. The Annville Stone Company sought an injunction against the Chocolate Corporation, enjoining them to cease

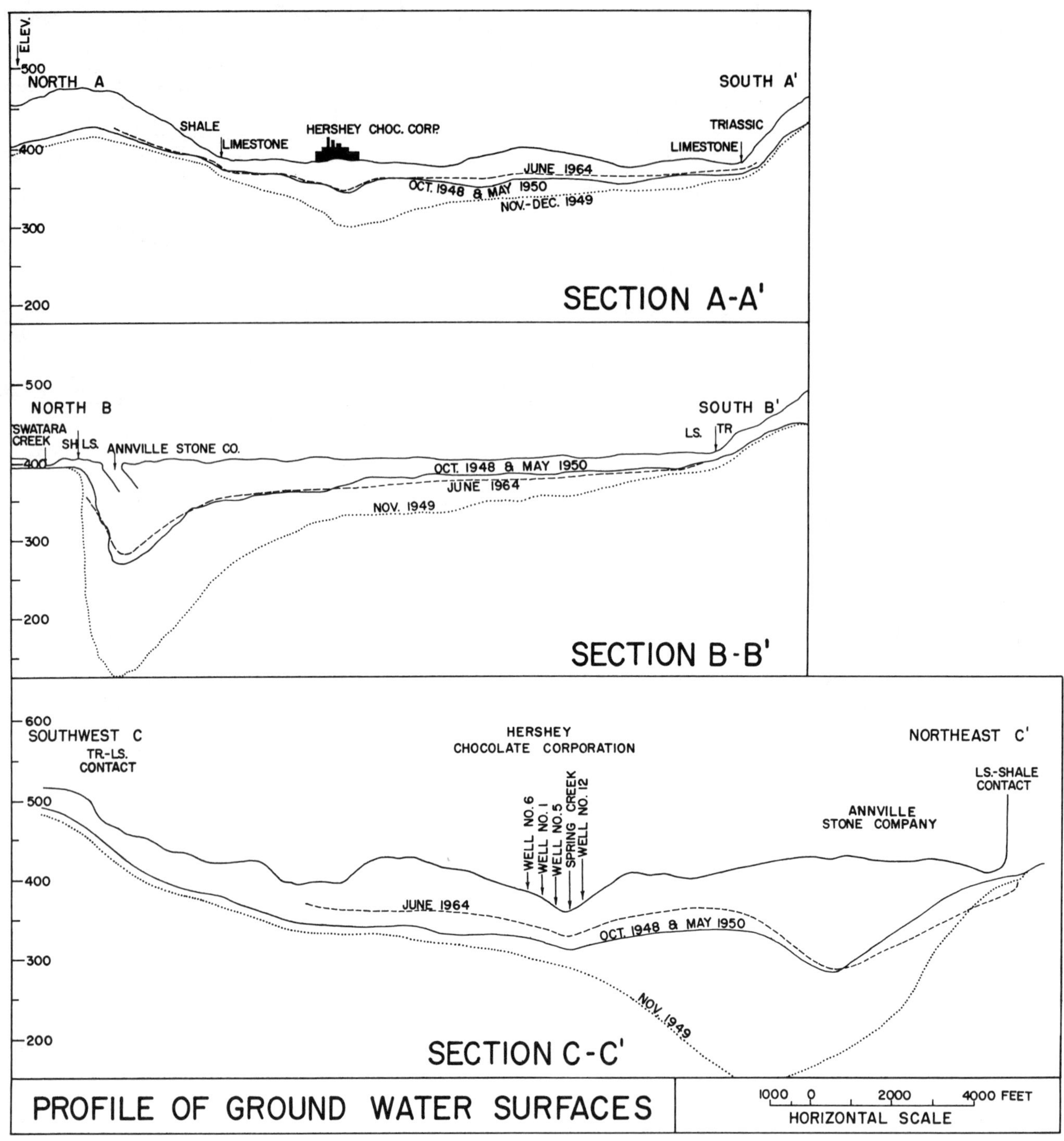

Figure 7. Profiles showing relationship of water levels in Hershey Valley during 1948, 1949, 1950, and 1964.

recharging water to the ground. The program of ground-water recharge was successfully defended, twice in the Dauphin County court and finally before the Supreme Court of Pennsylvania in Philadelphia. In the defense by Hershey Chocolate Corporation, the extensive geologic data were presented and provided the major basis for decision. The suit against the Hershey Chocolate Corporation was withdrawn by the Annville Stone Company in May 1950.

The burden of the arguments by plaintiff and defendant are embodied in the records of the Supreme Court of Pennsylvania, Middle District, May Term, 1950, No. 16. Volume 1 (268 p.) records all of the direct testimony and cross

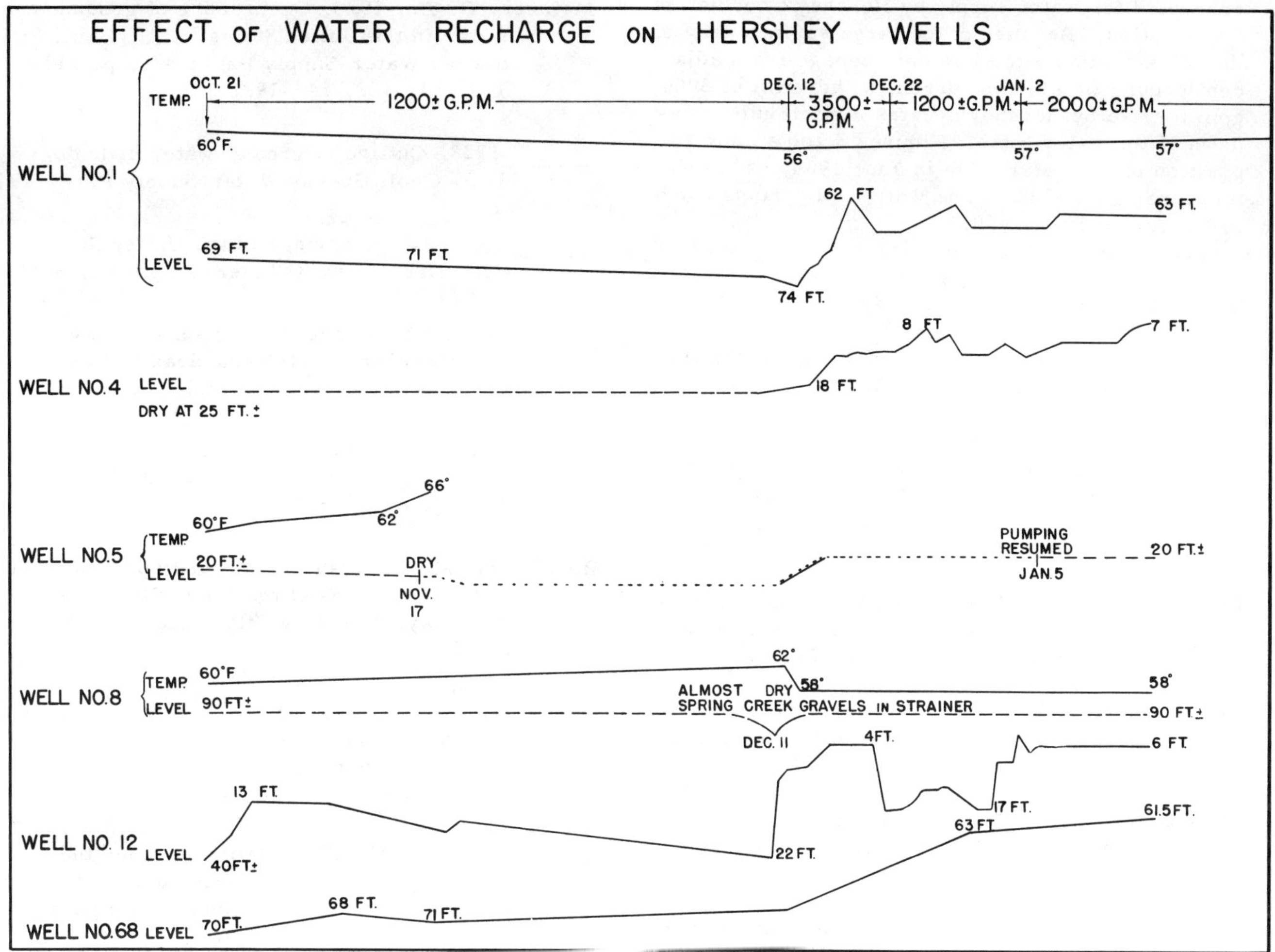

Figure 8. Effect of ground-water recharge on Hershey wells.

examination. Volume 2 (34 p.) and Volume 3 (32 p.) are briefs of the appellant and the appellee to the Supreme Court.

GROUTING PROGRAM

To reduce the burden of increasingly heavy pumping demands, which had risen from 6500 gpm to more than 8000 gpm, the Annville Stone Company engaged in a program of grouting in February and March 1950 to seal off the openings carrying water into their mine.

On the basis of the extensive geologic knowledge accumulated by Hershey Chocolate Corporation, that company strongly urged Annville to engage in a grouting program in 1949, prior to the dewatering program, but Annville refused. The 1950 program involved grout holes in a pattern that penetrated the cavernous contact zone between the dolomite and the uppermost part of the high-calcium Annville Limestone in the hanging wall (Fig. 3). The drawing shows the general pattern but not specific hole locations.

The success of this action, in addition to the recharge program, was demonstrated by a return to nearly normal conditions throughout the Hershey Valley beginning in late March 1950. Springs began to flow again. Water levels rose in wells that could once more be pumped. Spring Creek began to flow again. Sink holes no longer developed. The water-table map of May 1950 showed separate areas of ground-water depression around the Chocolate Corporation and the Stone Company mine once again. They were similar in size, shape, and location to those outlined in October 1948 (Fig. 4). The pumping costs of the mine were greatly reduced; the savings were sufficient to offset the cost of grouting within a short period.

DEVELOPMENTS SINCE 1950

In 1953 Hershey Chocolate Corporation purchased the Annville mine, then operated as a Division of the U.S. Steel Corporation. The mine was allowed to flood and is currently used as a reservoir for water supply by Hershey Chocolate Corporation. Because of the large surface area of the mine available for influent seepage from adjacent ground water, the current withdrawal of 3000 gpm for use by Hershey creates a minimum drawdown of the water table. Figures 3 and 4 show the position of the water table in June 1964, based on measurements of the same wells and springs used in the period 1948-1950.

REFERENCES CITED

Baldwin-Wiseman, W. R., 1905-1906, On the flow of underground water: Inst. Civ. Eng., Pr., v. 165, pt. 3.

Beede, J. W., 1910, The cycle of subterranean drainage as illustrated in the Bloomington quadrangle: Ind. Acad. Sci., Pr., p. 81-103.

Dole, R. B., 1906, Use of fluorescein in the study of underground waters: U. S. Geol. Survey Water-Supply Paper 160, p. 73-85.

Foose, R. M., 1953, Ground water behavior in the Hershey Valley, Pennsylvania: Geol. Soc. America Bull., v. 64, p. 623-645.

Hall, G. M., 1943, Ground water in southeastern Pennsylvania: Penna. Geol. Survey, Bull. W 2, p. 156-158, 196-198.

Horton, R. E., 1933, The role of infiltration in the hydrologic cycle: Am. Geophys. Union, Tr., p. 446-460.

Hoyt, W. G., and others, 1936, Studies of relations of rainfall and runoff in the United States: U.S. Geol. Survey Water-Supply Paper 772.

Lee, C. H., 1934, The interpretation of water-levels in wells and test-holes: Am. Geophys. Union, Tr., p. 540-554.

Lohman, S. W., 1941, Ground water resources of Pennsylvania: Penna. Geol. Survey, Bull. W 7, 32 p.

Meinzer, O. E., 1923, Occurrence of ground-water in the United States: U.S. Geol. Survey Water-Supply Paper 489, p. 111-116, 131-132, 149-192.

_______ 1923, Outline of ground-water hydrology: U.S. Geol. Survey Water-Supply Paper 494.

_______ 1927, Large springs in the United States: U.S. Geol. Survey Water-Supply Paper 557.

_______ 1936, Ebbing and flowing springs in ground-water resources of Shenandoah Valley: Virginia Geol. Survey, Bull. 45, p. 52-55.

Mitchelson, A. T., 1934, Underground storage by spreading water: Am. Geophys. Union, Tr., p. 622-623.

Rorabaugh, M. I., 1949, Progress report of ground-water resources of the Louisville area, Kentucky: Rept. by City of Louisville, p. 23.

Stringfield, V. T., 1935, The Piezometric surface of artesian water in the Florida Peninsula: U.S. Geol. Survey Water-Supply Paper, p. 524-529.

Theis, C. V., 1935, The relation between the lowering of the piezometric surface and the rate and duration of discharge of a well using ground-water storage: Am. Geophys. Union, Tr., p. 519-524.

_______ 1936, Ground-water in south-central Tennessee: U.S. Geol. Survey Water-Supply Paper 677, p. 33-46.

Unklesbay, A. G., and H. H. Cooper, Jr., 1945, Artificial recharge of artesian limestone at Orlando, Florida: Econ. Geol., v. 40, p. 95.

Veatch, A. C., 1906, Fluctuations of water level in wells: U.S. Geol. Survey Water-Supply Paper 155, p. 32-24, 60-61.

GROUND-WATER MANAGEMENT IN THE RAYMOND BASIN, CALIFORNIA

John F. Mann, Jr. (Consulting Geologist and Hydrologist, La Habra, California)

Abstract

The Raymond Basin, just northeast of Los Angeles, is an area of ground-water storage in alluvial fan deposits, flanked on the north by high mountains and on the south by a semi-permeable fault.

Because the Raymond Basin is an area in which there was early urban development made possible by intensive exploitation of local ground water, the effects of withdrawals in excess of replenishment were felt earlier than elsewhere in southern California. These conditions precipitated the first attempt in California to adjudicate the water rights of an entire ground-water basin.

The litigation resulted in a stipulated judgment under which pumping was cut back to the safe yield. Pumping rights were determined on the basis of maximum continuous five-year production prior to the filing of the action. The physical solution was rationalized by the invention of a new legal theory--mutual prescription. A Watermaster was appointed to enforce the provisions of the judgment.

Under the continuing jurisdiction of the Court, a re-study of the safe yield was ordered. The re-determination indicated that the safe yield had increased substantially and pumping rights were proportionately increased.

After 20 years of ground-water management, it is possible to conclude that the pioneering methods developed in the Raymond Basin have been effective in meeting the objectives of the local water users.

CONTENTS

ILLUSTRATIONS

INTRODUCTION

The Raymond Basin lies just a few miles to the northeast of the City of Los Angeles, and includes the City of Pasadena and all, or parts of, several smaller cities. The history of water development in this area is interesting and complex. The combination of substantial runoff from nearby high mountains, large alluvial volumes to absorb the flood flows, and a natural underground "dike" which forced the ground waters to the surface, resulted in perennial springs and surface flows which were utilized by early Indians. Some of these rising ground waters were developed by the padres of the San Gabriel Mission, and by white settlers in the early 1800's for such purposes as operating a grist mill, a sawmill, and a tannery, in addition to domestic uses (Blackburn, 1952). The earliest wells, just above the "dike," were artesian. As more wells were drilled in the basin above the "dike," and especially with the advent of the deep-well turbine pump, the rising waters were dried up and the artesian flows stopped. Alarmed by the progressive lowering of water levels in the basin, the City of Pasadena in 1937 sought, through the courts, to restrain pumping (Blackburn, 1961). The resulting litigation is referred to as Pasadena vs. Alhambra, or the Raymond Basin Adjudication.

From the standpoint of both physical and legal innovations, this is a landmark case. It was the first ground-water basin in California in which the Court reference procedure was used, the first in which the Division of Water Resources was appointed Watermaster to administer the decree, and the first basin in California in which essentially all the rights to pump ground water were adjudicated. Legally, the case is famous for developing the "doctrine of mutual prescription." At the time the action was filed, the pumpers consisted of true overliers who pumped their own water and used it on their own land, overlying appropriators such as municipalities and public utilities, and exporting appropriators who pumped water from the basin and transported it for use outside the basin. In the judgment, no recognition was given to priorities based on time or place of use. The trial court decreed that overdraft had existed for more than 5 years, and that each pumper had prescripted all the other pumpers. The safe yield was apportioned among the pumpers on the basis of extractions preceding the filing of the action. The judgment was affirmed by the California Supreme Court (33 Cal 2d 908), but one of the justices dissented vigorously. The shock with which some of the California legal fraternity greeted the doctrine of mutual prescription is reflected in the dissenting justice's opinion:

> I would say that the doctrine laid down in the majority opinion in the case at bar is based upon the philosophy of bureaucratic communism.

Control of ground water pumping in the Raymond Basin under the Watermaster has now been in effect for more than 20 years.

GEOLOGY

The Raymond Basin covers some 40 square miles in the northwestern portion of San Gabriel Valley, a large area of alluvial ground-water storage in the inland portion of coastal southern California. The Raymond Basin is separated from the main San Gabriel Valley by the Raymond fault which retards the southerly outflow of ground waters from the basin (Fig. 1). Those ground waters are

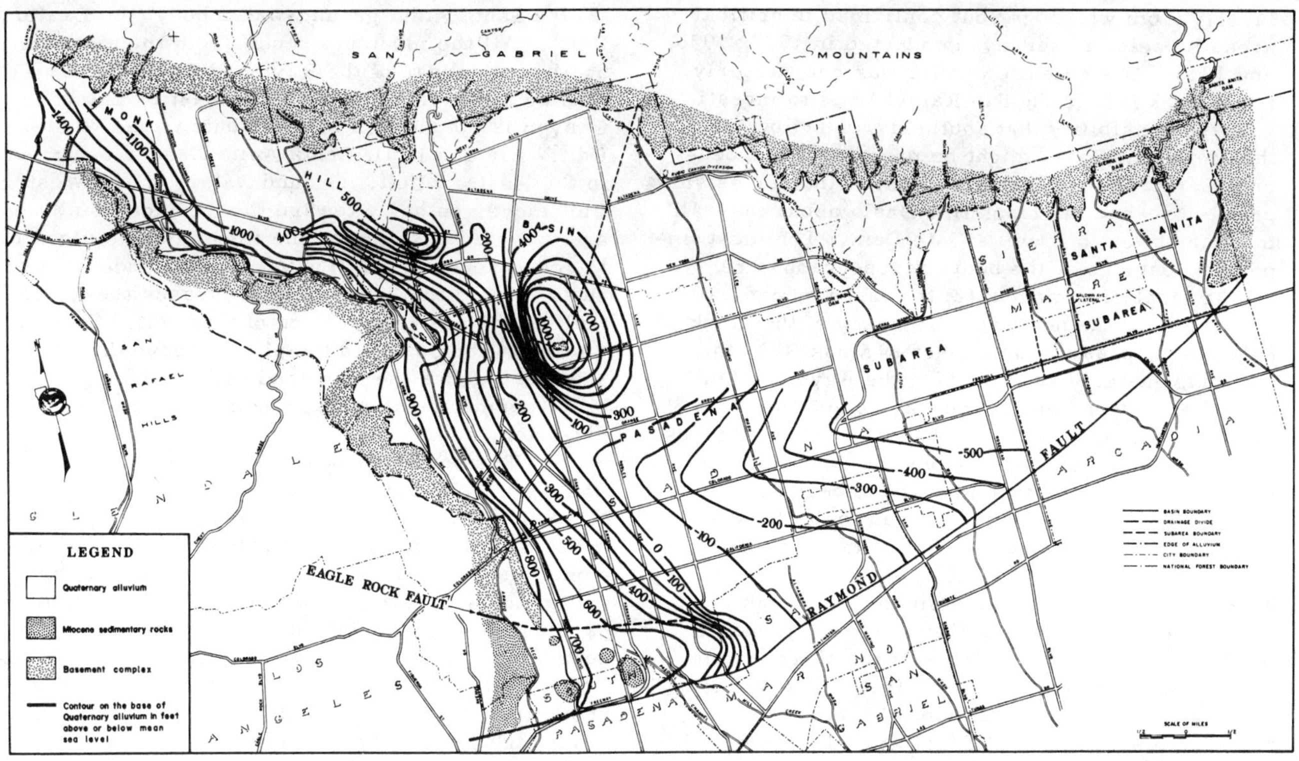

Figure 1. Geologic map and contours on the base of the Quaternary alluvium.

stored in thick alluvial deposits of fan origin which were laid down on an irregular bedrock topography.

Flanking the Raymond Basin to the north, and the source of the alluvial materials, are the high rugged San Gabriel Mountains consisting mainly of plutonic and metamorphic rocks, the so-called Basement Complex. The bold south front of those mountains is associated with the complex Sierra Madre fault zone, along which the Basement Complex has been dropped sharply. The Basement Complex is also exposed to the west of the Raymond Basin in the San Rafael Hills, which have only a few hundred feet of relief. Non-water-bearing rocks of Miocene age are exposed only near the southwest corner of the basin (Fig. 1). These consist of sandstones, shales, and conglomerates, and all have low porosity and permeability.

The Sierra Madre fault zone is one of the major zones along which the San Gabriel Mountains have been uplifted; it has been active since long prior to the deposition of the alluvial fan materials. Movements have continued through various episodes of the alluvial deposition, and near the mountain front, weathered alluvial gravels are in fault contact with the Basement Complex. The faulting does not affect the Recent alluvium.

The Eagle Rock fault forms the boundary between the Basement Complex and the Tertiary rocks near the southwest corner of the Raymond Basin (Fig. 1). To the east, the Eagle Rock fault is covered by alluvial materials, and the nature of its termination in that direction is not known. However, there is no evidence that movements on this fault have displaced any of the alluvial materials.

The important Raymond fault can be traced from several miles west of the Raymond Basin, along the south side of the Raymond Basin, and into the San Gabriel Mountains to the east (Fig. 1). Its surface trace is marked by a line of low hills, south-facing scarps, and sag ponds (Buwalda, 1940). Water levels stand 200 to 300 feet higher on the north side of the fault than on the south side. Near the southwest corner of the basin, where the up-faulted Tertiary rocks form hills 100 or more feet high, there are high dips and even some over-turning associated with the faulting. East of the Tertiary rock exposures the topographic expression of the faulting continues, but it is of a much more subdued nature. Where the active channels of Eaton and Santa Anita Creeks cross the fault, the topographic features have been removed by erosion and replaced by a featureless backfill of unfaulted Recent alluvium.

Almost the entire Raymond Basin is under-lain by Basement Complex, which is completely covered by alluvium except for the small exposure at Monk Hill (Fig. 1). The configuration of the bedrock surface shown on Figure 1 was derived largely from well logs, but confirmed in critical areas by seismic surveys conducted in 1937, 1938, and 1939. One seismic profile was run westerly from Monk Hill to the San Rafael Hills to investigate the possibility that southerly underflow from the Monk Hill Basin might be prevented by a bedrock barrier (Sonderegger, 1936). Another seismic profile crossed the Monk Hill Basin about one-half mile northwest of Devil's Gate Dam. The most important features of the bedrock topography are: (1) the Monk Hill mound; (2) the large buried valley starting near the northwest boundary of the Monk Hill Basin, passing to the west of Monk Hill, then swinging easterly and reaching the Raymond fault near the eastern edge of the Raymond Basin; and (3) the sharp drop-off east of the Tertiary outcrops. The below-sea-level elevations indicate that this buried erosional surface must have been developed at a time when the Raymond Basin fault block was standing relatively much higher than at present.

The present alluvial surface shows a generally smooth slope in a southerly to southeasterly direction. Near the northwest corner of the Raymond Basin the apices of the alluvial fans are at elevations of about 1800 feet; easterly along the mountain front the fan heads become lower, reaching elevations of 800 feet near the eastern corner of the basin. Minimum elevations of the alluvial surface are about 600 feet near the Raymond fault in the eastern part of the basin. The alluvial thickness is about 1100 feet where the main bedrock valley just west of Monk Hill. Except close to the bedrock outcrops surrounding the basin, the alluvial thicknesses are generally in excess of 700 feet.

Close to the mountains, the alluvial fan materials are expectably coarse, but bouldery materials extend all the way to the Raymond fault. Sandy layers are not common. Clays, resulting from the weathering of gravels, are abundantly distributed throughout the Raymond Basin, especially in the older gravels near the mountain front, and in the fans of the minor streams issuing from the mountains. Numerous clay and peaty layers are found just north of the Raymond fault, and are probably related to high water tables, ponds, and artesian leakage.

SOURCE AND MOVEMENT OF GROUND WATER

Recharge to the ground water of the Raymond Basin is derived from the percolation of stream flows originating in the mountains to the north, deep penetration of rain falling on the alluvial surfaces, and returns from delivered water used for irrigation or discharged to cesspools.

Two bodies of ground water are recognized: (1) the Arroyo Seco-Eaton ground-water body including the Monk Hill Basin and the Pasadena Subarea, which constitute the Western Unit; and (2) the Santa Anita ground-water body (the Eastern Unit). All the mountain runoff from the extreme northwest corner of the Raymond Basin and extending for several miles to the east of Eaton Canyon is considered to contribute to the Western Unit. Big and Little Santa Anita Creeks contribute to the Eastern Unit. Ground waters in the Western Unit move southerly toward the Raymond fault, but are diverted easterly by the impermeable block of Miocene rocks (Fig. 2). Most of the underground outflow from the Western Unit crosses the Raymond fault in the easterly portion of that unit. Ground waters in the Santa Anita Subarea move directly south and cross the Raymond fault, where permeability permits, without appreciable change of direction.

The southerly boundary of the Monk Hill Basin approximates the minimum cross-section of alluvial fill in the bedrock valley just west of Monk Hill. To the north of this boundary, water levels remain at appreciably higher levels than in areas to the south, suggesting retarded underflow. Lower permeability of alluvial materials near this boundary may contribute to the slowing of underflow. The easterly boundary of the Monk Hill Basin is arbitrary; the underlying alluvial materials are fairly tight, easterly components of water-level gradients are small, and underflow is considered minor.

The boundary between the Pasadena Subarea and the Santa Anita Subarea attempts to approximate the western boundary of ground waters derived from Santa Anita Creek. There is considered to be little underflow across this line because of limited permeability and unfavorable gradients. The position of the hydrologic boundary was influenced in part by a convenient nearby street.

In its functioning as a barrier to the movement of ground water, three conditions along the Raymond fault can be recognized: (1) the tight barrier of the Miocene rocks; (2) the leaking fault membrane; and (3) overspill through Recent alluvium. The topographically high exposures of Miocene rocks between the Eagle Rock and Raymond faults form a complete barrier to southward-moving ground water. Although there are several alluvial passes crossing this block, the water-table elevations indicate that the ground water in those passes is of local origin and creates a ground-water ridge from which there is northerly and easterly flow into the central Pasadena Subarea. East of the Miocene outcrops the underflow barrier is created by reduced permeability along the fault plane itself. The Raymond fault is a reverse fault with the north block up, and the fault plane dips at a high angle to nearly vertical. The barrier is caused by gouge, by the blocking of displaced aquifers by low permeability beds, or possibly by dragged tight layers. The barrier in the older alluvial materials is not completely effective, and there is considerable

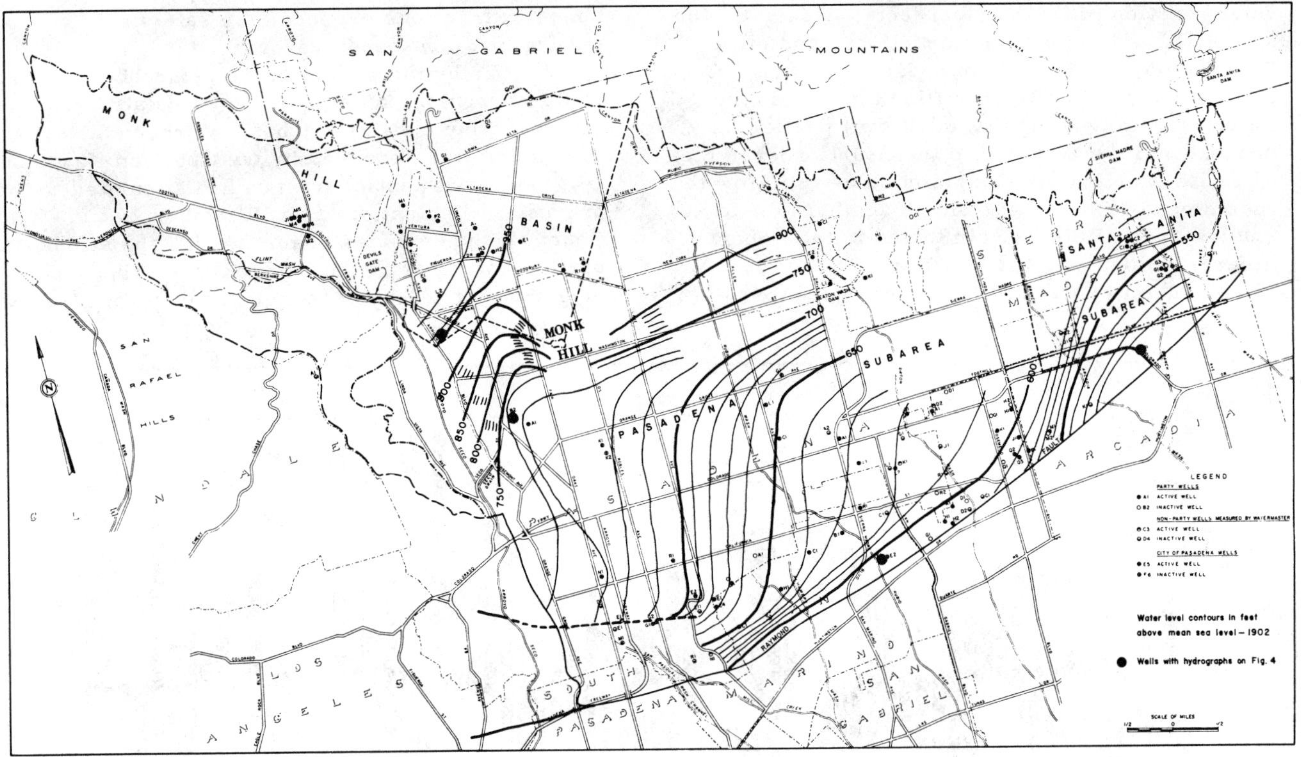

Figure 2. Water-level contours prior to heavy ground-water development.

leakage across it. At the Raymond fault, the Recent alluvium of Santa Anita Creek is about 80 feet thick, and there is opportunity for overspill through the Recent alluvium of ground water which rises because of the barrier effect in the older faulted alluvium. The Recent alluvium of Eaton Creek is so thin that subsurface overspill could only be minor.

FLUCTUATIONS OF WATER LEVELS

The general climatic trends in the Raymond Basin are shown on Figure 3. There was a general drought condition from 1916 to 1937, a wet period from 1937 to 1944, and a serious drought extending from 1944 with only minor relief through 1964. As shown in Figure 4a, the wells in the Monk Hill Basin show large fluctuations and good recoveries in wet periods. This is explained by strong recharge to the alluvium near the mountains, high rainfall (and therefore higher rainfall penetration) at the higher surface elevations prevailing in the Monk Hill Basin, and retarded ground-water outflow west of Monk Hill.

In the main Pasadena Subarea the recoveries during 1937-1944 are not so complete as in the Monk Hill Basin, and the continued recovery after 1944 is the result of the curtailment of pumping which started in 1944 (Fig. 4b). Just above the Raymond fault, where pressure conditions prevail and where local pumping, more than recharge, controls the water levels, the effect of the pumping curtailment is most evident (Fig. 4c). From 1944 to 1947 the recovery of static water levels was about 110 feet. In any year, the sharp annual fluctuations are indicative of the pressure conditions.

In the Santa Anita Subarea, the effects of direct recharge are shown as good recoveries, even in a single wet year (Fig. 4d). There was no initial curtailment of pumping in 1944, but as the drought progressed and water levels fell, a special provision of the judgment curtailed pumping in 1960-1961 and 1961-1962 (Fig. 5).

HISTORY OF THE LITIGATION

On September 23, 1937, the City of Pasadena filed a complaint against the City of Alhambra and 24 other named defendants in the Superior Court of Los Angeles County to quiet title to the water right within the Raymond Basin area. Because of the physical and legal complexities, an attempt was made to work out a "physical solution" to the problem. On January 31, 1939, most of the parties petitioned the Court to refer the matter to the Division of Water Resources as Referee for a determination of the physical facts, especially the safe yield and the annual extractions, and on February 8, 1939, such an order was made. On June 21, 1939, the Referee filed a preliminary report recommending that additional parties be brought into the action, and the Court ordered the Plaintiff to amend its complaint so as to follow that recommendation. The geologic and hydrologic studies by the Referee culminated in the Report of Referee, which was filed on July 12, 1943 (DWR, 1943). The Report of Referee concluded that the annual extractions exceeded the safe yield, resulting in a continued overdraft.

During the period of the reference the parties had been working on a proposed stipulated judgment, incorporating a water-exchange agreement. That agreement was approved by the Court on November 19, 1943. All parties but two had stipulated to the proposed judgment, and on November 24, 1943, the Court signed an order requiring that the stipulating parties comply with the terms of the proposed judgment during the pendency of the case. Following

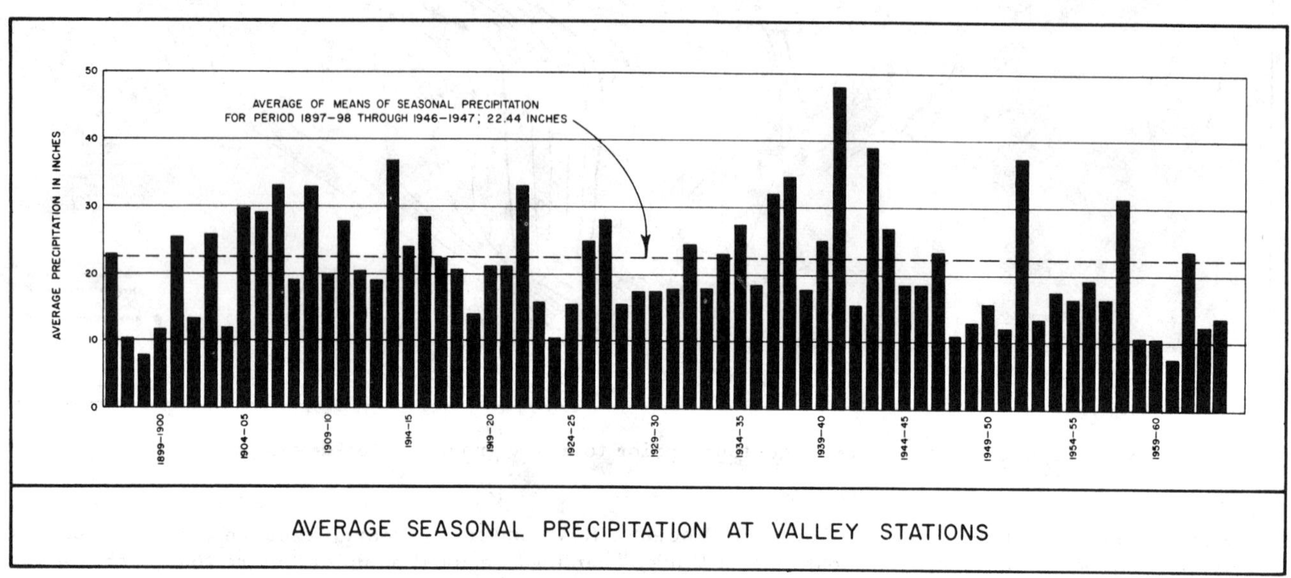

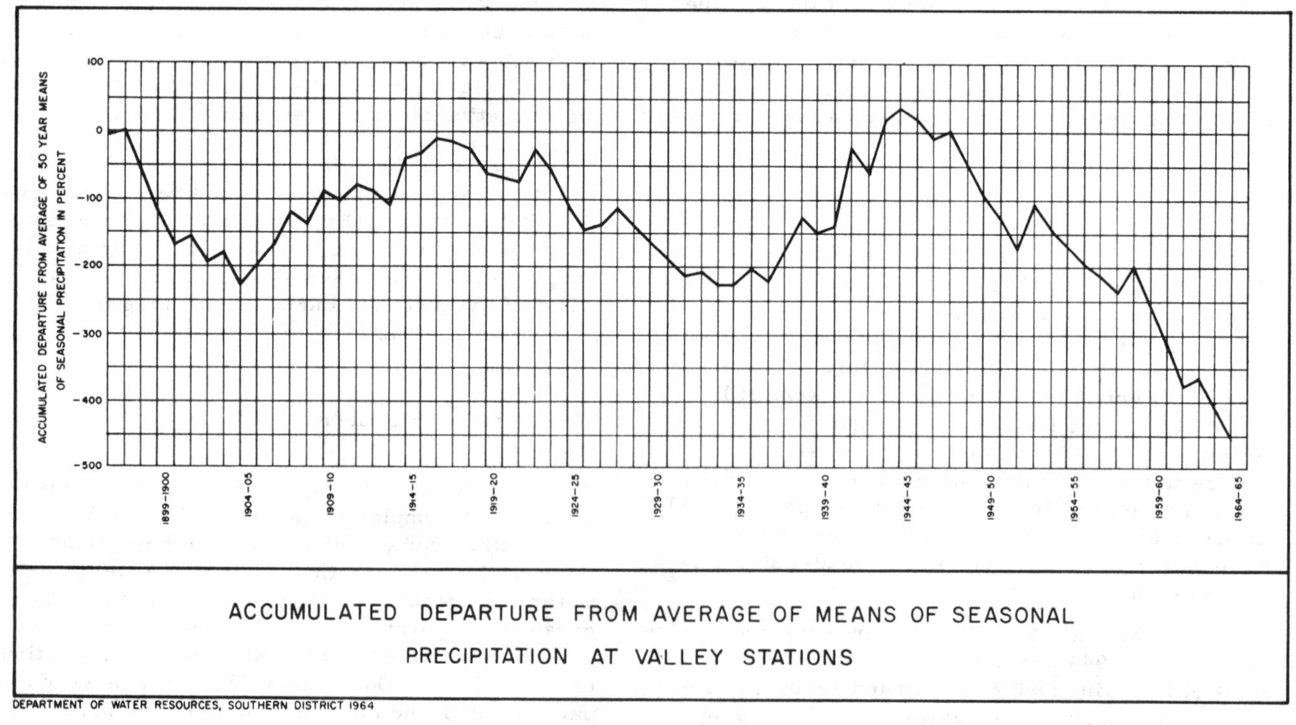

Figure 3. Average seasonal precipitation and accumulated departure from average.

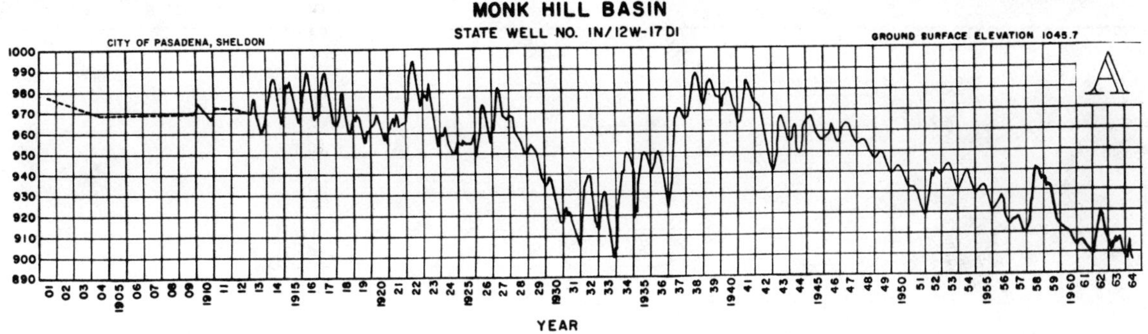

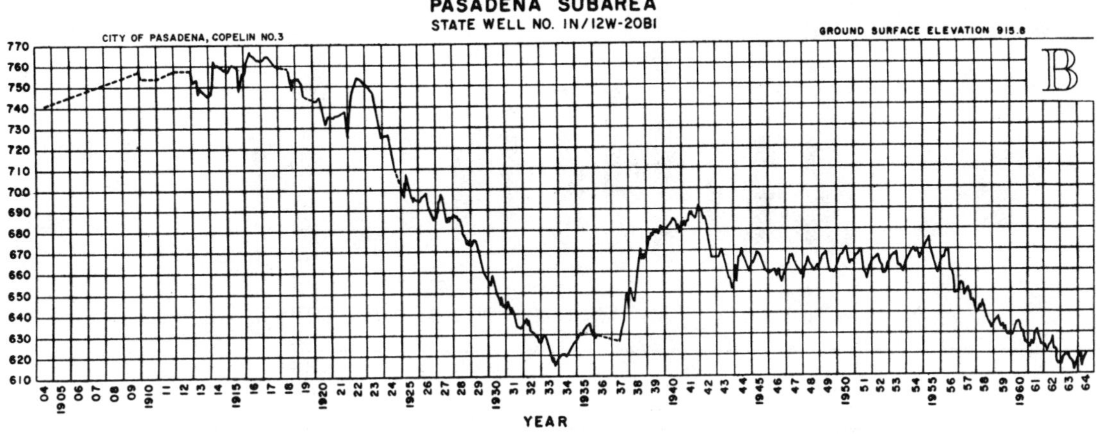

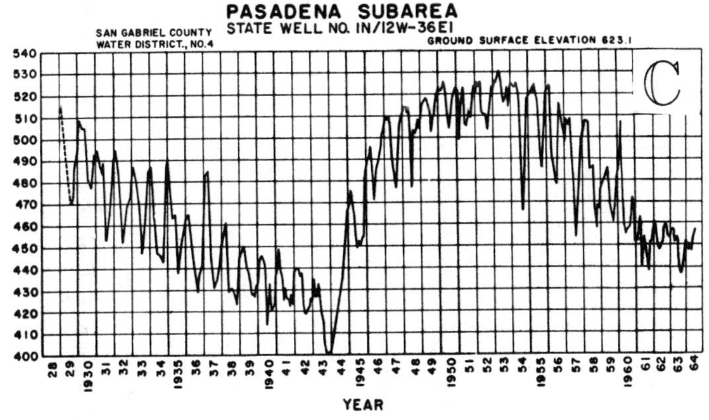

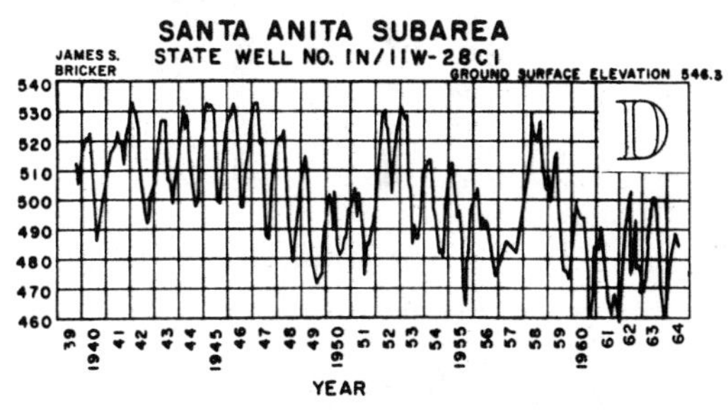

Figure 4. Representative hydrographs.

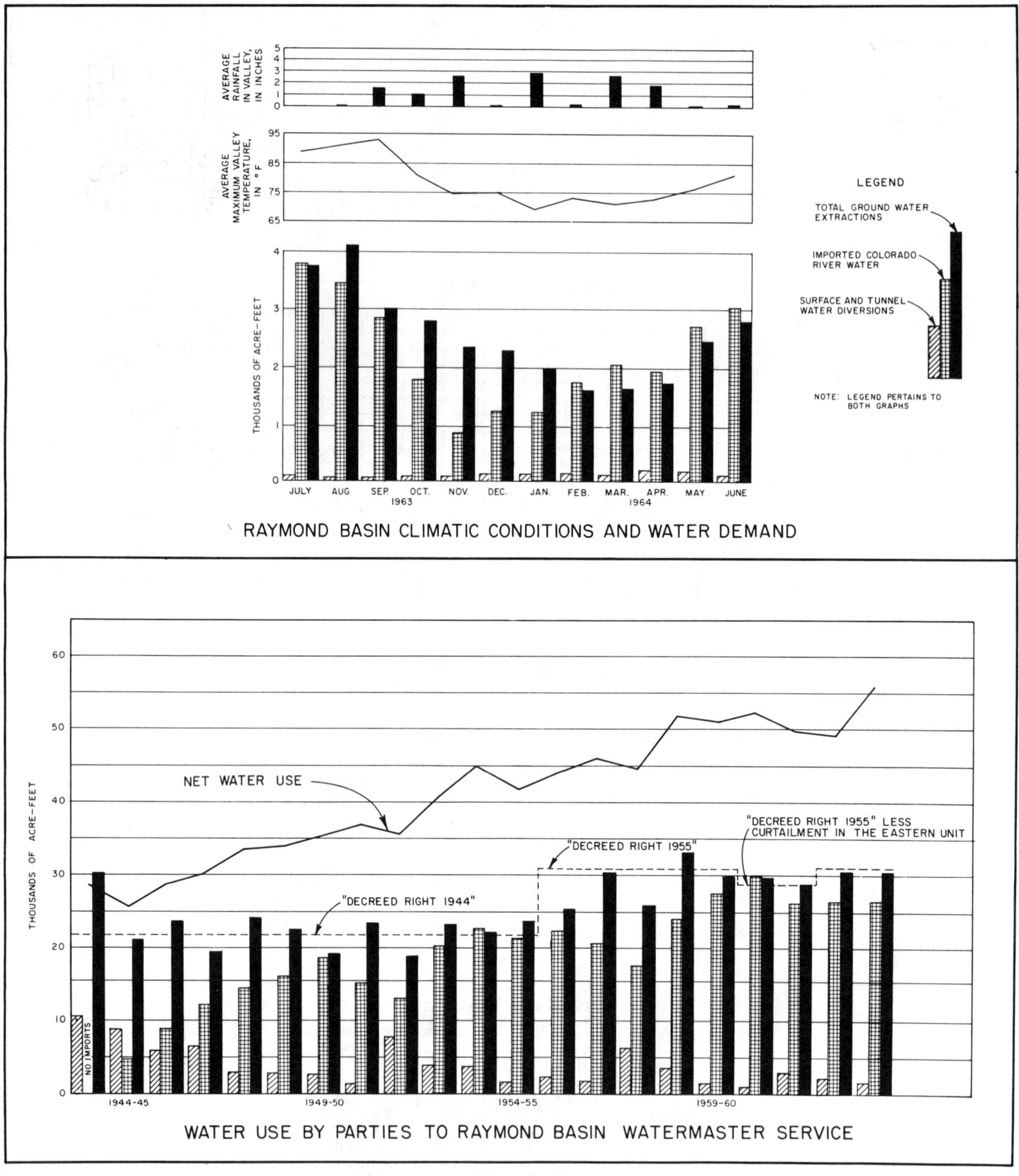

Figure 5. Climatic conditions, water demand, and water use by parties.

the provisions of the proposed judgment, the Court, on April 5, 1944, signed an order appointing the Division of Water Resources as Watermaster.

After a short trial, the Judgment was entered on December 23, 1944. Its most important provision was a required one-third cut-back in pumping in the Western Unit, effective as of July 1, 1944, except as to one party for whom the cut-back was to start July 1, 1945. One party appealed to the California Supreme Court, but the judgment was affirmed with minor modifications on June 3, 1949 (33 Cal 2d 908). A petition for a writ of certiorari to the United States Supreme Court was denied in 1950.

The terms of the judgment made it very difficult for the growing City of Sierra Madre, which had no source of imported water, to obtain its required water supplies. This situation led, on June 30, 1950, to the signing of an agreement between the City of Sierra Madre and the City of Arcadia, the only other pumper in the Santa Anita Subarea, which permitted the City of Sierra Madre to pump salvaged flood waters above its decreed right.

After six dry years of reduced pumping, during which water-level responses were better than expected, the City of Pasadena petitioned the Court, under its continuing jurisdiction, to order a review of the determination of the safe yield of the Raymond Basin area. On November 17, 1950, the Court appointed the Division of Water Resources as Referee for the review. The Report of Referee, filed October 5, 1954 (DWR, 1954), indicated a safe yield about 40 percent higher than the original figure, and on April 29, 1955, the Court signed a modification of the judgment which permitted increased pumping beginning July 1, 1955.

East Pasadena Water Company, Ltd., which had been added as a party Defendant in the original case and later dismissed, was not bound by the original judgment, and began pumping from one well in May 1955. The City of Pasadena, on April 26, 1960, filed a separate suit against this pumper and asked for an injunction. Negotiations were entered into and a stipulation for judgment was filed on March 26, 1965. The judgment provides for a waiver of pumping rights by the Defendant (except those that had been acquired by merger from another company with a decreed right), and an injunction against all pumping after December 31, 1966.

ORIGINAL SAFE-YIELD DETERMINATION

On June 9, 1939, the court made an Order of Reference to the Division of Water Resources, Department of Public Works, State of California for a determination of the physical facts, especially the safe yield. The resulting study was one of the earliest and most comprehensive studies of safe yield made in California (DWR, 1943).

The 29-year period from 1904-1905 through 1932-1933 was found to approximate closely the long-time mean precipitation. However, because of the many cultural changes in this urban area from 1904 to 1933, it was considered necessary to adopt a shorter, more recent period in which surface developments were much the same as in 1937, the year the lawsuit was filed. The shorter period selected was the 11-year period 1927-1928 through 1937-1938. The 11-year period had the further advantages of approximately normal precipitation, much more complete hydrologic data, and only a small net change of ground-water storage.

All inflow and outflow items were evaluated for the 11-year period and adjusted to the 29-year mean period. The Monk Hill Basin, Pasadena Subarea, and Santa Anita Subarea were handled separately. The inflow items evaluated were precipitation on the alluvial area, inflow from the mountains, and inflow from the hills. Measurements of mountain runoff went back only to 1913 and covered only 65 percent of the tributary mountain area. Correlations were used to extend and expand the existing records. Runoff from the hills was estimated from areas and precipitation. Long-time mean storm outflow from the Raymond Basin necessarily had to be related to the existing culture, and fortunately there were some gaging-station records during the 11-year period. The difference between the surface inflow and outflow was designated "local water remaining."

Detailed cultural surveys were made in 1938 and 1939, and to these acreages were assigned unit consumptive use values to determine total consumptive use. An independent study of consumptive use was based upon the differences between measured deliveries and the amounts of sewage exported from the basin or discharged to cesspools. Delivered water not derived from local wells was imported: (1) by the City of Pasadena from the San Gabriel River (since 1934); (2) a small amount from west of the Monk Hill Basin; and (3) a small amount by the City of South Pasadena northerly across the Raymond fault.

The change in storage during the 11-year period was based on water levels in 1927 and in 1938, and specific yield values. Two maps were prepared--contours of equal changes of water levels and contours of equal specific yield values averaged for depth in the zone of water-level change. The specific yield values had been developed mainly in an earlier study (DWR, 1934). Because of the complexity of underflow conditions at the Raymond fault, it was not possible to make a direct estimate of the underflow. It was calculated by taking total supply, subtracting disposal, and adjusting for change in storage.

The safe yield was determined to be 21,900 acre-feet per year, distributed as follows: Monk Hill Basin--6060; Pasadena Subarea--11,940; and Santa Anita Subarea--3900.

PROVISIONS OF THE JUDGMENT

The judgment declared that, under the then-existing conditions, the safe yield of the Eastern Unit was 3900 acre-feet per year and the safe yield of the Western Unit was 18,000 acre-feet per year. The taking by non-parties, upon whom the judgment is not binding, was declared to be 340 acre-feet per year for the Western Unit and 109 acre-feet per year for the Eastern Unit.

Diversions of surface-flow tributary to the Raymond Basin area were not treated in the same manner as ground-water pumpage. Such surface diversions, in the future, were to be limited to the maximum capacities of the diversion works as had existed at any time during the five-year period prior to October 1, 1937. Such maximum capacities for twelve specific diversion works were listed in cubic feet per second, and greater diversions were enjoined.

The most controversial and far-reaching provision is found in paragraph III:

That each and all of the rights of the parties hereto to pump water from wells or otherwise take water from the ground in said Raymond Basin area are of equal priority and of the same legal force and effect.

The "present unadjusted rights" were listed for two parties in the Eastern Unit, totaling 3791 acre-feet per year and for 26 parties in the Western Unit, totaling 25,608 acre-feet per year. The method used to calculate "present unadjusted rights" is not to be found in the judgment. However, a statement of what has come to be known as the "Raymond Basin formula" may be found in paragraph XIV of the Findings of Fact and Conclusions of Law:

. . . the highest continuous production of water for beneficial use in any five (5) year period prior to the filing of the complaint by each of the parties in each of said units, as to which there has been no cessation of use by it during any subsequent continuous five (5) year period . . .

Because the pumping of the "present unadjusted rights" in the Western Unit would have resulted in continued overdraft, each pumper was cut back proportionately to an amount designated as the "decreed right." In the Western Unit the "decreed right" was determined by reducing the "present unadjusted right" in the proportion that the safe yield of the Western Unit, less the nonparty pumping, bore to the aggregate of such rights of the parties in the Western Unit. This amounted to a cut-back of about 29 percent. Greater pumping than the "decreed right" was enjoined, except that in a single year, 120 percent of the "decreed right" was permitted, although over a 60-month period the pumping was restricted to 5 times the "decreed right." The "decreed rights" of all the parties in the Western Unit were listed in acre-feet per year, to the nearest acre-foot.

Because no overdraft was found in the Eastern Unit, there was no initial cut-back. As to the two parties, the City of Arcadia and the City of Sierra Madre, the pumping of 120 percent of the "decreed right" was permitted, but with the same 60-month reckoning. There was a specific limitation on the amount of water which could be pumped within one-half mile of the western boundary of the Eastern Unit and a further provision for reduction of the "decreed right" in the Eastern Unit from 3791 acre-feet per year to 3261 acre-feet per year in any year following a spring in which water levels were below an elevation of 500 feet in wells in a specified area near the southeastern corner of the Western Unit, and the water-table slope from this specified area was easterly into the Santa Anita Subarea.

The judgment appointed the Division of Water Resources as Watermaster, set up an advisory board of four members, and required each party to keep records of diversions, ground-water production, imports, exports, transfers, and water levels in wells. There were orders covering the responsibilities of the Watermaster, the financing of his activities, the preparation of annual reports, and appeals from Watermaster determinations. The judgment specifically denied making any determination regarding rights to imported water and intent to recapture such water after apreading. It incorporated an attached Raymond Basin Water Exchange Agreement of 1943 and gave legal protection to those parties exchanging water. An important provision was the retention of jurisdiction by the Court to review the safe yield, to protect against contamination, and to re-apportion rights forfeited or abandoned.

WATER EXCHANGE AGREEMENT

The Raymond Basin Area Water Exchange Agreement was signed by all but two of the parties during October and November 1943, and was approved by the Court on November 19, 1943. It became part of the judgment. The main objective of this agreement was to provide a means for any party to meet its needs by temporarily purchasing the rights released by another party. Each party agreed to participate in the Exchange Pool for a specified number of years--from 10 to 50. A party which offered to release water to the Exchange Pool for a certain year was required to specify the amount offered and the cost--equivalent to an average of its costs for all water (including imports). Water released in one unit could not be pumped in the other unit.

A special arrangement was made for the Monk Hill Basin. As to its taking from the Monk Hill Basin, the City of Pasadena was restricted to pumping the difference between the safe yield and the combined pumping of the non-parties and the six other parties in the Monk Hill Basin. Under certain conditions, the City of Pasadena, by virtue of this required shift of pumping from the higher Monk Hill Basin to the lower Pasadena Subarea, was allowed "undue costs" to be paid by the pumpers in the Monk Hill Basin.

Parties who desired to purchase water from the Exchange Pool were to file annual requests in advance. Requests were to be filled starting with the least expensive water offered, and the price to all Exchangees was the average price of all water taken from the Pool. Exchangors were considered to have pumped the water released and paid for. The Watermaster maintains an accounting of the water exchanged, collects from the Exchangees, and disburses money to the Exchangors (DWR, 1945W-1964W). The exchange agreement was amended in the spring of 1944 to include one additional party in the Monk Hill Basin.

SALVAGE OF WATER BY THE CITY OF SIERRA MADRE

In the Santa Anita Subarea, the only significant pumpers are the Cities of Arcadia and Sierra Madre. The City of Arcadia lies on both sides of the Raymond fault, and whereas pumping north of the fault is restricted, pumping south of the fault is unrestricted. As needed, the City of Arcadia imports ground water to its customers north of the fault. The City of Sierra Madre extends into the Pasadena Subarea, but has pumping rights only in the Santa Anita Subarea. Its rights under the judgment included a surface diversion of up to 6.0 cubic feet per second in Little Santa Anita Canyon, and a "decreed right" of 1264 acre-feet per year, which was later modified upward to 1764 acre-feet per year. However, when water levels drop and there is an easterly water-table slope from the Pasadena Subarea into the Santa Anita Subarea, the pumping right of the City of Sierra Madre is reduced to 1087 acre-feet. The reduced pumping allocation was in effect in 1960-1961 and 1961-1962 (Fig. 5) and was in effect in 1964-1965. To 1965, the City of Sierra Madre had not had available to it any imported waters.

In anticipation of future difficulties, the City of Sierra Madre on June 30, 1950 entered into an agreement with the City of Arcadia which permitted the City of Sierra Madre to spread certain flood waters within the Santa Anita Subarea and to recapture them, apart and in excess of its "decreed right." This agreement was approved by the Court on October 6, 1950, and the responsibility for calculating the net amount salvaged was assigned to the Watermaster (DWR, 1959).

The City of Sierra Madre has about 22 acres of spreading grounds on the west side of the lined channel of Sierra Madre Wash (Little Santa Anita Creek). The active production wells are just south of these spreading grounds. Water is diverted from the nearby channel of Sierra Madre Wash, from Santa Anita Creek to the east by means of a pipeline 1.2 miles long, and from storm drains carrying street runoff in the vicinity of the spreading grounds. The water is measured at the intake to the spreading grounds and at an overflow weir. The gross amount spread is reduced by two items: (1) that portion of the diversion from Santa Anita Creek which would have percolated above the Raymond fault, assuming the diversion had not been made; and (2) the increase in underflow across the Raymond fault which is attributable to the spreading.

For (1) an empirical relationship was set up relating discharge (in cubic feet per second) at the head of the percolation reach to percolation in the reach (in cubic feet per second). Data for this relationship have been obtained from streamflow measurements at both ends of the percolation reach. This water, calculated on a daily basis, is considered part of the natural safe yield. The determination of (2) involved monthly calculations of increased underflow resulting from increased water-table slopes across the Raymond fault. Deductions are taken for as long as the spread water remains unpumped. For the period 1951-1964, the City of Sierra Madre spread 11,792 acre-feet and was able to pump under a salvage credit a total of 4490 acre-feet.

In 1964, on behalf of the City of Sierra Madre, a proposal was made for the spreading and salvage of reclaimed sewage. This would involve some 1600 acre-feet per year developed by a water reclamation plant in the Pasadena Subarea. Several alternatives were suggested: (1) spreading in the Pasadena Subarea, pumping by the City of Sierra Madre in the Pasadena Subarea; (2) spreading in the Santa Anita Subarea, additional pumping by the City of Sierra Madre there; and (3) spreading in the Pasadena Subarea, pumping by the City of Arcadia there, with release of pumping rights in the Santa Anita Subarea by the City of Arcadia to the City of Sierra Madre. These suggestions are still under discussion.

REDETERMINATION OF SAFE YIELD

After a series of six dry years, during which time water levels had risen progressively, Plaintiff City of Pasadena petitioned the Court to order a review of the determination of the safe yield, and the motion was granted on November 17, 1950. This reference was also to the Division of Water Resources. The Report of Referee was completed in July 1954 (DWR, 1954).

The methods employed in the redetermination were much the same as used in the original study. Because of the availability of more complete hydrologic data, the original 29-year period (1904-1933) was replaced by a 30-year period (1919-1949), which was adopted as the long-time mean period. The original 11-year period (1927-1938), considered to represent cultural conditions as of 1937-1938, was replaced by the 12-year period 1938-1950, which more accurately reflected the cultural conditions as of 1951-1952.

The result of the redetermination was an increase of safe yield of about 40 percent (Table 1). The reasons for this increase can be seen in Table 2. The most obvious cause for the increase in safe yield is the sharp increase in imports, especially in the Pasadena Subarea. Another factor tending to increase the local safe yield is the reduction in exports, at least to the extent of the return waters. Items working in the direction of reducing the safe yield are the increases in surface outflow and sewage export. Although consumptive use remained relatively unchanged, there were greater delivery demands, with the result that a greater percentage of the delivered water got into the sewers. Furthermore, additional areas were sewered, and former cesspool recharge became sewage export. The increased surface outflow is related to the expansion of impervious areas that accompanied the further urbanization. The item labeled "decrease due to spreading" is an adjustment necessitated by the terms of the salvage agreement between the City of Arcadia and the City of Sierra Madre.

CONCLUSIONS

The Raymond Basin adjudication is but one event in a series through which the City of Pasadena sought to assure water supplies for future growth. In 1928, the City of Pasadena became one of the charter members of the Metropolitan Water District of Southern California, which was formed for the purpose of importing water from the Colorado River to coastal southern California. As early as 1923, that city had filed applications to appropriate flood waters of the San Gabriel River several miles to the east of the Raymond Basin. Those appropriations were protested vigorously by local interests and an agreement was negotiated in 1934 whereby the City of Pasadena, upon Colorado River water becoming available to it, would relinquish and abandon all rights to any of the waters of the San Gabriel River (Walters and others, 1961). The City of Pasadena imported water from the San Gabriel River during the period 1933-1941, but from 1941 to the present time all imports have been from the Colorado River.

The availability of supplemental water is a prime requisite to a physical solution of the Raymond Basin type. Where no supplemental water is available, an overdrafted ground water basin has three choices: (1) an immediate cut-back in pumping with curtailment of existing uses; (2) continued mining of water until the surface economy has to be abandoned; and (3) continued mining of water with a shift to higher water uses, such as agricultural to urban. Alternative (3) is often used in some basins with the expectation that the enlarged economic base generated by the mining of local ground water will demand (and be able to afford) future supplemental supplies.

The Exchange Pool can be a means of forcing those pumpers with growing demands to

Table 1. Comparison of 1938 and 1952 safe yields (acre-feet per year)

	Monk Hill Basin	Pasadena Subarea	Total Western Unit	Santa Anita Subarea	Total Raymond Basin
1938	6060	11940	18000	3900	21900
1952	7490	17990	25480	5290	30770
Percent Increase	23.6	50.7	41.6	35.6	40.5

Table 2. Changes in long-time mean items governing safe yield (acre-feet per year)

Item	Monk Hill Basin 1938	Monk Hill Basin 1952	Pasadena Subarea 1938	Pasadena Subarea 1952	Santa Anita Subarea 1938	Santa Anita Subarea 1952
Precipitation	12110	12100	29600	29600	4450	4450
Inflow from mountains	7720	5330	2770	2480	5680	5460
Inflow from hills	291	220	153	110	0	0
Inflow from Monk Hill Basin			5220	5680		
Subtotals	20111	17650	37743	37870	10130	9910
Surface diversions	1880	3640	1790	1780	1460	1140
Extractions	6380	6420	17570	12260	3380	2920
Imports	165	1060	2910	16400	0	810
Transfers in	0	640	4860	3530	0	0
Subtotals	8425	11760	27130	33970	4840	4870
TOTALS	28536	29410	64873	71840	14970	14780
Surface outflow	5430	6030	10980	14140	4140	4210
Exports	0	0	9520	4580	2030	0
Transfers out	4460	3010	0	640	403	530
Subtotals	4460	3010	9520	5220	2433	530
Sewage outflow	139	530	5776	8800	0	160
Consumptive use	10230	10130	25940	25050	3910	3770
Subtotals	10369	10660	31716	33850	3910	3930
Decrease due to spreading					0	200
TOTALS	20259	19700	52216	53210	10483	8670

arrange for receiving supplemental water, by making the costs of pumping quantities larger than assigned rights more than the costs of supplemental water. The Water Exchange Agreement was important in the earlier years of Watermaster operation, especially in the Monk Hill Basin. Colorado River water became available to the Monk Hill pumpers in 1955, but the quantity was somewhat lower than that of local supplies, and there was some reluctance to accept it. Under the terms of the agreement, however, a party with a connection to the Metropolitan Water District could get no water from the Exchange Pool unless undue hardship could be shown. After 1958 the Exchange Pool in the Western Unit ceased to be important; in 1963-1964, only one entity requested water from the Exchange Pool, in the amount of 30 acre-feet. In the Eastern Unit the amount exchanged in the 20 years of operation has averaged only about 100 acre-feet per year.

As illustrated by the activities of the City of Sierra Madre, the control of ground-water pumping tends to result in better conservation measures. The salvage of flood waters by the City of Sierra Madre may not exclusively involve waters that would otherwise waste to the ocean; some of those waters may have percolated to ground-water storage in San Gabriel Valley. In other ground-water basins, the proportion of water salvaged from unquestioned waste might be higher. The proposed reclamation of water from sewage completes an interesting cycle: (1) local disposal of sewage to individual cesspools; (2) expansion of sewerage facilities and connection to outfalls (tending to decrease the safe yield); and (3) local treatment of collected sewage with disposal of sludge to the outfalls, and local spreading of the effluent (tending to increase the safe yield).

In the more than 20 years that have elapsed since the start of ground-water management in the Raymond Basin, it appears that the pioneering approaches developed for that basin have been physically effective, the flare-up of water rights controversy has been minor, and the local entities are content with their solution.

In spite of the peace achieved locally, the Raymond Basin adjudication was not without external repercussions. The new doctrine of mutual prescription has brought strong reactions from supporters of classical concepts of water rights, and re-evaluations of those concepts (Towner, 1957; Hutchins, 1957; Sawyer, 1949). The Raymond Basin formula for calculating pumping rights has encouraged a race to build up prescriptive rights before adjudications are filed (Krieger and Banks, 1962). The Raymond Basin type of settlement has been completed in a second ground-water basin in southern California, it is almost completed in a third ground-water basin, and is currently being considered in several others.

ACKNOWLEDGMENTS

In the preparation of this paper, the author has necessarily drawn very heavily upon studies and reports prepared by the California Department of Water Resources in its dual role as Referee and as Watermaster. The department has kindly given permission to use Figures 3 and 5 and has made available the base maps for Figures 1 and 2, and the hydrographs on Figure 4. Through the courtesy of Duncan A. Blackburn, chief engineer and general manager of the Pasadena Water Department, access was gained to much information in the department's files.

REFERENCES CITED

Blackburn, D.A., 1952, Forty years in developing a water system: unpublished paper presented on October 29, 1952 at the California Section meeting, American Water Works Association.

________, 1961, The adjudication of the Raymond Basin: unpublished paper presented at the Southern California Coordinating Conference on November 21, 1961.

Buwalda, J. P., 1940, Geology of the Raymond Basin: private report submitted to the Pasadena Water Department.

Hutchins, W. A., 1957, California ground water: legal problems: 45 California Law Review 688-697.

Krieger, J. H., and H. O. Banks, 1962, Ground-water basin management: 50 California Law Review 56-77.

Sawyer, F. A., 1949, Water law: mutually prescriptive interests in underground water: 37 California Law Review 713-718.

Sonderegger, A. L., 1936, Report on ground-water conditions in Raymond Basin: private report submitted to Pasadena Water Department.

Towner, P. A., 1957, Mutual prescription--threat to vested rights to ground water: unpublished paper presented on December 12, 1957 at the convention of the Irrigation Districts Association of California at Long Beach.

Walters, E. H., I. R. Calvert, and M. A. Bengel, 1961, History of the San Gabriel River Water Committee, the San Gabriel River Spreading Corporation, and the San Gabriel Valley Protective Association: published by the Upper San Gabriel Valley Water Association.

The following reports by the California Department of Water Resources (and its predecessor, the Division of Water Resources) are cited as DWR, year. Watermaster reports are cited as DWR, yearW.

California State Department of Public Works, Division of Water Resources, 1934, South coastal basin investigation, geology and ground water storage capacity of valley fill, Bulletin 45.

________, 1943, Report of Referee, City of Pasadena vs. City of Alhambra, and others, No. Pasadena C-1323, Superior Court, Los Angeles County.

________, 1954, Report of Referee on a review of the determination of the safe yield of the Raymond Basin area, Los Angeles County, California. City of Pasadena vs. City of Alhambra, and others, No. Pasadena C-1323, Superior Court, Los Angeles County.

________, 1959, Report of Raymond Basin Watermaster on determinations of credit for water salvaged by the City of Sierra Madre in the Santa Anita Subarea, Raymond Basin, Los Angeles County, California.

________, 1945W to 1964W, Watermaster service in the Raymond Basin, Los Angeles County. Annual Reports.

FLAGG'S RESTAURANT AREA LANDSLIDE, PACIFIC PALISADES, CALIFORNIA[1]

Arthur B. Cleaves (Department of Earth Sciences, Washington University, St. Louis, Missouri)

Abstract

The Pacific Palisades (Flagg's Restaurant area) landslide of 1958 was only one of several local ones that occurred in that year. However, the area in general has a landslide record extending back over many years.

The toe of this slide spread across U.S. Route, Alternate 101, also known as the Coast Highway.

There are three general methods of correcting and/or controlling landslides: (1) avoidance; (2) the application of active measures where the driving force of the slide is opposed by a resistive force (cribbing, buttresses, retaining walls, and so on); and (3) application of passive measures such as drainage, unloading by excavation, total, or near the crown, and benching.

The treatment applied in this case, primarily because of possible legal conditions, was avoidance by relocation.

CONTENTS

ILLUSTRATIONS

INTRODUCTION

The history of landslide activity in the Pacific Palisades region along the coastline extends back many years. The best comprehensive information of the phenomena in this area is shown on a map--"Preliminary Map of Landslides in the Pacific Palisades Area, City of Los Angeles, California;" 1959 United States Geological Survey, Miscellaneous Geologic Investigations, Map I-284. This map was prepared by John T. McGill, and in the explanation, historic landslides are recorded as far back as 1884 with indications of even older slides. Data were obtained from field mapping, photographic interpretation, and records in the files of the City of Los Angeles and the State of California.

The slide of March 1958, which is the subject of this paper, is recorded by McGill. Figure 1 shows the site location of the Flagg's Restaurant slide. Because of the importance of, and heavy traffic on, this highway, it was necessary to reopen it as quickly as possible.

FLAGG'S RESTAURANT AREA LANDSLIDE

The name used above is for the purpose of identifying this particular slide from others in the

[1] Based on data obtained from the California Division of Highways.

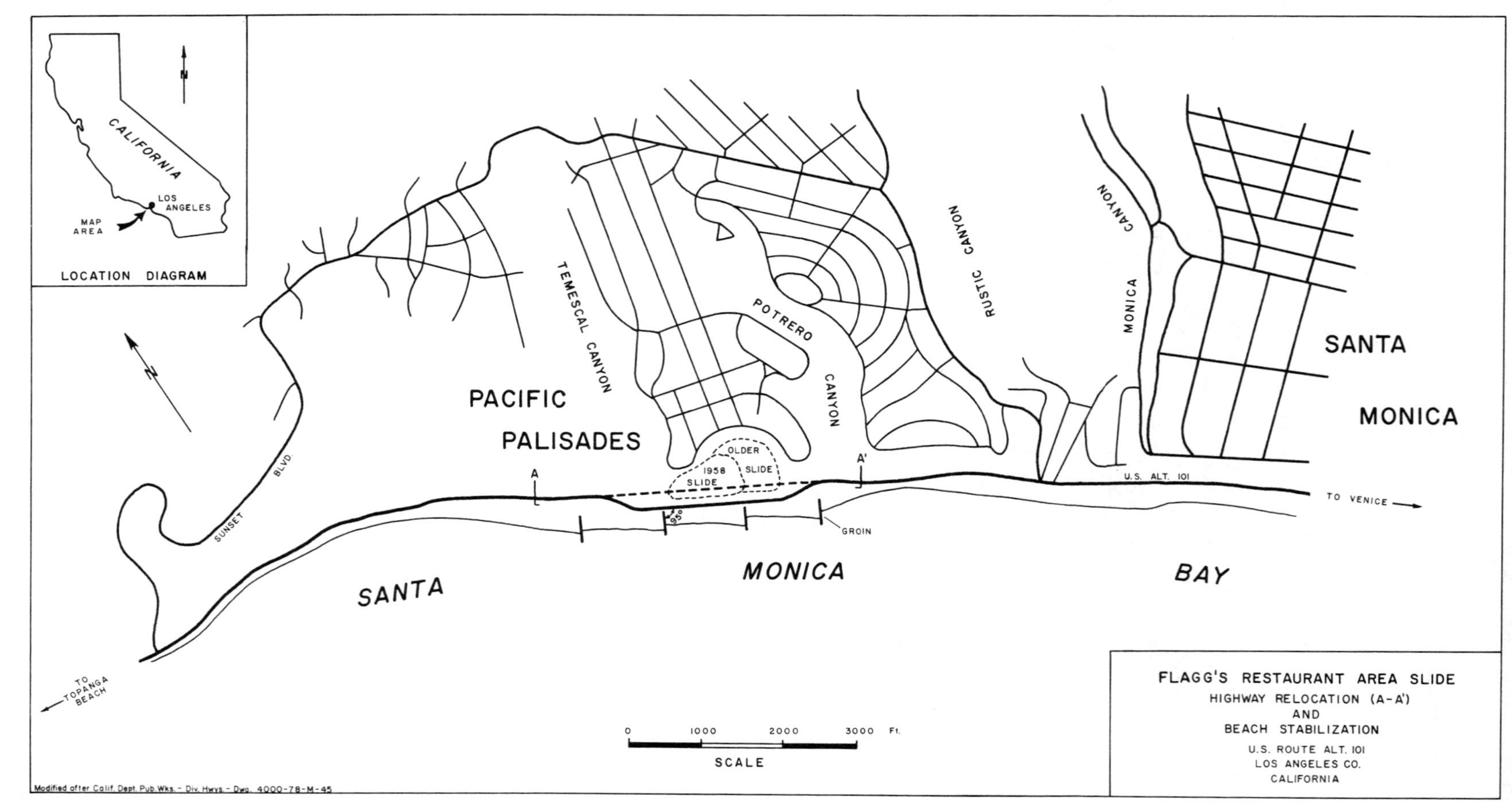

Figure 1. Location map of Flagg's Restaurant area landslide, Pacific Palisades, California.

vicinity. In Pl. 1, a corner of the restaurant is shown in the upper right, on the beach.

This slump type of landslide and others in the immediate vicinity occurred on March 27 to 31, 1958. The Flagg's Restaurant slide had an over-all width on the highway, near the toe, of about 950 feet, and the height, crown to toe, was roughly 225 feet. The movement took place in the precise area of an older slide that was approximately 1500 feet wide at the toe, and which probably antedated most of the homes on Via de Las Olas (street) above the crown. A portion of this street was dislocated downward.

Plate 1. Flagg's Restaurant slide looking west.

Heavier than normal rainfall had taken place in the days preceding the slide, and some ponding was observed on the backward rotated slide units (Pl. 2).

The landslide debris consisted of badly broken "stone" and poorly "cemented" clastic material in which the original bonding was probably never strong.

Relationship to Coast Highway

The landslide completely covered the Coast Highway, and in places buried it to depths of 20 and 25 feet. The field examination on April 15, 1958, only two weeks after the movement, revealed the slide mass was in a partially saturated condition. Limited local traffic was permitted over the toe of the slide. Near the west flank and about 20 feet above the original road, a steady seepage of water was observed.

Although no slide readjustments appeared to be occurring at the time of the field examination, they were anticipated as a normal condition. Several years later, a brief visit to the site confirmed this expectation. The principal area involved was Via de Las Olas, the street just above the slide crown. Along this street there were many expensive homes.

METHODS OF LANDSLIDE TREATMENT

The methods of slide treatment fall into three categories: (1) avoidance by relocation; (2) the application of active measures; and (3) the application of passive measures (Eckel and others, 1958; Cleaves, 1961).

The choice of the slide control, or corrective methods used, depends on variable factors which include: cost; possibility of relocation; physical properties of the slide materials, including grain size; drainage possibilities; and, of course, property values.

Avoidance, in the case of a highway, railroad, pipe-line or such utility, may involve the purchase of a new right-of-way, and in an urban area may be completely impracticable.

Active methods employing restraining devices such as retaining walls, cribbing, and buttresses work best in coarsely granular material. In any case, whether used in granular or silty and clayey materials, adequate drainage techniques should be included. In this respect care must be taken that the drainage method used (such as sand-filter perforated pipe drains) prevents clogging of voids by clay and silt-sized particles behind the restraining device.

Passive methods include drainage measures, unloading by excavation near the crown, benching, and similar activities. These are described in Eckel and others (1958) and Cleaves (1961).

LEGAL ASPECTS OF REOPENING THE ORIGINAL HIGHWAY

The right-of-way of the highway at the site of the Flagg's Restaurant landslide was very limited, hardly more than the width of the shoulder. This being the case, there was the possibility that removal of the debris at the toe of the slide could reactivate the slide movement. Should this occur, damage to properties along Via de Las Olas might develop.

The California Division of Highways had not done any excavation beyond it's right-of-way for many years. Consequently, at the time of the slide movement, they could not be held legally responsible. However, should the mass of debris covering the highway at the toe of the slide be removed by the Division of Highways, the argument could have been raised that the slide reactivation and possible damage to properties above the slide crown was the direct result of that work. If the slide had reached temporary stability, the removal of debris at the toe, by reducing the restraining forces, could have been a primary cause of reactivation.

Plate 2. Flagg's Restaurant slide looking east.

It was the author's opinion at the time that sidehill, horizontal, perforated-pipe drains might have stabilized the slide mass. This technique, successful in other areas in California, plus reshaping of the slide surface, would require co-operation, in the form of quit claims, by the property owners along Via de Las Olas, and probably also the administrative forces of Pacific Palisades.

Immediate reopening of the Coast Highway was of vital importance, and the legal arrangements referred to could, and probably would, have been involved and time consuming. Therefore instead of attempting stability, relocation of the road seaward was recommended.

RECOMMENDATIONS

Based on the recommendations, and with the permission and co-operation of the California Division of Beaches and Parks, the highway was relocated on the beach proper, and the landslide area was completely avoided.

NEW CONSTRUCTION

It is not the purpose of this paper to discuss in detail the features of the relocated highway, hence only the general features are mentioned.

The relocated highway consists of a four-lane, asphalt concrete roadway located on the original beach. The borrow material consisted of a fine-grained sandy soil with some silt and clay-sized particles. The upper 6 inches of the subgrade was compacted to 95 percent relative compaction by a California method that is comparable to the modified American Association of State Highway Officials compaction.

For highway protection against wave and long-shore current erosion, four heavy stone groins, rotated 5 degrees clockwise, were constructed. These were keyed to a seaward inclined, "special heavy" to "extra heavy" stone riprap layer 6 feet thick which rests on an inclined 3 foot thick gravel layer. Suitable drainage from the roadway and from the landward side of the right-of-way is carried through this structure.

An examination two years after construction indicated that the effectiveness of the groin- and riprap-protected road has been entirely successful.

REFERENCES CITED

Cleaves, A. B., 1961, Landslide investigations, a field handbook for use in highway location and design: U.S. Dept. Commerce, Bureau of Public Roads, Washington, D.C.

Eckel, E. B. and others, 1958, Landslides and engineering practice: Highway Research Board, Special Report No. 29, Washington, D.C.

PFEIFFER VERSUS GENERAL INSURANCE CORPORATION--LANDSLIDE DAMAGE TO INSURED DWELLING, ORINDA, CALIFORNIA, AND RELEVANT CASES

George A. Kiersch (Department of Geological Sciences, Cornell University, Ithaca, New York)

Abstract

The Pfeiffer case is illustrative of how geologic facts, events, and terminology can be central to the outcome of a controversy between an insured party and his insurance company. Gravity sliding occurred which damaged the dwelling of the insured. Geological testimony was used to determine causes; the possibility of recurrence; and the advisability of rebuilding at the site. In addition, such testimony was judicious on whether there was negligence on the part of the builder or the insurance company because they did not realize the geologic conditions would lead to large-scale gravity sliding. Geologic facts revealed that causes were visible and the landslide was foreseeable; therefore it could not be passed over by the insurance company as "an act of God."

The Pfeiffer case involved a decision concerning landslide insurance and the payment of damage to the insured. The court decision, which has since been precedent for other cases, was based on the critical geologic conditions and events that affected the site of the insured dwelling. The sole issue was "Does an insurance policy which insures a dwelling against the perils of landslide, include the foundation underlying the house, as well as the house structure itself?" Interestingly, this crucial point had not been clarified by a court decision prior to the Pfeiffer case. Insurance companies had customarily accepted no responsibility for the subsurface foundation part of the dwelling, and awarded damages for repair of the building only. The Pfeiffer's were awarded $31,000 in a judgment of August 1960 for repair of their dwelling to the conditions of the house and foundation prior to the landslide damage. Other important precedents in landslide insurance coverage concerned with geologic conditions are reviewed in this paper; all cases were decided uniformly in favor of homeowners on all major issues.

The Pfeiffer dwelling is situated on the slope of a northwest trending ridge of the Berkeley Hills at Orinda, California. Everywhere the area is underlain by rocks of the Orinda Formation which consist of alternating beds of siltstone, soft, fine-grained sandstone and conglomerate, clay shale and clays.

At the site, the natural slope is 20° to 40°; the underlying rock sequence dips up to 45° and more or less parallel to the surface. A mantle of soil and colluvium up to 10-feet thick overlies the soft, fractured and saturated Orinda beds. Gravity sliding, as determined by both surface and subsurface investigations, was caused by a series of geologic events, both ancient and current, which combined to create the slide of April 1958 that damaged Pfeiffer's dwelling. The numerous internal causes are described; no major external causes, such as earthquakes, were responsible.

CONTENTS

ILLUSTRATIONS

INTRODUCTION

The Robert J. Pfeiffer's negotiated purchase of a newly constructed home at No. 1 Easton Court in the suburban community of Orinda, California, in October 1957. They finalized purchase of the dwelling in December 1957, and an insurance policy was written by the General Insurance Corporation, which insured them against hazards of fire and an attached endorsement provided protection against natural events such as landslides. The General Insurance Corporation was also the representative of the contractor for the dwelling, Rana Builders, Inc., Lafayette, California. The community of Orinda is located in the hills some 6 miles east of the cities of Oakland and Berkeley.

Major gravity sliding occurred in the soil and bedrock directly upslope from the Pfeiffer dwelling at No. 1 Easton Court. Beginning on April 3, 1958, sliding took place on such a large scale that the Pfeiffer dwelling was partially distorted, twisted, and the subsurface foundation moved until two days later the house was unsafe for occupancy. Thus, on April 5th, the Contra Costa County officials ordered the Pfeiffer family to abandon their dwelling and move out. The sliding continued intermittently for several months thereafter, and the damage to the dwelling and building structures increased to a point where the garage section was demolished, the house structure was further damaged by twisting and shifting, and the foundation was weakened and unstable due to being fractured, deformed, and displaced.

Pfeiffer immediately reported the landslide event and damage to General Insurance Corporation and filed for reimbursement under terms of the special endorsement coverage of insurance policy. General Insurance Corporation subsequently refused to accept responsibility for damages to the dwelling, except for the repair of the house structure estimated to total $5000 to $8000. The insurance company refused to acknowledge responsibility for repair of the damage to the subsurface and the foundation of the housing structure which was fractured, deformed, and encompassed as part of an active landslide mass. Furthermore, General Insurance stated they would repair the damage to the housing structure only and were cancelling the insurance policy.

The action and attitude taken by General Insurance Corporation was contrary to Pfeiffer's interpretation of the policy which included an all-physical-loss coverage (<u>see</u> APPENDIX) and other endorsements which encompassed landslides. Consequently, the Pfeiffers filed suit and sought to recover the maximum amount set forth in the policy for landslide damage to their dwelling, namely $26,000, and an additional $5,000 which is the limit for additional living expenses incurred because they were forced to move from their home.

The trial of this case was held in United States District Court, San Francisco, before Judge J. Harris on May 4 and 5, 1960. The Pfeiffers were represented in court by Gilbert C. Wheat and Peter M. Kardel, Esq. of Lillick, Geary, Wheat, Adams and Charles, San Francisco, California. Oliver Merwin of Harding and Associates served as the soils engineering expert and the author served as the geological expert for the litigation. General Insurance Corporation was represented by Scott Conley, Esq. of Sedgwick, Detert, Moran and Arnold, San Francisco, California.

The Pfeiffer case involved a classic decision concerning landslide insurance and the payment of damage to the insured. The court decision, which has since been precedent for other cases, was based on the all importance of the geologic conditions and events that affected the site of the insured dwelling. The sole issue as defined at the trial was "Does an insurance policy which insures a dwelling against the perils of landslide, include the foundation and land underlying the house as well as the house structure itself?" Prior to the Pfeiffer case, this crucial point had not been clarified by a court decision; insurance companies had customarily paid the cost of damages to the building structure only, and had accepted no responsibility for the subsurface foundation part of the dwelling.

The importance of geologic conditions and a competent explanation of the interpretation for the court, as related to insurance coverage and landslides, is further demonstrated by summaries of several other precedent-making cases. For those readers who may wish a layman's explanation of certain legal terminology, an Appendix has been attached to the text which describes the principal types of insurance policies; important legal terms; and the causation of landslides as defined by the courts.

ACKNOWLEDGMENTS

Preparation of this summary on the Pfeiffer case and relevant landslide litigation involving insurance coverage has received the assistance of several interested parties. Messrs. Gilbert C. Wheat and Peter M. Kardel initially stimulated the author's interest in the matter of insurance claims as influenced by geologic conditions and R. J. Pfeiffer kindly granted permission to discuss his case.

Miss Sandra D. Glasgow, a graduate student at Cornell University, assisted the author greatly by assembling data for and preparing a draft of the material given in the sections on insurance and the Appendix.

GEOLOGIC SETTING

The community of Orinda lies in the Berkeley Hills of the Coast Ranges of California. The ranges are comprised of northwest-trending folded rocks that are cut by occasional faults and subsidiary structures. Throughout the Orinda area the long, steep-sided ridges and valleys are the dominant topographic features and are underlain by the Orinda Formation. Many ridges are the crest of a synclinal trough and the beds dip into the slopes, while others are anticlinal or fault controlled.

The folded Berkeley Hills are dissected by intermittent streams that generally flow in the northwest-trending valleys which are underlain by deeply eroded, soft clay shale and sandstone beds. The Orinda Formation and associated rocks were uplifted, folded, and faulted a few million years ago (Late Plio-Pleistocene time) and are still under active adjustment due to both tectonics and erosion. Elevations range from 600 to 1100 feet on the ridges; homes are constructed throughout the slopes and ridges as well as in the valleys.

The region has a mean annual precipitation of 26 inches per year, although heavy storms and prolonged periods of rainfall are frequent, such as the 13.8 inches of rainfall recorded during a 4-day period in October 1962, with 8.4 inches during one 24 hour period. The maximum precipitation for a rainy season is the 46.7 inches recorded for the 1957-1958 season--a period in which many landslides occurred throughout the Orinda Formation rocks (including Pfeiffer and Hughes described herein).

Orinda Formation

The Orinda Formation comprises the principal rocks throughout the Orinda community and occurs everywhere throughout the Pfeiffer property and vicinity. The formation has been studied with particular emphasis on its landslide potential by Radbruck and Weiler (1963) and the reader is referred to this paper for a detailed discussion and extensive bibliography.

Generally, the Orinda Formation consists of conglomerate, sandstone, and clayey shale with minor amounts of limestone, lignite, and tuff. Frequently, the beds are lenticular or gradational and change rapidly, such as from shale to conglomerate. Most of the rocks including the sandstone and conglomerate contain some clay which is commonly montmorillonite; this clay is reported to expand from 20 to 100 percent when saturated (Kachadoorian, 1956). The shales vary from silty clay to clayey fine sandstone and are very thin bedded or laminated.

The bedrock is commonly weathered to a depth of several feet, and on slopes is invariably overlain by several feet of colluvium. Weathered clayey shale generally consists of sandy silty clay with no visible inherent structures.

Landslide Conditions

The Orinda Formation is famous for its landslide topography and slide-prone soil and rock. The type of landslide which is most common and widespread in the Orinda Formation is a slump in the upper part and an earthflow in the lower part of the displaced mass. Many slump and earthflow slides in the Orinda Formation involve bedrock which is loosely consolidated and physically resembles a soil, as used by engineers, rather than a dense rock as used by geologists. Landslides in the formation occur as single, multiple, or coalescent slides according to the classification by Varnes (1958). The slide that damaged the Pfeiffer property is a multiple one and consisted of two well-defined slides and crowns with several smaller ones as shown on Figure 1. Coalescent slides are more characteristic of the Orinda Formation. They occur in the many basin-shaped ravines which are enlarged by the continuous creeping and slumping of their sides.

Some landslides in the Orinda Formation develop in one short period of time; they are subsequently stabilized by nature and do not move again. Historical records and personal observations, however, show that by far the greatest number of slides in the Orinda Formation move periodically and are reactivated over a period of years. True, a large movement may take place during an unusually wet season, but small increments of movement invariably occur during the drier seasons. Many soil slides move on the upper weathered surface of clayey bedrock. Less frequently the slide plane is in soft, clayey sandstone or conglomerate. The easiest slides to develop begin in clay shale and some of these move over sandstone and conglomerate beds according to Radbruck and Weiler (1963).

The depth of most of the landslides within the Orinda Formation range from 10 feet to more than 100 feet in depth. Most landslides show evidence of both slump and earthflow movement.

PFEIFFER PROPERTY AND LANDSLIDE

Geologic Features

The Pfeiffer property and vicinity is located on the western slope of a northwest-trending ridge of folded Orinda rocks. The dwelling at No. 1 Easton Court is situated on the side of the slope with a stream canyon below to the west and the ridge crest to the east and 250 feet higher in elevation. The principal topographic and geologic features of the Pfeiffer property and vicinity are shown on Figure 1; the surface is underlain by a mantle of soil and colluvium up to 10 feet thick.

The main rocks of the Orinda Formation in the vicinity of the site are described in the explanation, Figure 1. The rock sequence which dips up to 45° westward and strikes about N. 38° W. consists of alternating beds of siltstone, fine-grained sand or sandstone, fine-grained conglomerate, clay, and clay shale. The clay beds are yellow to grayish, soft and deeply weathered. The conglomerate is lenticular and interbedded with sandstone and clay; beds average 6 to 8 feet thick, are massive and weak to firmly cemented. The sandstone is fine-grained, soft and clayey; beds average 7 to 8 feet thick.

The tilted Orinda rocks at the site are dipping parallel to, but more steeply than, the natural slope. The landslide has developed somewhat parallel to the dip of the beds (Fig. 1), but has been influenced by a prominent joint set that trends N. 05° to 20° E. and dips 65° to 75° eastward. A second and less pronounced set trends N. 82° E. and dips 65° northward. The joints occur 1 to 3 feet apart and the joint planes are invariably coated with clay.

It is interesting to observe that an old landslide had been active at this site previously and is clearly recorded on aerial photographs taken in 1946. Furthermore the construction of Hall Drive thereafter by Contra Costa County undoubtedly exposed additional evidence of the ancient landslide. Such warnings were known and should have been utilized by the builder. In addition a slide in 1951-1952 on Hall Drive displaced part of the area within the slide mass of 1958 (Fig. 1).

Field Investigations

The landslide on the Pfeiffer property was initially investigated by representatives of Woodward-Clyde-Sherard and Associates, Oakland, and described in three memorandum reports submitted to Rana Builders, Inc., of May 16, June 13, and October 3, 1958. Raymond Lundgren, soils engineer, described the extent of the slide movement in the October report as interpreted from the data of four test borings (as shown on the geologic cross section Fig. 2). Lundgren included an item for stabilizing and repairing the subsurface foundation of the Pfeiffer dwelling in his cost estimate to repair the damage. The drill holes encountered a layer of firm, wet, yellow-gray silty clay at depth and was estimated to be the slide plane (Fig. 2). Underlying the clay is a moderately dense yellow-gray siltstone. The reports further concluded that the slide above Hall Drive was an independent slide from the active mass below the east side of Hall Drive; they recommended stabilizing the lower slide by excavation to below the slide plane and replacing it with a controlled backfill. The cost estimate for stabilizing the lower slide was $15,000 while the estimate for stabilizing both the lower and upper slides was $25,000.

In November 1959 the author undertook a geologic study of the Pfeiffer property and vicinity at the request of the owners and their legal counsel. This study included a review of earlier exploration and reports by Woodward-Clyde-Sherard and Associates, a study of the site conditions in the field, and the preparation of a plane table map of the property and vicinity as given on Figure 1. On the basis of the surface mapping and test borings the geologic cross section was constructed (Fig. 2). This field study served to delineate the need for and location of additional subsurface exploration (holes A, B, C). Subsequently the maps and drawings prepared were used as exhibits when appearing as an expert witness at court proceedings.

In April 1960 the firm of Richard S. Harding and Associates, San Francisco, drilled three test borings at the site under the supervision of Oliver Merwin, soils engineer. These test borings were located nearby the cross section alignment (Fig. 2).

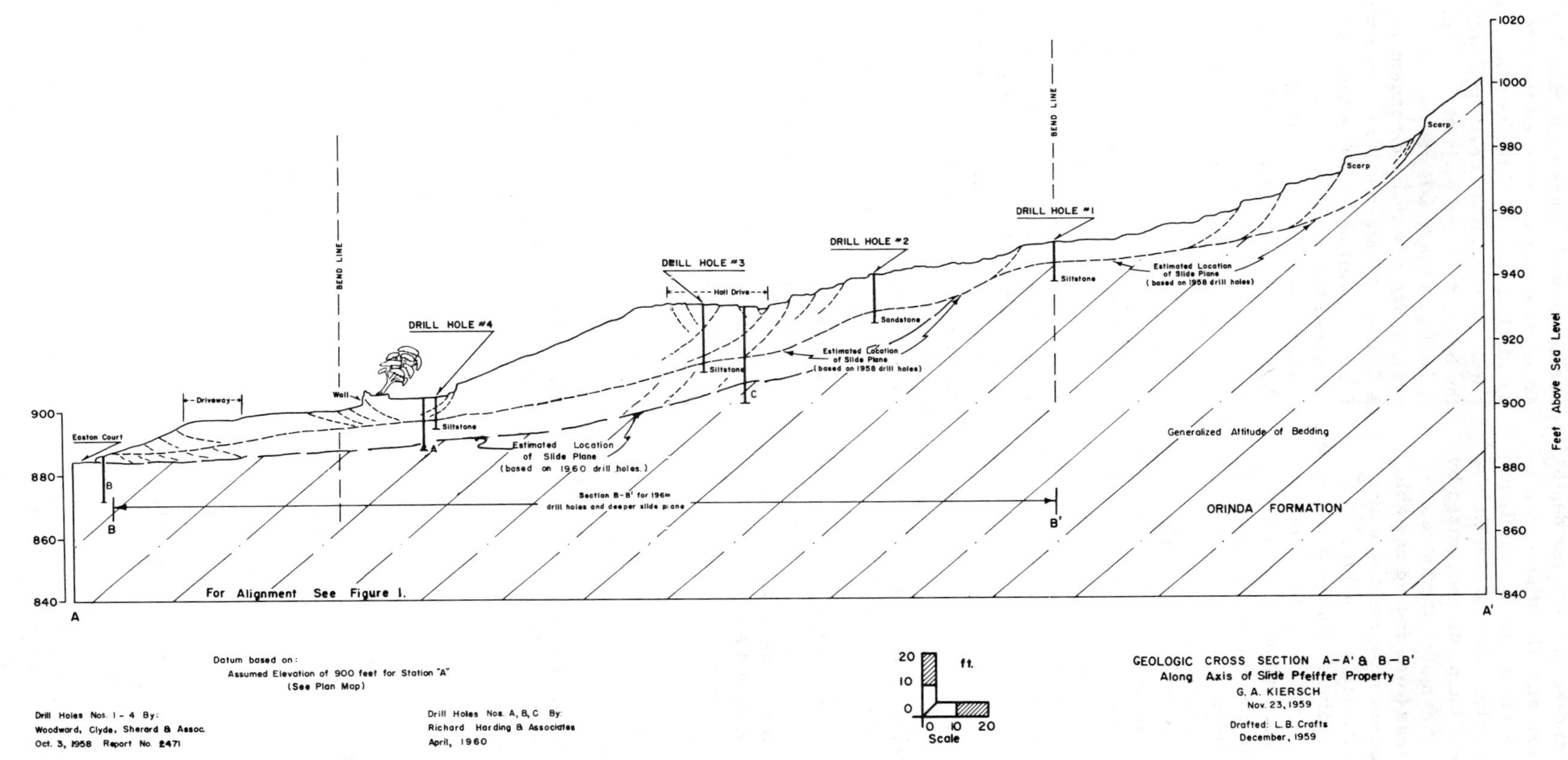

Figure 2. Geologic cross-section A-A' along the axis of slide through the Pfeiffer property.

The holes (A, B, C) were drilled to a greater depth than the holes of 1958. The significant new data realized by the 1960 test borings was the confirmation of a suspected second slide plane in the Orinda rocks located beneath the slide plane recognized by the 1958 drill holes. Furthermore, it was determined that the landslide mass was indeed unstable throughout its length as offset movement was observed to occur within individual test holes.

Geological data combined with the test boring information of 1958 and 1960 demonstrated that the upper and lower slide masses were connected and parts of one active landslide as shown on Figure 2. Consequently, the only means of protecting the Pfeiffer property against future slides would be an effective stabilization of the whole mass. It is interesting to note that a part of the slide mass was reactivated in 1962 during the unusual rainy season (Radbruck and Weiler, 1963), even though stabilization measures had been taken after the 1960 judgment awarded repair damages to Pfeiffer.

Geologic Causes of Slide

The multiple landslide which affected the Pfeiffer property and vicinity (Fig. 1) was caused by a series of geologic events, both ancient and current, which combined to create the sliding of April 1958. The numerous causes that contributed to sliding are: (1) The slide prone characteristics of the weathered Orinda rocks and soil, and the steep natural slope above the Pfeiffer property. (2) The soft bedrock sequence dips steeper than, and somewhat parallel to, the natural slope (20° to 40°). (3) Two sets of prominent joints fracture the Orinda beds and thereby enhance the tendency to slide downslope. One set dips roughly parallel to the sliding. The joint planes are invariably clay filled which further contributes to the susceptibility for sliding. (4) An old landslide was active within the limits of the 1958 slide mass. Aerial photographs of 1946 delineate this earlier slide. (5) The heavy inflow of surface runoff and the ground-water saturation of the underlying porous conglomerate and sandstone beds contributed to the tendency for sliding. The buoyancy effect due to hydrostatic pressure was further assisted in sliding by the abundance of clay seams and beds of clay shale throughout the bedrock. (6) Excessive runoff due to heavy storms created an abnormally high ground-water level in the bedrock during the spring of 1958 (46.7 inches of rain recorded during the 1957-1958 rainy season). Besides the hydrostatic pressure created, the high ground-water level contributed to weakening the clay shales of the bedrock sequence by lowering the cohesive and shear strength of the clays. (7) The swelling characteristics of the montmorillonite clay throughout the bedrock gives rise to pressures and heaving when wet, and shrinkage with cracking during dry seasons. (8) A small side canyon crossed the site more or less parallel to the dip of the bedrock. Consequently, small slump slides developed along the strike of the beds, assisted by joints and bedding planes, and moved into the small canyon. (9) The construction of Hall Drive with its embankment fill placed added weight on the steep natural slope directly above the Pfeiffer dwelling. (10) When the building contractor prepared the foundation site for the Pfeiffer dwelling, he cut away part of the toe of the natural slope. This contributed to further weakening the steep slope. (11) Unusual quantities of water infiltrated the hillside slope when a major water main of EBMUD (East Bay Municipal Utility District) was broken by the sliding across Hall Drive. (12) The progressive headward erosion of the multiple slide with its progressive slumping at the crown, placed an added dead weight of soil at the top of the over-all unstable landslide mass, which is over 560 feet long and 360 feet wide. This weight and the loss of friction within the mass due to the high ground-water level, combined with the inherent geologic conditions of the site to trigger a recurrence and movement of the full slide mass in April 1958.

The common external cause for triggering slides, earthquakes, was not active. No earthquakes were recorded during the period April 1 to 4, 1958 in northern California.

LANDSLIDES, INSURANCE AND GEOLOGY--A NARRATIVE

Judgment--Pfeiffer Case

The court on August 16, 1960, awarded the Pfeiffer's $23,000 to stabilize the land beneath the house and $8,000 to repair the house structure. The judgment specified that the $31,000 awarded to the insured was for restoring their dwelling to a safe condition and equivalent to the conditions of house and foundation prior to the landslide movement.

The case of Pfeiffer versus General Insurance Corp., N.D. Cal. S.D. 185 F. Supp. 605, (1960), revolved around the question of how to interpret the term "dwelling" in an insurance policy which included a provision to compensate for "all physical loss" and damage due to landslide. The Pfeiffer home was severely damaged by a landslide and the foundation was extensively damaged and caused to be unstable. The insurance policy covered payment for damages to the "dwelling" and did not go on to define whether that included the land and foundation as well as the structure itself. This ambiguity was at the heart of the court controversy. The court reasoned that "no amount of repairs to the present structure alone will cure the damage or replace the dwelling until the earth movement under the structure is stabilized" (Pfeiffer, 1960, p.608). Therefore, it is necessary that the land be included in the meaning of the word "dwelling" unless the policy is to be interpreted as illusory and gross injustice be suffered by the homeowners. The court declared that it was absurd to ask the owners to repair their house and garage on a mass of shifting earth where it appears certain that additional slides will occur, imperiling the occupants and causing them to suffer further damage.

As to the wish of the insurance company to cancel the policy, the court cites Harman versus American Casuality Company of Reading, Pennsylvania, C.D.Cal.S.D. 155 F.Supp.612, (1957) for support in its ruling that once the damage accrued, the liability of the insurance company becomes a contractual obligation; cancellation of the policy would not affect rights to coverage for events that have already occurred.

Pfeiffer, through geological and engineering testimony, established that the land under and about the house was not stable, nor would it stabilize itself. Furthermore, the land and foundation would require shoring up, and other engineering techniques and requirements to stabilize. Although testimony of the geologist disclosed that further slippage is imminent, no one could predict the precise moment when the event would take place. Geologic testimony clarified that the land beneath the house was unstable and was part of a slide known to exist in 1946 which is presently active. Although the slide may be momentarily halted, it will continue on in lesser or greater degree until eventually the movement culminates in a dramatic sequence and failure to the whole area affected by the initial slide. The hazards incident to maintaining a domicile for a family under such circumstances are obvious. Such circumstances would cause gas pipes and water and sewer connections to be severed and other incident damages to the premises, not to mention the danger of collapse, which would render them unlivable. Unless steps are taken by the insurance company to stabilize the land, as well as to repair the premises, the insurance policy is worthless to the homeowners (Pfeiffer, 1960, p. 609).

Relevance of Pfeiffer to Other Cases

The case which directly preceded Pfeiffer and lent the authority of precedence to the Pfeiffer opinion regarding cancellation of the policy concerned the famous landslide in the Portuguese Bend area, California, known as the Portuguese Bend Club. This area is southerly of Los Angeles and westerly of the City of San Pedro. Two fire insurance policies had been issued covering two family structures located in this area, when, in the fall of 1956, a massive and continuing land movement commenced affecting a large number of properties, including the two aforementioned. With full knowledge of the existing land movement, the company issued notices of cancellation of both policies. The question arose as to whether an insurance company could legally do this. The Harman (Supra) court held that liability, where a continuing loss had already commenced, could not be terminated or avoided. The insurance company would have to continue to cover the loss until the loss had occurred in its entirety or the cause of the loss had ceased.

Once the event insured against has occurred, there is nothing the insurance company can do to alter the policy; a contractual obligation has been formed which must be carried out with respect to a loss that is already in being. All that remains at that stage is a determination of the extent of damage.

The Harman case, like Pfeiffer, involved a policy which included an "all physical loss" provision (see APPENDIX) which was added to a fire insurance policy. A court will always endeavor to resolve doubt and ambiguities in the interpretation and construction of policies in favor of the insured homeowner (Harman, 1958, p. 614). This approach prevents insurance companies from taking an unjust and unfair advantage of the insured and thereby weaken the purpose for which the policies were issued. The recent development of the "all physical loss" provisions in insurance policies aids the court in its effort toward interpreting the policy in favor of property owners.

Harman held that to permit revocation of a fire policy which includes coverage against all physical loss while the contingency insured against is occurring would be fraud on the insured homeowner. This principle has been upheld to the present.

Pfeiffer Case as Precedent

Harman preceded Pfeiffer and, admittedly, the Harman court at that time (1957) had "been faced with the problem of determining the law without much aid from precedents." However, the case that directly followed Pfeiffer in this progression of the development of landslide insurance, the case of Hughes versus Potomac Insurance Company, had the authority of Pfeiffer to support it.

Not only could Pfeiffer be cited as precedent for the point of law in Hughes, but for the same type of landslide as well! The point of law at the heart of the case was again the interpretation of the term "dwelling" in the insurance policy. The defendant-insurance company, after having a judgment rendered against them in the trial court, appealed on the ground that the lower court erred in finding for Hughes, the homeowner. The insurance company hoped to have the decision reversed in the appellate court on the grounds that the policy is distinguishable from the Pfeiffer policy because an amendment to their policy refers to the "dwelling building" rather than to the "dwelling." Hughes pointed out that the basic policy uses only the term "dwelling," but the insurance company insisted that the amendment superceded the basic policy with use of the term "dwelling" and thus resolves any ambiguity inherent in this term. Here, again, the policy includes an "all physical loss" endorsement (see APPENDIX) which set forth the maximum sums for which the company would be liable "replacing the damaged parts of the building structure."

In addition to the question of the interpretation of the term "dwelling," the question of proximate cause (see APPENDIX) of the landslide was an issue in the Hughes case.

The Hughes were the owners of a home in Walnut Creek, California (5 miles east of Orinda) purchased in 1947. Beginning October 1, 1956, they were insured by the Potomac Insurance Company against all risks of physical loss of and damage to their dwelling, appurtenant private structures, household and personal property on the premises and additional living expenses caused by the loss of the dwelling. The policy excluded from coverage all losses caused by: "surface waters, flood waters, waves, tide or tidal wave, high water, or overflow of streams or bodies of water, all whether driven by wind or not."

The Hughes' home was located 30 feet from the banks of Las Trampas Creek which bounded their dwelling to the rear and on one side. Early on the morning of April 1, 1958, a shudder and shaking were heard and felt, and immediately thereafter the earth to the rear of and partially underlying the owner's house slid into the creek; their home now stood on the edge of, and partially overhanging a newly formed 30-foot cliff. This landslide resulted in the loss of a block of earth 30 feet wide and 100 feet long, and deprived the owners of subjacent and lateral support essential to the stability of their house.

When the Hughes filed for damages the insurance company denied all liability. The company claimed that the landslide had been caused by a risk specifically excluded by the policy; namely, all loss and damage caused by "flood" or "high water." The insurance company maintained that the evidence conclusively established that the homeowners' loss was due to one or both of these factors; hence that the trial court erred in awarding the Hughes any damages whatever.

The court differed with the insurance company's interpretation of the facts. The court decided that the record did not bear out the company's contention. The testimony of respondent Walter Hughes revealed that it had been raining steadily for about six weeks prior to the occurrence of the landslide. However, the water level in Las Trampas Creek was still 10 feet below the top of the bank at the time of the accident. Although the water continued to rise for two days thereafter, it never attained a greater height than two feet from the top of the bank. Under such circumstances, flood waters cannot be held the cause of the slide. There is no evidence that the creek ever overflowed its banks, either before or after the accident occurred. Therefore this was not a "flood" in the sense excluded by the policy and was evidence in favor of the homeowner.

On the issue of "high water," the testimony of two expert witnesses was introduced into evidence. Homeowners' witness, John Trantina, testified that in his opinion the landslide was caused by the build-up of ground-water pressure. He further testified that the ground water came from two sources; surface runoff and water which infiltrated laterally into the soil from the stream. Eventually the absorption of so much water into the soil resulted in an excessive hydrostatic pressure which caused the failure. Trantina testified that erosion from the stream itself contributed to the landslide only in the sense that it furnished some water to the soil and thus caused an increase in hydrostatic pressure.

The insurance company's expert, Hugh O'Neil, was of the opinion that the large quantity and high velocity of the flow of water in the creek eroded the banks and thus caused the landslide to occur. He also testified, however, that the rain might have been a contributing factor, since it could have weakened the soil and thus allowed it to be more easily eroded by the stream.

The court found, in view of the conflicting expert testimony, that there was ample evidence to support the lower court's finding; the landslide was caused solely by "the soaking of the land from heavy rains with consequent hydrostatic pressure and loosening of the soil." Even if the lower court could be deemed to have erred in finding that the landslide was in no way caused or contributed to by the water in the creek, the higher court ruled that the insurance company was not in any way prejudiced thereby because of the following legal principle:

In the law of insurance it has been held that "when two causes join in causing an injury, one of which is insured against, the insured (party) is covered by the policy...." Zimmerman versus Continental Life Insurance Company, 99 Cal.App. 723, 726, 729 p. 464 (1929).

The expert for the insurance company conceded that continued surface runoff may have contributed to the landslide by weakening the soil. Homeowner's expert testified that the slide was due to water pressure in the soil, and the stream contributed to the accident in only a minor degree. Since the insurance company at no time denied coverage for losses caused by the pressure of ground water, the court was satisfied that the rule of the Zimmerman case should control; coverage of the homeowner's loss was not barred by the "high water" exclusion.

The insurance company cited the Insurance Code, (section 523), which provides that if a peril is excluded by the insurance policy, and there is damage which could not have occurred without this excluded peril occurring, the insurance company is not liable for the loss; this is true even if the loss was not immediately or proximately caused by that excluded peril.

In the Hughes case, expert testimony indicated that the loss would have occurred even in the absence of the increased flow of the creek. (Flood damage was the peril excluded from that insurance policy.) Since geologic testimony established that the creek contributed to the slide in only a minor way, the inference is overwhelming that the accident would have happened even though the creek remained at a more normal level. Therefore, the trial court properly found that the homeowners could recover because the flood exclusion in the policy was not applicable to the geologic facts of the case.

The insurance company then tried to make the argument that the policy only covered the building structure or house, and did not insure the foundation of soil or ground underneath the house. The company quoted the policy headed "Property and Interests Covered:"

Coverage A--Dwelling: Dwelling building described in the declarations, including its additions and extensions, architects' fees, lawns, building equipment, fixtures and outdoor equipment pertaining to the service of the premises (if the property of the owner of the dwelling), while located on the premises of the described dwelling or temporarily elsewhere, and all lumber and materials on such premises or adjacent thereto incident to the construction, alteration or repair of such dwelling. TREES, SHRUBS, OR PLANTS ARE NOT COVERED.

This endorsement, the company contended, listed all the items to be covered; since no mention is made of the ground or soil underlying the building, the "dwelling" coverage is not applicable to a loss of land, even if the effect of this loss is to deprive the house of its foundation support. The company also claimed that part of the "all physical loss" dwelling endorsement, which refers to replacing the damaged parts of the "building structure," clearly indicated that coverage was to be limited to the building alone, and was not to extend to the underlying earth.

Hughes answered that the variation in terminology creates ambiguity and fails to clearly limit coverage to the building structure alone. They also stressed that since the policy specifically excluded trees, shrubs, or plants, that the company should have added an exclusion applicable to underlying ground had it so desired to limit the policy's coverage. Also, the "all physical loss" endorsement excludes from the replacement of the "building structure:"

excavations, underground flues and pipes, underground wiring and drains, and brick, stone, or concrete foundations, piers and other supports which are below the surface of the ground...

Therefore, argued Hughes, the only logical meaning to be given "other supports" is the support furnished by the ground itself and that includes the underlying soil and/or rock.

The case of Snapp versus State Farm Fire and Casualty Company, 206 C.A.2d 827, 24 Cal. Rptr. 44 (1962), is another example of evidence, dependent for its substantiation on geologic testimony. In that case, the homeowners' residence, and those of at least two of their neighbors, were founded on a man-made fill which, as subsequent investigation revealed, was poorly built. Due to the unstable condition of the fill and an unusually heavy rainfall, the landfill beneath the homeowners' residence began to move laterally during the term of the policy; this movement resulted in damage to the structure and its foundation. The as-contructed foundation, though apparently adequate to support the building prior to the land movement, proved inadequate once the earth movement commenced. The insurance company did not claim that the owners of the property were in any way personally responsible for the poor quality of the fill or they knew of its inherent danger and the potential inadequacy of the building foundations.

The insurance company contended that the policy did not cover the loss involved arguing that it was not a "fortuitous" event (see APPENDIX) and relied upon the lower court language which was: "Because of the instability of said fill, the earth movement which constituted the landslide was inevitable." However, the lower court also stated later: "Said landslide was and is a fortuitous event and is not a risk excluded by the terms and conditions of the contract of insurance entered into by the parties." This latter finding is supported by the evidence.

The lower court considered only damage done to the "habitable portion of the premises" and excluded damage to the "foundation." Further, they only considered the damage which occurred prior to the expiration date of the policy. The latter action is inconsistent with the lower court's finding with regard to the landslide, "this movement is still active and is without definite prospect of stabilization." All the experts who testified agreed that unless a way were devised to halt such movement or to protect the structure therefrom, the building eventually would be destroyed completely.

Providing sufficient information were available to geological experts, the high court states: "It is possible or probable that all earth movements might be forecast with accuracy. Further, after any movement has occurred, this event might be considered 'inevitable' with semantic correctness (see APPENDIX), but such 'inevitability' does not alter the facts at the time the insurance contract was made. In actuality, the event was only a contingency or risk that could conceivably occur within the term of the policy."

The issue of what date the insurance company was liable from arose in the lower court. The distinction was then made by the upper court that while the loss sustained up to a given date may have been "ascertainable" as the lower court decided, the real question was whether the liability of the insurance company was "terminable" on such date, or whether the company was liable for the "continuing damage or loss;" this, they said, was a legal rather than factual issue. Furthermore, to permit the company to terminate its liability while the fortuitous peril which materialized during the term of the policy was still active would not be in accord with previous precedents (such as Pfeiffer) or with the common understanding of the nature and purpose of insurance; it would defeat the very purpose for which the premiums were paid. Therefore, it is immaterial that the full extent of the damage may or may not be fully ascertained at the end of the policy period. The court concluded that the insurance company could not terminate its liability once the hazard insured against had occurred. The upper court reversed the lower court's decision and award the homeowners $25,000. In the Hughes holding the reason that such insurance is issued and purchased was stated as . . . "it is difficult to imagine how a homeowner could be expected to build a foundation so extensive as to protect against all possible land movements." Consequently, any argument advanced by an insurance company that the original foundation was weak or faulty in its construction and cannot adequately protect the premises from vertical or lateral earth movement, will not stand up in court.

CONCLUSION

The insured, the Robert J. Pfeiffer's, were given a judgment on August 16, 1960, by Judge Harris. He awarded them $23,000 to stabilize the land and foundation underneath the house and $8,000 to repair the house structure. The total of $31,000 awarded to the insured was for restoring their dwelling to a safe condition and equivalent to the conditions of house and foundation prior to the landslide movement.

Geologic testimony was central to the outcome of the controversy between Pfeiffer and the General Insurance Company to determine: the causes of gravity sliding; the possibility of recurrence; and the advisability of rebuilding at the site. Further testimony was judicious on whether there was negligence on the part of the builder or insurance company because they did not realize the geologic conditions would lead to large-scale gravity sliding. The multiple landslide that damaged the Pfeiffer dwelling was caused by a combination of geologic features and conditions inherent to the site plus the effects of hydrostatic pressure and a decreased shear strength of the eventual slide mass resulting from the high ground-water level created by excessive runoff.

The Pfeiffer case has since served as precedent in legal cases and the insured homeowner will recover for all damage to both his building and to the foundation and land that surrounds or is beneath the building. Furthermore, an insurance company will be liable for all continuing damage to the dwelling (this includes the building and its foundation of rock and/or soil) as long as such damage is due to vertical or lateral earth movements which are covered by the policy. Thus, an insurance company cannot cancel a policy and terminate its liability after sliding is in progress (the hazard insured against). Other important precedents in landslide coverage have been decided in favor of the homeowners on all major issues.

CASES CITED

The G. R. Booth, 171, U.S.450 (1898)

__________, in J. C. Bancroft Davis (Reporter), 1898, United States Reports: v. 171, p. 450-462. New York, Banks and Brothers, Law Publishers.

Harman versus American Casualty Company of Reading, Pennsylvania, C.D. Cal. S.D., 155 F.Supp. 612 (1957)

__________, in Federal Supplement, 1958, v. 155, p. 612-615. West Publ. Co., St. Paul, Minn.

Hughes versus Potomac Insurance Company, 18 Cal. Rptr. 650, 199 C.A. 2d 239 (1962)

__________, in Nankervis, Wm., Jr., (Reporter) 1962, Reports of cases determined in The District Court of Appeals, state of California: Appendix Calif. Supplement, v. 199 2nd series, p. 239-254. Bancroft-Whitney Co., San Francisco, Calif.

Pfeiffer versus General Insurance Corporation, N.D. Cal. S.D., 185 F. Supp.605 (1960)

__________, in Federal Supplement, 1960/1961, v. 185, p. 605-609. West Publ. Co., St. Paul, Minn.

Silver Eagle Company versus National Union Fire Insurance Company, 423 p. 2d 944 (Ore. App., 1967)

__________, in Pacific Reporter, 2d series, 1967, v. 423, p. 944-948. West Publ. Co., St. Paul, Minn.

Snapp versus State Farm Fire and Casualty Company, 24 Cal. Rptr. 44, 206 C.A. 2d 827, (1962)

__________, in Nankervis, Wm. Jr., (Reporter) 1962, Reports of cases determined in The District Court of Appeals, state of California: Appendix Calif. Supplement, v. 206 2d series, p. 827-834, Bancroft-Whitney Co., San Francisco.

Williams versus Southern Railway Company, 55 Tenn App 81, 396 S.W. 2d 98 (1965)

__________, in South Western Reporter, 2d series, 1966, v. 396, p. 98-102. West Publ. Co., St. Paul, Minn.

Zimmerman versus Continental Life Insurance Company, 99 Cal. App. 723, 279, p. 464 (1929)

__________, in Pacific Reporter, 1929, v. 279, p. 464-466. West Publ. Co., St. Paul, Minn.

REFERENCES CITED

Kachadoorian, Reuben, 1956, Engineering geology of the Warford Mesa subdivision, Orinda, California: U.S. Geol. Survey Open-file Report, Dec. 18, 1956, 14 p.

Long, John D., and Gregg, Davis W., 1965, Property and liability insurance handbook: Illinois, Richard D. Irwin, Inc., 1233 p.

Magee, John H., and Serbein, Oscar N., 1967, Property and liability insurance: Illinois, Richard D. Irwin, Inc., 929 p.

Michelbacher, G. F., 1957, Multiple-line insurance: New York, McGraw-Hill Book Company, Inc., 643 p.

Radbruck, Dorothy H. and Weiler, Louis M., 1963, Preliminary report on landslides in a part of the Orinda Formation, Contra Costa County, California: U.S. Geol. Survey, Open-file Report, 35 p., 1-map, 1-table.

Rodda, William H., 1956, Fire and property insurance: New Jersey, Prentice-Hall, Inc., 548 p.

Varnes, D. J., 1958, Landslide types and processes, in Eckel, E. B., Editor, Landslides and engineering practice: Natl. Research Council, Highway Research Board Spec. Report 29, p. 20-47.

Webster's New International Dictionary, 2nd Edition, 1956, Springfield, Massachusetts, Merriam Co. Publishers.

APPENDIX

INSURANCE POLICIES

Multiple-line Insurance

This type of insurance gradually modified the traditional "American system" of dividing insurance by subject matter; a multiple-line insurer occupies the entire file of insurance (outside life insurance) and can bring together in "package policies" various combinations of covers; it is insurance against loss caused by any one of several perils traditionally insured separately (Michelbacher, 1957, p. 17).

Underwriting

The underwriter is primarily responsible for the assumption of a certain sum or risk by way of insurance, made by his insurer. As the need for insurance expanded, the consequent necessity for large volumes of capital and regulatory authorities led to large corporate insurers and specialized functions for its underwriters. Those functions concern mainly the selection and rating of risks and continued observation and management of risks on the insurer's books. Basically the underwriter must determine which risks will be accepted by the insurer and the basis on which these risks will be insured (Magee and Serbein, 1967, p. 796).

Endorsement

The term "endorsement" developed from the practice of adding clauses to a contract after it had been issued, for the purpose of modifying the contract in some way, to restrict or enlarge the scope of coverage. It presumes a change in the conditions of the policy after preparation. Certain endorsements, such as those that extend the perils covered, may be lengthy, and attached when the contract is issued. Even here, the endorsement may be presumed to be an extension of the scope of the original contract, although prepared at the inception of the agreement and attached to the contract before delivery (Magee and Serbein, 1967, p. 179).

Specified Perils Policy

In such insurance policies, the perils insured against are named, and the insured and the insurance company can see just what coverage is provided. Rating and underwriting the policy are simplified because each peril is listed and known.

The "all risks" type of insurance coverage on the other hand, is difficult for insureds to understand because it does not list the perils insured against; and at times it is difficult for insurance companies to rate and to underwrite, since possible losses to a property from the unknown hazards cannot be anticipated with any degree of accuracy (Rodda, 1956, p. 185).

"All Risks" Endorsement

This type of endorsement was first made in 1950, and was called the "comprehensive dwelling endorsement;" it extends the coverage of the fire and extended coverage policy "to include direct loss resulting from physical damage by all other risks to the structures" described in the policy, subject to certain exclusions. It is broader than specified perils coverage because it covers many unknown and unpredictable losses (Rodda, 1956, p. 187).

"All Physical Loss"

This phrase is synonymous with "all risks" endorsement.

Dwelling Building(s) Special Form

This form is attached to the standard fire insurance policy and converts the fire policy into an "all risks" type of endorsement on a dwelling building. It insures against all risks of physical loss, except as excluded (Rodda, 1956, p. 193). The exclusions determine the coverage and a typical version of this form used in many states excludes loss to: "retaining walls not constituting part of a building when such loss is caused by landslide, water pressure, or earth movement" (among other exclusions). Use of this form to alter fire insurance policies is the most typical form of "landslide insurance" in operation today, and landslide coverage, as can be seen by the abovementioned exclusion, is often greatly limited. A separate landslide insurance policy is virtually nonexistant.

Responsibility for Determining Coverage

Under a specified perils policy, it is clearly the insured's responsibility to show that his loss was caused by one of the perils named. Under an "all risks" type of policy the insured still has the burden of proving his loss under the policy, but there is a presumption of coverage unless the insurance company can show that the exclusions in the policy apply to the loss. In case of doubt as to the exact cause of a loss, the insured is in a better position with an "all risks" type of coverage than he is with a specified perils type (Rodda, 1956, p. 185).

Earthquake Insurance

Policies covering earthquake insurance are written in different ways, depending upon the geographical area. Many west coast states write it as an endorsement to the standard fire policy. In other parts of the United States, earthquake insurance is written as a separate policy. In several areas dwellings can be insured against earthquakes by an endorsement to the standard fire policy where, in general, any damage caused to the land itself is excluded. This is relevant to landslides because the shaking and shearing caused by earthquakes occasionally results in serious damage to the land through the formation of faults and open fractures, hummocky topography and slumps, sand boils, springs, and so on. Thus earthquakes can indirectly precipitate landslides, and if so, the resulting damage would not be covered by such a policy (Long and Gregg, 1965, p. 107).

Exclusions

In an insurance policy, exclusions are those perils which are specifically excluded from coverage by the policy, and for which the insurance company is not held liable.

CAUSATION

Act of God

In lay usage it is an occurrence of nature that cannot be foreseen or prevented such as a storm, lightning, flood, or earthquake.

According to legal usage it is an inevitable accident; such an extraordinary interruption of the usual course of events that no experience, foresight, or care which might have reasonably been expected, could have foreseen or guarded against it, as lightning, tempests, and so one. No one is held liable for damage by an act of God, except under special contract. It includes every loss by a natural cause in which man, by act or negligence, has had no part (Webster, 1956).

Natural cause

As generally used, a "natural cause" is the usual, ordinary, regular force that produces a specific effect. It may be a "proximate cause" of some event, but if used geologically, generally means a cause occurring through natural processes as opposed to man-made artificial processes.

Proximate cause

According to legal usage, proximate cause is an ordinary, naturally explainable sequence which produces a specific result, and where no independent disturbing agencies are intervening. Proximate cause may be through man-made or natural agencies.

The following definition of proximate cause is frequently quoted in insurance cases:

> The question is not what cause was nearest in time or place to the catastrophe... The proximate cause is the efficient cause, the one that necessarily sets the other cause in operation. The causes that are merely incidental or instruments of a superior or controlling agency are not the proximate causes (or) the responsible ones. Though they may be nearer in time to the result, it is only when the causes are independent of each other that the nearest (cause) is charged with the disaster. (The G. R. Booth, 171 U.S. 450, 457, 1898).

Both Natural and Proximate Cause

Concepts of both natural and proximate cause necessitate geological considerations and testimony for placing liability on one of the parties. For example, large-scale blasting operations, miles away, may be a proximate (though not natural) cause for triggering a slide. Landslides may also result from the artificial removal of lateral support by earth and rock excavations. Such a case was Williams versus Southern Railway Company, 55 Tenn. App. 81, 396 S.W.2d 98 (1965) in which a railroad cut caused houses situated above the excavation to slide downhill and into the cut. The railroad company was held to have caused the landslide by its excavation, and was liable for property damage to the landowner.[1]

An earthquake may be a contributing natural cause of a landslide, but not its proximate cause. Similarly excess rain could contribute to the formation of a landslide, but not proximately cause it. A determination of causation (natural or proximate) is crucial to the question of who is liable for damages in a court of law; therein geologic testimony and considerations are decisive to the outcome of such litigation (see "direct cause").

Direct Cause

To be covered by the standard insurance contract, the loss must be directly caused by one of the perils covered by the contract. Efforts have been made to establish the causal relationship between a peril covered by the contract and the resulting damages. In the case of fire, for example, the fire must be the immediate or proximate cause of the loss, as distinguished from the remote cause. In a train of circumstances culminating in a loss, the proximate cause is held to be the efficient cause; that is the one that sets intervening agencies in motion. The courts have held that causes which are merely incidental to a superior or controlling agency are not proximate causes, even though they may be nearer in time to the result or have in some way contributed to it (Magee and Serbein, 1967, p. 115). Controversy as to cause of the landslide in the Hughes case, 1962, is an illustration of a court's decision involving the concept of proximate cause.

TERMS

Fortuitous

A happening by chance or accident; occurs unexpectedly, or without known cause, chance, (Webster, 1956).

Inevitable

An event that is not evitable or one incapable of being shunned; unavoidable (Webster, 1956).

Inevitable Accident

According to legal usage, it is an accident which is not foreseeable or to be prevented by due care or diligence; nearly equivalent to, though broader than, an "act of God" (Webster, 1956).

[1]Since the movement of Williams' land adjoining the railroad cut was the natural and proximite result of the removal of lateral support by the railroad, no negligence on the part of the railroad company had to be proven by them. According to law, every owner of land has an absolute right to naturally necessary lateral support from the adjoining soil. In this case the land moved of its own weight and pressure due to a man-made cause; the landowner did not contribute in any way to the slide as geologic testimony confirmed; it resulted naturally and proximately from the removal of lateral support. Williams recovered full damages.

MARYLAND STATE ROADS COMMISSION VERSUS CHAMPION BRICK COMPANY--CONDEMNATION OF CLAY-BEARING LANDS

Carl W. A. Supp, (Consulting Engineer - Geologist, Baltimore, Maryland)

Abstract

In 1959 the Maryland State Roads Commission condemned or isolated 55 acres of urban land for construction of a major highway interchange. The owner, Champion Brick Company, rejected a profferred settlement of $178,000, claiming that the taking diminished available deposits of brick clay to a 5-year supply, thus reducing market value of plant and related facilities by $3,500,000. The Commission contended that remaining clay reserves would outlast the residual economic plant life of 23 years, and denied consequential damages.

Pre-trial preparations included topographic mapping of the property; research on local Coastal Plain geology, clay technology, and brick-manufacturing processes; detailed subsurface exploration using various drilling and sampling techniques; laboratory testing (grain size analysis, Atterberg limits, etc.) of nearly 500 samples to determine suitability of remaining deposits by correlation with raw materials actually being used; and quantitative studies to determine depletion rate and volume of suitable and economically recoverable clay. These data substantiated the conclusion that at least 21 years of continued production were assured.

Annotated aerial photographs, geologic sections, and a three-dimensional geologic model were prepared as visual exhibits to facilitate comprehension of fundamental technical data by a lay jury.

Data accumulated by the investigation tend further to challenge the validity of current geologic nomenclature which recognizes three formations (Patapsco, Arundel, and Patuxent) as comprising the Potomac Group in the Baltimore area.

Co-ordinated preparation of legal, engineering, geological, and other technical aspects of the case were rewarded when in 1961 the jury awarded only token damages of $70,000 above the original state offer.

CONTENTS

ILLUSTRATIONS

INTRODUCTION

Issues of the Case

The case of Maryland State Roads Commission versus Champion Brick Company was unusually complex from a legal as well as a purely technical standpoint. The legal issues involved are summarized below by an excerpt from a paper by Carl H. Lehmann, Jr., (1962, p. 70-72), one of the attorneys who represented the State in the preparation of the case and the trial.

The brick company's land consisted of 172 acres, and was the site of subsurface clay deposits which were mined and used to make red face building bricks in the company's manufactory located 2.5 miles away. The State's acquisition of 55 acres of this tract for right-of-way purposes was carried out under Maryland's quick-taking law and was accompanied by the deposit of $178,000 as the estimated just compensation for this land. The company rejected this offer and asked instead for $3,500,000 on the ground that the taking destroyed the utility of the entire brickmaking business. This claim was later reduced to $1,500,000 in the opening statements made at the trial.

The company contended that its compensation should include damages resulting from the reduced amount and availability of its clay supply. It estimated that the plant had a life expectance of 23 years from the data of the taking (December 1959), whereas after the taking the plant could continue operations for only five years before the clay supply in its remaining property was depleted.

After five years, the company contended, equipment and remaining land would have no market value as a brick manufactory, and would have to be valued as disorganized land and buildings.

The question of prime importance was, therefore, can a landowner be compensated for damages to a parcel of land as a result of a taking from another parcel owned by him, and which, although not contiguous, is connected inseparably in the use of the former parcel?

The trial court held that the question of consequential damages should go to the jury, relying in its holding on Baetjer versus United States, 243 F.2d 391 (1944), cert. denied 323 U.S. 772. In substance the court's ruling in the Baetjer case was that damages could be awarded where it was proved that integrated use of the land existed, that the parcels were inseparable and not susceptible of being replaced, and that damage to one must necessarily damage the other. This involved proving unity of use, nonavailability of other suitable clay within an area capable of being used economically, and the diminution of the company's existing supply of clay below the 23-year period of the brick plant's life expectancy. This appears to be the first time in the history of Maryland's eminent domain law that the burden of proof of value has been placed on either party in a condemnation proceedings. Proof of these elements of the case called for extremely comprehensive and technical testimony regarding the artistry and history of brickmaking, and technical aspects of the brickmaking process, the geology of the area and site, the techniques of subsurface exploration, and interpretation of subsurface data.

Acknowledgments

The technical investigations, studies, and related work discussed herein were successfully completed only as a result of a concerted team effort. The writer wishes to express his appreciation for the assistance and cooperation offered by many individuals, particularly Joseph D. Buscher, Special Assistant Attorney General, Maryland State Roads Commission, and members of his legal staff; Special Attorneys Carl H. Lehmann, Jr., James S. Sfekas, and Eugene Ricks; LeRoy C. Moser, Chief, Right-of-Way Division, Maryland State Roads Commission, and members of his staff, Louis A. Yost, Jr., Sidney J. Ward, and Marion Silvestri; Benjamin E. Beavin, Sr., Consulting Engineer, Baltimore, Maryland, and members of his technical staff; Dr. Joseph T. Singewald, Jr. (deceased), formerly Maryland State Geologist; Dr. Maxwell M. Knechtel, U. S. Geological Survey; and Mr. Walter A. Weldon (deceased), Consulting Ceramist, Baltimore, Maryland.

SCOPE OF INVESTIGATION

General

The wide monetary divergence between the state's offer and the company's claim prompted the Commission to decide that a comprehensive geological and engineering investigation of the subject property would be a good investment. The writer, then associated with B. E. Beavin Company, Consulting Engineers of Baltimore, Maryland, was asked to undertake the project. Based upon reconnaissance studies, the following general phases of work were recommended and completed:

(1) Engineering studies--including boundary surveys, topographic and drainage mapping, and quantity estimates.

(2) Geological investigations--including preparation of geologic cross sections and construction of a geologic model.

(3) Laboratory testing program--to establish criteria for determining suitability or unsuitability of materials for face brick manufacture.

(4) Quantitative studies--pointed primarily toward developing a reliable calculation of how much (in terms of plant years) suitable and economically recoverable clay was still available in the remainder tract.

(5) Collaboration with counsel--to brief the State's attorneys on the engineering and geological technicalities involved so that they could effectively present the information in court.

(6) Preparation of report--a comprehensive, well-documented summary of all aspects of the investigation.

(7) Expert witness testimony in court.

Completion of this program required about a year, and many complex and interesting technical problems were encountered, of which some are briefly summarized in this paper.

Preliminary Engineering

Elevations and a precise traverse, referred to U. S. Coast and Geodetic Survey vertical and horizontal control, were carried into the property and tied to the control surveys for the highway interchange, then already under construction. A grid system of 500-foot squares, based on the Maryland State coordinate system, was laid out over the entire tract and permanently marked on the ground. This grid served as the framework for a boundary survey; preparation of a detailed topographic map (Fig. 1); and for location of test borings, test pits, and related features of interest.

STATE OF MARYLAND STATE ROADS COMMISSION	Benjamin E. Beavin Company Consulting Engineers Baltimore, Maryland	
Reviewed & Approved by: ______ Date: ______	Approved by: Benjamin E. Beavin Reg. Prof. Engr. (Md. No. 76) Date: Nov. 9, 1961	Reviewed and Approval Recommended by: Carl W. A. Supp Reg. Prof. Engr. (Md. No. 3772) Date: Nov. 9, 1961
Approved by: ______ Date: ______		
Approved by: ______ Date: ______		

STATE OF MARYLAND
STATE ROADS COMMISSION

STUDY OF PROPERTY
OF
CHAMPION BRICK COMPANY
AT
STEMMERS RUN, BALTIMORE COUNTY, MARYLAND
(MD. S.R.C. ITEM NO. 34248)

SCHEMATIC PLAN
SHOWING LOCATIONS OF CROSS SECTIONS
SCALE 1 INCH = 200 FEET

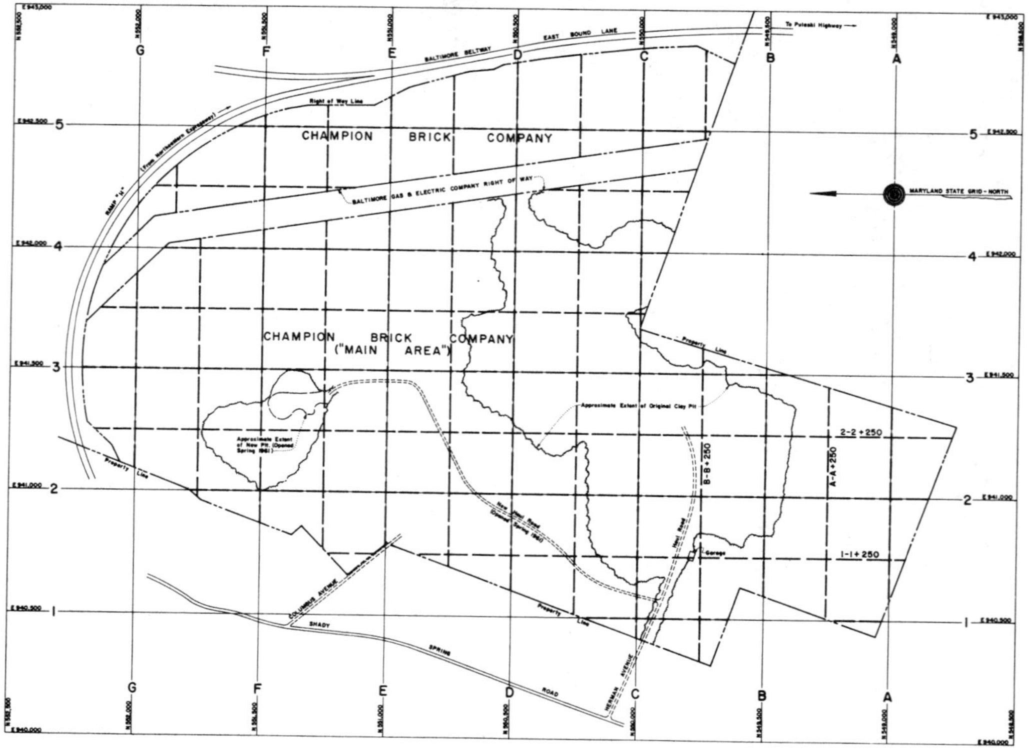

CHAMPION BRICK COMPANY

CHAMPION BRICK COMPANY ("MAIN AREA")

BALTIMORE GAS & ELECTRIC COMPANY RIGHT OF WAY

BALTIMORE BELTWAY EAST BOUND LANE

MARYLAND STATE GRID - NORTH

2-2+250

A-A+250

B-B+250

1-1+250

GENERAL NOTES

1. The bases for soil classifications according to the A.A.S.H.O. (American Association of State Highway Officals) and H.R.B. (Highway Research Board) Systems are presented in detail in the text of the Final Report by the Benjamin E. Beavin Company.
2. An asterisk* shown immediately above the soil classification indicates that the natural color (mainly gray) is such that the color when fired would probably not be acceptable.

LEGEND

- BA-12 (NUMBER) .... 36" Diameter Bucket Auger Boring Supervised by Benjamin E. Beavin Company
- RB-15 (NUMBER) .... 4" Conventional Rig Boring Supervised by Benjamin E. Beavin Company
- SRC-8 (NUMBER) .... Power Auger Boring by Maryland State Roads Commission (S.R.C. borings other than S R C-8 not shown on sections)
- 103 .... Power Auger Boring by Champion Brick Company
- (3) (1-4 inc.) .... Number of Composite Laboratory Samples (Prepared from Boring Samples)
- A-2-4(0) .... H.R.B.-A.A.S.H.O. Soil Classification. Number in parentheses indicates Group Index
- .... Boring on section line showing suitable clay
- .... Boring projected to section line showing suitable clay
- .... Boring projected to section line showing unsuitable material
- .... Boring on section line showing unsuitable material
- .... Suitable clay, economically available
- .... Suitable clay, Not economically available
- V .... Clay Deposit Number

INDEX OF EXHIBITS

	DESCRIPTION	NO. OF SHEETS
I	Map of Property of Champion Brick Company at Stemmers Run, Baltimore County, Maryland (Nov. 1960, Revised Aug. 1961)	1
II	Cross Sections of "Main Area" Showing Subsurface and Clay Suitability Data	8
III	Cross Sections of Area Between G.&E. Co. Transmission Line R/W and Beltway R/W Showing Subsurface and Suitability Data	2
IV	Cross Sections of "Main Area" for Determination of Total Net Quantity of Material Removed from Property	9
V	Study of a Practical Scheme for Drainage of Clay Pits	
	Plan of Ditch Layout and Deposits of Usable Clay Locations and ℄ Profiles of Drainage Ditches	3
VI	Grade Study of Haul Road from New Clay Pit (Opened Spring, 1961)	1

Figure 2. Schematic plan showing locations of subsurface cross sections.

Geologic sections were later developed along the same grid lines (Fig. 2).

A specially flown vertical aerial photograph, covering the subject project and vicinity, was enlarged to the same scale as the topographic map (1 in. = 100 ft.). Prior to the photographic flight, the control grid intersections were marked on the ground with crosses of white bunting. The images of these crosses were readily discernible on the enlarged photograph, enabling the grid to be quickly and accurately plotted. This photograph, suitably annotated, later served as a key exhibit.

Geological Considerations

The subject property, located in Baltimore County at the northeastern outskirts of Baltimore City, lies within and close to the western margin (the Fall Zone) of the Atlantic Coastal Plain. The location of the Fall Zone determines the effective head of navigation on the tributary streams of the Chesapeake Bay region. It is characterized by the availability of water power at the numerous places where falls and rapids occur as the streams descend from the relatively hard crystalline rocks of the Piedmont to the softer, more erodible, sediments of the Coastal Plain. In colonial times, these factors influenced the selection of sites of the larger towns which ultimately became the major cities of the region. At Baltimore, the availability of large quantities of good brick clay encouraged the early establishment of an extensive brick-making industry which left an indelible impression on the city's architecture.

As mapped by Berry and others (1929), the sedimentary formations occurring locally are (in descending order) the Patapsco, the Arundel, and the Patuxent, which comprise the Potomac Group (Cretaceous age). These three formations are considered to be separated by erosional unconformities. The Patuxent unquestionably rests unconformably on the old crystalline basement rocks which crop out in the Piedmont several miles to the west, and are known to be present at relatively shallow depth nearby the subject property.

Classical concepts of local Coastal Plain geology, as summarized by Berry and others (1929), indicated that the Patapsco, Arundel, and Patuxent are readily differentiated and laterally persistent formations. However, as early as 1902, Heinrich Ries, pioneer investigator of the Maryland Coastal Plain clays, stressed that clay deposits in these formations tend to be discontinous and to occur predominantly as lenses of varying extent (Ries, 1902). More recent work by others has established that Coastal Plain geology in the region is considerably more complex than previously interpreted. The author's experience, augmented by the studies made for this case, confirms the latter view.

In preparing the Coastal Plain portion of the geologic map for nearby Prince Georges County, C. W. Cooke (1951) did not attempt to separately map the Patapsco and Arundel. Later, M. M. Knechtel (1961) found it expedient to treat these "formations" without differentiation, as comprising the "non-marine division" of the Cretaceous sequence.

These considerations suggested the necessity for exploring the property in sufficient detail to avoid false assumptions or conclusions regarding the possibility of non-existent stratigraphic relationships and lateral continuity of the clay deposits.

Exploration and Classification of Clay Deposits

Drilling and sampling techniques. The drilling program was performed in two stages--a preliminary phase for general coverage, and a final phase designed to fill the gaps indicated by analysis of the earlier information and to develop additional data down to the "target" depth in areas where the earlier holes failed to do so. This "target" elevation at a given point was determined by the difference between ground surface elevation and the possible future floor elevation of a pit that would have not less than 1 percent natural drainage gradient toward a large box culvert (invert elevation 45) under the adjacent expressway.

At the time the drilling work was begun, the exact scope and nature of the testing program to follow could not be pre-determined. However, it was considered that the method chosen for the first stage should (1) Enable continuous, accurate, and uncontaminated sampling; (2) Provide relatively large quantities of sample; (3) Be relatively rapid and inexpensive; (4) Permit, if possible, in-situ examination of stratification and other subsurface features.

Continuous-flight power augering was considered and rejected for obvious reasons. The equipment selected was a truck-mounted "Calweld" bucket auger of a type commonly used to dig sanitary "dry wells." The rig has a telescoping "Kelly" bar which permits continuous operation to a depth of 40 feet; below that depth, 20-foot extensions must be added and uncoupled each time the bucket goes in and out of the hole. At depths of more than 80 feet, the operation becomes slow and unwieldy. Nevertheless, the overall progress (often over 200 feet of holes per day) was satisfactory.

It was correctly anticipated that some holes would encounter strata or pockets of water-bearing sand, and that further progress would be impossible with this auger because of a caving condition. Also, layers of dense limonite-hematite iron ore ranging from several inches to over two feet in thickness were occasionally encountered at interfaces between sand and clay. Where encountered at shallow depths the more massive layers were successfully dynamited; at greater depth, the hole was stopped. A 36-inch diameter bucket was used, and each "bite" represented about 1 to 1.5 feet of hole advance, so that sampling was done virtually on a foot-by-foot basis. Normally, composite 30-pound samples were taken from successive bucket loads to represent each 5 feet of hole. Using the bucket auger, 28 holes, totalling 1148 feet, were completed.

For the second and final boring operation, a combination of several conventional methods was used to overcome the numerous drilling and sampling problems caused by the variability of the subsurface materials and silty sands. These methods are summarized as follows:

(1) Auger borings. In instances where the new deeper boring was located adjacent to an existing boring, the repeat section was rapidly augered without taking samples. The boring was then continued to desired depth using either rotary or spoon drilling and sampling methods.

(2) Four-inch casing with spoon sampling. The conventional foundation procedures were modified so that the boring was advanced almost entirely by continuous spoon sampling rather than by intermittent washing. This modification enabled the recovery of an adequate sample and provided a detailed record of the strata.

(3) Four-inch casing and rotary drilling. In some highly consolidated, stiff clays which resisted penetration by the sampling spoon, a rotary core drill was used. A sawtooth bit was used and a continuous core of cohesive material was recovered. Wherever feasible, the use of "drilling mud" in lieu of casing eliminated the need for driving and pulling casing (see D. M. Greer, 1956).

Anticipating that every aspect of the exploration program might be (and was) subject to critical examination in court, competent supervision was maintained at all times; field logs and progress records were meticulously kept; and record photographs (color slides) were taken of every type of drilling and sampling operation for exhibit purposes. Nineteen "rig" borings, totalling 1210 linear feet, were completed during this final phase.

Miscellaneous sampling procedures. A prime purpose of the study was to compare physical characteristics of materials recovered from the borings with those of the materials being used for brick manufacture. The intent was to develop criteria to define the types or range of clay materials which would be suitable for that use. Therefore, several types of sampling procedures were devised to supplement the boring program.

A series of channel samples was taken at several locations along working faces of the company's original or "main" pit. Sampling was done in vertical increments of 5 feet, the intent being to obtain samples which could be tested and classified on an individual basis, and then combined to form a composite sample which would be representative of the materials actually mined and used. In the spring of 1961, the company opened up a new pit to the north of the "main" or "old" pit. Periodically, as this pit was enlarged, composite samples were taken here also, being representative of materials then being mined and used. Consequently, typically "suitable" materials from this "new pit" area were compared with those previously mined at the "old" pit.

Samples were also taken periodically from the plant stockpile and from materials at various stages of processing (for example, at the raw material hopper and the granulator). The stockpile samples were taken in composite fashion by digging through the layers from top to bottom; the others were taken at random.

It was considered essential to develop a record of reference sampling and testing of unfired or "green" brick taken from the production line at the plant. To determine, for subsequent purposes, a reasonably accurate value for the use factor (the number of cubic yards of suitable raw material, measured in place in the pit, required to produce 1000 or "M" unfired brick), the dimensions, volumes and weights of these sample bricks were carefully determined. They were also subjected to other tests as described hereafter.

It was known that the company would contend that the density of the clay in the "green" brick, resulting from the compactive effort of the extrusion process, was greater than the natural density in the pit. If true, this relationship would tend to increase the "use factor." To investigate this possibility, several "in-place" density determinations were made using the "sand-cone" method as defined by American Association of State Highway Officials Designation T-147. The resulting data are given in Figure 3. Comparison of the average natural dry density of 113.4 pcf with a value of 107.7 pcf for "green" brick, as subsequently determined (Fig. 5), completely repudiated the company's contention.

Laboratory testing and classification procedures. The state's investigation substantiated the company's claim that the deposits underlying the subject property are extremely variable and that much of the material--sands, silts, limonitic clays, and so forth--is worthless for brickmaking. Therefore, the problem of how to classify the materials sampled by the borings assumed critical importance.

Laboratory firing tests to determine the ceramic properties of a large number of samples would have been very time-consuming and costly. Also, it was apparent that the company would strongly object to this procedure with the contention that firing conditions in a laboratory kiln would not properly simulate the conditions in their tunnel

Figure 3. Natural density of suitable materials

Test No.	AASHO Group	Location	Density at nat. M.C. (lb./ft.^{3})	Nat. moist. cont. (%)	Dry density (lb./ft.^{3})
FD-1	A-6(11)	"Main" Area-TP-1, face of pit, about 40'W. of BA-1	126.7	12.7	112.2
FD-2	A-6(10)	New Pit (TP-4), west face of pit, upper bench	134.4	18.8	113.1
FD-3	A-7-6(12)	New pit (TP-4), north face, lower bench	135.2	17.7	114.8
		Averages	132.1	16.4	113.4

kilns. Obviously, it was not practicable to process test bricks through their plant.

After considerable research, the following line of reasoning was developed:

(1) The principal attributes of brick clay are related to particle size gradation and plasticity characteristics.

(2) These properties can be quantitatively determined by standard test procedures used in soils engineering--sieve and hydrometer anlysis, and Atterberg limits.

(3) The resulting data can be used to determine engineering classifications of soils according to one of the widely accepted systems. Of these, the American Association of State Highway Officials (AASHO) classification system was selected (Fig. 4). The basis of this system is given in PCA (1962).

(4) If reasonable correlations can be developed by parallel sampling and testing of "suitable" materials being consumed by the plant for brick production, the engineering classification may be used for evaluating the suitability of other materials (for example, samples from borings) for brick making purposes.

On this basis, over 500 samples from the test borings, working faces of the pits, plant stockpiles, and unfired or green brick were tested and classified in the laboratory. Good correlations were developed, thus enabling distinction between "suitable" and "unsuitable" materials. Although this distinction might be made visually by an experienced brick man, the procedure followed was objective and quantitative, rather than being based upon individual judgement. This consideration made a very favorable impression in court later.

From pre-trial discussion and interrogatory procedures, it was evident that the Company would attempt to discredit the use of an engineering soil classification system for evaluating the suitability of materials for brick making. Furthermore, it would be alleged that the types of tests involved were foreign to the ceramic and brick industry. Therefore, the literature was thoroughly searched for pertinent papers and precedent. H. Ries (1902) stressed the importance of determining grain size gradation in testing clays, and one of his papers appeared in a ceramic journal (1935). A pioneer paper by A. Atterberg (1911) became the basis for the widely used test procedure bearing his name. A review paper by E. A. Bauer (1959) concludes that Atterberg's work was directed toward investigating the plasticity of clays *per se*, and not exclusively for engineering purposes. Papers concerning the Atterberg limits procedures (Kinston, 1937) and the use of hydrometer analysis for grain size determination of soils (Bouyoucos, 1935) appeared in ceramic journals. And finally, substantiation of the applicability of AASHO test procedures to the testing of clays for ceramic purposes is described in a U. S. Bureau of Mines publication widely used in the brick industry (Klinefelter and Hamlin, 1957).

Interpretations and Conclusions

Correlation studies. An analysis of the voluminous data developed by the laboratory testing and classification work, as the basis for correlating the "suitable" and "unsuitable" materials, are shown on Figure 5.

It should be noted that brick samples Nos. 1 to 13 were obtained prior to the opening of the "new"

Group index charts.

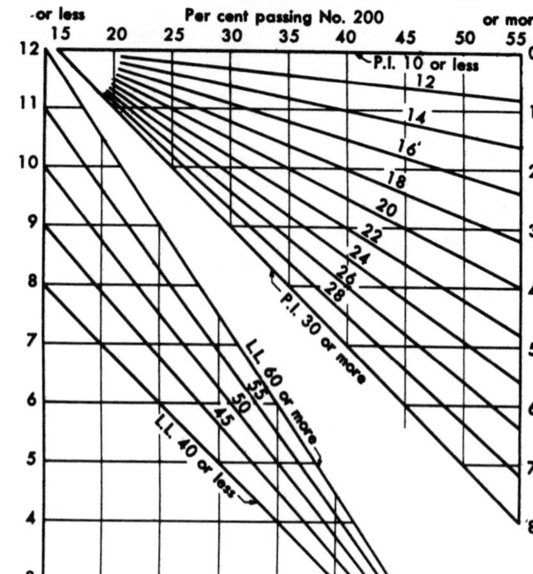

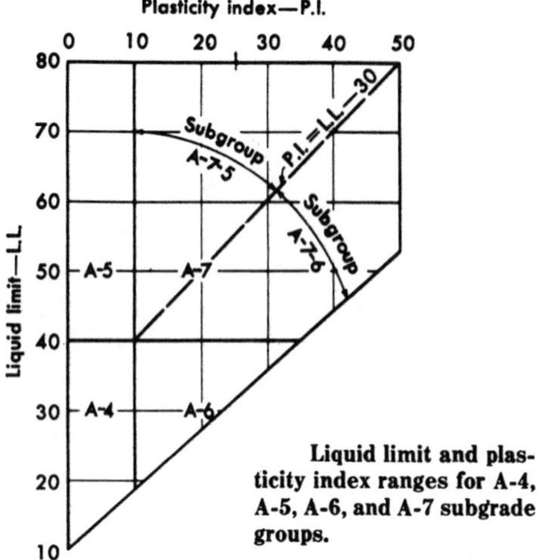

Liquid limit and plasticity index ranges for A-4, A-5, A-6, and A-7 subgrade groups.

Table 3. Classification of Highway Subgrade Materials (with Suggested Subgroups)

General classification	Granular materials (35 per cent or less of total sample passing No. 200)							Silt-clay materials (More than 35 per cent of total sample passing No. 200)			
Group classification	A-1		A-3	A-2				A-4	A-5	A-6	A-7
	A-1-a	A-1-b		A-2-4	A-2-5	A-2-6	A-2-7				A-7-5, A-7-6
Sieve analysis, per cent passing:											
No. 10	50 max.										
No. 40	30 max.	50 max.	51 min.								
No. 200	15 max.	25 max.	10 max.	35 max.	35 max.	35 max.	35 max.	36 min.	36 min.	36 min.	36 min.
Characteristics of fraction passing No. 40:											
Liquid limit				40 max.	41 min.	40 max.	41 min.	40 max.	41 min.	40 max.	41 min.
Plasticity index	6 max.		NP	10 max.	10 max.	11 min.	11 min.	10 max.	10 max.	11 min.	11 min.*
Group Index**	0		0	0		4 max.		8 max.	12 max.	16 max.	20 max.

Classification procedure: With required test data available, proceed from left to right on chart; correct group will be found by process of elimination. The first group from the left into which the test data will fit is the correct classification.

*P.I. of A-7-5 subgroup is equal to or less than L.L. minus 30. P.I. of A-7-6 subgroup is greater than L.L. minus 30 (see Fig. 4).

**See group index formula or Fig. 5 for method of calculation. Group index should be shown in parentheses after group symbol as: A-2-6(3), A-4(5), A-6(12), A-7-5(17), etc.

Figure 4. Classifications of soils and soil aggregate mixtures for highway construction purposes.

Figure 5. Classification test data and densities--unfired Champion brick

Brick sample no.	AASHO group	Atterberg limits L.L.	P.L.	P.I.	Mech. Anal. Gr. %	Sa. %	Si. %	Cl. %	Moist density lb. per cu. ft.	Moist cont. (%)	Dry density (lbs. per cu. ft.)
1										21.0	
2										23.4	
3										21.5	
4	*A-7-6(20)	58.5	23.3	35.2	0	10	36	54		23.1	
5										21.7	
6										23.6	
7	**A-7-6(20)	59.0	25.6	33.4	0	10	36	54		21.1	
8										23.6	
9										--	
10										23.0	
11										21.2	
12										23.3	
13										21.4	
Average	A-7-6(20)	58.8	24.5	34.2	0	10	36	54		23.3	
14	A-7-6(15)	48.7	24.7	24.0	0	18	41	41	--	18.8	--
15	A-7-6(16)	49.8	24.4	25.4	0	17	39	44	132.5	19.5	111.0
16	A-7-6(16)	48.9	24.8	24.1	0	18	40	42	133.7	19.7	111.8
17	A-7-6(16)	48.7	22.4	26.3	0	17	39	44	132.5	17.8	112.4
18	A-7-6(14)	46.5	23.4	23.1	0	18	39	32	131.5	18.9	110.7
19	A-7-6(16)	48.3	22.5	25.8	0	18	37	45	132.8	19.3	111.4
20	A-7-6(14)	46.9	23.7	23.2	0	18	37	45	131.0	19.3	109.7
21	A-7-6(17)	50.2	23.1	27.1	0	17	35	48	129.9	21.4	107.0
22	A-7-6(16)	50.0	23.9	26.1	0	16	36	48	129.9	21.2	107.2
23	A-7-6(17)	50.2	24.0	26.2	0	16	39	45	129.6	21.8	106.4
24	A-7-6(17)	50.2	23.6	26.6	0	16	37	47	127.4	21.3	105.0
25	A-7-6(17)	50.0	23.5	26.5	0	17	37	46	125.7	21.6	103.4
26	A-7-6(16)	50.1	24.1	26.0	0	16	36	48	126.1	21.7	103.7
27	A-7-6(16)	50.8	25.3	25.5	0	18	37	45	126.3	21.6	103.9
28	A-7-6(17)	51.4	24.3	27.1	0	17	36	47	127.4	20.6	105.6
29	A-7-6(17)	51.0	24.4	26.6	0	17	35	48	127.0	20.4	105.6
Average	A-7-6(16)	49.5	23.9	25.6	0	17	38	45	129.6	20.3	107.7
Total Average	A-7-6(18)	53.6	24.2	29.5	0	14	37	49	129.6	21.1	107.7

*Composite samples of the even numbered bricks
** " " " " " " "
Water contents of odd numbered samples taken one day later than even numbered samples and not included in determination of average M.C.

pit in the spring of 1961, and represent raw material from the "old" or original main pit. Nos. 14 to 29, sampled in the fall of 1961, represent raw material which came predominantly from the new pit. The test data confirm appreciable differences in the properties of the two groups of samples. Conclusions which may be drawn from the Figure 5 comparison are:

(1) Bricks made from "new pit" material have a higher sand content (17 versus 10 percent) and lower clay content (45 versus 54 percent) than bricks made from the "old pit" material.

(2) "New pit" bricks are made of less plastic clay (49.5 versus 58.8 L.L. and 25.6 versus 34.2 P.I.). This is consistent with the higher sand content and the fact that the material requires a

lower moisture content (20.3 versus 23.3 percent) to achieve a satisfactory degree of workability.

(3) The increased sand content and lower plasticity of the "new pit" material is reflected by the drop in the group index number from 20 to 16.

(4) Considerable variations in the texture of the blended raw materials can be tolerated.

Figure 6 summarizes the results of sampling at test pits (channel sampling locations) in the company's original or "old" pit. The data indicate that most of the materials sampled were A-7-6 clays with group index numbers in the range of 13 to 20.

The data for TP-1 indicate that 38.5 percent of A-7-6 (20) material can be mixed with 61.5 percent of A-6 (12) and A-6 (10) material to produce a blend or mixture of material which is still in the A-7-6 group but has a lower group index number of 14. This relationship affords an excellent example of the fact that "fat" clays are mixed with "lean" clays at every stage of Champion Brick Company's overall operation: for example, excavation from the full working face; loading, dumping, and spreading at the stockpile; excavation from the stockpile; and in processing through the plant.

Figure 7 shows the results of tests on samples from the "new" pit, and some interesting relationships are apparent. Data for composite samples BS-1 to BS-7 (TP-4) show that in this series when samples were taken at random on different dates as the face of the new pit advanced, about 72 percent of the material in the face was A-7-6 (12) and the remaining 28 percent was of lower plasticity A-6 (11) or lean clay. Therefore, these materials can and are blended in the routine operation to produce an acceptable mixture.

BS-6A and -7A (TP-4) also demonstrate that appreciable quantities of A-2-4 (silty sand) and A-4 (silt) materials occur in the face of the new pit and are mined and mixed with higher grade materials.

The "Check Strip" data (TP-4) further confirm the aforementioned relationships that about 45 percent is A-7-6 material along with 38 percent of A-6 and 17 percent of A-2-6 (clayey sand). This pit-run mixture as excavated at the full face is an A-6(6) or a "lean" clay. These facts explain the higher sand content and lower plasticity of the "green" bricks made from the "new pit" material.

Figure 8 reflects the data developed by sampling and testing raw material taken from the stockpile and at two stages of processing at the plant. Again, certain general relationships between the date of sampling, and source and texture of the material in the "green" brick produced are apparent (Fig. 5). For example, the 1960 stockpile, hopper, and granulator samples (material from "old pit") show relatively high plasticity (L.L. 47 and P.I. 26), high clay content (50 percent) and low sand content (8 percent). These indices are comparable to those for roughly contemporaneous "green" brick (Fig. 5), except that the plasticity is somewhat lower. The 1961 stockpile and hopper samples (material from "new pit") are lower in plasticity (L.L. 42 and P.I. 18) and

Figure 6. Results of sampling and testing at original (main) pit

Test pit no.	Date of sampling	Type and no. of sample	Stratum depth at pit face (ft.)	Stratum thickness or sample interval (ft.)	Percent of total thickness	AASHO soil group
TP-1	12-19-60	BS-1	0 - 5	5	38.5	A-7-6(20)
		BS-2	5 - 10	5	38.5	A-6(12)
		BS-3	10 - 13	3	23.0	A-6(10)
		Composite	0 - 13	13	100.0	A-7-6(14)
TP-2	12-19-60	BS-1	0 - 5	5	50.0	A-7-6(19)
		BS-2	5 - 10	5	50.0	A-7-6(17)
		Composite	0 - 10	10	100.0	A-7-6(18)
TP-3	12-19-60	BS-1	0 - 5	5	20.8	A-7-6(13)
		BS-2	5 - 10	5	20.8	A-7-6(13)
		BS-3	10 - 15	5	20.8	A-7-6(15)
		BS-4	15 - 20	5	20.8	A-7-6(15)
		BS-5	20 - 24	4	16.7	A-7-6(13)
		Composite	0 - 24	24	100.0	A-7-6(14)

Composite samples were prepared by thoroughly mixing materials from the individual samples in proportion to the thickness of the layer or sampling depth which they represented.

Figure 7. Results of sampling and testing at new pit (TP-4)

Type and no. of sample	Date of sampling	Stratum depth at pit face (ft.)	Stratum thickness or sample interval (ft.)	Percent of total thickness	AASHO soil group
BS-1 (Composite)	5-19-61	0 - 10	10	100.0	A-7-6(15)
BS-2 "	5-25-61	0 - 10	10	100.0	A-7-6(12)
BS-3 "	6-6-61	0 - 10	10	100.0	A-7-6(12)
BS-4 "	6-17-61	0 - 10	10	100.0	A-7-6(12)
BS-5 "	9-14-61	0 - 10	10	100.0	A-6(11)
BS-6 "	9-18-61	0 - 10	10	100.0	A-6(11)
BS-7 "	9-18-61	0 - 10	10	100.0	A-7-6(12)
BS-6A (Grab)	9-18-61	Pit Face	1	10	A-2-4(0)
BS-7A (Grab)	9-18-61	Pit Face	3 to 4	30 - 40	A-4(5)
"Check Strip"					
BS-1	9-28-61	0 - 3	3	20.7	A-6(3)
BS-2	"	3 - 5.5	2.5	17.2	A-2-6(1)
BS-3	"	5.5 - 6.5	1.0	6.9	A-6(10)
BS-4	"	6.5 - 8.0	1.5	10.3	A-6(9)
BS-5	"	8.0 - 14.5	6.5	44.9	A-7-6(10)
Composite	"	0 - 14.5	14.5	100.0	A-6(6)

1. Composite sample for "Check Strip" was prepared in laboratory.

2. BS-6A was a random grab sample taken from face immediately above Field Density Test 2 to illustrate presence of sand seams.

3. BS-7 was taken from sandy material piled up by power shovel and being mixed with more plastic materials.

Figure 8. Results of sampling and testing at stockpile and plant

Location and date of sampling	Sample no.	AASHO group	Atterberg limits			Mechanical analysis soil fractions in percent			
			L.L.	P.L.	P.I.	Gr.	Sa.	Si.	Cl
Stockpile									
12-20-60	SP-1 (1)	A-7-6(12)	41	20	21	0	8	44	48
	(2)	A-7-6(16)	46	20	26	1	9	46	44
12-20-60	SP-2 (1)	A-7-6(19)	53	21	32	0	13	42	45
	(2)	A-7-6(18)	52	21	31	4	9	41	46
12-20-60	SP-3 (1)	A-7-6(14)	44	22	22	2	8	39	51
	(2)	A-7-6(15)	46	21	25	2	8	36	54
	(Aver.)	A-7-6(16)	47	21	26	2	9	41	48
9-18-61	SP-4 (1)	A-7-6(11)	41	23	18	0	16	46	38
Plant hopper									
12-20-60	(1)	A-7-6(14)	45	22	23	1	8	36	55
	(2)	A-7-6(16)	48	22	26	0	5	42	53
	(Aver.)	A-7-6(15)	46	22	24		6	39	54
9-18-61	(1)	A-7-6(12)	43	24	19	0	16	38	46
Plant granulator									
12-20-60	(1)	A-7-6(17)	48	22	26	1	9	35	55
Total Averages		A-7-6(15)	46	22	24	1	10	40	49

All samples were either composite or "grab."

clay content (42 percent) and higher in sand content (16 percent), as is the "green" brick of comparable vintage.

Observation of the mining and material processing methods used by the Champion Brick Company, augmented by analysis of the classification test data as reviewed above, permits the following general conclusions to be drawn:

(1) In the pit, the mining excavation is planned so that the shovel alternately works in faces of "fat" and "lean" clay in order to supply the stockpile with the range of materials which will produce a mixture of reasonably uniform texture.

(2) Mixing and blending of the raw material occurs at a number of points in the normal excavation and processing cycle.

(3) Substantial quantities of "leaner" A-6 materials are excavated and used along with the "fatter" A-7-6 clays. Smaller amounts of sandy and silty materials (A-2-4, A-4, and A-5) are likewise used. The resulting blends will produce predominantly A-7-6 material as verified by the tests on unfired bricks. Although no bricks composed of A-6 material were found during the spot sampling, it is believed that they occur, and that A-6 material would be suitable.

Suitability criteria. Final determination of those materials deemed suitable for commercial brick production was made in accordance with the following criteria:

(1) Visual examination. Gray organic clays which would probably burn to a buff color, and materials containing undue amounts of gravel, iron ore, etc. were rejected on the basis of obvious physical defects.

(2) Physical classification tests. A-7-6 and A-6 materials, except those rejected because of visual characteristics, were tentatively considered to be usable. Large masses or thick layers of sands and silts (A-2, A-3, and A-4 materials) were rejected as unsuitable.

(3) Correlations with Champion Brick Company data. Wherever feasible, comparisons and correlations were made between test results of samples from the test borings and suitability determinations made by the Champion Brick Company on sample material from their borings at the same or nearby locations.

(4) Judgement. Where relatively thin seams or layers of obviously unsuitable materials (A-2-4 or A-4 sands and silts) occur in intimate association with large quantities of suitable materials, the former were frequently included in the determination of overall suitability. However, every effort was made to contemplate no greater degree of mixing and blending of unlike materials than is routine in the company operation.

Figure 9 presents a statistical analysis of the results obtained by applying the foregoing criteria. In terms of total footage penetrated by the borings, 68.5 percent of the materials considered suitable fall into the A-7-6 category; 27 percent are in the A-6 group; and the remaining 4.5 percent are A-5 and A-4.

The investigations concluded that the AASHO soil classification system may be effectively used in determining the suitability or unsuitability of materials for brick making. All of the basic studies discussed herein substantiate that the high ratio (about 2.5 to 1) of "fat" A-7-6 clays to "leaner" A-6 clays assures that the materials as classified are as suitable for the practical requirements of brick making. The small percentage of low-plasticity silts and sandy silts is considered unimportant, since larger percentages of such materials can and are used.

Subsurface cross sections. All of the boring logs, classification test data, and other related subsurface information were plotted on cross sectional alignments of the grid lines in two directions at 250-foot intervals (Fig. 2). Analysis and comparison of these data as they appeared on successive sections, enabled delineation of the approximate boundaries of lenses of "suitable" clay in three dimensions. The final format included features of both conventional profiles and geologic sections.

Also plotted on the cross sections were the limit lines beyond which otherwise suitable materials could not be recovered economically because of side design slopes of the pit, drainage limitations, or excessive overburden. Regarding the latter item, it was concluded that the Company could not mine clay profitably beyond a point where the ratio of overburden to clay exceeds 2 : 1. A search of the literature located one deposit in Kansas (Anonymous, 1959) where the ratio for a successful brick plant is 1 : 1.5. After the 2 :1 limiting ratio for mining was tentatively applied to the subject property, calculations from the cross sections determined that the resulting average ratio of stripping would be 1 : 2.45, a conservative value. Figure 2 illustrates a typical prepared subsurface section.

Calculation of Clay Reserves. Analysis of the plant operations, production records, and other factors, yielded the following basic data:

(1) Theoretically, 1.85 cu. yd. of raw clay (measured in situ) are required to make one thousand bricks.

(2) In practice, due to breakage, rejects, and miscellaneous losses, 2.10 cu. yd. are required per one thousand bricks.

(3) For an average annual production of 29 million bricks, the raw material requirement is about 61,000 cu. yd.

(4) The plant had continued operations during the year between date of R/W taking

Figure 9. Statistical analysis--composition of suitable material in terms of AASHO soil groups

BA and RB boring nos.	Suitable material vs. AASHO soil group									
	A-7		A-6		A-5		A-4		A-2-4	
	No. of samples	Total depth (ft.)	No. of samples	Total depth (ft.)	No. of samples	Total depth (ft.)	No. of samples	Total depth (ft.)	No. of samples	Total depth (ft.)
BA-1	2	10	6	30			2	10		
BA-3	1	5	1	5						
BA-5	8	40	4	20						
BA-6	2	10	1	5						
BA-7	2	11								
BA-8	1	5	3	15						
BA-9	6	30	2	10			2	10		
BA-10	2	10	1	5						
BA-11	5	25			1	15				
BA-12	1	5								
BA-13	1	5					1	5		
BA-14	3	15	2	10			1	3		
BA-16	2	10	1	3						
BA-17	5	25	2	10						
BA-19	1	5	6	28						
BA-25	3	15								
BA-28	4	20	3	15						
RB-2	2	10								
RB-3	2	12								
RB-4	10	40	1	5						
RB-6	2	9								
RB-8	1	7								
RB-11	5	27	3	14						
RB-13	1	5	2	12						
RB-14	7	34								
RB-15	11	50	2	6						
RB-16	6	31	1	5						
RB-17	4	20								
RB-18	2	9								
									0	0
Total suitable samples	102		41		1		6		Totals 150	
Percent of TOTAL	68.0		27.5		0.5		4.0		(100)	
Total footage of suitable material		500		198		5		28	Totals 731	
Percent of TOTAL		68.5		27.0		0.5		4.0	(100)	

All data shown above are based upon the final cross sections and suitability and soil classifications by Benjamin E. Beavin Company.

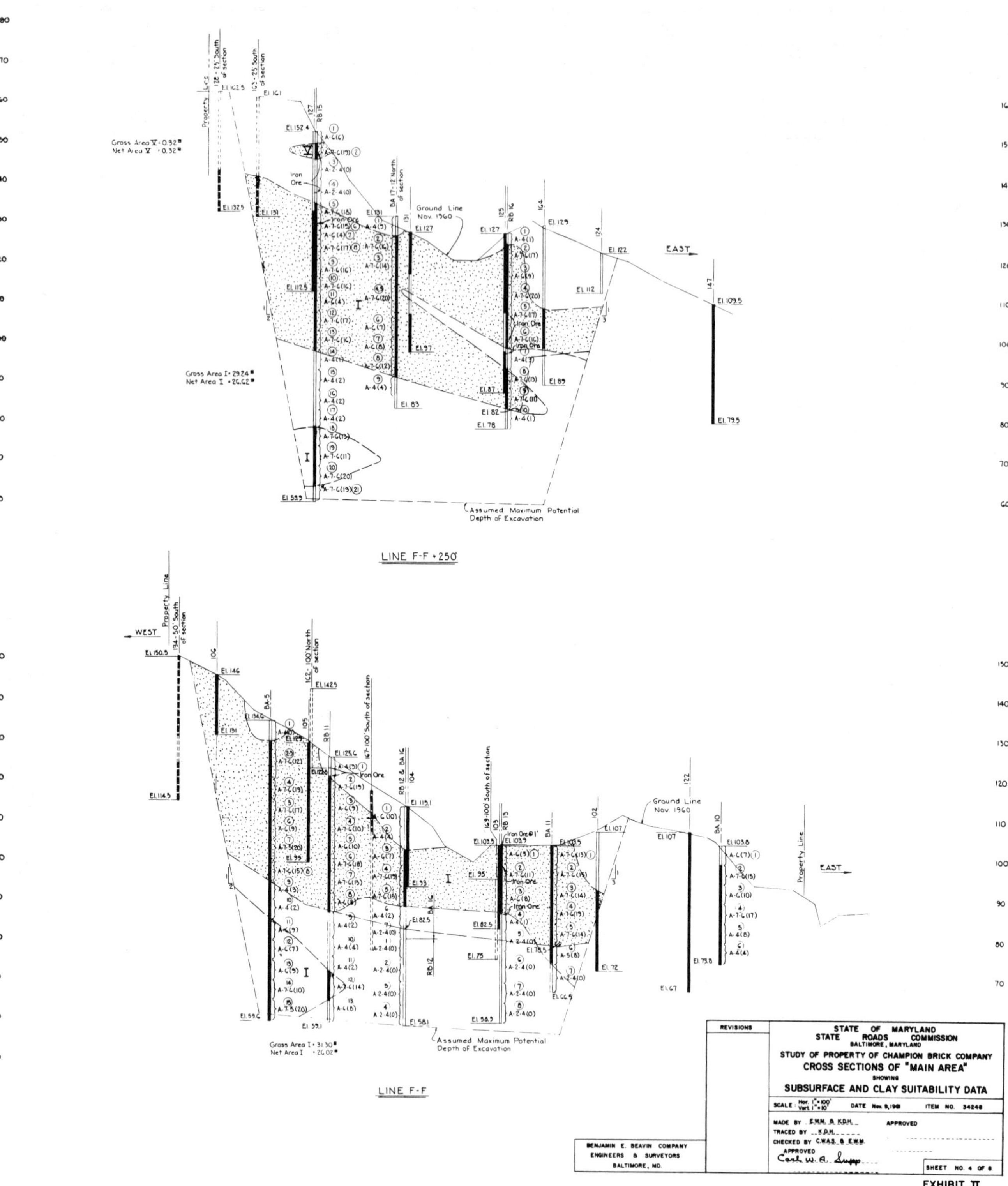

Figure 10. Typical subsurface cross sections, showing subsurface and clay suitability data.

(December, 1959) and date of estimate of remaining reserves November, 1960).

The total available volume of suitable and economically recoverable material was computed from the cross sections using the "average and area" method, a standard engineering practice (Peele, 1941). The volume so computed was 1,213,900 cu. yd., equivalent to 19.9 years of brick production. Adding the one year of plant operation prior to the estimate of reserves, it was concluded that the total reserves of clay available to the Champion Brick Company, as of the date of R/W taking (December, 1959) would assure a remaining life of manufacturing plant of not less than 20.9 years.

Preparation of Exhibits

Summary report. A comprehensive summary report (Supp, 1961) was completed shortly before the trial. It was accompanied by a volume of Appendices in which were compiled the field and laboratory logs of all borings, laboratory test and classification data, and related information.

The report served a number of important purposes:

(1) It organized an otherwise unwieldy mass of data into useful reference form for use by the State's attorneys before and during the trial.

(2) Copies were made available to the dependant's attorneys and may have dissuaded them from wasting efforts in areas where the State's case was obviously well prepared.

(3) The thoroughness of the preparation made a favorable impression upon the court and jury.

(4) A large body of technical and geological data was preserved for future reference.

Geologic model. Figure 11 shows the appearance of a 3-dimensional geologic model that was designed and built primarily to assist the court and lay jury in understanding and evaluating the practical significance of the almost intimidating mass of technical detail which was presented to them. It consists essentially of two series of interlocking plastic cross sections which were prepared by using the original subsurface sections as templates. Only the basic features of the reference sections are shown: the subsurface distribution of suitable clay, with the economically recoverable portions differentiated; limiting factors such as property and R/W lines, and side slopes and bottom

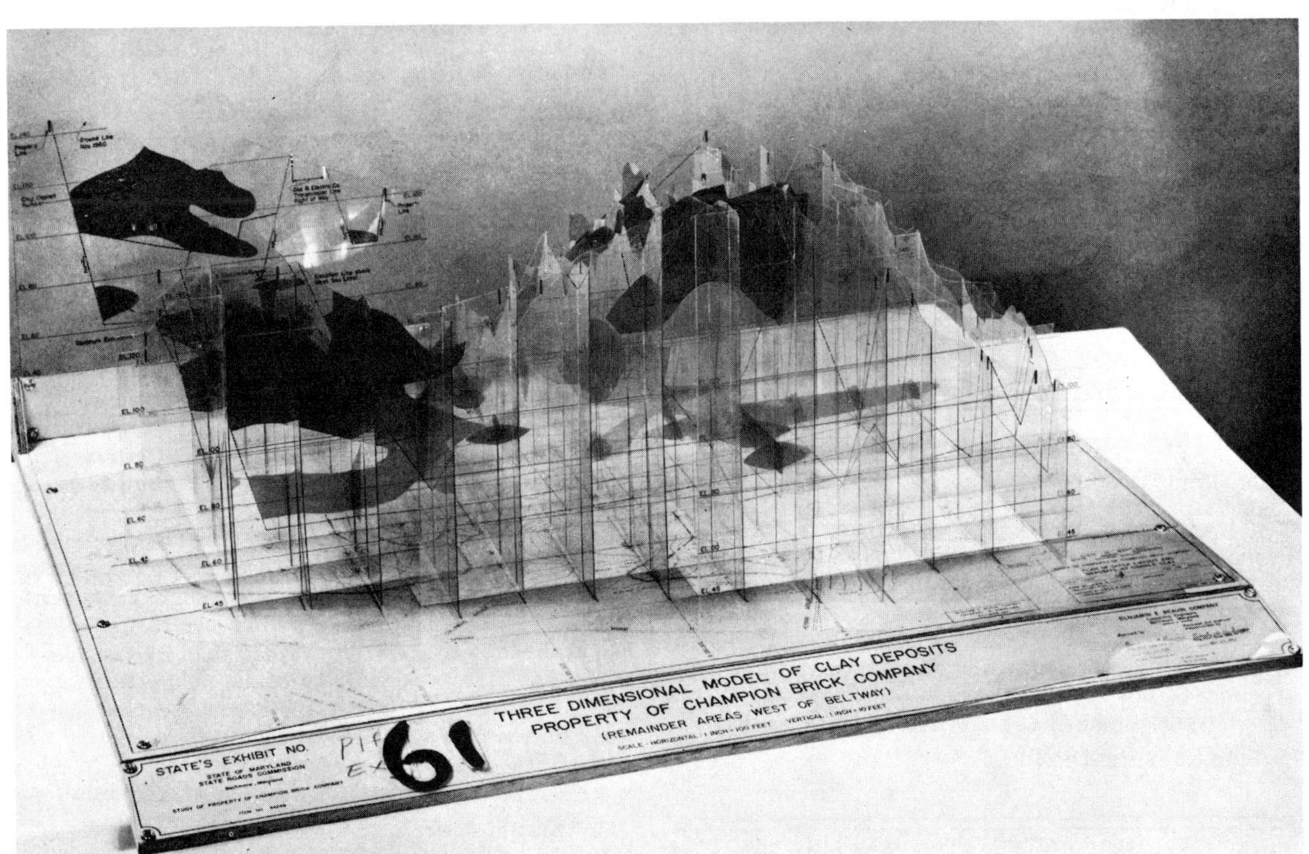

Figure 11. Photograph of three-dimensional geologic model.

of an "ultimate" pit; and existing ground profiles (as of November 1960).

The interlocking "egg-crate" type of construction permits the whole assembly to be collapsed into a flat bundle, and the individual sections are removable to permit comparison with the original detailed sections. The model rests on a plywood base upon which has been fixed a print of an annotated portion of the topographic map (Fig. 1). The individual sections are located on 250-foot centers (lin = 100-foot scale) and are in register with the grid lines on the base map.

The vertical scale is exaggerated, at ratio 10 : 1. Symbols and legend used on the model sections are explained on a removable hypothetical section set at the rear edge of the platform.

Other exhibits. Additional exhibits included:

(1) Large-scale (1 in. = 100 ft.) aerial photograph and topographic map (Fig. 1) of property.

(2) Complete set of suitably colored prints of subsurface cross sections (Fig. 10).

(3) Enlarged print of AASHO soil classification chart (Fig. 4).

(4) Soil testing equipment (liquid limit device and sand-density cone apparatus).

(5) Color photographs of exploration and testing procedures.

THE TRIAL

Duration

The case[1] came to trial before a jury in Baltimore County Circuit Court in November, 1961. It lasted some ten weeks and reputedly established a local record for technical complexity and "longevity."

Qualification Procedures

The Court allowed testimony from professional appraisers, valuation experts, and persons experienced in phases of brick manufacture, machinery and plant design, etc., without requiring that they have formal qualifications. Professional engineering registration was favorably regarded by the court, but did not become an issue.

Since the condemnee's counsel had prior knowledge through deposition and interrogatory procedures of the general scope of the author's intended testimony, his qualifications were subjected to critical review.

As is customary in condemnation cases, the property owner, in the person of the president of the Champion Brick Company, was permitted to testify.

Use of Exhibits

The company was in a position to make determinations of the suitability of materials for brick making by a more direct method than that used by the State. Several hundred power-auger borings were made in the ground adjacent to the subject property. Miniature briquettes were molded from all reasonably plastic materials recovered from these borings. They were then fired in the tunnel kiln and classified as "good" or "bad" on the basis of color, texture, shrinkage, dimensional distortion, etc. A key exhibit was then prepared by displaying these hundreds of briquettes in two racks. A companion and equally important exhibit consisted of a "peg model" intended to depict the subsurface distribution of suitable and unsuitable materials. The base consisted of a large sheet of plywood covered by a map of the property. Dowel sticks, cut to varying lengths according to a vertical scale and color-coded to correlate with the briquette exhibit, were set at the map locations of the borings.

The State's comparable key exhibits--maps, aerial photograph, subsurface sections, model, and summary report--have been previously described.

Expert Testimony

The company's case was directed toward establishment of a basis for consequential damages to the entire plant and related facilities due to alleged reduction of suitable clay reserves to a 5-year supply or less. Its succession of witnesses included experts in brick plant and equipment design, appraisal and valuation, and clay technology.

The company president testified at length regarding the drilling program and ceramic testing procedures which formed the basis for the briquette exhibit and the "peg model." The effectiveness of his testimony was impaired by the fact that all of this work, including construction of the model, was performed by his own forces under his personal supervision, rather than by outside experts. Also, the validity of the model was further challenged by the State's demonstration that many of the borings were of insufficient depth; some borings were incorrectly located; and there were technical flaws in its construction and in the correlations within the briquette exhibit. Since the model was based upon the same data used in computing a reserve clay supply of less than 5 years, these points were of major importance. It

[1]Recorded in the archives of Baltimore County as Condemnation Docket #4, Folio 160, Case #741.

was also apparent that the model was difficult for the jury to understand.

The prime objective of the state's case, in rebuttal, was to deny consequential damages by proving that the remaining clay reserves were adequate to support the plant for its remaining useful life. The author's testimony, which lasted one week, covered all of the matters discussed in this paper. Extensive use was made of the exhibits in substantiating his conclusion that not less than 21 years of remaining plant life were assurred. The three-dimensional model was unquestionably effective in providing the jury with a readily understandable representation of the practical meaning of much of the previous highly technical testimony.

Another witness for the State, a professional valuation engineer with considerable experience in the clay industry, testified to a remaining clay supply of from 20 to 80 years. This had the effect of emphasizing the conservatism of the author's estimate.

Verdict

After one and one-half days of deliberation, the jury's award to the Champion Brick Company was $247,752, or $69,752 above the state's original offer, as compared to the original company claim of $3,500,000. The basis of the monetary award was not susceptible to analysis, but the length of the jury's deliberation strongly suggested that it was determined by compromise between varying views.

CONCLUSION

It is considered appropriate to quote here conclusions which were reached by one of the attorneys for the State in this case (Lehmann, 1962). From the viewpoint of the condemnor, the effort spent in ten weeks of arduous trial and approximately two years of painstaking preparation of this case was well justified. If conclusions can be drawn from this case and applied generally to right-of-way condemnation work, the writer believes they may be the following. This case emphasizes the value of expertly prepared visual aids, particularly to present technical data which might otherwise be unintelligible to lay persons.
Further, the case illustrates that no condemnation situation is too complicated or technical to be entrusted to a jury, and that if a jury retires to deliberate confused by technical gobbledegook or overwhelmed by complicated scientific data, it is the fault of counsel. The presentation of the evidence is a responsibility that trial counsel alone must assume. Thus, adequate preparation with emphasis on an orderly presentation and lucid demonstration of the facts will always be the best insurance that counsel can have a correct and fair verdict.

The foregoing quotation reflects the opinions and philosophy of an experienced condemnation attorney. Most professional persons would concur with his comments regarding the value of expert preparation and presentation of technical testimony. Some might disagree with his appraisal of the merits of a jury trial, and argue to good effect that complex technical cases should be heard before a highly qualified judge (without jury), or before a board of experts. In Maryland, this question is answered by the law which requires that property condemnation cases be tried before judge and jury in the Circuit Court of the County in which the subject property is situated.

REFERENCES CITED

Anonymous, 1959, Acme's Kanopolis plant produces wide size and color range: Brick and Clay Record, v. 8, 42 p.

Atterberg, A., 1911, Uber die physikalische bodenuntersuchung, und über die plastizitat der tone: Internationale Mitteilungen für Bodenkunde, v. 1.

Bauer, E. A., 1959, History and development of the Atterberg limits tests: Symposium on Atterberg limits, Sixty-Second Annual Meeting ASTM, June 1959, ASTM Spec. Tech. Pub. No. 254.

Berry, E. W., Mathews, E. B., Watson, E. H., 1929, Baltimore County: Maryland Geological Survey, Baltimore, The Johns Hopkins Press, p. 200-207, 270-276.

Bouyoucos, G. J., 1935, Hydrometer method for making mechanical analysis of soils: Am. Ceramic Soc. Bull. v. 14, no. 8, p. 259-262.

Cooke, C. W., and Cloos, E., 1951, Geologic map of Prince Georges County and the District of Columbia: Baltimore, Maryland Department of Geology, Mines and Water Resources.

Greer, D. M., 1956, Use of drilling mud and modern methods for better soil test borings: Consulting Engineer, v. 8, no. 4, p. 70-74.

Kinston, C. S., 1937, A study of the Atterberg Plasticity Method: Am. Ceramic Soc. Trans., v. 16, 472 p.

Klinefelter, T. A., and Hamlin, H. P., 1957, Syllabus of clay testing: U. S. Bureau of Mines Bull. 565, Government Printing Office, 67 p.

Knechtel, M. M., Hamlin, H. P., Hosterman, J. W., and Carroll, C., 1961, Physical properties of nonmarine Cretaceous clays in the Maryland Coastal Plain: Maryland Department of Geology, Mines and Water Resources Bull. 23.

Lehmann, C. H., Jr., 1962, The Champion Brick Case: A study of the proof of value: in Selected problems relating to highway laws, 1962: Highway Research Board Spec. Report 76, Nat. Acad. of Sciences, Nat. Research Council, Washington, D.C., p. 70-77.

Mathews, E. B., 1925, Map of Baltimore County and Baltimore City showing the geological formations: Baltimore, Maryland Geological Survey.

Peele, R., 1941, Mining Engineer's Handbook: v. I, 3rd ed., New York, John Wiley & Sons.

Portland Cement Association, 1956, PCA soil primer: Chicago, Portland Cement Association, 86 p.

Ries, H., 1902, Report on the clays of Maryland: Maryland Geological Survey, v. IV, pt. III, Baltimore, The Johns Hopkins Press, 214 p.

________ 1935, Geology and clay research: Am. Ceramic Soc. Bull. v. 14, no. 9, p. 279-290.

Supp, C. W. A., 1961, Report on the clay reserves of the Champion Brick Company property at Stemmer's Run, Baltimore County, Maryland: Benjamin E. Beavin Co., Consulting Engineers, Baltimore, Maryland. (unpublished report, Md. S. R. C. Item No. 34248), 2 vol., November 1961.

Engineering Geology Case Histories

Number 8

Engineering Seismology: The Works of Man

Edited by

WM. MANSFIELD ADAMS

Prepared for the Division on Engineering Geology

of

The Geological Society of America

1970

Reprinted 1973

Published by

THE GEOLOGICAL SOCIETY OF AMERICA, INC.

3300 Penrose Place

Boulder, Colorado 80301

Printed in the United States of America

The printing of this volume has been

made possible

through the Henry R. Aldrich

Publication Fund.

This Number of the

Case Histories is Dedicated

to the Memory of

Parker D. Trask,

Editor of Three Previous Case Histories

CONTENTS

FOREWORD

The geologist and the geophysicist have always been aware of the environment as it is the subject of their continual study. The objective of such study is frequently to actively modify, not just passively observe, the environment. Most spectacular of the interactions of man with his geological environment is underground nuclear explosions. For the engineering seismologist (or seismological engineer) underground nuclear explosions present a new tool-for excavating harbors or canals; a new problem - of predicting and measuring high intensity seismic vibrations; and even a new environment - by the creation of novel minerals and isotopes. Man can at last probe the interior of the Earth with vibrations from a controlled source of known origin time and energy level. An understanding, or at least an appreciation, of underground nuclear explosions is requisite for the engineering seismologist today. Therefore, the first two papers in this number are studies of the seismic energy released by underground explosions. The first study covers the free-field zone, before interface effects, within tens of feet of chemical explosions being used to model some of the seismic features of an underground nuclear explosion: the second study covers the 300-meter to 20-kilometer range.

Man is aware not only of the significance of thresholds in the environment, such as critical mass, but also of the sensitivity of the environment to changes in boundary conditions.

One such boundary condition is for a point--injection or withdrawal of fluid from a well. The well operation has occasionally been associated geographically with seismic activity. Most pertinent is to ascertain if the association is causal, not just coincidental. For understanding, and ultimately prediction and control, study of possible causal mechanisms must be explored. Three papers in this number are devoted to the earthquakes which occurred in the vicinity of the well being used for disposal of effluents from the Rocky Mountain Arsenal. One paper compares the effects of waterflooding an oil field.

Another such boundary condition is for a line - the damming of rivers. Changing such a relatively minor boundary has permitted the impoundment of more mass than could have been moved otherwise, even with nuclear explosives. Indeed, the weight of the water impounded by such a dam becomes comparable to the level of stress in the crust of the Earth. These induced stresses have sometimes exceeded the strength of the crust and shocks have resulted. One paper in this number is devoted to local earthquakes caused by the filling of reservoirs.

Not only has man been actively modifying his environment, but also he has been pressed to live in areas having high exposure to natural hazards, such as earthquakes and tsunamis (sometimes erroneously called tidal waves). One paper in this group is devoted to analyzing and estimating the additional risk entailed by construction in a coastal environment subject to tsunami inundation.

Hopefully the engineering seismologist will benefit notably from the data, methods, or results in these case histories. The reader interested in additional details should contact the author directly, as I have taken the liberty of condensing and revising the papers.

Reports which might be appropriate for consideration for inclusion in a future Case Histories Number on Engineering Seismology should be brought to the attention of the Chairman of the Topical Committee on Engineering Seismology, currently Dean S. Carder.

Wm. Mansfield Adams
Editor of this Number
February 1970

University of Hawaii
Honolulu, Hawaii

PREFACE

This report on Engineering Seismology Case Histories is a supplement to a series of similar reports on Engineering Geology which first appeared in May 1957 under the editorship of Parker D. Trask. The purpose of this and other reports which are to follow is to present the seismological aspects of projects having general interest to the civil engineer and the engineering geologist. Topics covered in this first issue are in two categories: ground effects from nuclear and other explosions, and association of local earthquakes with the works of man such as reservoir loading and downhole injection of fluids. Reports in the latter category are in large part drawn from published material appearing in journals of limited circulation; however considering current interest in this topic, inclusion of pioneering case histories and a summary of current work in one volume seems appropriate. The same argument applies to pioneering efforts in predicting ground effects from large explosions. Work on these and other topics of interest to the engineering geologist are scheduled for inclusion in a subsequent issue of Engineering Seismology Case Histories.

Dean S. Carder, Chairman
Topical Committee on Engineering Seismology

QUASI-PERMANENT DEFORMATION IN THE ELASTIC ZONE NEAR EXPLOSIONS

Wm. Mansfield Adams (University of Hawaii, Honolulu, Hawaii)

Abstract

The low frequency (<10 cycles per second) content of body waves from chemical explosions has been studied. The traces recorded on an FM magnetic tape recording system have been reduced for instrumental effects. The corrected data have a coherent but peculiar distribution. The shape of the distribution suggests an hypothesis about the nature and behavior of the medium. Not all facts are consistent with the proposed hypothesis. This may be due to low level of the energy spectrum considered and the nature of the FM recording system used. Digital recording, permitting greater accuracy, is suggested.

CONTENTS

ILLUSTRATIONS

INTRODUCTION

The decoupling of underground explosions (Adams and Allen, 1961) and the effect of the medium surround the explosion (Adams and Swift, 1961) are difficult and expensive aspects of nuclear explosions to study.

However, for these aspects, chemical explosions may be considered as models of the nuclear explosions. The dominant frequency of small nuclear explosions, say 10 kilotons, is in the band of less than 10 cycles per second, so the low frequency content of the chemical explosions must be considered. There is no adequate analytical theory, even for the free-field region close to the source without interface effects, hence an experimental effort seems appropriate. No experiment specifically for this purpose has been conducted, but data available from two other experiments appeared suitable for analysis. These

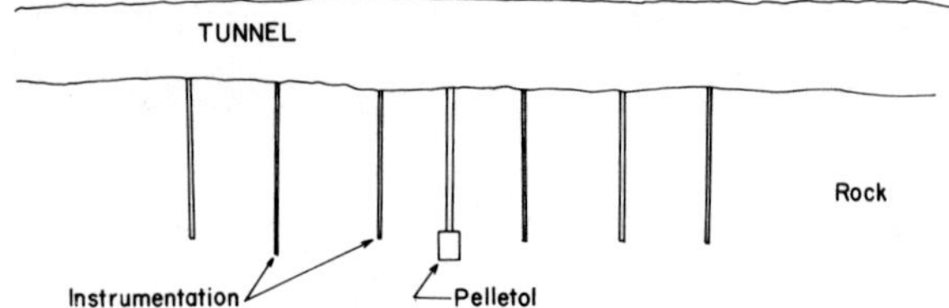

Fig. 1. Experimental Environment

were chemical explosives shot in an experimental array (Fig. 1) a hundred feet below the floor of a tunnel, with instruments for radial particle velocity at nominal distances of 35, 50, and 80 feet. The charge sizes ranged from 100 to 1000 pounds of pelletol and occurred in two media: Oak Spring Tuff at the Nevada Test Site, and salt in the Carey Salt Mine of the Winnfield Salt Dome near Winnfield, Louisiana. These shots had been initially planned for a study of the higher frequency content (30 to 60 cycles per second) (Adams and Swift, 1961); here the low-frequency means below ten cycles per second. This may be approximated by the position at 25 milliseconds after arrival; this position is called the quasi-permanent displacement.

The purpose of this report is to describe an experimental estimation of the relative low-frequency content in tamped chemical explosions. Consider the situation surrounding the shot a few milliseconds after detonation (Fig. 2). The initially cylindrical charges has formed an approximately spherical cavity which is surrounded by a crushed region. Radiating vertical plane cracks are dominant in the cracked zone. The envelope of the cracks is considered somewhat arbitrarily to define the start of the elastic zone. The low frequencies should be observed in the elastic zone to avoid the non-linear effects and heterogeneity of the inelastic region. The quasi-permanent displacements in the elastic zone are considered to be the deformation which occurs in the elastic zone for short times--while the cavity is in the process of cooling down, lowering the interior pressure, and thus slowly reducing the displacements in the elastic zone before a significant cooling can occur, allowing the pressure in the cavity to drop. This quasi-permanent displacement approximates the low frequency content. Quasi-permanent displacement should be easier to measure than the low-frequency content of the signal.

Fig. 2. Post-shot Condition of Cavity Region

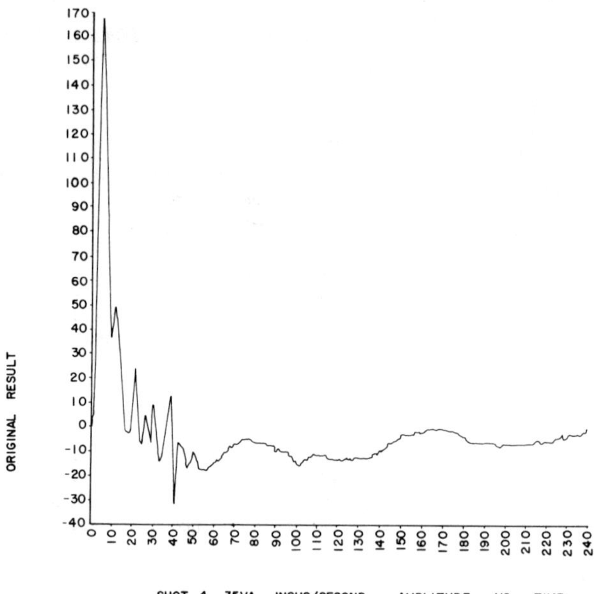

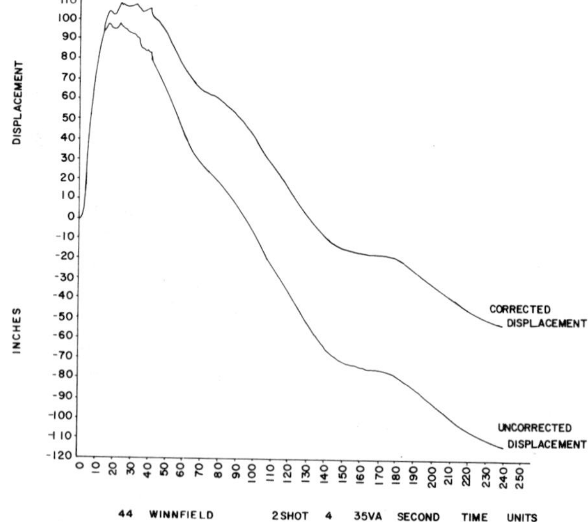

Fig. 3. Relative velocity (inches per second) versus time (milliseconds).

The first way to determine the quasi-permanent displacement is to assume that the material is incompressible and distribute the newly created cavity volume in a shell at a distance from the shock point. Assume that the displacement in the elastic zone is not significantly relaxed by the pressure drop in the cavity as it cools but is sustained by the compressive strength of the inelastic region. To measure the newly created void, one may re-enter the shot point and determine the volume. Two methods have been used: one is to mine back into the shot point; the other is to drill into the shot point and insert a measured quantity of sand or brine. These methods give comparable results which indicate that for salt with 910 feet overburden about 0.062 cubic foot of volume is created for each pound of pelletol used. (No definitive results were obtained on the tuff experiment.) The volume so determined is considered to be distributed in a shell at a given distance of instrumentation.

The other procedure, which does not require the assumption of non-relaxation, is to observe the particle motion in the elastic zone and from such measurements determine the displacement during the times of interest. As there does not exist a suitable displacement gage, a velocity gage has been used and the observation integrated. Figure 3 shows the typical observed velocity recording. In Figure 4 is shown the uncorrected displacement (integrated relative velocity) and the corrected displacement (reduced for the motion of the inertial mass using drift correction for the preliminary integrations). Note that the position of the ground about 25 milliseconds after detonation of the shot is negative, corresponding to an inward collapse.

Table 1 presents the quasi-permanent displacement in mils. The data are presented in graphical form in Figures 5 and 6 for halite and tuff, respectively. The peculiar distribution of the data suggested the back-to-back log-log plot. The dashed line corresponds to the displacement caused by distribution of 0.062 cubic foot per pound of pelletol at the respective radius. Although this value is derived from data in halite, it is also shown on the tuff figure to permit comparison of the data on the two figures.

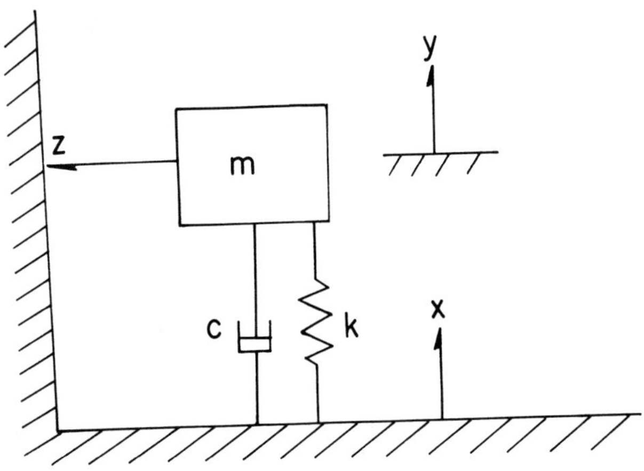

Fig. 4. Integrated relative velocity (inches) versus time (milliseconds). Upper trace is described in text.

There are three significant features of these graphs:

1. The data are peculiarly distributed. Instead of being Gaussian about an inverse square decay line for positive amplitudes, the data scatter about two inverse square decay lines, one for positive and one for negative amplitudes.
2. The tuff to salt ratio of the data in the radial range of 50 to 100 feet, is about 0.2 or less. This is considerably at variance with the value of 1.6 reported earlier for the frequency range of 30 to 60 cycles per second (Adams and Swift, 1961).
3. The noticeable break in slope of the plotted data, at 100 feet in halite, and from 80

TABLE I

Shot	Instrument	Distance Feet	QUASI-PERMANENT DISPLACEMENT (MILS)
COWBOY			
7	2.2- 2V	50.3	+ 50
7	2.2- 1V	79.4	+ 42.5
7	2.2- 3V	49.0	+112.5
7	2.2- 4V	80.6	- 9
9	2.3- 1V	79.6	20
9	2.3- 2V	50.5	42
9	2.3- 4V	79.7	- 4
9	2.3-17VB	368.8	?
11	2.4-15V	208.2	?
11	2.4- 1V	77.5	+ 42.5
11	2.1-17VB	477.7	- 0.8
16	1.2- 3V	173	+ 1.25
16	1.2- 4V	157	0
16	1.3- 8V	207	+ .25
16	2.1-17VB	431	- .8
17	1.2- 4V	167	+ 1.25
17	1.3- 7V	99	4
17	1.3- 8V	116	12
HOBO			
1	50VA	51.1	-152
1	35VB	35.0	+ 65
1	50VB	51.1	-145
1	80V	78.2	0
1	196V	195.4	+ 1
2	35VA	36.0	- 55
2	50VA	52.1	- 8
2	35VB	36.0	-105
2	50VB	52.1	- 30
2	80VB	76.5	- 5
2	80VA	76.5	+ 7.5
2	375VA	369.6	0
2	375VB	369.6	+ 1
3	80V1	81.5	+ 1.5
3	375V1	373.7	+ 2.5
3	35V	37.2	12
3	50V	50.6	14
3	80V2	81.5	- 7.5
3	375V2	373.7	+ 1.2
4	35VA	34.9	-106
4	35VB	34.9	- 52
4	50VA	47.9	- 10
4	50VB	47.9	+ 7
4	5V	14.4	572
4	290VA	289.4	1.0
4	290VB	289.4	3

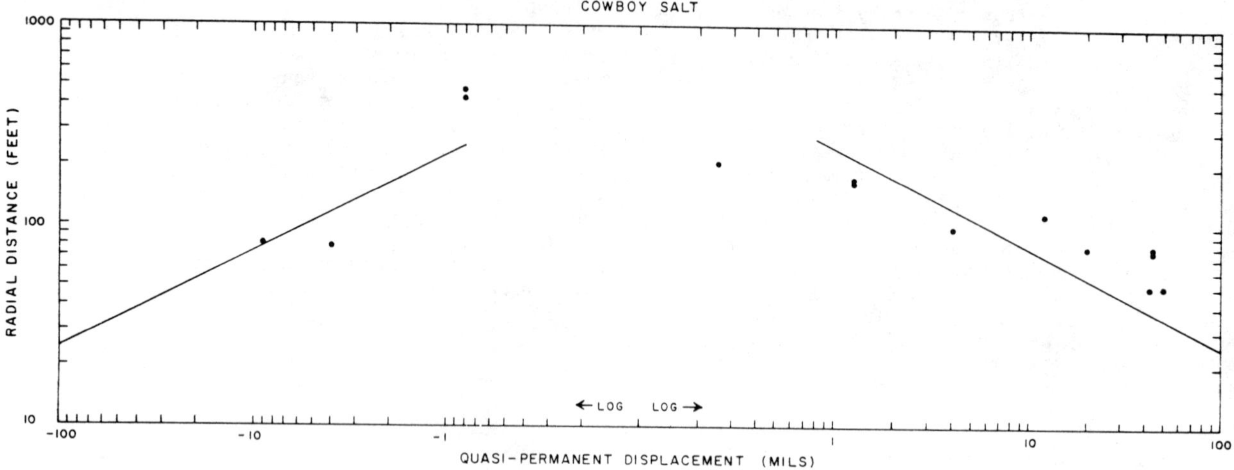

Fig. 5. Displacement (reduced integrated velocity, in mils) versus radial distance from a shotpoint in feet. Halite matrix.

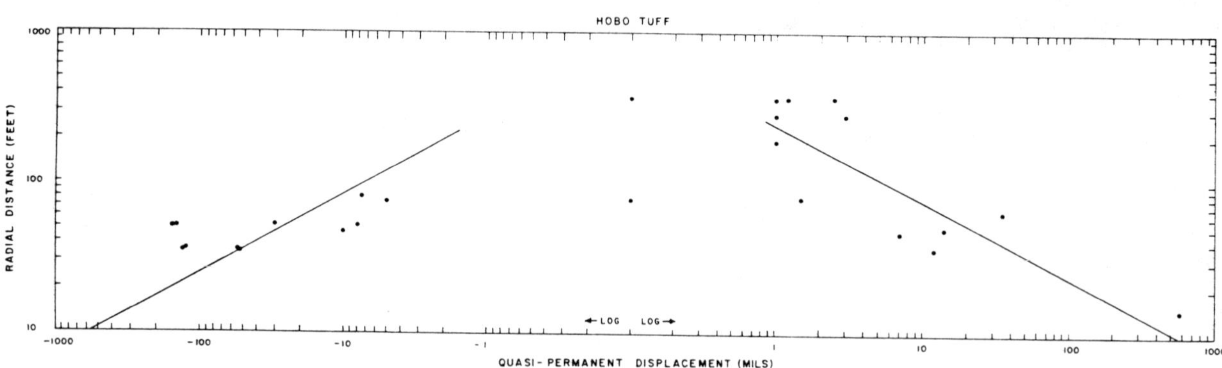

Fig. 6. Displacement (reduced integrated velocity, in mils) versus radial distance from shotpoint in feet. Tuff matrix.

to 100 feet in tuff, may determine the effective elastic radius.

DISCUSSION ON INSTRUMENTATION

Recording for the shots in the tuff was done for each geophone on two different tape transports, providing backup. These used FM modulation with ±40% deviation. In Figure 7, we see two traces played back from the two tape recorders for the same geophone. The differences are considered to be due to errors in the record and playback functions. They cause significant differences in the integrated results. The data in Figure 8 are considered more typical. In order to have some comparison with the variation shown by the two traces (Fig. 7), we also present two traces recorded on different instruments on diametrically opposite positions about 1000 pounds in salt. The data are presented in Figure 9 and show variation comparable to that due to different record systems.

We now consider the error between identical FM record systems recording the same seismometer. Assuming a 40% deviation, and a maximum cumulative flutter of 0.50% peak-to-peak, then the maximum record error would be 2.5 x .50% or 1.25%,

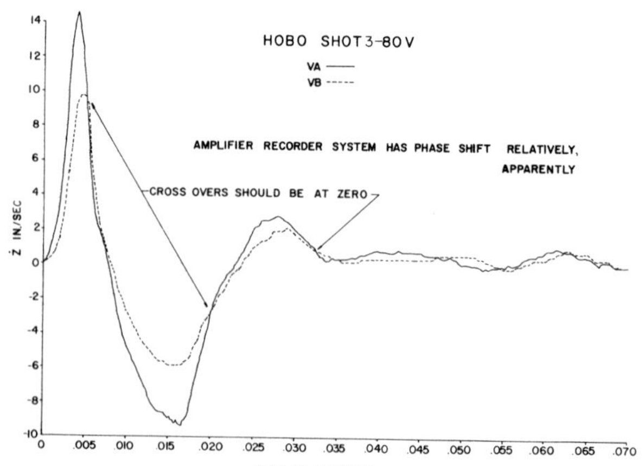

Fig. 7. Extreme differences in two traces played back from two tape recorders recording the same geophone. Tuff matrix.

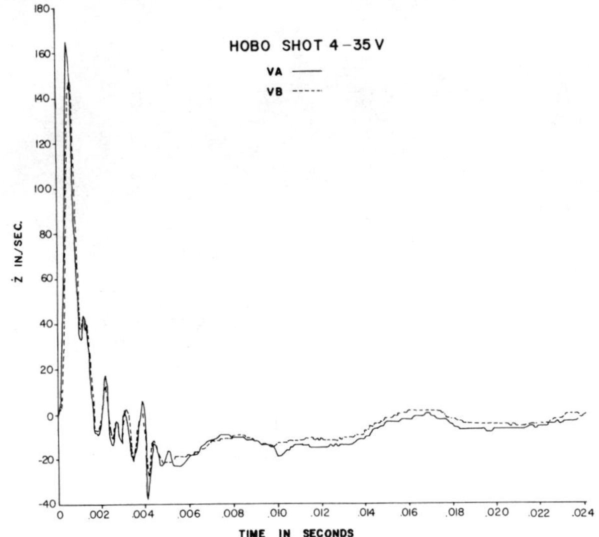

Fig. 8. Typical differences in two traces played back from two tape recorders recording the same geophone. Tuff matrix.

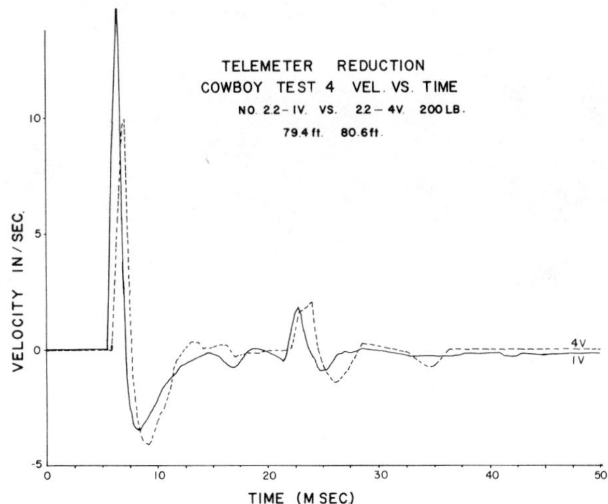

Fig. 9. Typical differences in two traces recorded on the same tape recorder at approximately the same distance from the same shot. Halite matrix.

with the maximum playback error being the same, 1.25% for a total maximum of 2.5%. Such an error is permissible if actually random.

DISCUSSION ON RESULTS

The data in Figures 5 and 6 show an unusual distribution, necessitating the unusual back-to-back log-log graphs. These unusual results suggest an alternative model for the ground motion close-in to a shotpoint.

Consider that the shotpoint is surrounded by material having linear compressive strength but no tensile strength, e.g. rubble. Assume that the explosion sets the material into sinusoidal oscillation with an amplitude that decreases with distance. An observation is then taken at some random time within a fixed time interval after the detonation (or arrival time).

The density function for a sinusoidal path is bimodal, with peaks at the two extremities. Thus an observation is most probable at the extremities. This interpretation produces, however, a mean value of zero at all distances. For twenty-five milliseconds after the arrival time, this does not appear probable.

REFERENCES CITED

Adams, W. M. and L. M. Swift 1961. The effect of shotpoint medium on seismic coupling. *Geophysics*. Vol. XXVI. p. 765-771.

Adams, W. M. and C. DeWitt Allen 1961. Seismic decoupling for explosions in spherical underground cavities. *Geophysics*. Vol. XXVI. p. 772-799.

SEISMIC GROUND EFFECTS FROM NUCLEAR EXPLOSIONS

Dean S. Carder (Environmental Science Services Administration/
Earth Science Laboratories, San Francisco, California)

Abstract

Some of the results of measuring surface effects from many nuclear explosions in the Nevada Test Site (NTS) and elsewhere, which may be of interest to Civil Defense and to the scientific public, are given and discussed. The distance range are in the elastic zone and measurements are confined in general to the surface within the radii from 0.3 to 20 kilometers. However, projections are made to greater distances.

CONTENTS

ILLUSTRATIONS

INTRODUCTION

The instruments used in this distance range are the standard strong-motion seismographs which have been used for years by the U. S. Coast & Geodetic Survey for measuring strong earthquake motion (Carder and Cloud, 1956). They are direct recording seismographs consisting of simply constructed compound pendulums damped by permanent magnets, together with timing clocks and 30 centimeter photographic tape cameras. Direct recording is accomplished by a focused light beam reflected to the photographic tape from mirrors attached to the pendulums near the axes of oscillation. These instruments are capable of recording ground motion in the acceleration range from

0.001 g or less to 3.0 g, and transitory ground displacements of 15 centimeters or less. For this purpose, two sets of seismometers are required; one set having pendulum periods substantially less than the periods of expected ground motion, and the other with periods substantially greater than the ground periods. Each set measures ground motion in the vertical and two horizontal directions. To accomplish this purpose, the periods of the accelerometer (short-period) pendulums range from 0.03 to 0.15 seconds, depending on the distances from shot zero, and the displacement meter (long-period) pendulums are from 2 to 3 seconds with magnifications from 0.5 to 10. At distances approaching 20 kilometers, magnifications of the displacement type seismographs are as high as 120.

PREDICTIONS OF EXPECTED GROUND EFFECTS

In the preparation of instrumentation for a given shot, an estimate of expected ground acceleration and amplitude are considered invaluable to insure that the records will be fully contained on the tapes yet large enough to be measurable. For this, data from earlier explosions were utilized. The Rainier explosion of 19 September 1957 was the first underground nuclear explosion to receive wide attention. In the preparation of suitable gain settings for this explosion, Mr. W. K. Cloud and I took advantage of two sources of information:

1. On February 21, 1957, a 10-ton high explosive shot was fired by the Geological Survey in the tuff near the Rainier site, and on April 5, 1957, a 50-ton shot was fired in nearly the same location. Both of these shots were recorded at the Boulder City and Eureka, Nevada, Coast & Geodetic Survey teleseismic stations, at respective distances of 180 and 260 kilometers. The latter shot was recorded by a strong-motion seismograph on the surface, 95 meters from the explosives charge. (Fig. 1 shows the station layout and the resulting record.) The Eureka and Boulder City data indicate that ground amplitudes were observed to vary as the 1/2 to the 1.0 power of the explosives charge.
2. In 1945 and 1946, seismic measurements of ground effects from controlled explosions of ammunition dumps on the U. S. Navy Proving Grounds near Arco, Idaho, were made by the Coast & Geodetic Survey (Carder, 1948, and Carder and Cloud, 1959). From the latter investigation, attenuation with distance as the 2.0 power seemed to give reasonable fits to maximum ground accelerations and transitory ground displacements.

Using information just cited and putting Y as the equivalent weight of the explosive in kilotons, R the distance in kilometers, a_g, the acceleration in units of gravity, and A_{cm} the zero-to-peak ground displacement in centimeters.

$$a_2/a_1 = A_2/A_1 = (Y_2/Y_1)^{0.75} \quad (R_1/R_2)^2$$

and more specifically by using information gained from the 50-ton explosion and applying to a generalized formula, we have

$$a_g = 0.17\ Y^{0.75}\ R^{-2}$$

and

$$A_{cm} = 0.6\ Y^{0.75}\ R^{-2}$$

With the use of these formulas, ground motions resulting from the Rainier explosion of 19 September 1957 were satisfactorily predicted.

However, for better results, certain revisions are made, and after the Hardtack Phase II series of September and October, 1958, the generalized formulas for maximum ground acceleration and transitory ground displacements in the range from 0.4 to 20 kilometers were considered to lie within the envelope represented by

$$a_g = 0.1\ Y^{0.75}\ R^{-2} \quad (1a)$$

$$a_g = 0.3\ Y^{0.75}\ R^{-2} \quad (1b)$$

$$A_{cm} = 0.3\ Y^{0.75}\ R^{-1} \times 10^{-0.013R} \quad (2)$$

if the station is on crystalline rock, or more simply

$$A_{cm} = Y^{0.75}\ R^{-1.7} \quad (3)$$

Y is the yield in kilotons (907 metric tons) equivalent of TNT, and the shot in every case is in bedded tuff. These formulas were later found to be applicable, with certain revisions, out to sev-

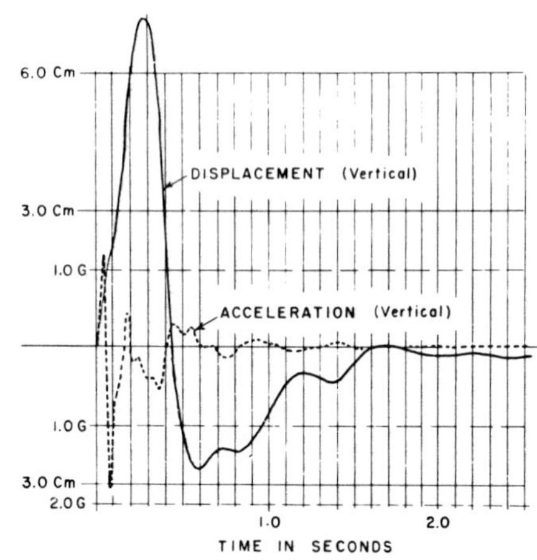

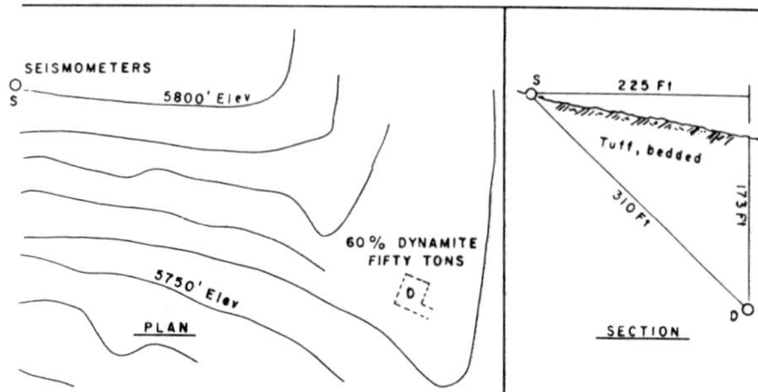

Fig. 1. Location of a strong-motion seismograph station, 50-ton high-explosive shot, and a tracing of the resulting record. From Carder and Cloud (1959).

eral hundred kilometers and for yields up to 1200 kilotons (Cloud and Carder, 1969). The revisions will be discussed later.

GROUND ACCELERATIONS

Figure 2 is a plot of the envelope of equation 1 showing available data (through 1963) on underground explosions in tuff in the Nevada Test Site (NTS) scaled to the equivalent of 1 kiloton. The shots represented in this plot varied from 13 tons to 19 kilotons. The formula is not applicable to ground frequencies greater than 15 cycles per second. A replot of the same envelope as in equation 1 shows different data (Fig. 3). The circles represent maximum accelerations from the Bilby shot (200 kilotons) scaled to 1 kiloton on the 0.75 law, and the triangles represent Gnome data scaled from 3.1 kilotons.

In this discussion, the Bilby data will be considered first. This shot was fired more than 600 meters beneath the floor of a desert valley. Although the floor of the valley is covered by 300 or more meters of alluvium, the shot itself was in tuff, as in the Rainier and Hardtack series. The stations, however, were in general on different terrain and the distance versus acceleration results could be expected to follow a slightly different pattern. Nevertheless, as indicated in Fig. 3, equation 1 predicted ground accelerations from Bilby with fair accuracy.

An attempt was made to refine equation 1 using Rainier and Hardtack data and the general formula

$$a_g = CY^xR^{-y}.$$

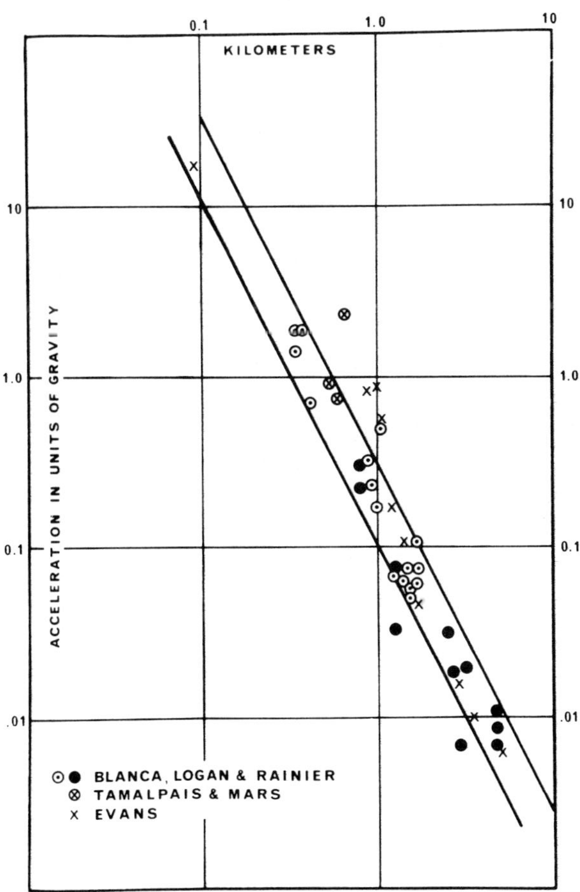

Fig. 2. Acceleration versus distance scaled to 1 kt yield, from early NTS underground shots. Filled circles, stations on crystalline rock. From Carder and Mickey (1962).

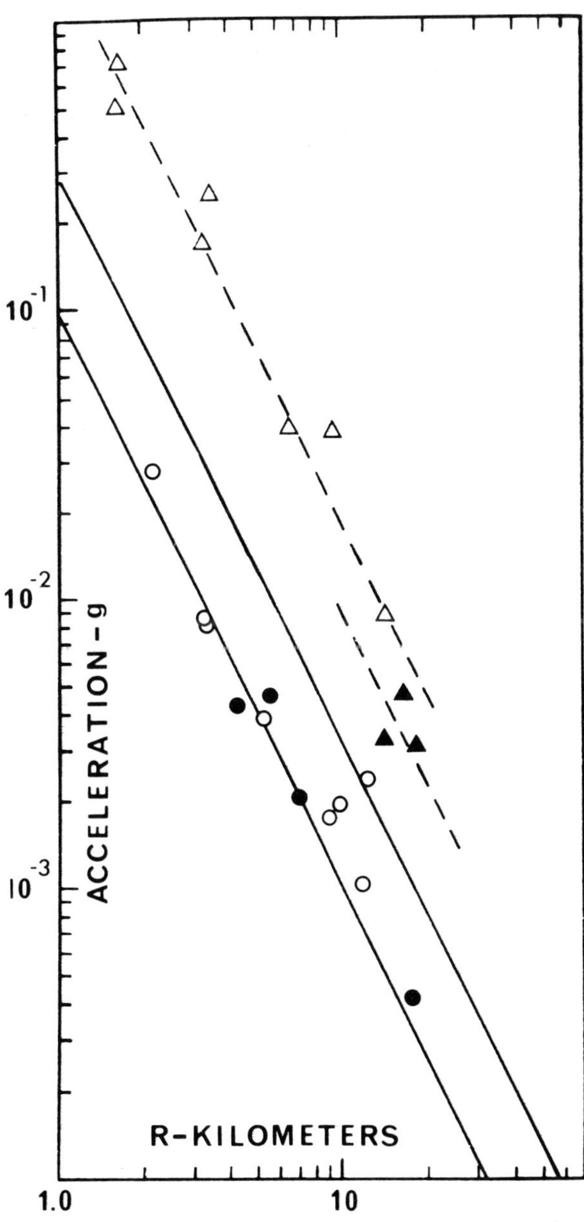

Fig. 3. Accelerations from Bilby (circles) and Gnome (triangles) scaled to 1 kt yield. Filled circles on rock. Filled triangles underground. Solid lines same as for Figure 2.

It appears that x is a function of distance having the value of 0.6 at 1 kilometer and about 0.8 at 10 kilometers, and that y is related to yield by the formulas

$$y = 0.32\ (7\text{-}\log Y).$$

Replacing the exponents in equation 1 by x and y defined above, and applying them to the Bilby shot gives us

$$a_g = 4.8\ R^{-1.4}.$$

This is slightly higher than the maximum observed values and a factor of 2 above the median. Mickey's (1964) general equations give

$$a_g = 7.2\ R^{-1.4}$$

for a shot in tuff corresponding to Bilby,

and

$$a_g = 3\ R^{-1.4}$$

for a shot in alluvium. In actuality the source conditions probably lie between tuff and alluvium. We will, therefore, bring forward the generalized formula

$$a_g = CY^xR^{-y} \qquad (4)$$

where $x = 0.6\ R^{0.12}$; and $y = 0.3\ (7\text{-}\log Y)$; and C has values according to the source medium as follows:

granite, C = 0.5
tuff, C = 0.2
alluvium, C 0.08 (5)
bedded salt, C = 2.0 (surface recording)
bedded salt, C = 1.0 (underground recording)

The first three values of C are based on Mickey's general results of his study of seismic wave propogation from underground nuclear explosions in the NTS since September 1961 (Mickey, 1964). The last two values are applicable to the Gnome shot in a highly different environment.

The NTS is in an intermontane desert fault block area. Absorption of seismic waves in the media near the surface is relatively high. The Gnome shot, on the other hand, is in a salt bed 360 meters beneath the surface and in layered media. Absorption to seismic waves associated with the higher frequencies, 5 to 10 cycles per second, is relatively low.

The Gnome site is located about 40 kilometers southeast of Carlsbad, New Mexico, near the western boundary of the Great Plains which extend to the east and northeast. The underlying rock consists of relatively flat lying sediments, largely limestone, with a salt member about 300 meters thick. The Western Cordillera begins a few tens of kilometers to the west. This setting has importance here because it is found that seismic wave speeds and seismic energy distribution in the two physiographic environments are highly dissimilar. The travel-times over the first 1500 kilometers toward the northwest, through the upper mantle under the Cordillera, is slower by as much as 12 seconds, than over a like distance under the plains to the east and northwast. Furthermore, it is found that seismic energy absorption under the plains is lower by one or two orders of magnitude. In other words, the upper mantle and crustal layers under the plains serve as a wave guide to great distances. This may account in part for the relatively high value of the parameter C in equations 1 and 4 when applied to the Gnome shot. Only two yield levels applicable to the Gnome site are available: the pre-Gnome series of 0.09, 0.375, and 3.1 tons and the Gnome shot of 3.1 kilotons. Pre-Gnome accelerations were associated with high frequency waves of 30 to 80 cycles per second, and the Gnome with frequencies of 5 to 12 cycles per second. Nevertheless, equation 4 scales to pre-Gnome accelerations within a factor of 2, but pre-Gnome accelerations within themselves do not follow this rule.

The variable exponents x and y in equation 4 have been explained as follows: as the yield y increases, maximum accelerations are associated with lower frequencies which attenuate with distance at a lower rate. Equation 4 probably is more applicable to acceleration propagation in the NTS for low yields and short distances than equation 1. Nevertheless, as explained earlier, use of equation 1 with the constant term as defined by C above may be used to predict ground accelerations from underground nuclear explosions in the NTS with fair accuracy even to high yields and relatively long distances.

GROUND PARTICLE DISPLACEMENTS

Ground displacements resulting from underground explosions, or in fact, from any other disturbance, are more difficult to predict. They are more dependent on the nature of the ground underlying the receiving station than are ground aceelerations. It can be shown that a variety of scaling laws may give adequate results if the distance range is confined to 20 kilometers. Application of equation 3 predicts with fair accuracy displacements resulting from most of the nuclear explosions fired in the NTS. Fair accuracy is defined here as within plus or minus a half order of magnitude. No other equation gives better results because of the scatter of data. However, to predict maximum ground amplitudes, a slightly higher yield factor seems appropriate. We will test the formula

$$A_{cm} = Y^{0.85}\ R^{-1.7} \qquad (5)$$

in two environments.

Bilby data from Mickey's (1964) report have been scaled to 1 kiloton by this law (Fig. 4). The Gnome and pre-Gnome data (Carder, et al., 1959, 1962) are scaled by the same law (Fig. 5). Scalings in these two illustrations are for yields from 0.09 ton of high explosives to 3,100 tons equivalent in the same environment, and from 13 tons to 200,000 tons equivalent in another environment.

It is apparent that the environment of the shot point is immaterial from the standpoint of ground particle displacement (Figs. 4 and 5). Ground amplitudes in a range to 17 kilometers are about as great from shots in tuff as from a shot in a salt bed. However, frequencies from the Gnome shot are at least twice those from most of the NTS shots. The latter are 1 to 2 cycles per second, and the former to 6.5 kilometers are about 4 cycles per second.

The following observations should be made relative to ground particle amplitudes from underground explosions in the kiloton range or higher:

1. If the source is in crystalline rock, ground frequencies and associated energies received on the surface are higher. Shots in tuff or alluvium are associated with relatively long periods--1 to 2 seconds--and with longer periods from higher yields.

2. Ground amplitudes on deep alluvium from shots in tuff or alluvium are generally higher by a factor of 2 to 4 or more, than ground amplitudes on crystalline rock.

GROUND PARTICLE VELOCITIES

The Coast and Geodetic Survey strong-motion instruments employed to monitor nuclear explosions in the distance range to 20 kilometers do not in general record ground particle velocities directly. However, these velocities may be obtained indirectly by integrating the accelerations or differentiating the displacements. Both operations have been attempted on a limited basis yielding almost identical results.

GROUND EFFECTS FROM AIR BURSTS

Ground particle accelerations and displacements from a shot in air but near the ground generally are about an order of magnitude less than from similar shots that are fully contained underground. In this case, damage from a shock wave in air probably exceeds possible damage caused by seismic waves from tee same source. The energy input into the ground from shallow cratering shots is about the same as surface or near surface air bursts.

SEISMIC ENERGIES FROM UNDERGROUND EXPLOSIONS

Seismic energies from underground explosions depend on the yield and the nature of the underground. The portion of the energy release which enters the ground as seismic waves is usually from 0.2 to 0.4 percent of the total yield.

DAMAGE FROM UNDERGROUND EXPLOSIONS

Damage from underground explosions is a function of acceleration, ground amplitude, frequency and duration. More correctly, it is considered a

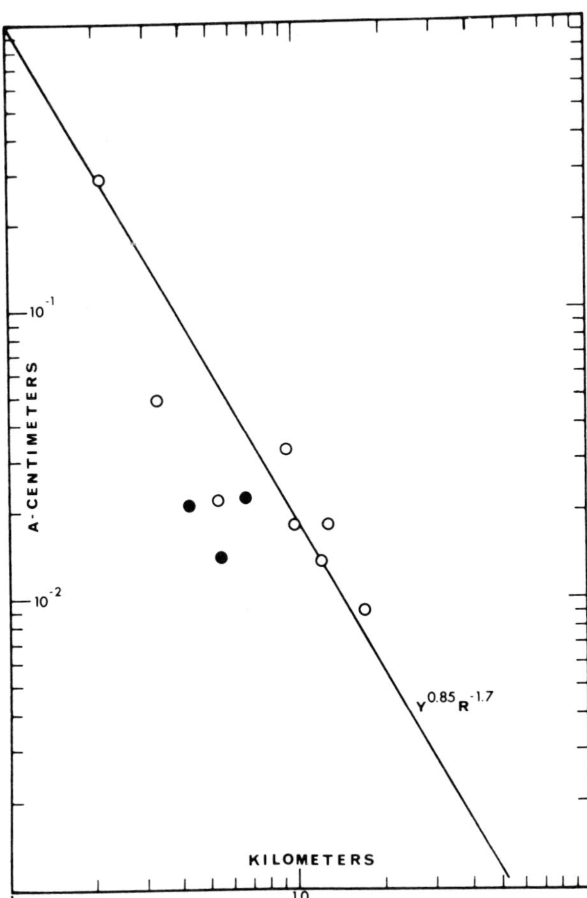

Fig. 4. Bilby displacement data scaled to 1 kt yield. Filled circles on rock.

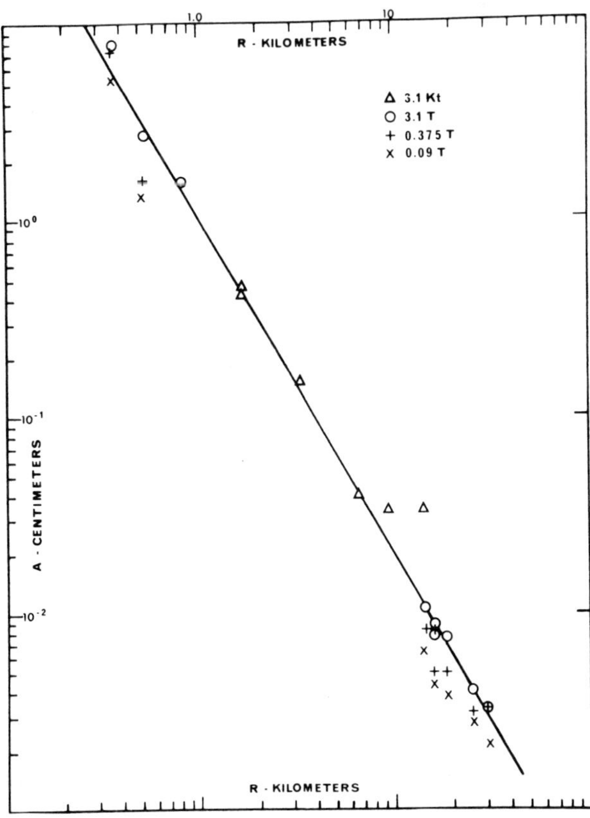

Fig. 5. Gnome and pre-Gnome displacement data scaled to 1 kt yield.

function of ground particle velocity and duration since ground particle velocity is a measure of energy.

According to the U.S. Bureau of Mines, the upper limit of the safe zone is 5 centimeters per second. Minor damage begins about 13 centimeters per second. According to other investigators (Langefors, et al., 1958; Crandell, 1949; Edwards and Northwood, 1960) structure damage begins with ground particle velocities of about 10 centimeters per second. For safety purposes, the criteria for minor damage begins at 6 centimeters per second and has a maximum duration of 2 cycles. Criteria for earthquakes are somewhat less than this because of the longer duration of earthquake vibrations with consequent possibilities of resonance.

Figure 6 shows generalized estimates of damage as functions of distance and yield based on expected ground particle velocities of 6 centimeters per second for the threshold of minor damage.

These estimates are based on the formula

$$V_{cm/sec} = 2\pi\, Y^{0.85}\, R^{-n}\, T^{-1}$$

where T = 1 second and n = 1.73 (which is generally applicable in the NTS), and T = 0.5 seconds and n = 1.6 (presumed applicable in the Great Plains). (Carder, 1962.)

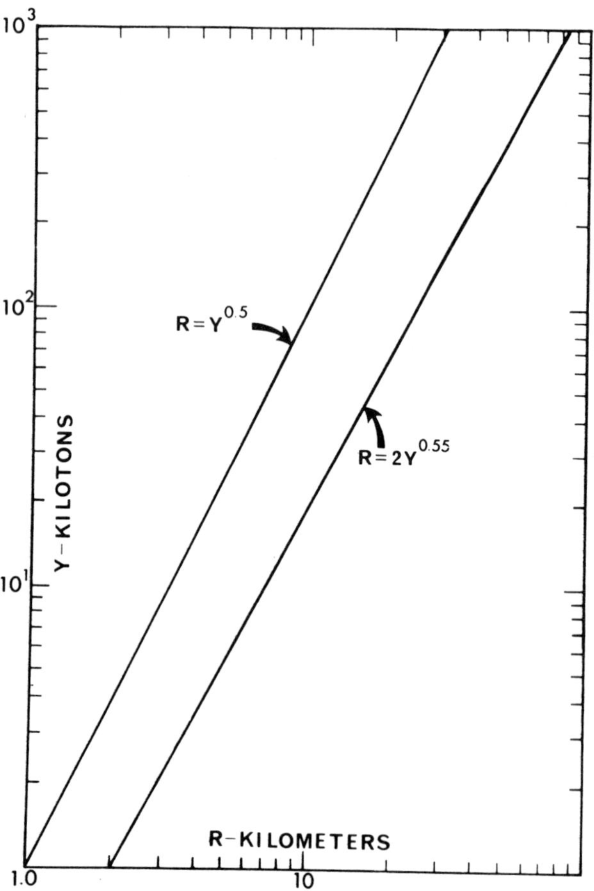

Fig. 6. Damage threshold from underground explosions; left - Western Cordillera; right - Great Plains. Damage distance to left of solid line.

CONCLUSIONS

If any energy source is a fully contained underground nuclear explosion:

1. Structure damage is believed to be a function of ground particle velocity. The threshold of minor damage is believed associated with ground velocities of 6 centimeters per second. Using this criterion, the outer limit in kilometers of minor damage is associated with yield by the formula

$$R_{km} = Y^{0.5}$$

where Y is the yield in kilotons of a fully contained buried shot and the source area is the intermountain U.S. (e.g. the NTS). East of the Rocky Mountains the damage threshold is more nearly

$$R_{km} = 2Y^{0.55}$$

2. Ground particle acceleration follow the general formula

$$a_g = C\, Y^{0.75}\, R^{-2}$$

where a_g is in units of gravity, and C is a parameter depending on the source medium. Estimating values of C for various media are defined in equation 5.

This scale has been applied to yields of 3.1 kilotons near the Gnome site, and from 13 tons to 200 kilotons in the NTS. The distance range is from 0.4 kilometers to 17 kilometers for 1 kiloton shots and from 1 kilometer or more to 20 kilometers for yields of 5 kilotons or more. For shots in the megaton range, it is found fairly applicable to distances of 200 kilometers. A more generalized formula with variable exponents is given in the text. It is applicable for low yields and short distances.

3. Maximum zero-to-peak ground particle displacements are usually scattered by an order of magnitude. The formula

$$A_{cm} = Y^{0.85}\, R^{-1.73}$$

is applicable to yields of from 0.09 tons to 3.1 kilotons near the Gnome site and from 13 tons to 200 kilotons in the NTS.

4. If the shot is an air burst, near or on the ground, use Y/7 in place of Y in the above formulas. This applies to seismic effects only and not to effects from a shock wave in the air.

5. Maximum amplitudes are associated with longer periods on alluvium than on rock, and longer periods result from higher yields.

Formulas in items 2 and 3 when extrapolated to the equivalent of the two megaton underground explosions Boxcar and Benham came within a half order of predicting ground effects resulting from these explosions in the distance range to 200 kilometers and beyond (Cloud and Carder, 1969). However, for ground accelerations in the NTS, the formula

$$a_g = 0.1\ Y^{0.85}\ R^{-1.73}$$

would give better results.

Acknowledgements are due to W. V. Mickey of the Coast and Geodetic Survey for data on the Bilby shot and ground velocity information; also to W. K. Cloud of the Seismological Field Survey, Coast and Geodetic Survey for his assistance in developing initial formulas.

REFERENCES CITED

Carder, D. S. and Cloud W. K. 1959. Surface motion from large underground explosions. *J. Geoph. Res.* 64, pp. 1471-87.

Carder, D. S., Murphy, L. M., Cloud, W. K., and Pearce, T. H. 1960. Seismic data from natural phenomena and high-explosive tests near Carlsbad, New Mexico. *Internal Report to the U.C. Lawrence Radiation Laboratory*. January, 1960

Carder, D. S., Murphy, L. M., Pearce, T. H., Mickey, W. V., and Cloud, W. K. 1961. Ground motions from underground explosions. Final Report WT-1741. *Operation Hardtack, Phase II,* March, 1961.

Carder, D. S. and Mickey, W. V. 1962. Ground effects from underground explosions. *Bull. Seism. Soc. Am.* 52, pp. 67-75, January 1962.

Carder, D. S. 1962. Ground effects from the GNOME and LOGAN explosions. *Bull Seism. Soc. Am.* 52, pp. 1047-1056.

Cloud, W. K. and Carder, D. S. 1969. Ground effects from the BOXCAR and BENHAM nuclear explosions. *Bull. Seism. Soc. Am.* 59, December, 1969.

Crandell, F. J. 1949. Ground vibrations due to blasting and its effect upon structures. *Journal of the Boston Society of Civil Engineers,* Vol. 36, No. 2, April 1949.

Duvall, W. I. and Fogelson, E. E. 1962. Review of criteria for estimating damage to residences from blasting vibrations. *Bureau of Mines Report of Investigations* 5968.

Edwards, A. T. and Northwood, T. D. 1960. Experimental studies of the effects of blasting on structures. *The Engineer,* Vol. 210, September 30, 1960.

Frantti, G. E. 1963. Seismic energy from ripple-fired explosions. *Earthquake Notes,* Vol. XXXIV, No. 2, June, 1963.

Mickey, W. V. 1964. Seismic wave propagation. *Internal Report, Coast and Geodetic Survey,* March, 1964.

Thoenen, J. R. and Windes, S. L. 1942. Seismic effects of quarry blasting. *Bureau of Mines Bulletin* 442.

PRESSURE INJECTION DISPOSAL WELL, ROCKY MOUNTAIN ARSENAL, DENVER, COLORADO*

Louis J. Scopel

Abstract

The pressure injection disposal well completed for the U. S. Corps of Engineers at the Rocky Mountain Arsenal, 10 miles northeast of Denver, exemplifies a revolutionary concept in industrial waste disposal. This method is one which has been used by the oil industry for many years. The well has two major distinctions: it is the deepest well drilled in the Denver Basin, and it may provide a precedent for solving waste disposal problems which are anticipated in the future industrial growth of the Denver area.

The well was drilled on the east flank and near the axis of the Denver Basin and penetrated Tertiary through pre-Pennsylvanian sediments. Drilling was completed at a depth of 12,045 feet in Precambrian gneiss. The sediments from Tertiary through Pennsylvanian are in normal sequence. They show only minor variations in thickness and lithology from sections in adjacent areas. The occurrence in this part of the Denver Basin of pre-Pennsylvanian sediments, possibly Ordovician in age, was unexpected and may shed valuable light on the paleography of the Denver area.

Lost circulation was a major drilling problem in the Paleozoic sediments and in the Precambrian gneiss. Slow penetration rates resulting from the induration of the Paleozoic sediments necessitated the use of special hard-formation drilling bits. Core information shows that the Permian and Pennsylvanian sediments are fractured and have a rock matrix of low porosity and permeability. Drill-stem tests indicate that the Paleozoic sediments contain low-pressure reservoirs.

CONTENTS

*Reprinted, with permission and editing, from *The Mountain Geologist,* V. 1, No. 1, pp. 35-42.

ILLUSTRATIONS

INTRODUCTION

The pressure injection disposal well which was completed at the Rocky Mountain Arsenal for the U. S. Corps of Engineers, NW/4 NE/4 sec. 26, T. 2 S., R. 67 W., is an example of a continuation of the new trend in waste disposal. The concept of subterranean waste-water disposal is not unique--the oil industry has been practicing this method for many years. The well is one of the first to be drilled in this area, however, for the sole purpose of industrial waste disposal rather than for the disposal of oil field saline waters.

Three aspects of the well distinguish it from others drilled in the area: (1) it is the deepest well drilled in the Denver Basin, (2) it penetrated hitherto unsuspected sediments of pre-Pennsylvanian age, and (3) it is the pace-setter for the Denver industrial area and pioneers a possible solution to a problem of future industrial growth and development.

This paper will consider certain geological and mechanical factors which are significant in a well drilled for waste disposal. Among other factors to be discussed are:

1. The thickness, extent, and impervious character of the sediments that would serve to restrain contaminated fluids within reservoirs;
2. The permeability and porosity of rocks into which fluids may be injected;
3. Drilling and casing procedures which will prevent contamination of potable water horizons and prevent damage to potential or actual oil and gas producing formations.

HISTORY OF DISPOSAL PROBLEM

The Rocky Mountain Arsenal began manufacturing chemical mechanisms for national defense on a large scale in 1942. A detrimental by-product was contaminated water. The Chemical Corps decided that the simplest way to eliminate all but the solid waste was to collect the solution in dirt tanks and spray the liquid into the atmosphere, thus causing the water portion to evaporate.

The waste water seeped into the subsurface, contaminated the potable ground water which spread from the Arsenal grounds into adjacent areas, endangered crops, and became detrimental to the water supply (Gahr, 1961; Walker, 1961). The Chemical Corps then decided to water-proof the reservoirs, and again proceeded to evaporate the waste water. The accumulation of waste was greater than the evaporation of the water, and an over-abundance of waste was accumulated.

The Chemical Corps and the Corps of Engineers further studied the problem of disposing of the waste economically. Three main methods of surface disposal were considered: (1) heat, (2) chemical, and (3) electrolysis. Each of these methods had its "bad" points.

The engineers then turned their attention to the subsurface--if they could drill a well to the Precambrian basement and find a suitable reservoir in which to dispose of the waste water, the initial expense would be high, but the cost of maintenance would be low, and the double handling of the waste by the other three methods would be eliminated. It was decided to drill the experimental well on the Arsenal grounds, near the axis of the Denver Basin. There, the maximum thickness of sedimentary rocks would be present so that the engineers would have a better chance of finding a suitable reservoir (Fig. 1).

The U. S. Army Corps of Engineers, Omaha District, commissioned the firm of E. A. Polumbus, Jr., and Associates, Inc., to design the well, supervise the drilling and completion, provide the necessary engineering and geological services, and manage the project. The writer, as an associate, was the Project Geologist and was responsible for all geological aspects of the operation. Another geological associate was G. R. Downs, who contributed materially to the initial design of the project and acted in an advisory capacity throughout the operation. John Neighbours represented the Chemical Corps and James Zeltinger was the field representative of the Corps of Engineers.

REGIONAL GEOLOGY

Strata which are exposed along the mountain front dip steeply eastward. On the east side of the Denver Basin, the strata dip gently westward from the plains of Nebraska and eastern Colorado.

The rocks of the basin are separated from the rocks of the mountain area by a system of faults which extends discontinuously from Canyon City northward to the Wyoming-Colorado state boundary and beyond. Figure 2 is a cross section which shows the sediments penetrated in the Arsenal well as

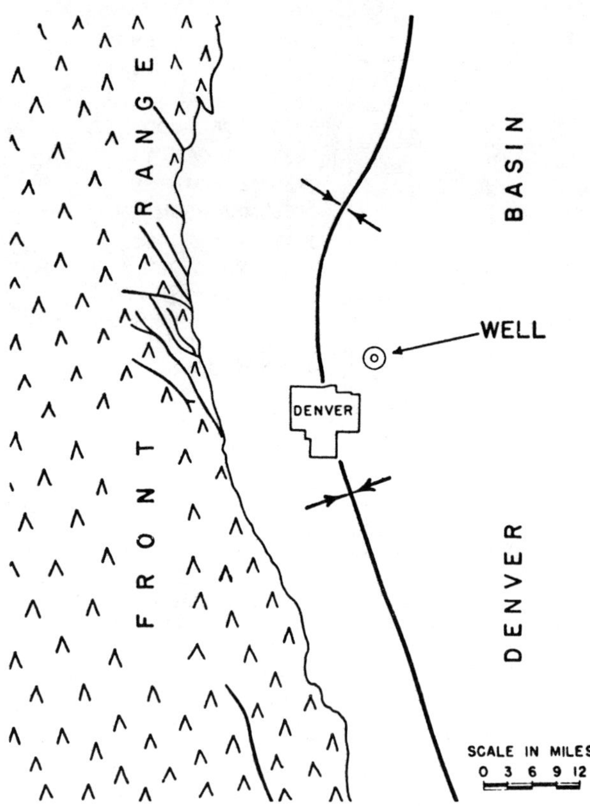

Fig. 1. The disposal well is located on the Rocky Mountain Arsenal grounds, 12 miles northeast of Denver and 20 miles east of the mountain front. The well is near the structural axis of the Denver Basin.

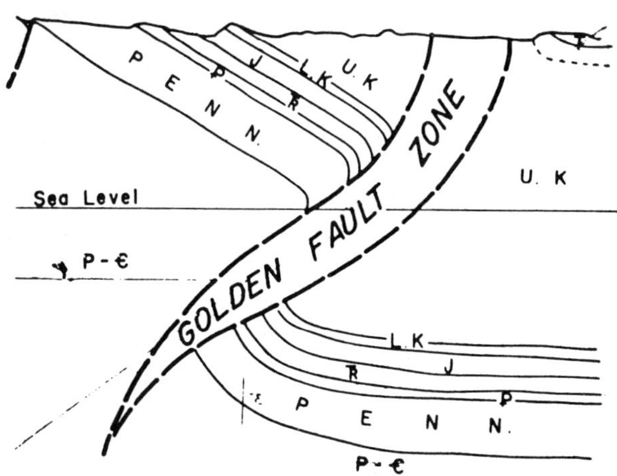

Fig. 2. Cross section of the fault zone just southwest of Denver (after C. M. Boos and M. F. Boos, 1957).

they are exposed at the mountain front west of Denver. The sediments at the surface dip eastward and are disconnected from the same sediments in the subsurface by the Golden fault zone. Stratigraphic displacement along this fault, in places, exceeds 10,000 feet. The formations range in age from Pleistocene to Precambrian. No pre-Pennsylvanian sediments are present at the mountain front west of Denver, but were penetrated in the Arsenal well.

PRESSURE INJECTION DISPOSAL WELL

The well, spudded March 10, 1961, was drilled to a total depth of 12,045 feet. Drilling operations were completed on September 11, 1961. The thickness and character of formations penetrated in the Arsenal well are described below and shown in Figure 3.

Pleistocene and Upper Cretaceous Rocks

Surface sediments to a depth of 460 feet consist of dune sands and gravels of early Pleistocene age which overlie unconsolidated sandstones and dark gray shales of the Upper Cretaceous Arapahoe Forma-

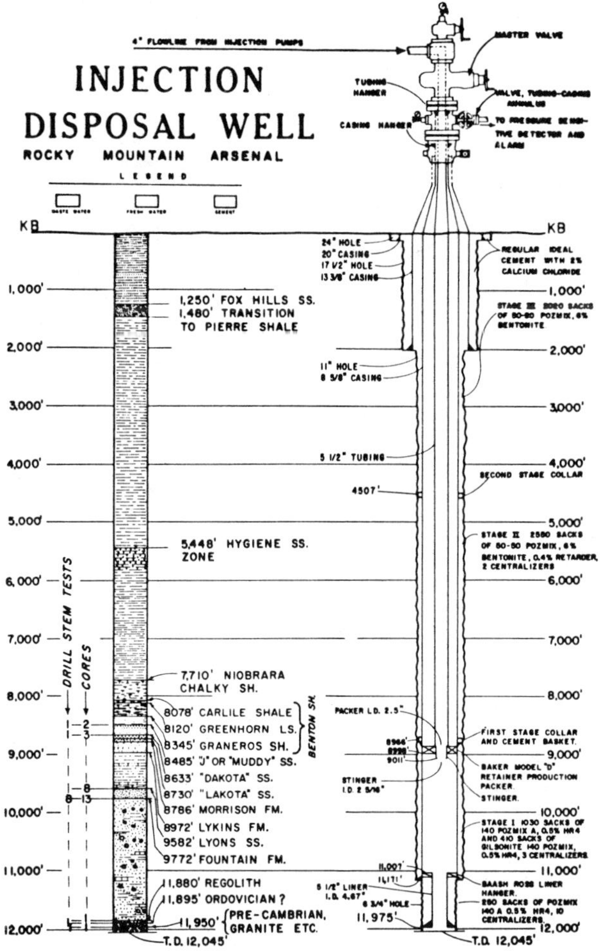

Figure 3

tion. Below these beds is the Upper Cretaceous Laramie Formation which consists of 1,020 feet of alternating fine-grained, gray glauconitic sandstones, and dark gray shales with coal beds. The Fox Hills Sandstone which underlies the Laramie Formation consists of 230 feet of gray, fine-grained sandstone.

Surface gravels and the sandstones through the Fox Hills are important sources of potable water in the vicinity of the Rocky Mountain Arsenal. In order to protect these reservoirs from contamination of waste fluids to be injected in the deeper reservoir, 13-3/8-inch casing was run and cemented from the surface to 2,005 feet. It was set in the upper portion of the Pierre Shale.

The underlying 6,230 feet of dark gray indurated Pierre Shale, 368 feet of Niobrara speckled shale and limestone, and 407 feet of the Benton Group (Carlile Shale, Greenhorn Limestone, and Graneros Shale) are impervious and are considered to constitute an excellent cover for the permeable rocks into which waste material is injected. The lower portion of the diagrammatic log (Fig. 3) is enlarged and is shown on Figure 4.

Lower Cretaceous Rocks

The first possible objective for fluid injection was the "J" Sandstone which is a principal oil producer of the Denver Basin and which underlies the Benton Group. The upper 25 feet of the sandstone was slightly oil-stained. Two cores and one drill-stem test were taken. The cores cut a total of 62 feet of "J" Sandstone and "reworked" sandstone and shale. The drill-stem test, which covered the total "J" Sandstone interval, was open one hour and recovered only 200 feet of slightly gas-cut mud. Due to the low permeability and the slight oil stain, this sand is of no commercial significance either as an oil sand or as an injection reservoir in the Rocky Mountain Arsenal well.

The next potential reservoirs were the "Dakota" and "Lakota" Sandstones of Early Cretaceous age. The "Dakota" Sandstone consists of 70 feet of sandstone, "reworked" sandstone and dark gray shale. Three cores were cut in the "Dakota" which represent 53 feet cut with 96 percent recovery. The permeability was less than one millidarcy, except for one fractured foot which had 3.5 millidarcys permeability. The average porosity was 4.6 percent.

The "Lakota" consists of 55 feet of white, hard, quartzitic sandstone. A drill-stem test covered both the "Dakota" and the "Lakota" interval. The tool was open one hour and the recovery was 400 feet of slightly gas-cut mud. The test verified the interpretation based on core and sample cuttings and indicated that the sandstones of the Dakota Group were of very poor reservoir quality.

Jurassic and Triassic Rocks

The Jurassic Morrison Formation underlies the "Lakota" Sandstone. The Morrison consists of 186 feet of varicolored shales and thin beds of freshwater limestones, white, hard sandstones, and anhydrite. Triassic sediments are represented by the Lykins Formation which underlies the Morrison and consists of 610 feet of red silty shale with 15 feet of fine, hard, quartzitic sandstone at the top of the formation, and 80 feet of anhydrite and red shale near the base. Jurassic and Triassic rocks do not have sufficient permeability to be favorable reservoirs.

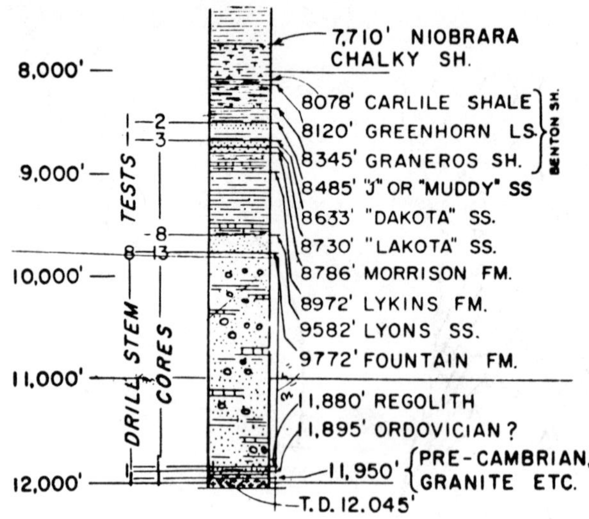

Fig. 4. Log of lowermost part of the disposal well.

Permian Rocks

The Permian Lyons Sandstone underlies the Lykins Formation and consists of 190 feet of fine-grained, highly fractured, orange, cross-bedded, quartzitic sandstone. Eight cores were cut in the Lyons. The cored intervals represent 95 feet of Lyons Sandstone cut with 90 percent recovered. Permeability of the sandstone is less than one millidarcy. The average porosity is 3.0 percent. While drilling and coring the Lyons Sandstone, much lost circulation was encountered.

The Lyons Sandstone is considered to be an important potential reservoir for the injection of waste fluids even though the formation matrix had less than one millidarcy permeability. Lyons cores showed intense fracturing (Fig. 5) and the lost circulation may be indicative of an extensive network of fractures away from the well-bore that would carry the waste fluid into more porous and permeable sandstones.

Pennsylvanian and Older Paleozoic Rocks

The Pennsylvanian Fountain Formation underlies the Lyons Sandstone. At the Arsenal well, the Fountain is 2,110 feet thick. It contains arkosic conglomerates, siltstones, maroon shale, and a few thin beds of limestone. The top 110 feet of the formation consist of fine-grained feldspathic sandstones and maroon shales, possible equivalent to the lower Satanka tongue (Ingleside Formation) in northeastern Colorado. Below this interval, the Fountain Formation is primarily composed of coarse-grained arkosic conglomerate. The cement in the clastics is calcareous, or consists of red and green shale or kaolinite.

Eleven cores were cut in the Fountain Formation by the conventional diamond core method, and three Bowen junk basket cores were cut. Eight drill-stem tests were made. The cored intervals represent 426.5 feet cut, and 85 percent recovered. Permeabilities of these cores ranged from less than one millidarcy to as much as 3.7 millidarcys. The average porosity was 3.3 percent. Figure 6 illustrates the significant typical fractures which are present in the upper 630 feet of the Fountain Formation. Little or no fracturing is evident below this depth although lost circulation difficulties were encountered.

Of the eight drill-stem tests taken in the Fountain Formation, three were failures, and five recovered only water cushion and mud. The upper one-third of the Fountain Formation is considered to be an important potential reservoir for the injection of waste fluids even though the formation matrix exhibits very low porosity. The fracturing and lost circulation in this section, as in the Lyons Sandstone, may be indicative of an extensive network of fractures.

When the drilling had proceeded to a point near the base of the Fountain, the hole condition became critical because of the prolonged exposure of the Upper Cretaceous shales to the drilling filtrate and because of the extensive period of lost circulation in the Lyons Sandstone and the Fountain Formation which caused drastic fluctuations of chemical properties in the hole. It was decided to run the 8-5/8-inch casing immediately. The casing was run and cemented from total depth to the surface employing a three-stage cementing procedure. This string of casing will give additional protection to the shallow potable water sands. The well was then drilled ahead. The pre-Pennsylvanian portion of the log is shown in Figure 7.

A regolith or fossil soil, 15 feet thick, underlies the Fountain Formation (Figure 7). It represents possibly five distinct soils, each from one to five feet thick. The four upper "soils" are composed primarily of weathered maroon to dark-brown shales which have white, leached, irregular areas. The bottom foot of the regolith is dark reddish brown, highly fractured quartzite with white to light-green irregular areas.

A series of sedimentary rocks of Ordovician? or Cambrian? age lies between the regolith and the Precambrian surface. They are described as follows:

20 feet of quartz conglomerate, composed of angular clear and orange quartz;
13 feet of purple shale, fissile;
15 feet of dolomite, coarsely crystalline, pink to white, and contains dark green and gray-green chert fragments;
5 feet of purple shale.

These sediments are believed to be of Early Ordovician or Cambrian age. They may possibly be a wedge edge of the sediments of the Colorado Springs trough. They are not known in the mountain front west of Denver and their presence may make necessary a revision of Paleozoic paleogeographic maps. G. L. Scott of the U. S. Geological Survey and J. G. Mitchell of Amstrat believe these sediments to be Cambrian.

Precambrian

The above-described sediments overlie 20 feet of bright green weathered schist which contains brown to copper-colored mica and kaolinite. The pre-Pennsylvanian sediments and the Precambrian schist were not cored.

The Precambrian schist is immediately above highly fractured hornblende granite gneiss which contains pegmatite intrusions. The top eight-foot section of the gneiss was cored. Hedge and Walthall (1963) have dated the gneiss to be 1350×10^6 years old. A drill-stem test was taken from the basal portion of the Fountain Formation across the pre-Pennsylvanian sediments and into the top eight feet of the Precambrian gneiss. The tool was open two hours and thirty minutes, and 2,000 feet of water

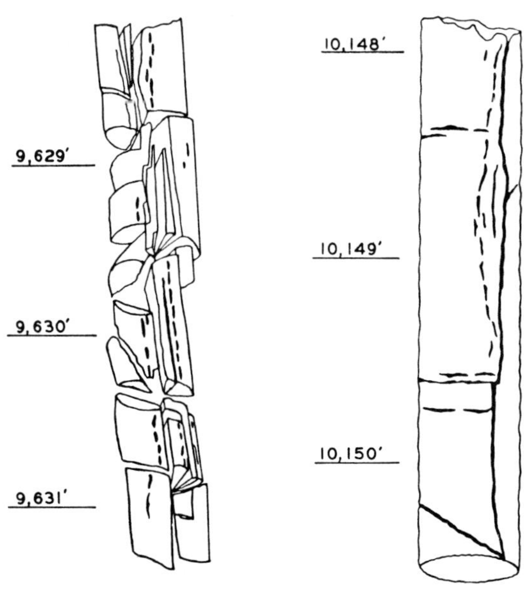

Fig. 5. A typical segment of the vertically fractured, indurated Lyons Sandstone. Nearly horizontal fractures are also present.

Fig. 6. Fractures in the upper 630 feet of the Fountain Formation.

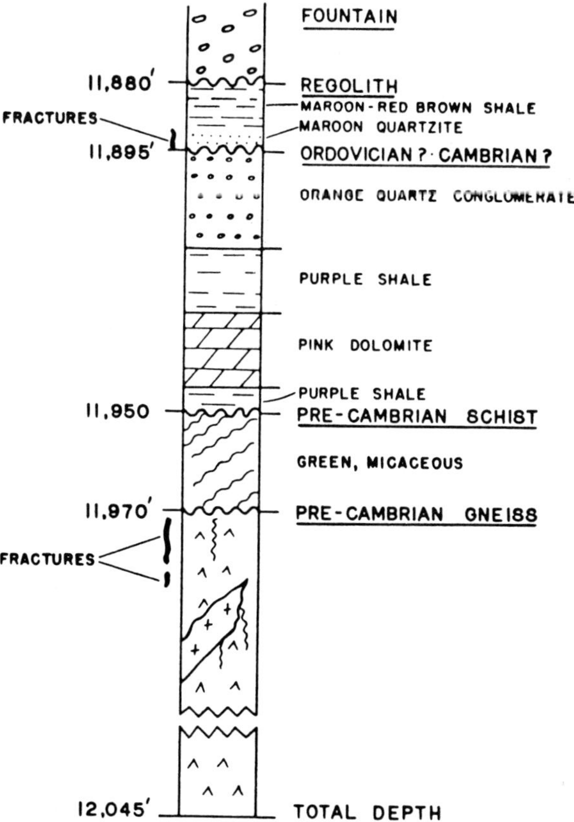

Fig. 7. Log of pre-Pennsylvanian portion of the disposal well.

cushion and 5,400 feet of salt water were recovered on the test. Fluid recovery of the drill-stem test indicated that this interval was more favorable for an immediate potential injection reservoir than any other zones in the well.

The well was then drilled an additional 60 feet into the Precambrian gneiss to the total depth of 12,045 feet. Short injection tests, conducted in the fractured Precambrian rocks, lead to setting a 5-1/2-inch liner from the top of the Precambrian gneiss into the basal portion of the 8-5/8-inch casing. The initial injection test indicated that this reservoir would accept more than 400 gallons per minute at less than 2,000 pounds surface pressure. This rate of injection was adequate to fulfill the disposal requirements.

DISPOSAL DATA

A summary of the disposal reservoir requirements and results of injection in the Rocky Mountain Arsenal well (Mechem and Garrett, 1963) is as follows:

"1. The interval between 11,975 feet and 12,045 feet has accepted over 30 million gallons of waste fluid as of July 18, 1962.

2. The well was accepting fluid at the rate of 200 gpm at a surface pressure of 550 psig.......

3. The impervious massive shale body, essentially comprising the upper 8,000 feet of the geologic section, is considered to constitute an excellent cover for the deeper potential reservoir rocks.

4. Reservoir evaluation techniques have indicated that no natural resources of commercial or domestic value are contained in this lower part of the disposal well.

The original objective of the Rocky Mountain Arsenal disposal well was to identify, select, and evaluate prospective injection reservoirs capable of accepting waste fluids at rates to 800 gpm under a maximum surface injection pressure of 2,000 psig. The well, as completed in the fractured Precambrian interval from 11,975 feet to 12,045 feet, is regarded as fulfilling this objective."

REFERENCES CITED

Boos, C. M. and Boos, M. F. 1957. Tectonics of eastern flank and foothills of Front Range, Colorado. *Am. Assoc. Petroleum Geologists Bull.*, V. 41, pp. 2603-2676.

Gahr, W. M. 1961. Contamination of ground water, vicinity of Denver (abstract). *Symposium on Water Improvement*, Am. Assoc. Adv. Sci., pp. 9-20.

Hedge, C. E. and Walthall, F. G. 1963. Radiogenic strontium-87 as an index of geologic processes. *Science*, V. 140, pp. 1214-1217.

Mechem, O. E. and Garrett, J. H. 1963. Deep injection disposal well for liquid toxic waste. *Proc. Am. Soc. Civil Eng., Jour. Construction Division*, pp. 111-121.

Scopel, L. J. 1961. Pressure injection disposal well, Rocky Mountain Arsenal, Denver, Colorado Area (abstract). *Program of Ann. Meeting*, Am. Assoc. Adv. Sci., Sec. E, p. 17.

Walker, T. R. 1961. Ground-water contamination in the Rocky Mountain Arsenal area, Denver, Colorado. *Geol. Soc. America Bull.*, V. 72, pp. 489-494.

THE DENVER AREA EARTHQUAKES AND THE ROCKY MOUNTAIN ARSENAL DISPOSAL WELL*

David M. Evans

Abstract

During 1961, a deep well was drilled at the Rocky Mountain Arsenal northeast of Denver, Colorado, to dispose of contaminated waste water. The well is bottomed in 75 feet of highly fractured Precambrian gneiss. Pressure injection of waste water into the fractured Precambrian rock was begun in March 1962. Since the start of fluid injection, 710 Denver-area earthquakes have been recorded. The majority of these earthquakes had epicenters within a five-mile radius of the Arsenal well. The volume of fluid and pressure of fluid injection appears to be directly related to the frequency of earthquakes. Evidence also suggests that rock movement is due to the increase of fluid pressure within the fractured reservoir and that open fractures may exist at depths greater than previously considered possible.

CONTENTS

ILLUSTRATIONS

*Reprinted, with permission and editing, from *The Mountain Geologist*, V. 3, No. 1, pp. 23-36.

INTRODUCTION

Products for chemical warfare have been manufactured on a large scale under the direction of the Chemical Corps of the U. S. Army at the Rocky Mountain Arsenal since 1942. A by-product of this operation is contaminated waste water and, until 1961, this waste water was disposed of by evaporation from dirt reservoirs.

When it was determined that Arsenal waste water was contaminating the local ground-water supply and endangering crops (Gahr, 1961; Walker, 1961), the Chemical Corps tried evaporation of the contaminated waste from water-tight reservoirs. This proved unsuccessful. The Chemical Corps and the Corps of Engineers then decided to drill an injection disposal well for the purpose of disposing of the contaminated waste water.

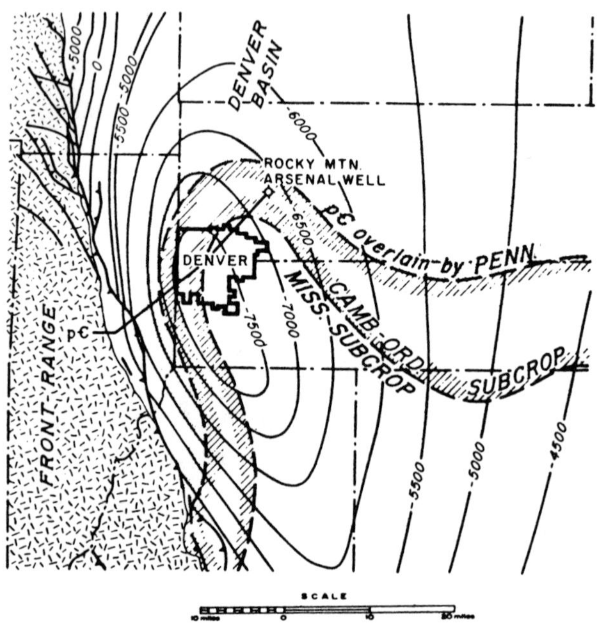

Fig. 1. Structural map of a portion of the Denver-Julesburg Basin (after Anderman and Ackman, 1963), showing the location of the Rocky Mountain Arsenal well.

The well was located and drilled in the NW/4 NE/4 sec. 26, T. 2 S., R. 67 W. (39°-51.5' N., 104°-51' W.), Adams County, Colorado. It was spudded 10 March 1961 and completed at a total depth of 12,045 feet 11 September 1961.

REGIONAL GEOLOGY

The Rocky Mountain Arsenal disposal well is located on the gently dipping east flank of the Denver-Julesburg basin, just a few miles west of the axis of the basin. As indicated in Figure 1, the Arsenal well is located in a region of the subcrop of Cambro-Ordovician rocks, near the area where these rocks are truncated by Pennsylvanian sediments.

Figure 1 is a structural map of a portion of the Denver-Julesburg Basin in the vicinity of the Rocky Mountain Arsenal well after Anderman and Ackman (1963). Figure 2 is a cross-section (after M. F. and C. M. Boos and H. H. Odiorne) which shows the subsurface geology from the Arsenal well to the outcrop of Precambrian granite gneiss west of Denver.

The granite gneiss is identified as the Mount Morrison Formation by C. M. and M. F. Boos (1957), who describe typical Mount Morrison granite as medium to fine grained, pink to tan, and delicately gneissic. Parts of the granitic gneiss are permeated with ill-defined pegmatite.

Approximately 13,000 feet of structural relief exists between the top of the Precambrian in the Arsenal well and the Precambrian outcrop west of Denver.

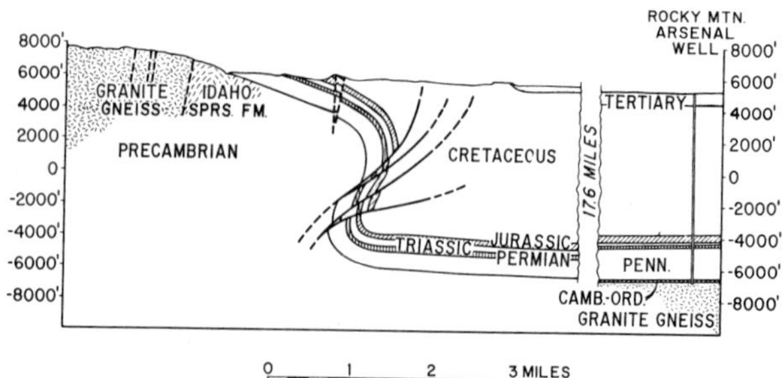

Fig. 2. Cross-section showing the subsurface geology from the Arsenal well to the outcrop of Precambrian granite gneiss west of Denver (after M. F. and C. F. Boos, and H. H. Odiorne). The line of cross-section is shown in Figure 1.

TESTING OF THE WELL

A drill stem test was taken of the basal Fountain Formation, the pre-Pennsylvanian rocks, and Precambrian rocks from the bottom of the 8-5/8-inch casing at 11,171 feet to the total depth of 11,985 feet. Recovery was 5,400 feet of salt water, in addition to 2,000 feet of water cushion, in 156 minutes. Ninety-three-minute final shut-in pressure was 4,128 pounds, measured at 11,002 feet. Density of the water was 1.05 grams per cubic centimeter.

The well was drilled ahead to 12,045 feet where it was completed in Precambrian gneiss. Considerable lost circulation was experienced while coring, testing, and drilling the Precambrian gneiss from 11,970 to 12,045 feet.

A 5-1/2-inch liner was cemented five feet into the Precambrian gneiss from the bottom 64 feet of the 8-5/8-inch casing. Five-and-one-half-inch tubing was run to 9,011 feet to complete the well.

During November and December 1961 a conventional oil field pump was run in the well, and pumping tests were conducted. After pumping 1,100 barrels of water, a quantity in excess of the amount of fluid that had been lost into the formation during drilling operations, the well pumped down and fluid recovery became negligible. It was concluded, at the time of testing, that fluid recovery was from fractures. It was further believed that as fluid was withdrawn from these fractures, they were squeezed shut by compressive forces which restricted fluid entry into the well bore.

Pressure injection tests were conducted on the well during January 1962 to determine the rates and injection pressures at which the Precambrian would take the fluid. As a result of these tests, it was noticed that calculations of the drainage radius and formation capacity increased as fluid was injected (see Calhoun, 1953, for more on reservoir calculations).

As a result of the testing program, it was concluded that the formation would take fresh water at 400 gallons per minute under 650 pounds pressure, and that the reservoir consisted of fractures which expanded as additional volumes of fluid were injected.

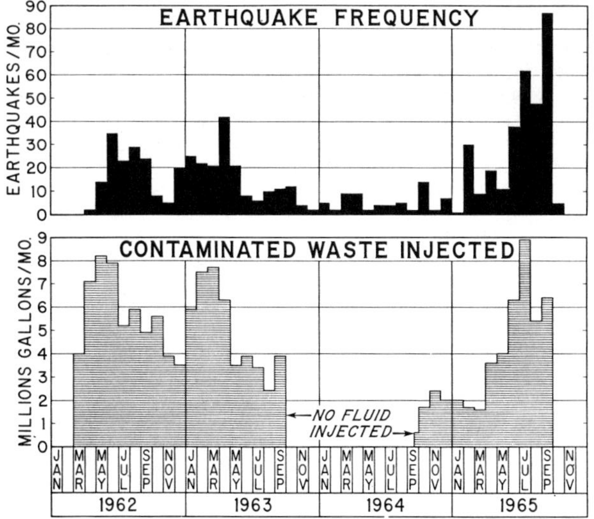

Fig. 3. Upper half: number of earthquakes per month recorded in the Denver area.
Lower half: monthly volume of contaminated waste water injected into the Arsenal well.

THE PRESSURE INJECTION PROGRAM

Contaminated waste from the Arsenal plants is first collected and allowed to settle in a two-hundred-million-gallon waste-settling basin that is sealed with an asphaltic membrane to prevent seepage. It is then flocculated and clarified. Next it is filtered to less than 20 parts per million of suspended solids less than 5 microns in diameter. It is sterilized and monitered for bacteria, then pumped into the well. Four 130-horsepower positive-displacement electric pumps are available. Normally, two or three pumps are used.

The first contaminated waste was injected into the well during March 1962, when 4.2 million gallons of waste were injected into the well. The monthly volume of waste injected into the well is shown in the lower half of Figure 3. During the first year of operation, considerable trouble was experienced with the filter plane with the result that the injection well was often shut down for a few days or weeks at a time. From March 1962 until September 1963 the maximum injection pressure is reported to have been about 550 pounds, with a fluid injection rate of 200 gallons per minute.

At the end of September 1963 the injection well was shut down, and no fluid was injected until operations were resumed 17 September 1964. During the shut-down period, surface evaporation, from the settling basin, was sufficient to handle the plant output.

From 17 September 1964 until the end of March 1965, injection operations were resumed by gravity discharge into the well. No well-head pressure was necessary to inject the maximum of 2.4 million gallons of waste per month into the well.

Beginning in April 1965 larger quantities of fluid were injected. The filter plant operated efficiently, and fluid was usually injected 16 to 24 hours daily. During April and May a maximum pump pressure of 800 pounds was required. From June to the end of September 1965 a maximum pressure of 1,050 pounds was required to inject 300 gallons per minute into the well.

THE DENVER EARTHQUAKES

The U. S. Coast and Geodetic Survey reports that on 7 November 1882 an earthquake was felt in the Denver, Louisville, Georgetown, and S. E. Wyoming area (Wang, 1965). From that date until April 1962 no earthquake epicenters were recorded in the Denver area by either the U. S. Coast and Geodetic Survey or by the Regis College Seismological Observatory, located ten miles southeast of the Rocky Mountain Arsenal well (Joseph V. Downey, personal communication, 1965).

During the period from April 1962 to the end of September 1965, 710 earthquakes were recorded with epicenters in the vicinity of the Arsenal at the Cecil H. Green Observatory, Bergen Park, Colorado, operated by the Colorado School of Mines (Pan, 1964; Wang, 1965; Jones, 1965; Mines Magazine, 1965).

The total number of earthquakes reported in the Denver area is plotted in the upper half of Figure 5. The magnitude of the earthquakes reported range from 0.7 to 4.3 on the Richter scale. Table 1 lists the earthquakes in Colorado of magnitude 3 and larger, according to the U. S. Coast and Geodetic Survey reports (Wang, 1965). Wang (1965) calculated the epicenters and hypocenters of the 1963-65 Denver earthquakes, and the results of his calculations are shown in Figure 4.

The majority of the earthquake epicenters are within a five-mile radius of the well. All epicenters calculated from four or more recording stations are within seven miles of the well.

Wang (1965) calculated the best-fitting plane passing through the zone of hypocenters calculated from

Table 1. COLORADO EARTHQUAKES, MAGNITUDE 3 AND LARGER
(From U. S. Coast and Geodetic Survey Reports)

No.	Year	Mon.	Day	Greenwich Mean Time Hour**	Min.	Sec.	Longitude (N)	Latitude (W)	Depth (km)	Magnitude	Felt Area
1.	1882	Nov.	7				?	?	?	?	Denver, Louisville, Georgetown and S. E. Wyoming
2.	1960	Oct.	11	01	05	30.5	38.3°	107.6°		5.5	Cimarron, Lake City, Montrose, Ophir, Ouray, Placerville, Powderhorn, Ridgeway, Telluride
3.	1961	Nov.	27	00	55	45.7	39.0°	106.1°	(33)	?	Fairplay, Hartsel, Leadville, Jefferson, Buena Vista, Alma
4.	1962	June	18	00	46	05.0	?	?	?	3.1	Denver, Derby, Henderson
5.	1962	Aug.	7	00	51	00.0	?	?	?	3.0	Denver, Derby, Henderson
6.	1962	Dec.	4	17	49	59.4	39.8°	104.7°	(33)	3.6	Denver, Jefferson, Adams
7.	1962	Dec.	5	13	48	00.4	39.9°	104.6°	(33)	3.8	Denver, Pueblo
8.	1963	Jan.	30	23	05	09.6	39.8°	104.6°	(33)	3.2	Denver, Boulder
9.	1963	April	8	00	03	59.1	39.6°	104.9°	(33)	3.2	Denver, Derby, Henderson
10.	1963	April	24	22	29	35.7	39.7°	104.8°	(33)	3.2	Denver, Derby, Henderson
11.	1963	May	25	10	44	38.1	39.8°	104.7°	(33)	3.5	Denver, Derby, Henderson
*12.	1963	June	5	00	13	56.6	39°52.5'	104°55.5'	2.6	3.0	Denver, Derby, Henderson
13.	1963	July	2	08	02	54.1	39.8°	104.7°	(15)	3.0	Denver, Boulder
14.	1963	July	28	13	18	47.0	?	?	?	3.1	Denver, Derby, Henderson
*15.	1965	Jan.	5	01	05	31.2	39°54.7'	105°15.8'	15.6	?	Western Denver, Louisville
16.	1965	Feb.	16	20	17	54.0	39.9°	105.1°	5	4.6	Western Denver, Louisville
*17.	1965	Feb.	16	22	21	46.9	39°52.4'	104°55.2'	8.3	3.7	Western Denver, Louisville
18.	1965	July	31	13***	41	43.0	39.7°	104.9°	5	4.6	Western Denver, Louisville
19.	1965	Sept.	14	22***	46	24.0	39.9°	104.6°	5	4.7	Western Denver, Louisville
20.	1965	Sept.	29	18***	59	56.0	39.8°	105.1°	5	4.7	Western Denver, Louisville
21.	1965	Sept.	29	19***	20	41.0	39.7°	104.9°	5	4.6	Western Denver, Louisville

* Determined from local seismogram data.
** For Mountain Standard Time, subtract 7 hours.
*** For Mountain Daylight Savings Time, subtract 6 hours.

(Wang, 1965, with additions)

four or more recording stations. He concluded that this plane might be a fault along which movement was taking place. The plane dips to the east, and passes beneath the arsenal well at a depth of about six and one-half miles (Figure 4).

ROCK MOVEMENT AND EARTHQUAKES

An attempt has been made to develop a method of estimating, directly from seismograms, wave energy radiated during an earthquake. Using the formula favored by Tocher (1964) and Richter (1958), the elastic wave energy of a magnitude 3 earthquake could be provided by dropping a 100 foot cube of rock a distance of a few feet.

Admittedly, the formula applies to distant earthquakes and is not routinely applicable to large numbers of earthquakes, but it does suggest that the Denver earthquakes may be caused by relatively minor rock movements.

PRESSURE INJECTION AND EARTHQUAKE FREQUENCY

Pressure injection began in March 1962. The first two earthquakes with epicenters in the Arsenal area were recorded during April 1962.

The lower half of Figure 5 is a graph of the monthly volume of waste injected into the Arsenal well. The total number of earthquakes recorded in the Arsenal area is plotted each month in the upper half of the graph.

During the initial injection period from March 1962 to the end of September 1963, the injection program was often shut down for repairs to the filter plant. During this period there does not appear to be a direct month by month correlation. However, the high injection months of April, May, and June 1962 seem to correlate with the high earthquake frequency months of June, July, and August. The high injection months of February and March 1963 may correlate with the high earthquake month of April.

The period of no injection from September 1963 to September 1964 coincides with a period of minimum earthquake frequency. The period of low volume injection by gravity flow, from September 1964 to April 1965, is characterized by two months (October and February) of greater earthquake frequency than experienced during the preceeding year.

The most direct correlation of fluid injection with earthquake frequency is during the months of June through September 1965. This period was characterized by the pumping of 300 gallons per minute, 16 to 24 hours a day, at pressures of from 800 to 1,050 pounds.

A review of the injection program reveals that there have been five characteristic periods of injection into the well (Figure 5):

1. April 1962-April 1963: High injection at medium pressure.
2. May 1963-September 1963: Medium injection.
3. October 1963-September 1964: No injection.
4. September 1964-March 1965: Low injection at zero pressure (gravity).
5. April 1965-September 1965: High injection at high pressure.

The average numbers of earthquakes per month are plotted in Figure 5 above the average volumes of fluid injected per month for each of these five periods. The injection for March 1962 is not used in the average because the exact day when injection was started is not known.

Figure 5 indicates that there is a direct correlation between average monthly injection and earthquake frequency when an injection program is carried out for a period of five months.

Bardwell (1966) has prepared a statistical analysis that suggests that a mathematical relationship exists

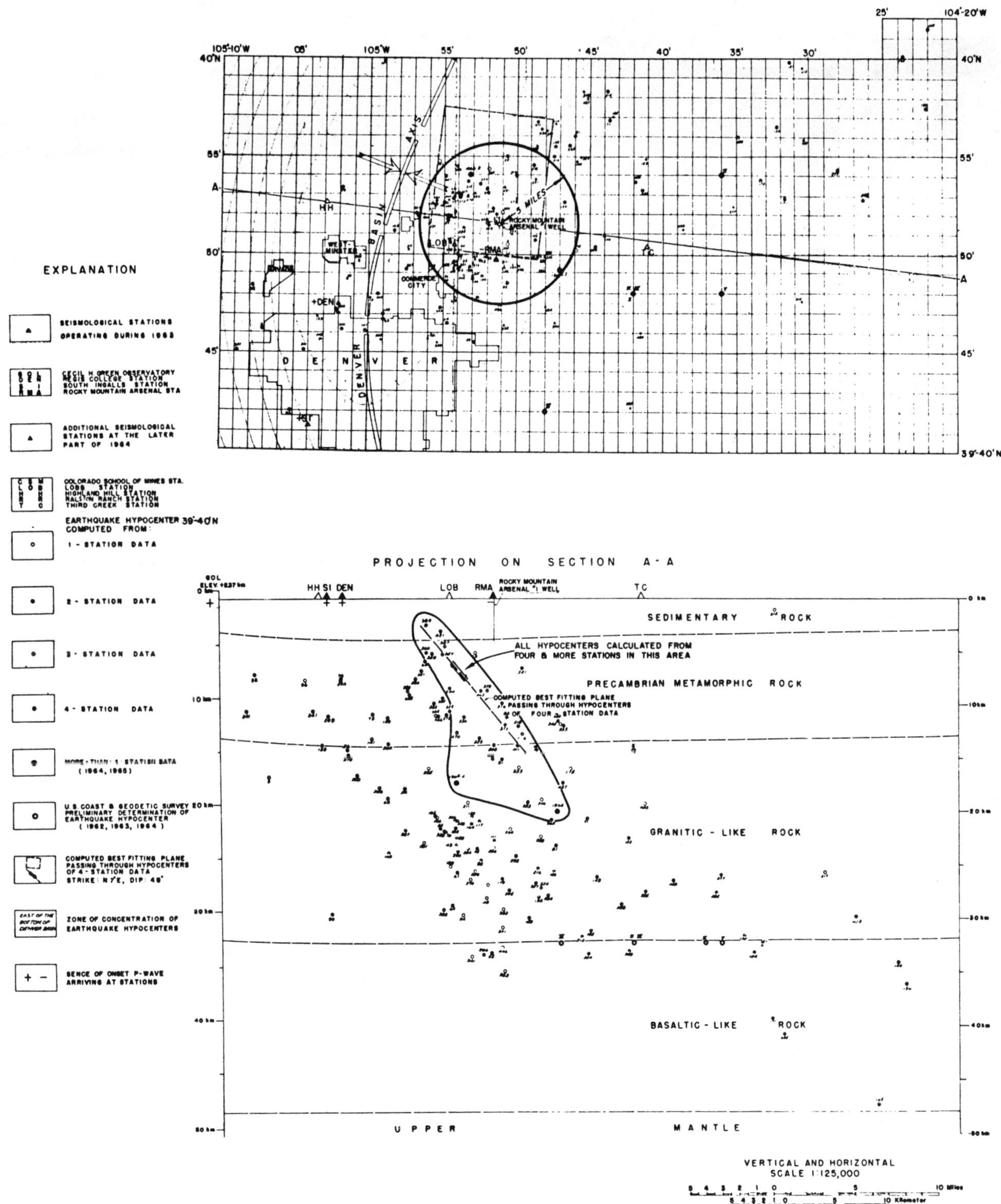

Fig. 4. Earthquake hypocenters during 1963-64 from local seismological stations in the Denver area (after Wang, 1965). All epicenters calculated from four or more recording stations are within seven miles of the Arsenal well. All hypocenters calculated from four or more recording stations are within area indicated on section A - A.

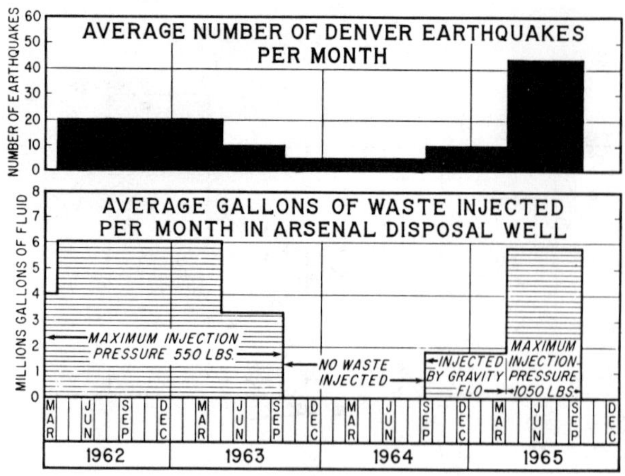

Fig. 5. Earthquake frequency - waste injection relationships during five characteristic periods.

between the Denver earthquakes and the volume of contaminated waste injected into the Arsenal well.

EFFECT OF EARTHQUAKES ON INJECTION PRESSURE

The wellhead-pressure injection charts were not available for the years 1962 and 1963. Only the earthquakes of magnitude 3 or larger were checked against the pressure injection charts for 1963. These earthquakes are listed in Table 1.

Three pumps were maintaining a pressure of 725 pounds when the September 14 earthquake occurred. There was no change in injection at the time of the event.

Allowing for a few minutes time discrepancy between the chart time and recorded time of the earthquakes, the two earthquakes of 29 September may have affected the injection pressure. During the first earthquake, at 12:59 P.M. MDST, the pressure recording needle on the pressure chart jumped from 970 pounds to 940 pounds and also repeated a ten minute time interval on the chart. During the second earthquake, at 1:20 P.M. MDST, the pressure dropped from 960 pounds to 780 pounds. Whether this 180-pound pressure drop was due to the earthquake or to the slowing up of one of the pumps is not known.

FLUID PRESSURE AND THE ARSENAL EARTHQUAKES

The evidence gained from drilling and testing the Arsenal disposal well indicates that the Precambrian reservoir is composed of a highly fractured granite gneiss which is substantially impermeable. The fractures are almost vertical, and porosity of the reservoir is formed by these fractures. The evidence indicates that as fluid was pumped out of the reservoir the fractures closed, and as fluid was injected into the reservoir the fractures opened. In other words, the pumping and injection tests indicated that rock movement occurred as fluid was withdrawn or injected at relatively low pressures.

The pressure-depth relations of the Precambrian reservoir, showing hydrostatic and lithostatic pressure variations with depth, are shown in Figure 6. These data were determined from the drill stem test. As shown on the chart, the observed pressure of the Precambrian reservoir is almost 900 pounds less than hydrostatic pressure.

Hubbert and Rubey (1959) devised a simple and adequate means of reducing by the required amount the frictional resistance to the sliding of large overthrust blocks down very gentle slopes. This arises from the circumstance that the weight of such a block is jointly supported by solid stress and the pressure of interstitial fluids. As the fluid pressure approaches the lithostatic pressure, corresponding to a flotation of the overburden, the shear stress required to move the block approaches zero.

If high fluid pressures reduce frictional resistance and permit rocks to slide down very gentle slopes, it follows that, as fluid pressure is decreased, frictional resistance between blocks of rock is increased, thus permitting them to come to rest on increasingly steep slopes. The steeper the slope upon which a block of rock is at rest, the lower the required raise in fluid pressure necessary to produce movement.

In the case of the Precambrian reservoir beneath the Arsenal well, these rocks were at equilibrium on high-angle fracture planes with a fluid pressure of 900 pounds less than hydrostatic pressure before injection began.

As fluid was injected into the Precambrian reservoir, the fluid pressure adjacent to the well bore rose, and the frictional resistance along the fracture planes was thereby reduced. When, finally, enough fluid pressure was exerted over a large enough area movement took place. The elastic wave energy released was recorded as an earthquake.

Since the formation fluid pressure is 900 pounds sub-hydrostatic, merely filling the hole with contaminated waste (mostly salt water) raises the formation pressure 900 pounds, or to the equivalent of hydrostatic pressure. Any applied injection pressure above that of gravity flow brings about an increase in pressure resulting in a total in excess of hydrostatic pressure. For example, an injection pressure of 1,000 pounds would raise the reservoir pressure adjacent to the well bore 1,900 pounds, or by the amount necessary to bring the pressure to hydrostatic (by filling the hole) plus 1,000 pounds.

Apparently a rise in fluid pressure within the Precambrian reservoir of from 900 to 1,900 pounds is sufficient to allow movement to take place.

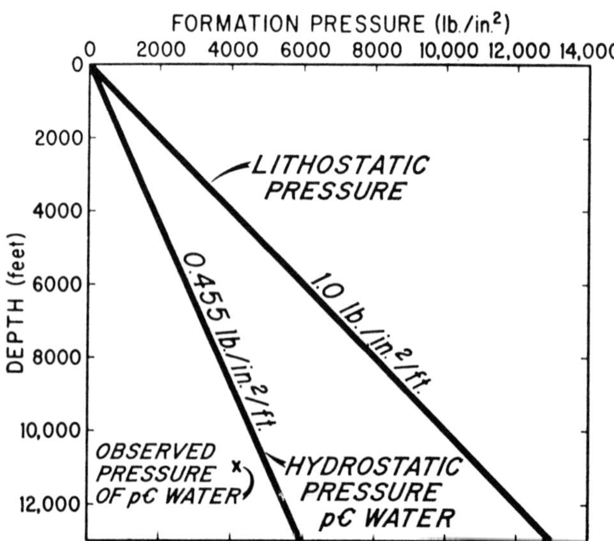

Fig. 6. Pressure-depth relations, Precambrian reservoir, Rocky Mountain Arsenal Disposal well.

OPEN FRACTURES

The hypocenters in the Arsenal area plotted from data derived from four or more recording stations indicate that movement is taking place beneath the Arsenal well at depths of from 1-1/2 to 12 miles. If the Precambrian fracture system extends to a depth of 12 miles, then fluid pressure could be transmitted to this depth by moderate surface injection pressure as long as the fracture system was open for the transmission of this pressure.

Secor (1965) concluded that open fractures can occur to great depths even with only moderately high fluid pressure-overburden weight ratios. It appears possible that high-angle, open fractures may be present beneath the Arsenal well at great depths with much lower fluid pressure-overburden weight ratios than has formerly been considered possible.

Almost 150 million gallons of contaminated waste had been injected into the Arsenal well by the end of September 1965. Since this amount of water would be enough to fill four continuous 1/16-inch fractures each seven miles long (the maximum distance of epicenters from the well located by four or more recording stations) and five miles deep, it can be seen that a relatively small area is being affected by the injection program.

TIME LAG BETWEEN FLUID INJECTION AND EARTHQUAKES

The correlation of fluid injected with earthquake frequency (Figure 3) suggests that a time lag exists between the two. Bardwell (1966) notes that the frequency of Denver earthquakes appears to lag injected waste by approximately one to four months. This phenomenon is probably the same as that described by Serafim and del Campo (1965). They describe the observed time lag between water levels in reservoirs and the pressures measured in the foundations of dams, and ascribe this to an unsteady state of percolation through open joints in the rock mass due to the opening and closing of these passages resulting from internal and externally applied pressures.

The time lag between waste injected in the Arsenal well and earthquake frequency is probably due to an unsteady state of percolation through fractures in the Precambrian reservoir due to the opening and closing of these fractures resulting from the applied fluid pressure of the injected waste. The delayed application of this pressure at a distance from the well bore is believed to trigger the movement recorded as an earthquake.

EARTHQUAKES DURING SHUT-DOWN PERIOD

In considering the earthquake frequency during the year the injection well was shut down, it is unfortunate that neither periodic bottom-hole pressure tests nor checks of the fluid level in the hole were run. Had these measurements been made, then speculation as to how long it took for the bottom-hole pressure to decline would have been unnecessary.

By the end of September 1963, 102.3 million gallons of fluid had been injected into the well. It is believed that this injection had raised the fluid pressure in the reservoir for some distance surrounding the well bore. During the shut-down period this elevated pressure was equalizing throughout the reservoir and at increasing distances from the well bore. As this fluid pressure reduced the frictional resistance in fractures farther away from the well, movement occurred, and small earthquakes were the result.

CONCLUSIONS

The Rocky Mountain Arsenal Pressure Injection Disposal well was drilled for the purpose of disposing of contaminated waste water, which is a by-product of chemical warfare products manufactured at the Arsenal.

During the month following the initiation of injection of waste water into the almost vertically fractured Precambrian rocks there were two earthquakes with epicenters in the Arsenal area.

A summary of the evidence relating the Arsenal injection program with the earthquakes is:

1. The first earthquakes observed during the present century with epicenters in the Denver area were recorded during the month following the initiation of the Arsenal injection program.
2. Since the initiation of the injection program in March 1962, 150 million gallons of waste have been injected into the Arsenal well, and there have been 710 earthquakes (to 1 October 1965).
3. The majority of the earthquake epicenters are located within five miles of the Arsenal well. All epicenters determined from four-or-more station data are within seven miles of the well.
4. There is evidence that the earthquake activity is taking place along a plane that dips eastward and passes beneath the Arsenal well at a depth of 6.5 miles (Wang, 1965).
5. When the Arsenal injection program is considered on the basis of high, medium, low, or no injection, there is a correlation between the fluid injected and earthquake frequency.
6. The best correlation of earthquake frequency with fluid injected occurred during July, August, and September 1965, when relatively large amounts of fluid were injected at higher pressures and for longer periods of time than previously.
7. A statistical analysis (Bardwell, 1966) is cited that suggests a mathematical relationship between Arsenal earthquakes and volumes of waste injected into the Arsenal well.

The volume of fluid injected appears to be affecting the Precambrian reservoir only for a limited distance from the well bore, and rough estimates of the energy released by a single earthquake suggest that relatively minor rock movement is involved.

The Precambrian reservoir receiving the Arsenal waste is highly fractured granite gneiss of very low permeability. The fractures are nearly vertical. The fracture porosity of the reservoir is filled with salt water. Reservoir pressure is 900 pounds sub-hydrostatic.

It appears that movement is taking place in this fractured reservoir as a result of the injection of water at pressures from 900 to 1,950 pounds greater than reservoir pressure.

Hubbert and Rubey (1959) point out that rock masses in fluid-filled reservoirs are supported by solid stress and the pressure of interstitial fluids. As fluid pressure approaches lithostatic pressure, the shear stress required to move rock masses down very gentle slopes approaches zero.

It appears that these principles offer an explanation of the rock movement in the Arsenal reservoir. The highly fractured rocks of the reservoir are at rest on steep slopes under a condition of sub-hydrostatic fluid pressure. As the fluid pressure is raised within the reservoir, frictional resistance along fracture planes is reduced and, eventually, movement takes place. The elastic wave energy released is recorded as an earthquake.

If earthquake hypocenters indicate the point at which movement is taking place and injected fluid is triggering this movement, then there is evidence that open fractures exist at depths of 12 miles under conditions of lower fluid pressure-overburden weight ratios than has formerly been considered possible (Secor, 1965). It is believed that the high angle of this fracture system is an important factor. Because the fractures are almost vertical, only a small part of the lithostatic pressure is acting to force the fractures closed, and they can remain open under conditions of lower fluid pressure, and at greater depth than if they were horizontal or inclined at a lower angle.

The time lag between fluid injection and earthquake frequency is believed to be due to the unsteady state of percolation of fluid through the fractures in the reservoir due to the opening and closing of these passages resulting from the applied pressure of the injected waste.

It is believed that as fluid continues to be injected into this reservoir fluid pressure will be increased at greater distances from the well bore, and rock movement will be occurring at ever increasing distances.

In the present case it is believed that a stable situation in this Precambrian reservoir is being made unstable by the application of fluid pressure. However, it is interesting to speculate that the principle of increasing fluid pressure to release elastic wave energy might have an application in the subject of earthquake modification. That is, it might someday be possible to relieve the stresses along some fault zone in urban areas by increasing the fluid pressures along the zone using a series of injection wells. The accumulated areas might thus be released at will in a series of non-damaging earthquakes instead of eventually resulting in one large event that might cause a major disaster.

ACKNOWLEDGMENTS

The author is especially obligated to Ben H. Parker, who critically read the manuscript and made many valuable suggestions. Thanks are also due to Lt. Col. Martin J. Burke, Jr., Commanding Officer of the Rocky Mountain Arsenal, Lt. Col. William J. Tisdale, Director of Industrial Operations, and to the Arsenal Engineering Department, for aid in compiling the well and injection data; John C. Hollister and Maurice W. Major of the Colorado School of Mines Geophysical Engineering Department, for help in compiling the earthquake data; Joseph V. Downey, S. J., Director of the Regis College Seismological Observatory, for earthquake data prior to 1962; George E. Bardwell, for undertaking the statistical analysis, presented elsewhere in this issue; L. Trowbridge Grose and David T. Snow for information concerning the role of fluid pressure in jointing; Harry C. Kent and David A. Moore for critically reading the manuscript; and to John A. Rathbone and Charles C. Works for suggestions concerning the statistical analysis.

REFERENCES CITED

Anderman, G. G. and Ackman, E. J. 1963. Structure of the Denver-Julesburg Basin and surrounding areas: in *Rocky Mtn. Assoc. Geologists Guidebook to the Geology of the northern Denver Basin and adjacent uplifts.*

Anonymous, 1965. Geophysical observatory. *Mines Mag.*, V. 55, No. 5, pp. 24-26.

Bardwell, G. E. 1966. Some statistical features of the relationship between Rocky Mountain Arsenal waste disposal and frequency of earthquakes. *The Mountain Geologist*, V. 3, No. 1, pp. 37-42.

Boos, C. M. and Boos, M. F. 1957. Tectonics of eastern flank and foothills of the Front Range, Colorado. *Am. Assoc. Petroleum Geologists Bull.*, V. 41, pp. 2603-2676.

Calhoun, J. C. 1953. Reservoir rocks and rock-fluid systems, Part II. *Fundamentals of Reservoir Engineering.* Norman, University of Okla. Press.

Gahr, W. M. 1961. Contamination of ground water, vicinity of Denver (abstract). *Symposium on Water Improvement*, Am. Assoc. Adv. Sci., pp. 9-20.

Hedge, C. E. and Walthall, F. G. 1963. Radiogenic Strontium-87 as an index of geologic processes. *Science*, V. 140, pp. 1214-1217.

Hubbert, M. K. and Rubey, W. W. 1959. Role of fluid pressure in mechanics of overthrust faulting: Pt. I, Mechanics of fluid-filled porous solids and its application to overthrust faulting. *Geol. Soc. America Bull.*, V. 70, pp. 115-166.

_______, 1961. Role of fluid pressure in mechanics of overthrust faulting, a reply. *Geol. Soc. America Bull.*, V. 72, pp. 1445-1452.

Jones, P. B. 1965. Derby area earthquakes, 1964. Colorado School of Mines, Dept. of Geophysics, unpublished report.

Mechem, O. E. and Garrett, J. H. 1963. Deep injection disposal well for liquid toxic waste. *Proc. Am. Soc. Civil Eng.*, Jour. Construction Division, pp. 111-121.

Pan, Poh-Hsi 1963. The 1962 earthquakes and microearthquakes near Derby, Colorado. Unpublished M. Sc. thesis, Colorado School of Mines.

Richter, C. F. 1958. Elementary seismology, San Francisco, W. H. Freeman & Co.

Scopel, L. J. 1964. Pressure injection disposal well, Rocky Mountain Arsenal, Denver, Colorado. *The Mountain Geologist*, V. 1, No. 1, pp. 35-42. (See also the preceding article.)

Secor, D. T., Jr. 1965. Role of fluid pressure in jointing. *Am. Jour. Sci.*, Vol. 263, pp. 633-646.

Serafim, J. L. and del Campo, A. 1965. Interstitial pressures on rock foundations of dams. *Jour. of the Soil Mechanics and Foundations Div., Proc. Am. Soc. of Civil Engrs.*, V. 91, No. SM5.

Tocher, D. 1964. Earthquakes and rating scales. *Geotimes*, V. 8, No. 8, pp. 15-20.

Walker, T. R. 1961. Ground water contamination in the Rocky Mountain Arsenal area, Denver, Colorado. *Geol. Soc. America Bull.*, V. 72, pp. 489-494.

Wang, Yung-liang 1965. Local hypocenter determination in linearly varying layers applied to earthquakes in the Denver area. Unpublished D. Sc. thesis, Colorado School of Mines.

SOME STATISTICAL FEATURES OF THE RELATIONSHIP BETWEEN ROCKY MOUNTAIN ARSENAL WASTE DISPOSAL AND FREQUENCY OF EARTHQUAKES*

George E. Bardwell

Abstract

Aspects of the statistical problem of relating frequency of earthquakes in the Denver area to injection of waste water by the Rocky Mountain Arsenal are discussed. Behavior of cumulative time series for earthquake frequency and volume of water are analyzed. A regression-correlation analysis leads to an exponential model approximating the relationship between these two variables. Also considered are statistical tests of the differences in mean number of earthquakes per month for three periods of waste injection.

CONTENTS

ILLUSTRATIONS

INTRODUCTION

Evans has proposed a theoretical construct for a relationship between the injection of waste water by the Rocky Mountain Arsenal and the occurrence of earthquakes in the Denver (Derby) area. This relationship suggests a number of interesting statistical overtones, some of which are examined in this paper.

The analysis here is confined exclusively to volumes of waste injected by the Arsenal and the number of recorded earthquakes by month from March 1962 through October 1965 (Table 1). These data are not derived, of course, from controlled experimentation, and because of this, have some undesirable characteristics for purposes of statistical analysis. Moreover, there is uncertain precision in the location of the epicenters of the earthquakes as well as their magnitude is measured on the Richter scale. On the other hand, figures for volumes of waste injected and number of earthquakes are quite reliable. However, waste injection pressures have varied considerably during certain periods between 1962 and 1965 which add another dimension to the data.

*Reprinted, with permission and editing, from *The Mountain Geologist*, V. 3, No. 1, pp. 37-42.

Table 1. Waste Injected by Rocky Mountain Arsenal and Frequency of Recorded Earthquakes in the Denver Area, November 1882 - October 1965.

Month & Year	Gallons of Waste Injected (millions)	Number of Earthquakes	Earthquakes with Magnitude 3 and over (Richter Scale)
Nov. 1882			Felt Denver, Louisville, Georgetown and S. E. Wyoming
Mar. 1962	4.2	-	
April	7.2	2	
May	8.4	12	
June	8.0	35	18 June - 3.1
July	5.2	23	
Aug.	6.0	29	7 Aug. - 3.0
Sept.	5.0	24	
Oct.	5.6	8	
Nov.	4.0	6	
Dec.	3.6	20	4 Dec. - 3.6; 5 Dec. 3.8
Jan. 1963	6.0	25	30 Jan. - 3.2
Feb.	7.6	22	
Mar.	7.8	21	
April	6.4	42	8 Apr. - 3.2; 24 Apr. - 3.2
May	3.6	21	25 May - 3.5
June	4.0	8	5 June - 3.0
July	3.4	6	2 July - 3.0; 28 July - 3.1
Aug.	2.4	10	
Sept.	3.9	11	
Oct.	0	12	
Nov.	0	4	
Dec.	0	2	
Jan. 1964	0	5	
Feb.	0	2	
Mar.	0	9	
April	0	9	
May	0	2	
June	0	4	
July	0	4	
Aug.	0	5	
Sept.	0.6	2	
Oct.	1.8	14	
Nov.	2.4	2	
Dec.	2.0	7	
Jan. 1965	2.0	1	
Feb.	1.7	30	Two, 16 Feb., each of magnitude 4.1
Mar.	1.6	9	
April	3.6	19	
May	4.0	11	
June	6.4	38	
July	8.9	62	31 July - 4.0
Aug.	5.4	48	
Sept.	6.4	87	Two, 14 Sept. each of magnitude 4.1
Oct.	3.8	5	Three, 29 Sept. each of magnitude 4.1

Sources: Rocky Mountain Arsenal; Dept. of Geophysics, Colorado School of Mines; U. S. Coast & Geodetic Survey.

CUMULATIVE TIME SERIES

Volumes of waste water cumulated on a monthly basis are compared to corresponding monthly cumulative frequencies of earthquakes for the period March 1962 to October 1965 in Figure 1. For example, the cumulative volume for April 1962 is obtained by adding the volume for March 1962 (4.2 million gallons) to the volume for April 1962 (7.2 million gallons), giving a cumulative total for April 1962 of 11.4 million gallons, etc. Cumulative totals for number of earthquakes are computed in an identical manner.

The patterns of behavior of each of the series shown in Figure 1 are strikingly similar. Examination of Figure 1 suggests the following characteristics:

1. Frequency of earthquakes lags volumes of injected waste by approximately one to four months. For example, the frequency of earthquakes from March 1962 to June 1962 exhibits a tendency to lag behind injected volumes by about three months. The number of earthquakes during this period became appreciable in May and June 1962, two to three months following significant injection volumes. In September 1963, injection of waste was discontinued until September 1964. Yet, the curve representing frequency of earthquakes does not reflect this change in waste volumes until November 1963.

The period March 1965 to June 1965, also exhibits this lag characteristic. In April 1965, waste was injected at appreciably higher pressures than in previous months, yet the apparent effects of these higher injection volumes do not register until July 1965.

2. A "carry over" or residual influence from one month to the next of injection volumes on frequency of earthquakes. This characteristic is illustrated during the period of no waste injection from October 1963 through August 1964. This period is also characterized by a low, but persistent, level in the number of earthquakes, suggesting that the influence of injected waste during the preceding months carried over into the period of no injection. One might question whether a characteristic collateral to this carry over effect is the reduced level of frequency of earthquake immediately following disturbances having relatively large magnitudes. The behavior of the data in this respect is shown in Table 1. While there is a hint of such a relationship in these data, this evidence is confounded by a number of influences, and appears inconclusive.

3. Number of earthquakes tends to increase with higher injection pressures. Examination of the period March 1965 to October 1965 shows an appreciable monthly increase in the number of earthquakes over all preceding time periods. During this period waste was injected under substantially greater pressures than at any previous time. Yet the monthly volumes of water injected during this period are similar to the monthly injection volumes in 1962 and 1963.

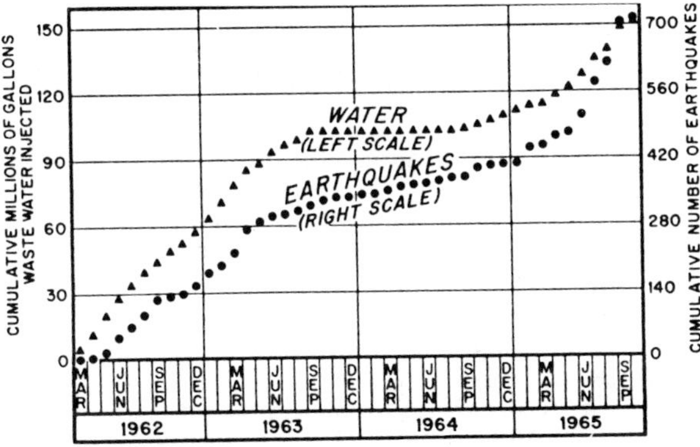

Fig. 1. Cumulative volumes of injected waste water compared to cumulative number of earthquakes, March 1962 to October 1965.

MODEL RELATING INJECTION VOLUMES TO FREQUENCY OF EARTHQUAKES

A simple mathematical model relating frequency of earthquakes to injection has been fitted to the data in Table 1 grouped in three-month intervals excluding the first and last month figures. The analysis above suggests that month-to-month frequencies of earthquakes are dependent upon one another to an unknown extent. Moreover, there appears to be about a three-month lag in the series for frequency of earthquakes behind the series for injection volumes. These dependence and lag characteristics of the data are undoubtedly far more complicated than pictured here, but they do suggest that grouping of the data by three-month intervals suppresses, to some extent, these undesirable properties for purposes of statistical analysis. Table 2 shows the injected volumes and frequency of earthquakes grouped into successive three-month intervals. Data for March 1962 and October 1965 have been omitted from the analysis to give an integral number of three-month periods.

Waste volume in millions of gallons versus the logarithm of frequency of earthquakes in each three-month interval is shown as a scatter diagram in Figure 2. The points in Figure 2 are generally scattered along a linear path showing that large waste volumes are associated with large numbers of earthquakes and small frequency of earthquakes accompany low waste volumes. The equation of this

Table 2. Volume of Waste and Number of Earthquakes--3-month intervals, April 1962 to Sept. 1965.

Year	Month	Volume of Waste (millions of gallons)	Number of Earthquakes
1962	April - June	23.6	49
	July - Sept.	16.2	76
	Oct. - Dec.	13.2	34
1963	Jan. - March	21.4	68
	April - June	14.0	71
	July - Sept.	9.7	27
	Oct. - Dec.	0	18
1964	Jan. - March	0	16
	April - June	0	15
	July - Sept.	0.6	11
	Oct. - Dec.	6.2	23
1965	Jan. - March	5.3	40
	April - June	14.0	68
	July - Sept.	20.7	197
	TOTAL	144.9	713

linear path, found using the method of least squares (Cramér, 1945; Mood and Graybill, 1963; Croxton and Camden, 1955) is given by

$$\log E_i = 1.207 + 0.0350\, W_i ,$$

where E_i denotes the number of earthquakes in three-month interval, i, and W_i is the volume of injected waste in millions of gallons in the same three-month interval, i. This equation is equivalent to

$$E_i = 16.1\,(1.0825)^{W_i}.$$

This formula states that the frequency of earthquakes is approximately exponentially related to volumes of waste. For example, if in two three-month periods the volume of waste for the second period is double that for the first period, the number of earthquakes in the second period is expected to be about $(1.0825)^2$ - 1.172 times that for the first period or about 17 percent greater.

Estimates from this model are subject to variation. This is to be expected, since the model itself is derived from data subject to considerable variation. Figure 2 shows the upper and lower 95 percent confidence belts. If certain mathematical assumptions are assumed to hold, these belts provide practical limits to the error in the estimates. It may be asserted that in repeated observations of the number of earthquakes at a given injection volume, 95 percent of such observations are expected to lie within the upper and lower 95 percent confidence belts.

The confidence belts exhibit considerable spread. This is not surprising since the number of earthquakes during the period October 1964 to September 1965 has a very large variance, apparently due to substantial changes in injection pressure of the waste during this period.

The extent to which there is a linear dependence of the logarithm of E_i and W_i is, roughly speaking, measured by their correlation coefficient (Cramér, 1945; Mood and Graybill, 1963; Croxton and Camden, 1955). For the data above, the correlation coefficient is 0.85. Theoretically, the absolute value of the correlation coefficient is always less than or equal to 1, where values near 1 correspond to greater linear dependence than where the absolute value is near zero. Since the correlation coefficient is computed from sample data, it, too, is subject to variability. With 14 observations, it may be asserted that the probability that a correlation coefficient of 0.85 could have arisen by mere chance, when actually there was no linear dependence between the log E_i and W_i, is about 5 in 1000.

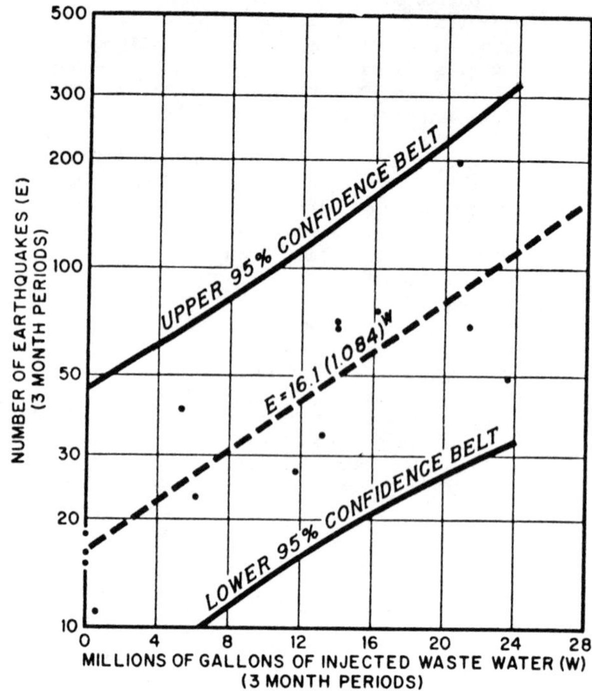

Fig. 2. Scatter diagram showing relationship of volume of injected waste to logarithm of number of earthquakes (3-month intervals, April 1962 to October 1965), semi-logarithm scale.

MEAN MONTHLY FREQUENCY OF EARTHQUAKES

Visual inspection of Table 1 suggests that the

Table 3. Comparison of Time Periods:

March 1962 to September 1963
October 1963 to August 1964
September 1964 to October 1965

in Gallons of Waste per Month and Number of Earthquakes per Month

Time Period	Number of Months	Mean Number of Gallons of Waste per Month (millions)	Mean Number of Earthquakes per Month	Variance in Number of Earthquakes
1. (March 1962 - Sept. 1963)	19	5.38	17.1	11.1
2. (Oct. 1963 - Aug. 1964)	11	0	5.3	3.2
3. (Sept. 1964 - Oct. 1965)	14	3.61	23.9	25.2

Table 4. Test of Difference Between Mean Number of Earthquakes per Month for 3 Time Periods.

Time Periods Compared	Large Sample Test	Exact Test
Period 1 vs. Period 2	Significant difference at 0.001 level	Significant difference at 0.005 level
Period 1 vs. Period 3	No significant difference at 0.30 level	No significant difference at 0.30 level
Period 2 vs. Period 3	Significant difference at 0.009 level	Significant difference at 0.05 level

mean monthly number of earthquakes in the three periods: March 1962 to September 1963, October 1963 to August 1964, and September 1964 to October 1965, vary markedly from one another. To test this observation, two statistical analyses have been made, one based upon so-called large sample theory, the other based upon exact distributions (Cramér, 1945; Mood and Graybill, 1963; Croxton and Camden, 1955). Table 3 shows certain statistical estimates for each of these three time periods. The hypothesis that there is, in fact, no difference in the mean number of earthquakes is tested against the alternative hypothesis that a difference does exist. The results of these tests are shown in Table 4. Thus, for example, the hypothesis that there is no difference in the mean number of earthquakes between period 1 and period 2 is rejected by both tests. Moreover, it may be asserted that as between period 1 and period 2 the probability of such a difference between the means, as that observed, arising by random fluctuation is about 1 in 1000. This shows that there is some assurance that the difference in means is real.

It is noteworthy that these tests failed to detect any statistically significant difference in frequency of earthquakes per month between period 1 and period 3. A different result might have been expected since some point has been made of the fact that the higher pressures of injected volumes of waste in period 3 would contribute to a substantially larger number of earthquakes per month. Part of the explanation is provided by the size of the variances shown in the last column of Table 3. The variance in the number of earthquakes in both periods 1 and 3 are very large. From a statistical point of view, this means that the difference between means must be very substantial to compensate for the variability between monthly figures.

LIMITATIONS OF THE ANALYSIS

The foregoing analysis should be considered a tentative one. Many factors such as pressures of injected waste, location of epicenters, magnitudes of earthquakes, geophysical constraints, should be taken into account in a more thorough-going analysis of the phenomenon under consideration. Even under most favorable circumstances there are severe limitations in the data for purposes of statistical analysis, and it is likely that we may have to be satisfied with something less than a definitive answer. No claim is made that the implications of the above analysis are invariant.

There is always present the temptation to conclude that a cause and effect relationship exists between two factors when confronted with data as compelling as that presented here. All sciences are replete with unhappy instances where this temptation preempted sober precaution.

ACKNOWLEDGMENT

The author extends his appreciation to Mr. Dennis Howard for his able assistance with many of the computations.

REFERENCES CITED

Cramér, H., 1945. *Mathematical Methods of Statistics.* Princeton University Press (seventh printing, 1957).

Croxton, F. E. and Camden, D. S. 1955. *Applied General Statistics.* Second Edition: Prentice-Hall.

Mood, A. M. and Graybill, F. A. 1963. *Introduction to the Theory of Statistics.* Second Edition: McGraw-Hill.

RELATIONSHIP OF EFFECT OF WATERFLOODING OF THE RANGELY OIL FIELD ON SEISMICITY*

R. C. Munson

Abstract

Waterflooding of the Rangely Oil Field of Colorado started in December 1957 in an effort to recover the maximum amount of oil from the reservoir. After 10 years of waterflooding, fluid production still exceeds fluid injection. As of October 1967 the deficit was 92 million barrels of fluid.

A seismic study using seismograms from the Uinta Basin Seismological Observatory near Vernal, Utah, was made of the Rangely area. The epicenters of the earthquakes centered on the reservoir and extended beyond its limits. No prominent secondary area of activity was found.

A study of the net fluid injected into the reservoir and the number of earthquakes that occurred disclosed no positive time correlation between these two parameters.

CONTENTS

ILLUSTRATIONS

*Condensation by the Editor of a thesis entitled "An Investigation of the Seismicity in the Vicinity of Rangely, Colorado", MS., Colorado School of Mines, 1968.

INTRODUCTION

The Rocky Mountain Arsenal Well and the northeast Denver earthquakes have generated great interest in determining whether there is any correlation between fluid injection into the ground and earthquake activity. The large number of earthquakes with their increasing magnitude prompted more detailed investigations of the seismicity in northeast Denver. The Colorado School of Mines now has extensive instrumentation in the area to monitor future events. W. W. Rubey, consultant for the U.S. Army investigating the Denver earthquakes, recommended studying other regions where fluid injection might effect the rate of earthquake activity. Another location at which fluid injection and earthquakes are found together is the Rangely Oil Field. This investigation of that area was undertaken to see whether any further information on fluid injection in relation to earthquakes could be obtained.

THE RANGELY OIL FIELD

The Rangely Oil Field, T 2 N, R 102 & 103 W, Rio Blanco County, Colorado, is the most prolific field in the Rocky Mountain Region. It produced a cumulative total of about 393 million barrels of oil prior to 1 October 1967 (written communication, Murphy, F. J., 11/28/67).

The reservoir is an anticlinal entrapment of the top 400 - 500 feet of the Permian-Pennsylvanian Weber Sandstone, Figure 1. The reservoir is about 12 miles long and 5 miles wide, with the longitudinal axis trending northwest-southeast. A normal fault with a northeast-southwest strike and a throw of about 50 feet cuts the longitudinal axis about one-third of the distance from the southeast end. The fault has no apparent effect on fluid migration in the reservoir (Bleakley, 1964). The position of the fault is shown in Figure 1. There a few smaller faults southeast of the reservoir limits.

The California Company (now Chevron Oil Company, Western Division) discovered the reservoir in 1933 but due to the remoteness of the area they did not develop the field at that time. During World War II oil became a premium product and the expansion of the Rangely field was started. In October 1957, the 45 operating interests and the 450 royalty interests unitized the field under the management of the California Oil Company. In December 1957, waterflooding was started in an effort to recover the maximum amount of oil possible from the reservoir.

Of the 482 wells in the field, 98 wells were converted to water injection by January 1964. Of these 98 water-injection wells there were 75 wells on the periphery of the reservoir, 16 wells in the 5-spot pattern on the east flank, and 7 wells in the gas cap

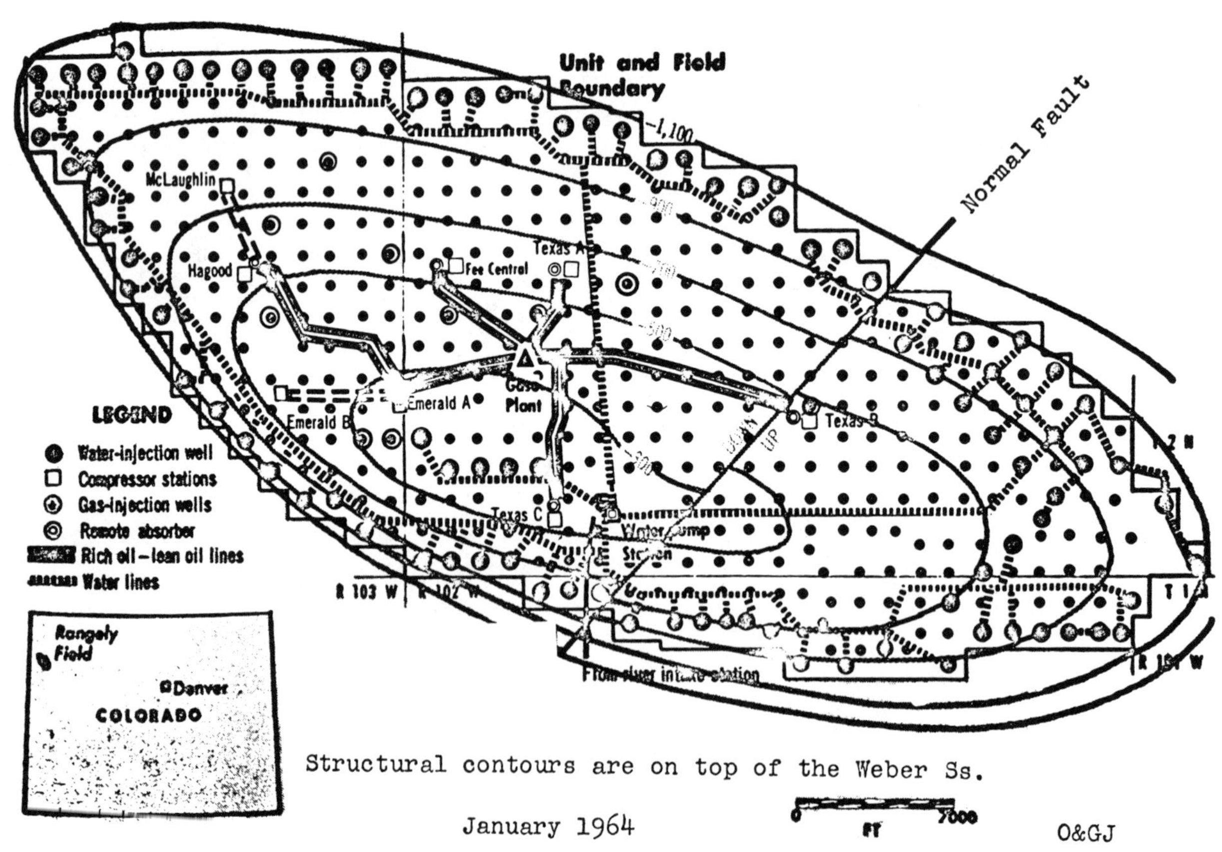

Fig. 1. Rangely Oil Field, Rangely, Colorado.

area. The original gas cap contained no residual saturation of the pore spaces. Therefore, migration of the oil into the gas cap area would result only in loss of that quantity of oil required to "wet" the reservoir rock. For the prevention of this needless loss, the gas wells in the cap were converted to water injection. As the oil production of the peripheral wells becomes too low economically, the exploitation of the wells is converted to the water injection method. The number of injection wells has increased steadily through the years, reaching a total of 158 in July 1967 (Chevron Oil Company, 1967). The wells in the field and the water injection pipelines are shown in Figure 1.

Water for injection is obtained from the very muddy White River, from water wells in the Entrada and Navajo formations, and from the Weber reservoir during oil production. Before injection the water is chemically prepared by the company's water treatment plant and is made crystal clear with a turbidity of less than 0.1 parts per million (Bleakley, 1964). The water is distributed through 12- and 14-inch cement-lined pipes to the east, north, northwest, and south portions of the field and then distributed to the individual injection wells through 3-inch cement-lined pipes (Figure 1).

The capacity of the 10 injection pumps is 210,000 barrels per day at a pressure of 1,200 pounds per square inch. The water injection, maintained at the full capacity of the water treatment plant, results in an oil production of about 1.3 million barrels per month.

Production is from the Weber Sandstone. The wells have an average depth of 6,700 feet. The producing interval is from 100 to 700 feet thick. The 98 injection wells were taking in excess of 160,000 barrels of water per day by January 1964 (Bleakley, 1964). Only 5 peripheral wells were then producing oil at profitable rates, and they will be exploited with the fluid-injection method when the water cuts become too high.

In the period 1933 to December 1957, before waterflooding, the total oil production was 236 million barrels. Data on the oil/(oil+water) ratio are inaccessible for the time prior to unitization in October 1957. However, data given in Appendix I of the original thesis since January 1958 indicate that the ratio was probably greater than 0.97 during the preunitization production (Figure 2). It is concluded that the total volume of fluid withdrawn prior to December 1957 was not more than 245 million barrels.

Since December 1957, 161 million barrels of oil and 76 million barrels of water have been produced. These combined to give a total of 237 million barrels of fluid removed from the reservoir since the field was unitized. During this same time 390 million barrels of water were injected. Therefore a net volume of 153 million barrels of fluid have been added to the reservoir in this time.

Comparison of the 245 million barrels of fluid removed prior to December 1957 with the 153 million barrels of fluid added since that date shows a deficit of 92 million barrels of fluid.

At the present rate of 3.1 million barrels of fluid injected per month more than that removed, in 30 months the fluid in the reservoir will equal the amount contained when the reservoir was discovered in 1933. It will be interesting to observe what, if any, change will occur in the earthquake activity at Rangely when this equilibrium is attained and then exceeded.

SEISMOGRAPH RECORDS AND INTERPRETATION

The only seismograph station close to the Rangely Field is the Uinta Basin Seismological Observatory (UBSO) located near Vernal, Utah. This portion of the study is based on selected records from that station. The observatory was constructed for the Department of Defense under the Vela-Uniform Project. It was established in November 1962 and is operated by the Geotech Division of Teledyne Industries Incorporated. UBSO (latitude 40° 19' 18" N, longitude 109° 34' 07" W, Elevation + 1600 meters) is about 10 miles south of Vernal, Utah. The Rangely reservoir is located between 50 and 80 kilometers away at an angle whose tangent is between 0.2 and 0.5 south of east from the observatory as shown in Figure 3.

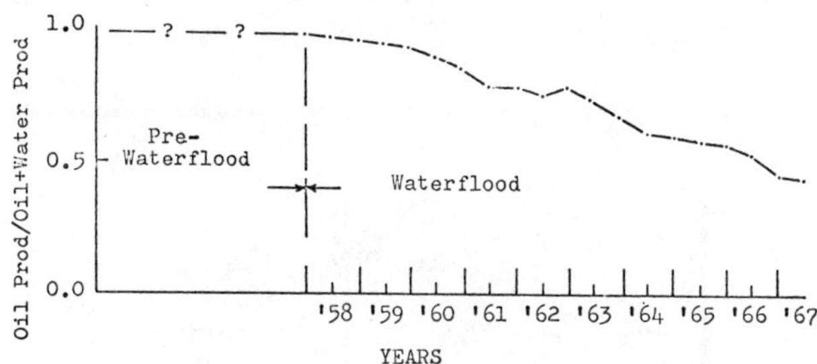

Fig. 2. Graph showing that part of fluid production that is oil.

Ten vertical and two horizontal Johnson-Matheson short-period seismometers are arranged in a star-type array as shown in Figure 4. The array is about 3 kilometers on a side with a maximum difference in elevation between the seismometers of 102 feet. The normalized response curve of the seismographs is shown in Figure 5. The seismographs, initially operated at a magnification of 200,000, were later increased to 600,000. During the periods of high wind and microseisms, it is impossible to identify the smaller Rangely events due to the high noise level.

The seismograms from the short-period seismograph array were used to determine the number and the location of those local earthquakes that have their epicenters in the vicinity of Rangely. The epicenters of the earthquakes were determined by computing the azimuth and distance from the observatory to the epicenter. The azimuth was found by one of two methods. The first and most precise method of determining azimuth was based on the difference in the arrival time of the Pg phase across the seismometer array. This is a geometric solution of the configuration of the array. As a matter of convenience, earthquakes were assigned to one of five easily identifiable azimuthal sectors based on the sequence of the Pg arrivals at the various seismometers as shown in Figure 6. The geometry of the UBSO seismometer array was so resolved that the azimuth was determined in 5 1/2° intervals for the Pg phase arrivals. For the listing of the azimuth on the data sheets, only the whole degree was used so that exceptional accuracy would not be implied.

The second method of determining the azimuth was used when the Pg arrivals on the vertical seismometers were masked by noise or made uncertain in other ways. The azimuth could sometimes be determined by scaling the double amplitude of the Pg arrival on the N/S and E/W traces. The tangent of the angle south of east was then determined by dividing the N/S amplitude by the E/W amplitude. The azimuth from north was found by converting the tangent value to degrees and adding 90°. In cases where the azimuth was determined by both methods, the azimuth agreed within 3° of each other. Both results are listed in Appendix II of the original thesis.

Of the two methods for determining azimuth, the use of Pg arrival time across the array is considered to be the more accurate method.

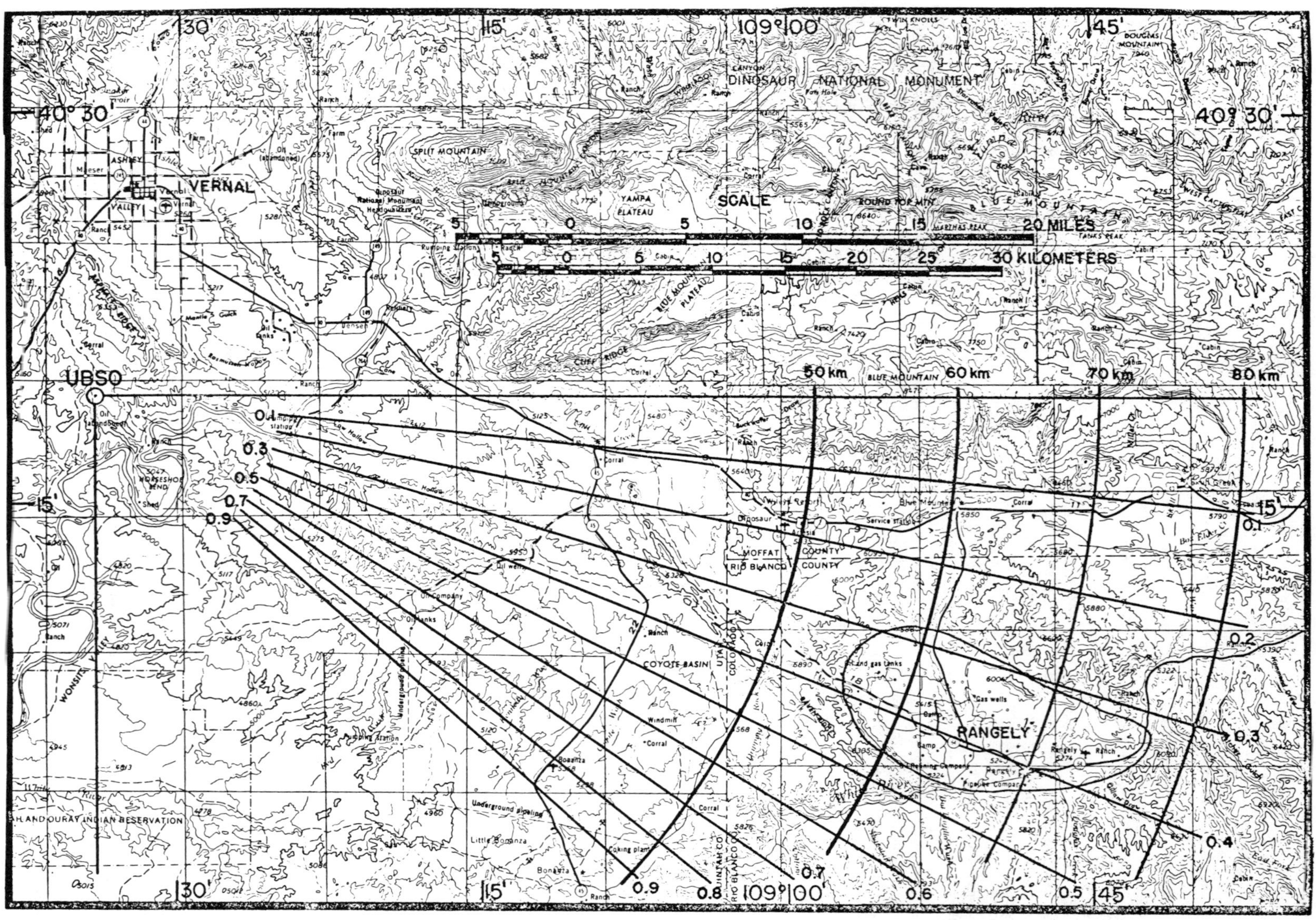

Fig. 3. Uinta Basin Seismological Observatory - Vernal, Utah - Rangely, Colorado.

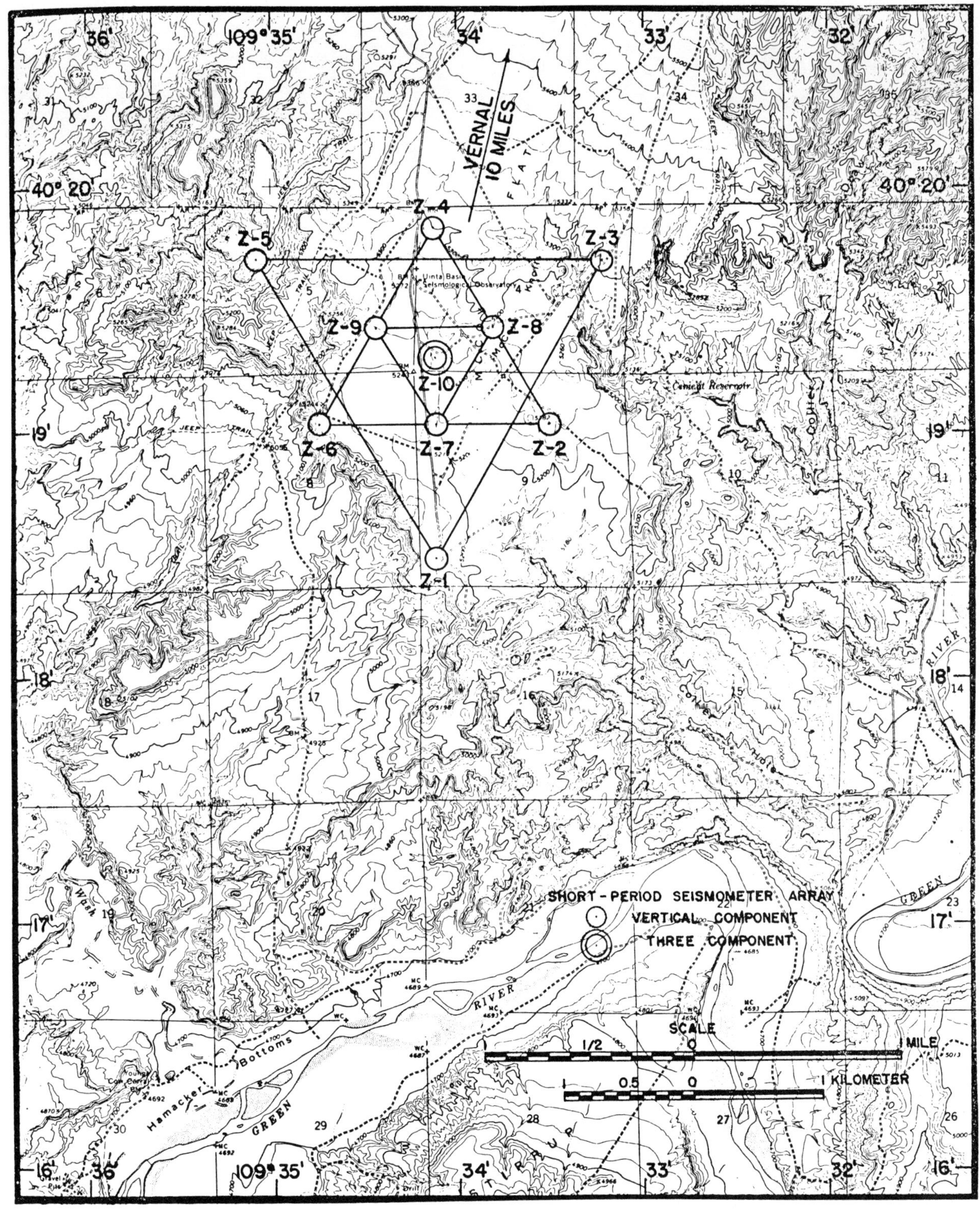

Fig. 4. Star-type seismometer array at UBSO.

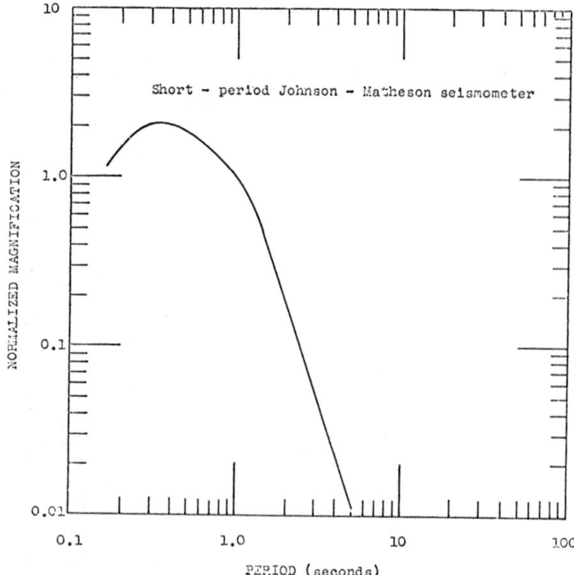

Fig. 5. Normalized frequency response of UBSO seismograph.

The distance from UBSO to the epicenter was found by using the Sg - Pg phase interval, from the three-component seismograph at the center of the array, and the Jefferys-Bullen local travel time curves. The Sg arrival was best recorded on the N/S seismograph.

A typical seismogram of a Rangely earthquake is shown in Figure 7. This figure shows the earliest arrival of the Pg phase on trace Z-2, followed closely by the arrivals on traces Z-3 and Z-1. This is the time difference that was converted to azimuth in determining the epicenter location.

The distinctive character of the Pn, Sg, Sn, and surface waves made the Rangely earthquakes readily recognizable. However, many of the recognizable Rangely earthquakes were of such small magnitude that an accurate epicenter determination could not be made. Pg and P* are very difficult to identify on events of magnitudes less than 0.5.

The ground displacement for the earthquakes was determined from the seismograph films using the formula:

$$A = \frac{\alpha \times 10^3}{K\ g_t} = \frac{\alpha \times 10^3}{600\ (1.6)} = \frac{\alpha}{0.96}$$

where A = peak-to-peak ground displacement in millimicrons, of the Pg phase.

α = peak-to-peak pulse amplitude of the Pg phase in millimeters when the seismograph film is enlarged 10 times.

K = magnification in thousands, at the calibration frequency.

g_t = correction factor applied to obtain true magnification at the signal frequency; obtained from the seismograph frequency-response curve.

With the use of the ground displacement, A, the approximate magnitude of the earthquake can be determined from Richter's magnitude formula, $M = \log_{10}A - \log_{10}A_0$. The ground displacement must first be converted to the A used in Richter's formula. This step is done by changing the actual ground displacement from millimicrons to millimeters, increasing A by a

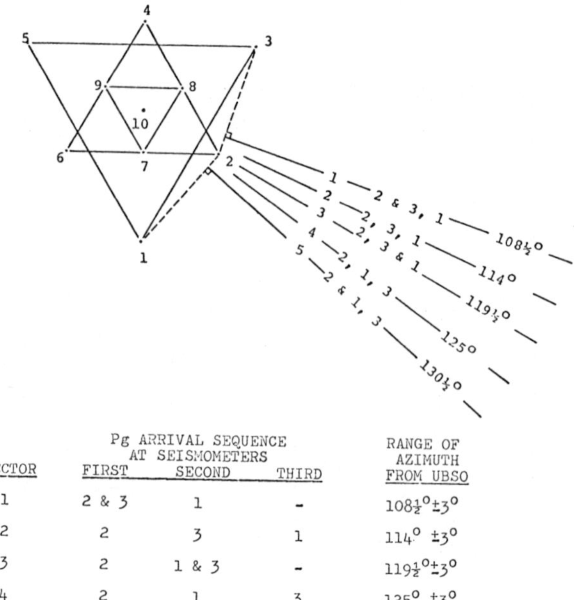

SECTOR	Pg ARRIVAL SEQUENCE AT SEISMOMETERS FIRST	SECOND	THIRD	RANGE OF AZIMUTH FROM UBSO
1	2 & 3	1	-	108½°±3°
2	2	3	1	114° ±3°
3	2	1 & 3	-	119½°±3°
4	2	1	3	125° ±3°
5	2 & 1	3	-	130½°±3°

Fig. 6. Azimuth determination from the geometric solution of the array configuration.

factor of 4, and multiplying this value by 2800 to convert it to the maximum trace amplitude that would have been recorded on the torsion seismometers, which Richter used in establishing his relationship. The factor of 4 is the average ratio of the maximum trace amplitude to the Pg amplitude. The constant 2800 is the magnification of the instruments that were used by Richter to derive the formula. The value of the term $-\log_{10}A_0$ (minus $\log_{10}A_0$) is obtained directly from his table for various distances to the epicenter. The majority of the epicenters were about 65 kilometers from UBSO. At this distance the value of $-\log_{10}A_0 = +2.8$. Substituting these values into Richter's formula results in the following conversion formula:

$$M = \log_{10} 2800(4A \times 10^{-6}) - \log_{10}A_0 \text{ or}$$

$$M = \log_{10} 0.0112A + 2.8$$

for a ground displacement of 10 millimicrons the magnitude is deduced thus:

$$M = \log_{10} 0.0112(10) + 2.8 = 1.85.$$

The Pg phase arrivals of the smallest events were too small, equal to or less than 0.1 millimeter, to obtain peak-to-peak amplitude measurements. Thus, they were designated an amplitude of 0.1 millimeter arbitrarily. This designation gives a magnitude of -0.15 as the lower limit of detectability for the earthquakes. Amplitudes ranging from 0.2 millimeter and greater could be measured accurately.

FLUID INJECTION IN RELATION TO EARTHQUAKES

An ideal investigation would require the seismograph records prior to waterflooding, during the initial stages of the waterflood, and after the waterflood had reached its present proportions. This ideal investigation could not be undertaken because the waterflood started in December 1957 and the observatory did not become operational until November 1962.

The observatory bulletins available in the Colorado School of Mines Geophysics Library start with June 1963 and continue to the present with the omission of the volumes for March 1964. These bulletins give four years of data to assist in this investigation. The volumes prior to June 1963 and for March 1964 were requested from the Defense Documentation Center, Cameron Station, Alexandria, Virginia, but the supply was already exhausted. It was found in reading the seismograph films that few of the smaller events are recorded in the UBSO bulletins; and when several earthquakes occur together, only the largest event is recorded. It was also found that in the earlier bulletins only the largest events were identified and none of the small earthquakes were listed. Therefore the number of Rangely earthquakes in the monthly bulletins is a relative number rather than a true figure.

Two methods were used in conducting the investigation. First the net fluid injected into the reservoir was compared with the number of Rangely earthquakes listed in the UBSO bulletins. This method used a large 4-year sample of data of questionable precision from the bulletins. It is probable, however, that the bulletins reflect the general level of seismic activity. The second method was based on using a small sample of data of much greater precision in comparing the net fluid injected with the earthquake frequency. The small sample periods were separated by 4 years in time in hopes that the large increase in the rate of net fluid injected between the two periods would be reflected by a substantial increase in the earthquake activity.

A comparison of fluid injection with the general level of seismic activity, as determined from the UBSO bulletins, is shown in Figure 8. In this figure the net amount of water injected per month from December 1957 to the present is plotted on the left, with the number of earthquakes per month recorded in the observatory bulletins shown below the injection graph. There exists two pronounced peaks, P_1 and P_2, on the left-hand graphs, separated by 12 months in time. It is impossible to use this time interval and correlate a significant number of the other peaks or valleys to substantiate a 12-month lapse between fluid injection and the resulting earthquakes. The graph on the right of Figure 8 shows the same data after the monthly values have been averaged over a year. The small number of events in 1963 probably represents failure to list small local events in the initial bulletins. They were very conspicuous by their absence. No other period showed such a low level of activity. Otherwise there was no positive correlation between the two parameters.

A more precise comparison of injection and seismicity is constrained by the format of the seismic data. The 16-millimeter film records from the seismographs are available from the Seismic Data Laboratory in

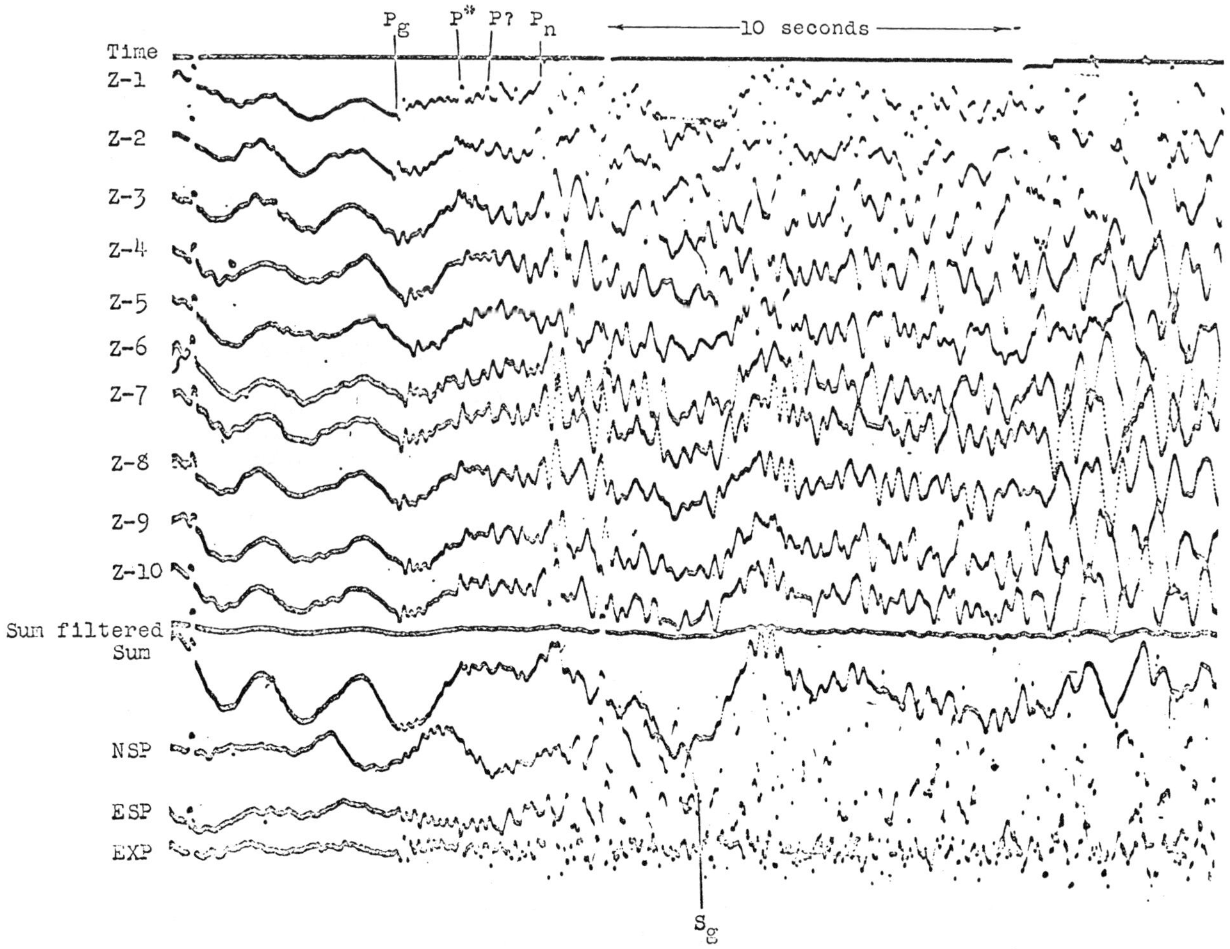

Fig. 7. Rangely earthquakes as recorded at UBSO.

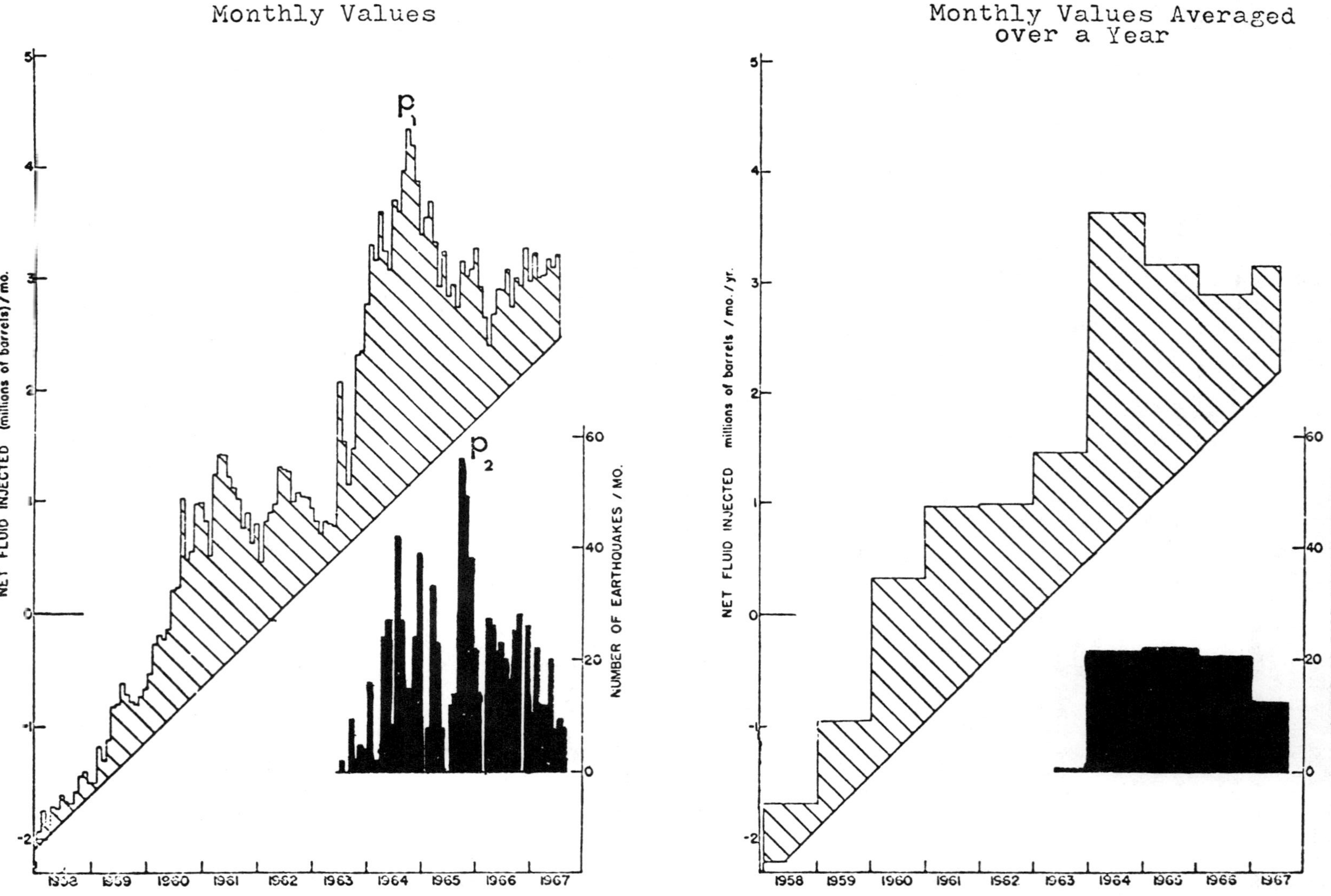

Fig. 8. Comparison of rate of net fluid injected with earthquake frequency.

Alexandria, Virginia. Although 16-millimeter film is an excellent format for storing records, it is a tedious, eye-straining, time-consuming format for searching records to identify specific events, such as those occurring at Rangely. It was therefore felt to be impractical to view all of the films from November 1962 to the present because the additional information obtained, over that received from selected small sample periods, would not warrant the investment of such an enormous amount of time. This decision was supported by the results from two sample periods. The first sample period selected was November-December 1962 so that the earliest seismic information available would be used in the investigation. During this interval 284 Rangely events were identified with 185 of these earthquakes large enough for accurate epicenter determinations. The second sample interval was the three-month period, October-November-December 1966. One hundred and seventy-five Rangely earthquakes were identified with 107 of these events large enough for epicenter determinations. All of the Rangely earthquakes that were identified in the two sample periods are tabulated in Appendix II of the original thesis.

The largest Rangely earthquake that occurred during the sample periods was of a magnitude 2.6. However, there have been much larger events recorded. The two largest Rangely events listed in the UBSO bulletins are of a magnitude 4.5 on July 6, 1966, and of a magnitude 4.4 on February 18, 1967.

The $\log_{10} N$ versus M curve is shown in Figure 9. The formula $\log_{10} N = 3.06 - 1.03M$ describes this curve in which N = the number of earthquakes of magnitude M. The slope of the curve is significantly steeper than that determined for the northeast Denver earthquakes. The northeast Denver earthquakes closely agree with the formula used by Richter, i.e., $\log_{10} N = C - 0.8M$. If either of the two largest earthquakes, whose magnitudes were 4.4 and 4.5, had occurred during the sampling periods, the slope of the curve would have been about 0.8 for the Rangely events.

Of the larger earthquakes whose epicenters were determined, 79% were within the reservoir area and the remainder were within 15 kilometers of the reservoir. Figure 10 shows the concentration of the epicenters with respect to the reservoir for the two individual sample periods, and Figure 11 shows the totals for the two sample periods. Figures 10 and 11 emphasize the fact that there is no predominant secondary area in which the earthquakes occur. The seismic activity gradually diminishes as the distance from the reservoir gradually increases.

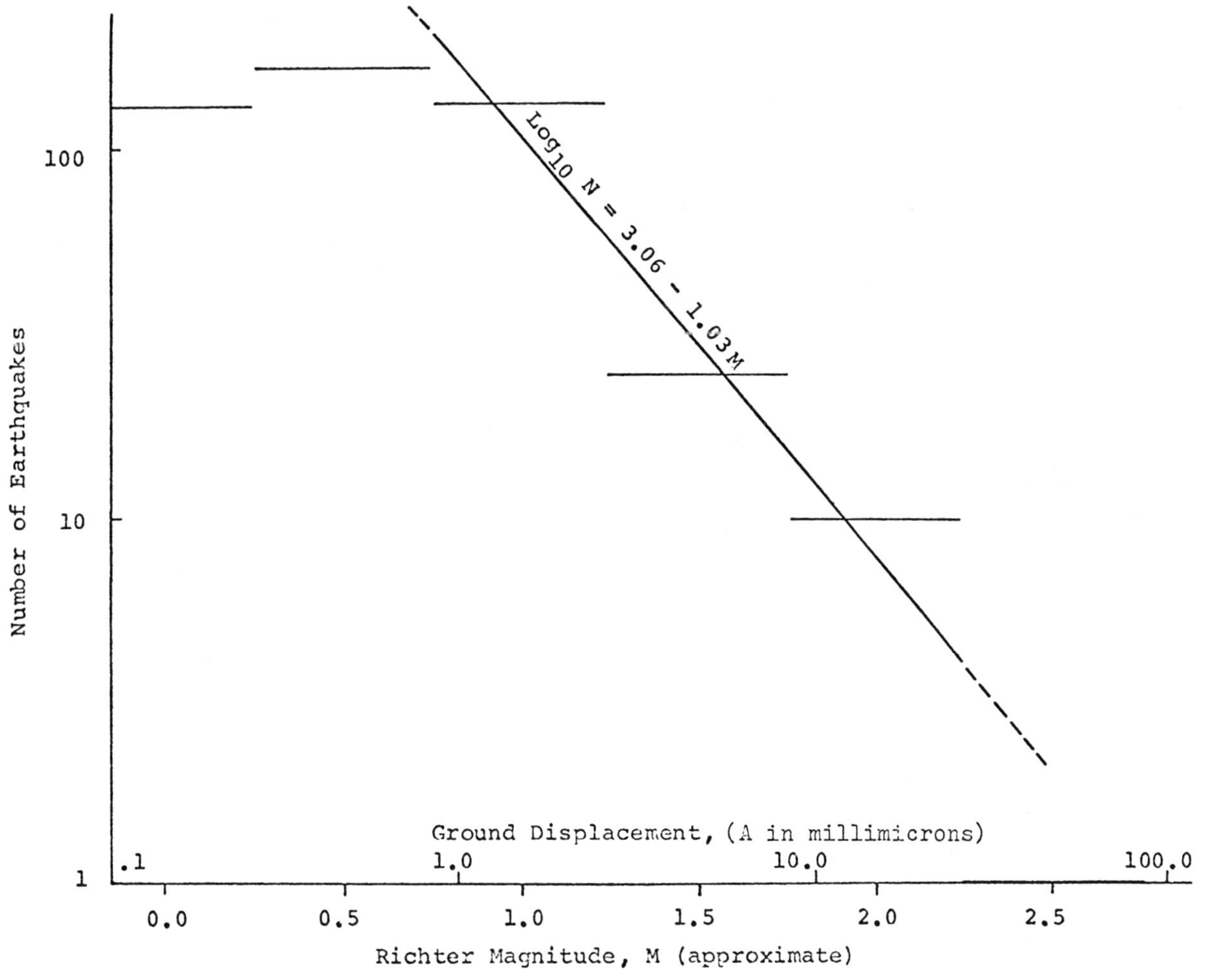

Fig. 9. Log_{10} N vs M.

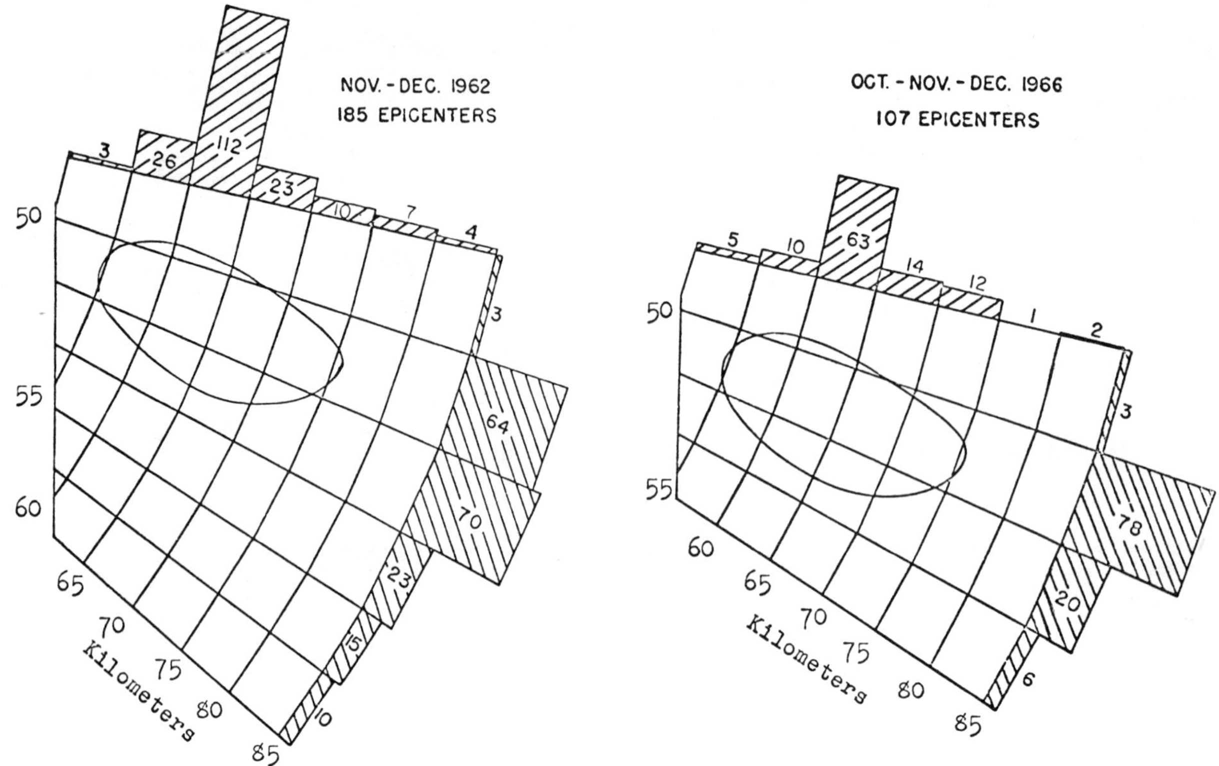

Fig. 10. Location of earthquake epicenters with respect to the Rangely Reservoir for the two sample periods.

The rate of net fluid injection into the reservoir increased about threefold between 1962 and 1966, yet the earthquake frequency reduced from 142 per month to 58 per month. This reduction in earthquake frequency is further emphasized by the seismograph magnification increasing from 200,000 to 600,000 in the same interval. It is concluded that, while the seismic activity is continuous at Rangely, there is no positive correlation in time between the net fluid injected and the number of earthquakes recorded.

Regardless of the net amount of fluid injected per month per year, the number of earthquakes per month per year, as listed in the UBSO bulletins, remains at about 20 events. In two of the three months, for which comparison was possible, between two and three times as many earthquakes were identified from the records as were listed in the bulletins. It is therefore assumed that the bulletins provide only a gross index to the seismic activity. There were no bulletins published in 1962. The tabulation of this information follows.

Table 1. Summary of Results from the Two Sample Periods.

Year	Month	Mag.	No. of Earthquakes Bulletin	Munson	Net Fluid Injected 10^6 barrels per month
1962	Nov.	200,000	NA	171	1
	Dec.	400,000	NA	114	1
1966	Oct.	600,000	21	59	3
	Nov.	600,000	6	41	3
	Dec.	600,000	26	75	3

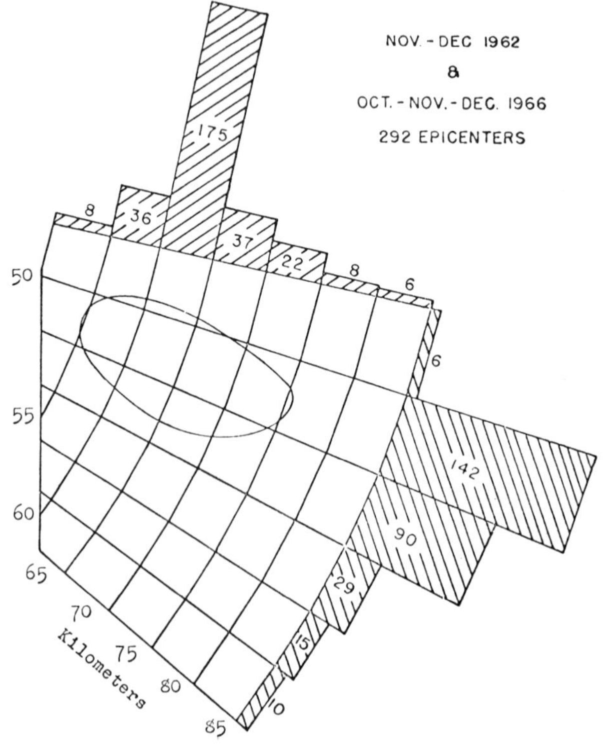

Fig. 11. Location of earthquake epicenters with respect to the Rangely Reservoir for the total of the two sample periods.

CONCLUSIONS

In considering the Weber reservoir from 1933 to the present, there has been a total of 482 million barrels of fluid removed and 390 million barrels of fluid injected. Hence, because of steady removal there are 92 million barrels of fluid less as of October 1967. Therefore, although fluid injection has been in operation for 10 years, the net result is fluid withdrawal.

There are definitely earthquakes occurring in the Rangely oil field. The extent of the epicenters covers an area larger than that bounded by the reservoir limits, and there is no predominant secondary area in which the earthquakes originate.

The Rangely data in the UBSO bulletins cannot be used as a precise reference but may be used to show the general seismic activity.

No positive correlation in time can be determined between the fluid injection and the frequency of earthquakes in the Rangely oil field. Neither the data from the UBSO bulletins nor the data from the two short sample periods disclosed a positive correlation in time when compared with the net fluid injected.

This investigation does not reveal that the earthquake activity is directly caused by fluid injection.

REFERENCES CITED

Allen, D. D. 1963. High-volume pumping in the Rangely Weber sand unit. *Jour. of Petroleum Technology*, V. 15, pp. 823-827.

Bleakley, W. B. 1957. Rangely is unitized. *Oil and Gas Jour.*, V. 55, p. 70.

_________ 1958. Rangely field ready for big flood. *Oil and Gas Jour.*, V. 56, p. 50.

_________ 1959. Rangely flood moving in high gear. *Oil and Gas Jour.*, V. 57, p. 93.

_________ 1960. Rangely flood moves ahead gasoline plant gains satellites. *Oil and Gas Jour.*, V. 58, p. 110.

_________ 1964. Colorado's biggest field engineered for maximum recovery. *Oil and Gas Jour.*, V. 62, pp. 108-116.

Campbell, Graham S. 1955. Weber pool of Rangely Field, Colorado. *R.M.A.G.-I.A.P.G. Guidebook to the Geology of Northwest Colorado*, pp. 99-100.

Evans, David M. 1967. Man-made earthquakes - a progress report. *Geotimes*, V. 12, No. 6, pp. 19-20.

Hembree, M. R. 1961. Rangely field. *R.M.A.G. Oil and Gas Field Volume*, Colorado-Nebraska-1961, pp. 224-225.

Herrin, Eugene and Taggart, James 1962. Regional variations in P_n velocity and their effect on the location of epicenters. *Seismol. Soc. America Bull.*, V. 52, pp. 1937-1046.

Peterson, V. E. 1955. Future production from the Mancos shale, Rangely Field, Rio Blanco County, Colorado. *R.M.A.G.-I.A.P.G. Guidebook to the Geology of Northwest Colorado*. pp. 101-105.

Pickering, W. Y. and Dorn, C. L. 1947. Rangely oil field, Rio Blanco County, Colorado. *A.A.P.G. Symposium, Structure of Typical American Oil Fields*, V. III.

Richter, C. F. 1958. *Elementary Seismology*. San Francisco, California, W. H. Freeman and Company, Inc., 738 pp.

Ritzma, H. R. 1955. The Rangely boom. *R.M.A.G.-I.A.P.G. Guidebook to the Geology of Northwest Colorado*, p. 100.

Ryall, Alan and Stewart, D. J. 1963. Travel times and amplitudes from nuclear explosions, Nevada Test Site to Ordway, Colorado. *Jour. Geophys. Research*, V. 68.

Smith, W. R. 1961. Rangely field, Colorado - a large fluid injection operation. *R.M.A.G. Oil and Gas Field Volume*, Colorado-Nebraska-1961, pp. 272-278.

White, J. E. 1965. *Seismic Waves: Radiation, Transmission, and Attenuation*. New York City, New York, McGraw-Hill Book Company, 302 pp.

RESERVOIR LOADING AND LOCAL EARTHQUAKES

Dean S. Carder (U. S. Department of Commerce, Environmental Science Services Administration)

Abstract

Shortly after Lake Mead (the reservoir impounded by Hoover Dam) began to fill, local earthquakes were felt. The earthquakes reached a culmination in a magnitude 5 earthquake about a year after the reservoir had filled to 80 percent capacity. For a number of years thereafter, small local earthquakes showed a close correlation in numbers and energy release with seasonal peak loads. In due time the correlation was as close with unloading as with loading. Sometimes there was no direct correlation at all.

In recent years many instances of local earthquakes attributed to crustal unbalance have been noted. Noteworthy is the earthquake activity near Denver and associated downhole injections of waste chemicals; and earthquakes associated with reservoir loading at Kariba in Africa, Kremasta and Marathon in Greece, Vajont in Italy, Monteynard and Grandval in France, and Koyna in India. In one or two instances the earthquakes may be a direct result of settlement under load. More likely they result from a triggering of stored strain energy. The crustal block is considered to be strained to near yield but in delicate equilibrium. This equilibrium may be upset by a sudden shifting of the load, or by lubrication or build-up of fluid pressure within fractures in the rock. This decreases frictional resistance between the walls of fractures. Many newly created reservoirs have no association with local earthquakes because there are no ready-made conditions in the crustal block containing the reservoir for the production of earthquakes.

CONTENTS

ILLUSTRATIONS

INTRODUCTION

Recent earthquake activities attributed to artificial unbalance of natural forces in various localities have revived interest in an investigation of similar nature which was begun more than 30 years ago in the Lake Mead region of Nevada-Arizona.

Lake Mead is the reservoir impounded by Hoover Dam (formerly Boulder Dam). Until a few years ago, prior to construction of Kariba Dam in Rhodesia, it was the world's largest artificial lake. In mid-1936, about a year after the reservoir had started to fill, a number of earthquakes were felt immediately following and accompanying the peak load of the year. It was not known at the time whether this was the result of loading or whether prior to this time earthquakes were considered as blasts because of construction. There is no evidence that earthquakes were felt prior to construction which in itself is inconclusive because the area was relatively uninhabited at the time.

In an effort to gain some understanding of earthquake activity local to the area, the U. S. Bureau of Reclamation requested assistance of the U. S. Coast and Geodetic Survey (C&GS). Early efforts began in 1937 with the installation of three strong-motion seismographs in or near the dam and in 1938, a Wood Anderson seismograph at Boulder City. In 1940, a C&GS field party financed by the Bureau of Reclamation was established to install and operate a quadripartite net of seismographs about the lake and to interpret the records. Operational details and test results of the first four years of operation are covered in earlier papers (Mead and Carder, 1941 and Carder, 1945).

Eventually 12 years of records were collected from this net. Several hundred small earthquakes were located, most of them near the lower basin of this reservoir. Several thousand earthquakes were not located but were recorded by the seismograph at Boulder City. Focal depths of most of the earthquakes near Boulder City probably did not exceed 6 kilometers.

Originally, the method of locating local earthquakes was to assume homogeneity in the crustal layers and to assign certain speeds to *P* and *S* waves assuming spherical wave fronts. Later it was found that homogeneity of the surface rocks and spherical wave fronts were gross oversimplifications.

In February 1946, 700 surplus land mines containing a total of 4000 pounds of TNT were exploded in the end of an old prospect hole driven 80 feet horizontally into the granitic face of the Black Mountains about ten miles east of Boulder City. A 6.35 kilometer per second medium underlying 3 kilometers of a 3.5 kilometer per second medium formed the best fit of the resulting seismic data. Later, seismic waves from the NTS nuclear explosions were observed to cross under the net at a speed of about 6.15 kilometers per second. However, by using this more realistic model, still an oversimplification, it was found that epicentral locations within the net were practically the same. The extreme shift toward the edge of the net was no more than 2 kilometers.

Locations of local earthquakes from late 1940 through 1947 are reproduced from Carder and Small (1948) in Figure 1. Additional locations since 1947 follow the same general pattern and are not reproduced.

Figure 1 is a partial fulfillment of the principal objective of the Lake Mead seismological program, from the viewpoint of the Bureau of Reclamation, which was to locate the larger local earthquakes and thus determine the position of potentially active faults in relation to the dam. The influence of reservoir loading on local earthquake activity was a secondary objective. However, from the viewpoint of the Engineering and Scientific Community, this relationship is of high interest.

RESERVOIR LOADING AND LOCAL SEISMICITY

Local seismicity will be defined here as the count of earthquakes occurring in the immediate vicinity of Lake Mead and at focal distances not exceeding 80 kilometers from Boulder City with an added factor of energy released by the earthquakes.

Figure 2 is a revision of an illustration I presented at the Ottawa meeting of the Eastern Section, Seismological Society of America, in 1951. In this figure, energy estimates are on the basis of one of two methods; (1) For earthquakes recorded by the Pasadena net which had a magnitude estimate, the following formula is used: $\log E = 11.8 + 1.5M$, where $\underline{E}$ is the energy release in ergs and $\underline{M}$ is the Richter magnitude. This is a revision of the earlier estimate. Records from the local net of these earthquakes were usually off scale. (2) For smaller earthquakes yielding measurable trace amplitudes,

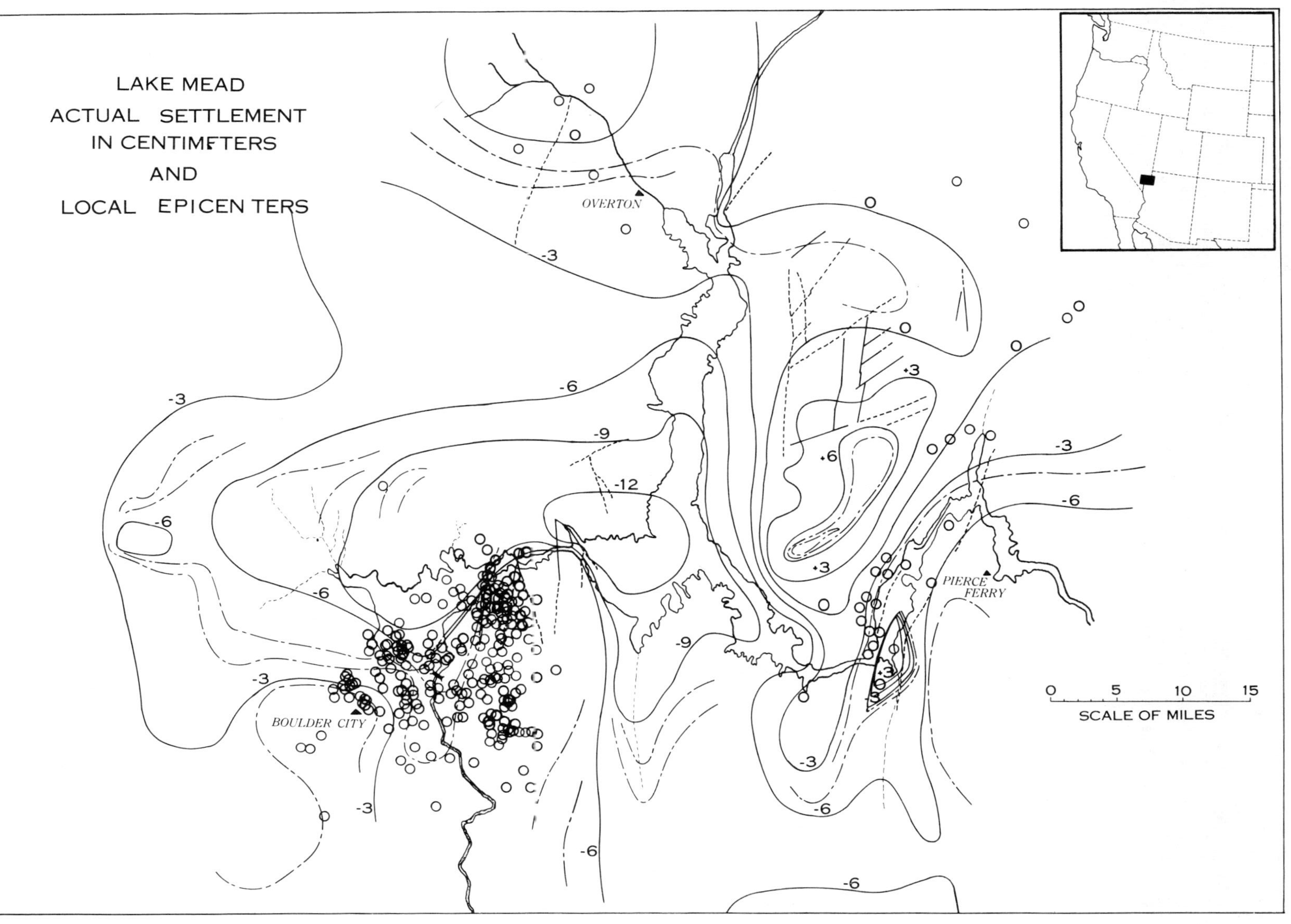

Fig. 1. Lake Mead.

energy was usually estimated by direct calculation using methods outlined in an earlier paper (Carder, 1962).

Several premises may be deduced from Figure 2.

(1) A single earthquake in May 1939 released as much or more energy than all the other local earthquakes combined.

(2) There is strong evidence that the May 1939 series had some correlation with the load peak of the previous year with a triggering effect resulting from the rising water of 1939.

(3) For the same reason, the energy and frequency peak during the summer of 1942 may have been associated with the all-time maximum of the previous year, but with triggering by the load of the year 1942.

(4) Prior to 1945, a close ass ation existed between the energy peak and the load peak of the given year.

(5) In 1945 and later, energy peaks were at times associated with load minima, and energy maxima were not always associated with frequency maxima.

(6) Leslie F. Bailey, seismologist for the ESSA/Earth Sciences Laboratories, examined statistically the energy release in reference to reservoir loading by making running averages of seismic energy and reservoir loading over the years 1938 through 1949. He found, nevertheless, a close correlation between seismic energy and loading over the entire period.

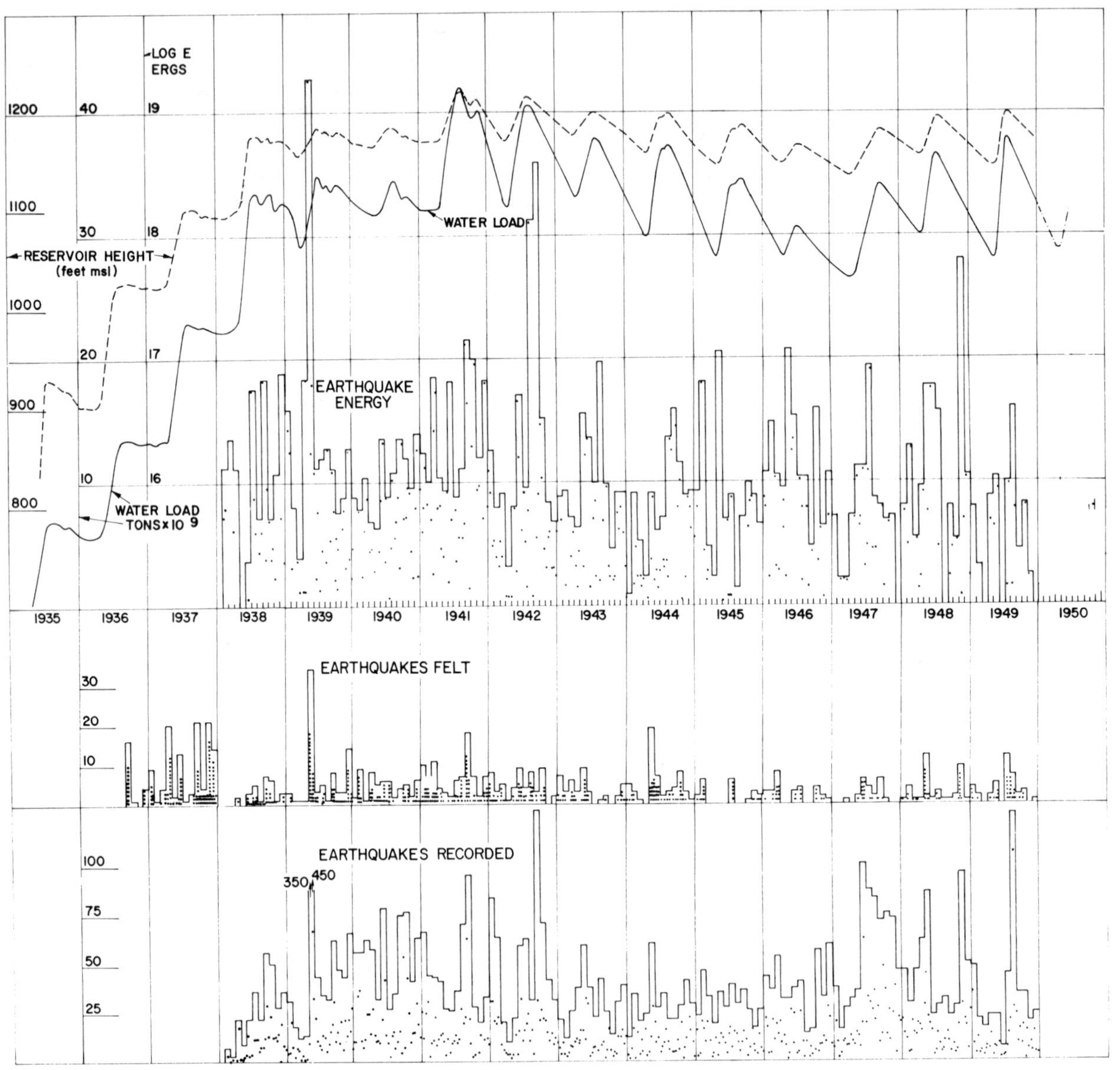

Fig. 2. Lake Mead earthquake and hydrographic chart, 1935 to 1949. The dots refer to ten-day periods.

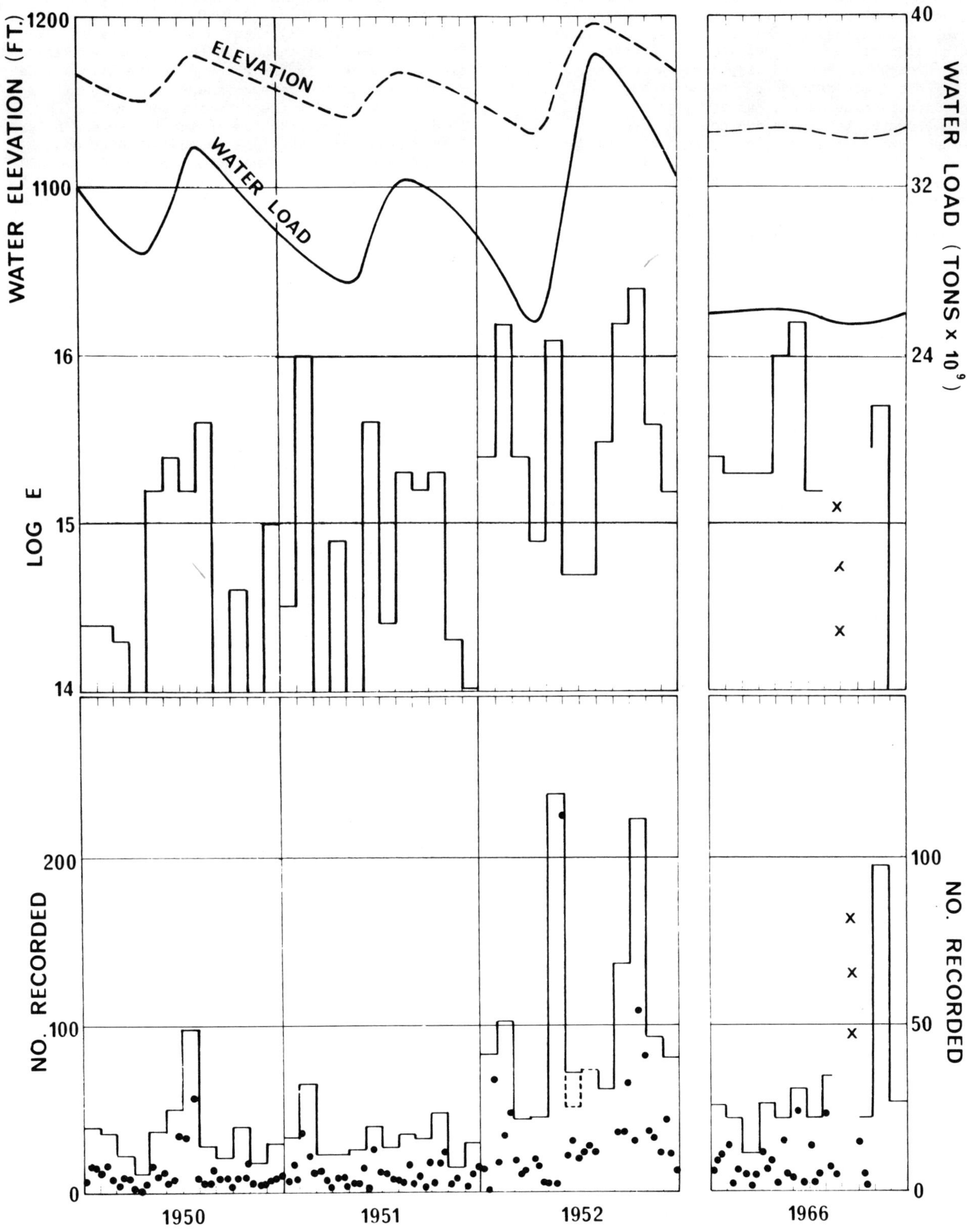

Fig. 3. Lake Mead earthquake and hydrographic chart, 1950 to 1952 and 1966. The 1966 energy data are uncertain and the August-September hiatus is a result of the Utah-Nevada magnitude-6 earthquake of August and its aftershock.

The Boulder City records for 1950, 1951, and 1952 were recently examined for the double purpose of extending the study of the relationship of seismicity to reservoir loading over an additional three years, and to obtain an estimate of the numerical count versus magnitude of the local microearthquakes. A continuation of local seismicity and reservoir loading through these years is presented in Figure 3. The seismic data were obtained from the Boulder City records alone, and for lack of reliable documents, felt data are not included. Water-load data represented in Figure 3 are based on reservoir capacity as of 1935 and are in actuality somewhat higher than available water because accumulation of silt over the years has materially reduced the reservoir capacity. However, the actual seasonal load fluctuation is nearly as represented, and the total would be even higher than represented here.

From the evidence of Figure 3, if earlier data were not available, there would be little or no reason to suspect that any relation existed between reservoir loading and seismic energy.

For a possible explanation, it should be recalled that surveys by level parties of the C&GS were made in 1935 as the reservoir began to fill and again in 1940-41 when the reservoir load was at about 80 percent capacity. It was observed that areas of greatest settlement as a result of the added load were not necessarily the areas of greatest seismicity which is understandable considering the complexity of the region. Areas of monzonite or pre-Cambrian gneisses and granites crossed by level lines were positive areas where little or no settlement occurred. It was concluded that many of the local earthquakes were caused by down faulting of the crustal block occupied by the lower lobe of the Lake against the granitic masses to the southeast and southwest. It was mentioned that this settlement was probably a renewal, on a small scale, of pre-Pleistocene activity under the stimulus of a suddenly added load of 10^{10} tons of water and silt.

The same level lines were resurveyed in 1950. The trend of settlement, apparent under loading observed between 1935 and 1940-41, did not continue into 1950. Evidently then, seismicity resulting from adjustment of the crust to loading was satisfied during the first few years of the life of the reservoir, and thereafter, there was an equal probability of adjustment to seasonal unloading. It should be noted that beginning in the mid 60's, there is no such thing as seasonal loading and unloading because of regulation by Glen Canyon and other upstream dams.

FREQUENCY VERSUS MAGNITUDE, BOULDER CITY AREA

As mentioned earlier, a count of local earthquakes was made by re-examining Boulder City records for 1950, 1951, and 1952. Events with an *S* minus *P* interval greater than 8 seconds were not considered and those greater than 3 seconds only if from the direction of the reservoir. Magnitudes as low as -0.5 were recognizable, but only for nearby events and on quiet days.

The plots in Figure 4 are from 1950 and 1951 data; and Figure 5 is a fusion of the two with the 1952 data included. Graphs in Figures 4 and 5 are plots of the well known equation $\log N = a - bM$, where N is the cumulative number of all events of magnitude M or greater. The most reliable evaluation of the b value is from the mid portion of the data because events of magnitudes less than say 0.3 are often lost in the noise if they are much more remote than 20 kilometers or so and the larger events are relatively infrequent, or magnitude estimates were uncertain. If this relationship has physical meaning, one may expect a magnitude 5 earthquake in the reservoir area once each 30 years or so. One such event occurred about 30 years prior to this paper.

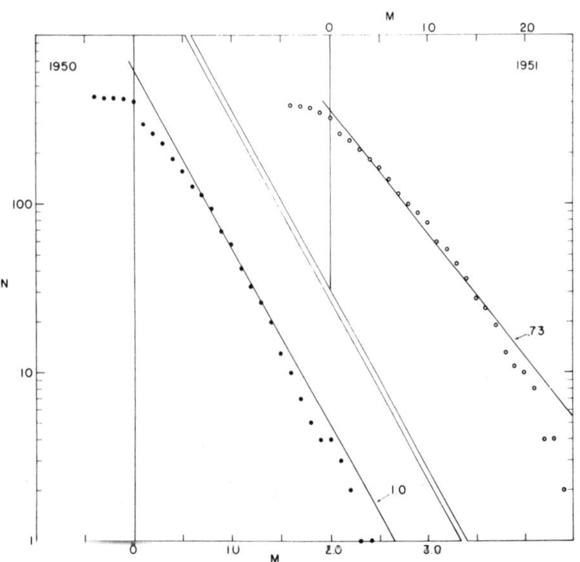

Fig. 4. Magnitude versus number of earthquakes local to Boulder City, 1950 and 1951.

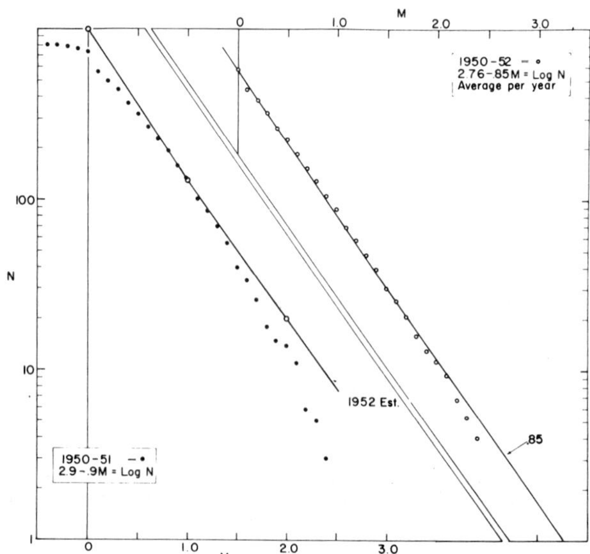

Fig. 5. Magnitude versus number of earthquakes local to Boulder City: Cumulative 1950-1951 and 1952.

OTHER PROJECTS

Bureau of Reclamation Projects

Under the sponsorship of the Bureau of Reclamation, the C&GS was requested to investigate local seismicity near a number of other dam sites. These sites include Shasta and Oroville in northern California; Grand Coulee in central Washington; Hungry Horse in northern Montana; Glen Canyon in northern Arizona; Flaming Gorge in northern Utah; and San Luis in central California. All seismograph installations, except that at Grand Coulee, preceded reservoir filling.

At Shasta, a swarm of local earthquakes was recorded during the first year of reservoir loading. The stronger ones had magnitudes of about 3.0 on the Richter scale and were located a few miles southeast of the reservoir. Some of them were felt. Very few earthquakes, if any, which could be attributed to reservoir loading, have occurred in the area since.

At every other installation cited, the relation of reservoir loading to local earthquakes in each case is more or less negative. This is particularly true of the reservoir behind Grand Coulee Dam because any depression of the crust under load was probably by flexure. Very few local earthquakes were recorded during a seven-year period of observation.

Earthquakes and Downwell Injection

In February 1962, waste chemicals were disposed of by injecting them into the pre-Cambrian basement through a 12,045-feet deep well located on the Rocky Mountain Arsenal a few miles northeast of Denver. A few weeks after injections began, sharp earthquakes were felt in inhabited areas near the Arsenal.

In November 1965, David Evans, a consulting geologist in the Denver area, announced in a televised interview that the earthquakes were directly related to the injection processes. For supporting evidence he used a correlation between numbers of earthquakes recorded by nearby seismographs and the volume of fluid injected. He hypothesized that the earthquakes resulted from increased fluid pressures within the cracks and instertices in the pre-Cambrian basement near the bottom of the well, and supported his argument with a theory of overthrusting advanced by Hubbert and Rubey (1959).

Shortly thereafter, the problem was studied in detail by a number of organizations including the U. S. Geological Survey (GS) and the Colorado School of Mines. Preliminary results are published in the GS Open-File Report (Healy, 1966). Evans' (1966) early report includes a graphical presentation by months of the volume of injected fluids and the numbers of earthquakes recorded by the Bergen Park seismograph located 32 miles southwest of the Arsenal. Figures 6 and 7 are similar presentations from an energy viewpoint. Injection energy is in the form $Ei = K(Pw + Ph)V$ where Pw is the well head pressure, V is the volume of injected fluids, and Ph is the difference in hydrostatic pressure of the fluid at ground level and at the supposed pre-injection level. K is a constant depending on the choice of units. Well head pressure was taken from charts supplied by the Corps of Engineers. Earthquake frequency and magnitude data were taken from the Open-File Report and from readings supplied subsequently by Maurice Major and Ruth Simon of the Colorado School of Mines. Figure 6 was presented in Omaha before a panel of consultants assembled by the Corps of Engineers (1966). Other reports before that panel considered a close correlation between seismicity and injection pressure, rather than volume.

Injection processes were discontinued in February 1966. Relatively strong earthquakes nevertheless continued to be felt. In fact in August 1967, a magnitude 5.5 earthquake, located in the general area of the arsenal probably released as much energy as all the rest of the local earthquakes combined.

The Rangely Oil Field in northwestern Colorado is probably the only other area where an apparent correlation between earthquakes and fluid injection exists. In this case, the volume of fluid removal and injection (water flooding for the purpose of increasing production) far exceeds that of the Rocky Mountain Arsenal. The earthquakes are being monitored by the Uinta Basin Seismological Observatory, a ten-unit array located 50 to 80 kilometers west northwest of the Rangely Field. Recordings started in November 1962, several years after the field had started operation. R. C. Munson (1968), investigating the seismicity of this field, came to the conclusion that there was no relation between the earthquakes and water flooding because peak flooding and peak earthquake activity did not correlate. However, in view of similar instances in which an apparent time lag of somewhat more than a year was indicated, the lack of correlation does not necessarily rule out the possibility that the earthquakes are in some way associated with oil field operations.

Earthquakes and Loading from Flood Waters

Newspapers in Memphis, Tennessee, May 7, 1927, contained accounts of an earthquake which occurred in the flood-swept Mississippi Valley at 2:34 a.m. on May 7th. It was reported that two distinct tremors shook Memphis and surrounding territory in rapid succession. Window panes were shattered, dishes broken, etc., but no serious damage resulted. Reports indicated that the shock was felt as far south as Sikeston, Missouri, throughout Mississippi, and western Tennessee. This event supports a study by McGinnis (1965) from which he concluded that earthquake frequency increases with an increase in water mass culminating in flood stages.

Tidal Unloading and Alaska Earthquakes

Eduard Berg (1965) in a study of tidal effects on earthquakes concluded that the unloading of the crust by large ocean tides may be considered as a triggering mechanism for the Alaska earthquake of March, 1964 and major aftershocks. He considered that the stress system is compatible with that deduced from fault plane solutions and land uplift and subsidence which occurred during the main shock.

Marathon and Kremasta Lakes in Greece and Vajont Dam in Italy

A. G. Galanopoulos (1966) reported a correlation of small to moderate earthquakes felt in Attica over the period 1931 to 1965 with peak loadings of Marathon Lake. The lake is relatively small, capacity 4×10^7 m^3, and the earthquake epicenters were located as far as 20 kilometers from the lake, yet most of them occurred at the time when the lake was overflowing. Since 1958, local microearthquakes recorded by Wiechert seismographs

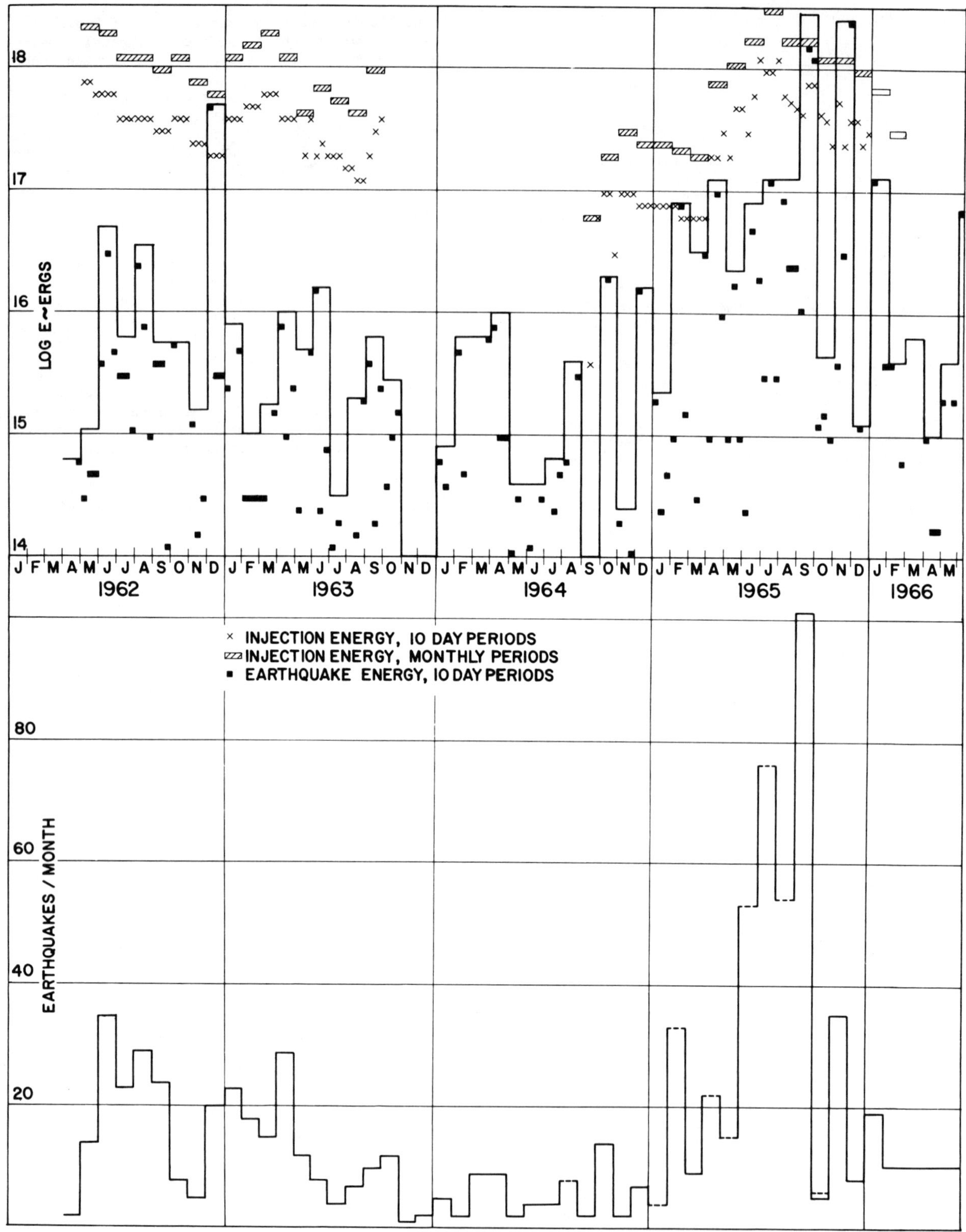

Fig. 6. Injection energy (upper graph) and number of earthquakes per month (lower graph) versus time.

indicate a like correspondence of these earthquakes with seasonal peak loads. In the same publication, Galanopoulos depicts (a) a spurt of earthquake activity corresponding with the filling of Kremasta Lake, capacity 5 x 10^9 m^3, culminating in a magnitude 6 to 6 1/4 shock shortly before the peak load of 1966; and (b) a numerical count of microearthquakes recorded near Vajont Dam and the seasonal fluctuations of the Vajont reservoir over a 4-year period (1960-1963). In the latter case, he uses Pietro Caloi's data (Caloi, 1966) from which the correlation of numerical peaks of microearthquakes with seasonal load peaks appears definite.

Snowy Mountain and Sydney Water Board Projects, Australia

In 1958, these authorities installed networks of seismographs for the principal purpose of locating potentially active faults in relation to hydroelectric and water storage projects under their control. A moderate amount of seismic activity has been reported, most of it located at some distance from the reservoir sites. (Jaeger and Read, 1969). However, the filling of Lake Eucumbene was associated with moderate seismic activity which began not at high load but at relatively high water, (Jaeger,personal communication, June 1969). Lake Eucumbene is a new reservoir located in the Snowy Mountains in southeastern Australia. The peak load is about 2.6 x 10^9 tons.

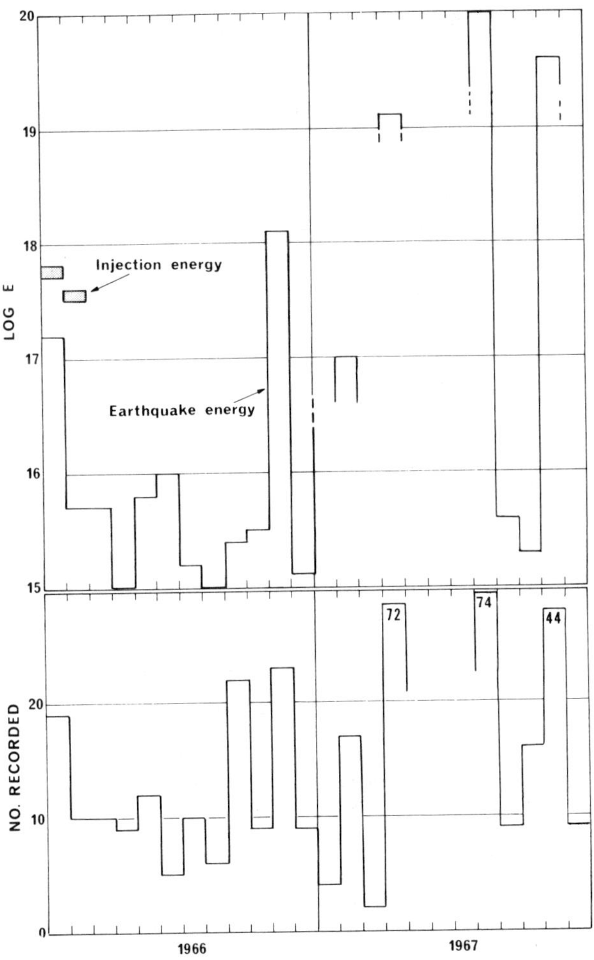

Fig. 7. Earthquake and injection data, Rocky Mountain Arsenal, 1966 and 1967. Earthquake data were supplied by Ruth Simon, Colorado School of Mines.

Lake Kariba

The Zambezi River was dammed in the Kariba Gorge (Rhodesia-Zambia) in December 1958. The ultimate capacity of Lake Kariba, the impounded reservoir, is 16 x 10^{10} m^3 which is about 4 times that of Lake Mead; and at that time, the largest load placed on the earth's surface by human action. Its ultimate length is 250 kilometers and its ultimate depth at the dam face is 119 m (Gough and Gough; 1962, 1968). The seismographic record of local earthquake activity began in May 1959, about 5 months after closure of the river and at the beginning of the peak load for the year. Thereafter hundreds of earthquakes of magnitudes upward from 1.9 were located in the east and deepest end of the reservoir area with a culmination of three magnitude 5.5 to 5.8 earthquakes occurring in September 1963 after the reservoir had reached somewhat more than 50 percent capacity. During this time, the association of seismicity with the 5 seasonal load peaks was definite. Many faults, at least one of them known to be active, are located in the area. The settlement of the basin of the reservoir was about 20 to 30 cm (D. I. Gough, personal commun.). The resulting elastic strain energy should be more than sufficient to account for the energy released by the earthquakes.

Koyna Reservoir (Lake Shivajisagar)

This reservoir, impounded by the Koyna Dam is located on the peninsula shield of India about 120 kilometers south of Poona. The area was generally considered aseismic, except that swarms of small tremors have been noted at times in scattered localities (Guha, et al., 1966). However, as filling of the reservoir progressed a number of earthquakes were felt, and by mid-September 1967, upward of a hundred epicenters had been located within the reservoir area. These earthquakes were relatively small and were attributed by some to the reservoir loading. On 13 September 1967 two magnitude 5 to 5.5 earthquakes caused minor damage, and on 11 December 1967 (10 December 22:51 GMT) a magnitude 6.5 earthquake practically leveled the nearby city and caused extensive damage to the dam. This earthquake killed about 180 people and injured about 2200 (Housner,1969). The epicenter was located within a few kilometers of the dam. Reported maximum acceleration at the dam exceeded 0.50 g at high frequency (Berg, 1968). Within the following 14 days, 6 aftershocks having magnitudes of 5.5 to 6.2 occurred in the general area. The focal depths varied downward to 20 kilometers. Views as to the relationship of these earthquakes - or specifically the 11 December event which is considered as the Koyna earthquake - are varied. Nine short papers appearing in an Indian journal (Geophysical Abstracts, 1969) are in general agreement that the earthquake was tectonic with a veiled suggestion that some triggering action may have taken place. It was emphasized that 11 other reservoirs located on the

Deccan peninsula had no association with local earthquakes, and that the focal depths were too great to be other than tectonic. A hint of a N-S fault following the course of the Koyna River was suggested. A focal depth of 30 kilometers for the main shock was suggested but also a suggestion that a relatively shallow earthquake may have immediately triggered the larger one which had the greater focal depth.

DISCUSSION

In the foregoing presentation, any relation of crustal disturbances resulting from human action or natural phenomena of like nature to local earthquake situations is based solely on statistical evidence. For this reason, there is doubt among some that any such relation exists. In support of these doubts, case histories similar to the ones cited here are relatively few. Many reservoirs where there is little or no association with local earthquakes have been impounded. Some of these reservoirs approach Lake Kariba in volume. These doubts may in particular apply to the devastating Koyna earthquake of December 1967 because energy reportedly released from this earthquake was far in excess of the loading of the nearby reservoir. Be this as it may and with the assumption that such a relationship may exist, the conditions where this may be possible will now be discussed.

As pointed out earlier, the Lake Mead and Kariba earthquakes could be a result of energy release from settlement along pre-existing faults. According to level surveys, the lower basin of Lake Mead had settled about 10 centimeters over the period from 1935 to 1940, during which time a water and silt load of 10^{10} tons had been added, and that settlement was spasmodic along brittle fractures. Energy of settlement could alone account for seismic energy released by the observed earthquakes which has the equivalence of 1.5 magnitude 5 earthquakes, a reasonable approximation. As mentioned, a like argument could possibly apply to the Kariba situation. With that, this argument ceases to apply because the reservoirs in Greece and India and downhole injections in Colorado are of themselves inadequate to be directly responsible for the attributed earthquakes. In any case including Lakes Mead and Kariba, assuming that the earthquakes are directly or indirectly a result of fluctuations in surficial crustal loading, a ready-made crustal structure, such as a crustal block containing potentially active faults, is a necessary condition. Additional loading may then reactivate these faults by sheer added weight as mentioned, or more likely by the condition that the crustal section is under high strain and in delicate equilibrium. Then (1) a relatively small added load may possibly upset this equilibrium with the resulting release of energy in the form of earthquakes; or, (2) fluids forced into fissures thus destroying the frictional resistance between these walls; or (3) as a correlary to the above, fluid pressure in pores, joints, and other discontinuities in the rock mass, (Lane, 1969; Hubbart and Ruby, 1959; Evans, 1966; Jaeger, op. cit.). In condition (3) water level rather than water load is of importance.

In contrast, if a ready-made crustal structure and other relevant conditions do not exist, such as is supposed with a large majority of reservoirs, local earthquakes will not result.

In almost every instance where local earthquakes have been attributed to crustal loading or to some other unbalance, a gradual buildup of seismic activity has been observed, with a culmination into one or more relatively strong shocks. As examples: at Kremasta, the main shock occurred near the seasonal peak four months after preliminary shocks were noticed; at Lake Mead, it occurred nearly three years after initial shocks were noticed, and one year after the plateau load of the lake had been reached; at Koyna, it occurred four years after initial shocks were noted and 4.3 years after the lake had filled; and near Denver, the strongest earthquake occurred 1.5 years after injections had ceased and 5.5 years after they had begun.

ADDENDUM

The following is a resume of a paper presented by J. P. Rothé at the 4th World Conference on Earthquake Engineering in Santiago de Chile, January 13-19, 1969. This résumé is credited to Teledyne SSD newsletter of April 1969. Professor Rothé adds two reservoirs to the list of possible earthquake generators. His overall conclusions are much the same as mine.

Earthquakes and Reservoir Loadings

In a paper presented at 4WCEE, J. P. Rothé, University of Strasbourg Professor, discusses the circumstances surrounding increased earthquake activity near six reservoirs in different parts of the world: Monteynard Dam (France), Grandval Dam (France), Hoover Dam (Arizona-Nevada), Kariba Dam (South Africa), Kremasta Dam (Greece), and Koyna Dam (India). The author draws a number of conclusions:

1. The earthquake activity is particularly clear when the reservoir is deeper than 100 meters. The height of the water seems to be more important than the total volume of the reservoir.
2. In many cases, contrary to natural seismicity, the strongest shocks follow numerous foreshocks, the frequency of the foreshocks increases progressively during a certain period; the more tectonically quiescent the area, the longer this period is.
3. The seismic activity generally increases at the time of the first filling and, after having reached a maximum, gradually disappears in the course of a few years.
4. The accident at the Malpasset Dam, in France, on December 2, 1959, brought to light the risk - underestimated at this time - of a sudden breaking of the rock support of the arch dam by the action of the interstitial strengths. Consequently it is important to study the underground water flow and the internal stresses which are created.
5. Some particular geological conditions are necessary for the release of the shocks (the nature of the strata, faults, diaclases, etc.); of course, in many cases, the filling of reservoirs did not bring any notable seismic activity.
6. The age of the concerned faults does not seem to be relevant. In the reported examples, there appear strata with an old orogeny and a very low seismicity (African Shield, Deccan), as well as areas with a recent orogeny and a

notable seismicity (Greece, Alps, Pyrénées, Rocky Mountains). Even if the faults have been quiescent during a long geological time, their activity can be revived under the weight of the water reservoir and the interstitial stresses.

ACKNOWLEDGMENTS

Acknowledgments are due the U. S. Bureau of Reclamation for support of seismological programs and the use of data in reference to the Boulder Canyon and various other hydroelectric projects; the U. S. Army Corps of Engineers, Omaha Division, for supplying data on downhole injections in the Rocky Mountain Arsenal; and the Coast and Geodetic Survey for support in collection and interpretation of records. Individual acknowledgments are due Leslie F. Bailey for statistical assistance in interpreting Lake Mead data; Maurice Major and Ruth Simon for supplying Derby earthquake data as recorded by the Bergen Park seismograph; Professor Gough for supplying information on Lake Kariba and associated earthquakes; David Evans and Don Tocher for supplying literature on the Koyna Dam and other projects; and A. G. Galanopoulos for reprints of his papers on the seismicity in Greece and Italy.

REFERENCES CITED

Berg, Eduard. Triggering of the Alaskan Earthquake of March 28, 1964 and major aftershocks by low ocean tide loads. Undated report to Air Force Office of Scientific Research.

Caloi, P. 1966. L'evento del Vajont nei suoi aspetti geodinamici. *Ann. de Geof.*, 65, pp. 1-74.

Carder, D. S. 1945. Seismic investigations in the Boulder Dam area 1940-1944, and the influence of reservoir loading on local earthquake activity. *Bull. Seis. Soc. Am.*, 35, pp. 175-192.

Carder D. S. 1962. Ground effects from Gnome and Logan explosions. *Bull. Seis. Soc. Am.*, 52, pp. 1047-1056.

Carder, D. S. and Small, J. B. 1948. Level divergences, seismic activity, and reservoir loading in the Lake Mead area, Nevada and Arizona. *Trans. Am. Geophys. Union*, 29, pp. 767-771.

Earthquake Engineering Research Institute 1968. *Newsletter*, 2, p. 1.

Evans, David M. 1966. The Denver area earthquakes and the Rocky Mountain Arsenal disposal well. *The Mountain Geologist*, 3, pp. 23-36.

Evans, David M. 1966. Man-made earthquakes in Denver. *Geotimes*, 10, pp. 11-18.

Evans, David M. 1966. Man-made earthquakes--a progress report. *Geotimes*, 12, pp. 19-20.

Galanopoulos, A. G. 1967. The influence of the fluctuations of Marathon Lake elevation on local earthquake activity in the Attica Basin area. *Annals. Geol. des Pays Helleniques*, 18, pp. 281-306.

Gough, D. I. and Gough, W. I. 1962. Geophysical investigations at Lake Kariba. *Internationales Symposium uber rezente Erdkrustenbewegungen* vom 21. bis 26. Mai 1962 in Leipzig, DDR, 441-447, Akademic-Verlag, Berlin.

Gough, D. I. 1967. Letter to Lynn R. Sykes; and 1968, Letter to D. S. Carder.

Guha, S. K., Gosavi, P. D., Padale, J. G., and Marwadi, S. C. 1966. Crustal disturbance in the Shivajisager Lake area of the Koyna hydroelectric project. (Maharashtra, India) *Third Symposium on Earthquake Engineering*. Roorkee, India, November 4-6, 1966, pp. 399-416.

Healy, J. H. and others 1966. Geophysical and geological investigations relating to earthquakes in the Denver area, Colorado. *U. S. Geological Survey open-file report.*

Housner, G. W. 1969. The seismic events at Koyna Dam. *Eleventh Symposium on Rock Mechanics*. Berkeley, California.

Hubbert, M. King and Rubey, W. W. 1959. Role of fluid pressure in mechanics of overthrust faulting. *Bull. Geol. Soc. Am.*, 70, pp. 115-166.

Jaeger, J. C. and Reed, Leslie, 1969. Seismicity of Southeast Australia. *Am. Geophys. Un.* Geophysical Monograph 13, pp. 145-147.

Lane, K. S. 1969. Engineering problems due to fluid pressure in rock. *Eleventh Symposium on Rock Mechanics*. Berkeley, California.

McGinnis, L. D. 1965. Earthquakes and crustal movement as related to water load in the Mississippi Valley Region. *Illinois State Geol. Survey* Circular 344.

Mead, J. C. and Carder, D. S. 1941. Seismic investigations in the Boulder Dam area in 1940. *Bull. Seismol. Soc. Am.*, 31, pp. 321-324.

Munson, Robert C. 1968. An investigation of the seismicity in the vicinity of Rangely, Colorado. *Masters Thesis*. Colorado School of Mines.

Rothé, J. P. 1969. Earthquakes and reservoir loadings. *Proc. 4th World Conference on Earthquake Engineering*. Santiago, Chile.

U. S. Army, Corps of Engineers, Omaha District 1966. Report of investigations, injection well-earthquake relationship, Rocky Mountain Arsenal.

U. S. Geological Survey 1969. Geophysical Abstracts No. 271. pp. 1113-1114. 9 papers on the Koyna Earthquake by Indian geophysicists appearing in pages 1 to 36, *Indian Geophysical Journal*, V. 5, No. 1, 1968, are abstracted.

TSUNAMI EFFECTS AND RISK AT KAHUKU POINT, OAHU, HAWAII*

Wm. Mansfield Adams (University of Hawaii, Honolulu, Hawaii)

Abstract

Historical tsunami inundation data for the Kahuku Point area on Oahu, Hawaii, are compiled or deduced from comparison with Hilo, Hawaii. The data are considered in two parts: North Pacific tsunamis and South American tsunamis. The South American tsunamis are about 1/3 as large at Kahuku Point as they are at Hilo: North Pacific tsunamis are about the same at both locations. The maximum North Pacific inundation is 27 feet; the maximum South American inundation is 9 feet.

The distribution of inundation along the shoreline area of the point from Kawela Bay to Kahuku is rather uniform--being within 20 percent of the average.

The historic tsunami data are ranked by size. Assuming a Poisson distribution and independence between the observations, a graphical relationship for the frequency of occurrences is found. The extreme tsunami height encountered for a given duration and at a given level of accepted risk is graphically determined. The graphical presentation is modified to permit easier use. For example, accepting a risk of one in twenty for 40 years, results in a tsunami design height of 43 feet. By accepting greater risk, a one in ten chance, the tsunami design level is reduced to 35 feet. The tsunami design level determined by the risk level actually accepted could define the height of the first floor. In the previous example of a risk of one in twenty, this height would be 17 feet (43 minus 26, the ground elevation of the structure); for a risk of one in ten, only 9 feet.

The probable sequence of events at Kahuku Point for an Aleutian Tsunami is:

1. Knowledge of the generating earthquake and the possible tsunami is reported to Kahuku Point at least two hours before the tsunami arrives.
2. The Kahuku Point area receives tsunami inundation highs relative to most parts of Oahu.
3. The inundation is of the form of a tidal bore.
4. Several inundations occur.
5. The surf breaks at or inland of the present beach line during periods of inundation.
6. The currents and waves at Kahuku Point are unusual for several days after the tsunami. The currents are much stronger than normal. No one should enter the water.

There is the possibility that protective structures against the potential tsunami damage at Kahuku Point can be built. These include armor, breakwaters, sloping walls, etc. A similar problem and situation occurs at Hilo, Hawaii. Recent studies of the Hilo situation indicate that the esthetic benefits of a natural environment are more important than the physical protection provided by the designed breakwater. Every breakwater design proposed is very expensive and severely mars the vistas.

By analogy, the findings at Hilo, Hawaii, apply to Kahuku Point on Oahu. Any protective structure would have to be so large as to blot out the seascape. The expense of any such protective structure is better invested in improving the design and construction of the buildings.

*This paper is based on five reports developed by the author under contract to Del E. Webb Inc. Del E. Webb Inc. has graciously agreed to publication.

Two "Rules-of-Thumb" which have evolved for design against tsunami damage are:

"Build up off the ground."

"Orient shear walls perpendicular to the wavefronts."

For Kahuku Point, the direction from which the waves arrive will be about 10° east of north, regardless of the direction to the source of the tsunami. Hence shear walls should be aligned to N 10° E.

The columns in the first floor of any structure should be streamlined, because they will endure large rising currents relative to ebbing currents.

CONTENTS

ILLUSTRATIONS

HISTORICAL TSUNAMI DATA FOR THE KAHUKU POINT AREA, OAHU, HAWAII

Introduction

In order to evaluate the risk of any proposed project, a prediction of the future is necessary. Such a prediction may be either deterministic, based on knowledge of the physics of the phenomenon, or statistic, based on knowledge of the history of the phenomenon. For a phenomenon as complex as tsunamis, the basis for predicting future events is historical data. The use of historical data is fraught with danger because of the human tendency to be imprecise, to make a tale more interesting by elaboration and variation, and to legendize extreme occurrences. These frailities have been considered in gathering the data presented here. All values are based on concurrent documentation whenever possible. All values have been reduced to a standard datum--mean sea level, except where noted.

The historical tsunami data directly available at the Kahuku Point area are for the tsunamis of 1946, 1952, 1957, and 1960. Maximum run-ups for these four tsunamis are presented (see Table 1.1) as measured above mean sea level. Run-up is defined to be the maximum elevation at which water inundation was seen or was evident on the land.

Estimates for other tsunamis may be obtained from comparison of Kahuku Point with a place having relatively plentiful and well documented tsunami data. Hilo, Hawaii, is such a place (Cox, 1964).

The tsunami run-up observed at Hilo has been extrapolated to Kahuku Point. The ratio of the 1960 run-up at Kahuku Point to that at Hilo is about 1/3.* This ratio was used to adjust run-up data due to tsunamis originating in South America. The ratio of the sum of 1946, 1952, and 1957 run-ups at Kahuku Point to the sum of the respective tsunami run-ups at Hilo is approximately 1. Hence the data due to tsunami run-up originating from earthquakes in the North Pacific may be used interchangeably between Hilo and Kahuku Point.

Chronological Presentation of Historical Tsunami Data

The Kahuku Point observed run-ups and the data extrapolated from Hilo (see Table 1.2) are shown in the upper part of Figure 1.1. The tsunami data for Hilo,

Table 1.1

Tsunami	Runup
1946	27
1952	10
1957	23
1960	9

*(At Hilo, the upper decile value has been used instead of the actual maximum. The intensive study of the Hilo conditions is considered to reveal extraordinary extreme values, not directly comparable with values elsewhere.)

on which the extrapolation is based, are given in the lower part of the figure for comparison. The period of time covered is 141 years. Hilo was instrumented to record fluctuations in the sea level for only 27 years of this time; these years were from 1927 to 1932 and December 1946 to present.

This chronological presentation of the historical tsunami data deduced for the Kahuku Point area immediately shows the time periods during which tide gauges have been operating. Except for 1938, no tsunami less than two feet was noticed except by instrumentation. Evidently, two feet is a reasonable value for an average threshold for visually observing a tsunami in Hilo Bay area. The threshold for visually observing a tsunami at Kahuku Point is not known, but is higher because of the ambient background of surf.

Inspection of the chronological distribution of tsunamis might also seem to indicate a slight emphasis for large values to have occurred recently. The collection process is considered to be unbiased, hence the possibility must be considered that earthquakes have recently become more frequent.

A study of the seismicity of the earth made by Dr. H. Benioff concluded that a notable increase in earthquake frequency occurred about 1949. This increased seismicity might be considered in explaining the several large values of run-up observed in the past two decades. However, detailed inspection of the data show that the recent large events are primarily located in the North Pacific. Prior to 1930, the large events were dominantly South American. There is no explanation for any such trend.

Geographical Presentation of Historical Tsunami Data

The maximum historical tsunami inundation in the Kahuku Point area was caused by the 1946 earthquake for the North Pacific earthquakes, and by the 1960 earthquake for the South American earthquakes. Only these two will be considered for geographical comparisons of tsunami inundation.

The coastline of the Kahuku Point area is shown in Figure 1.2. The observed tsunami inundations due to the 1946 tsunami are given in the upper half of the figure with subjective interpolation between data points. The two observed tsunami inundations due to the 1960 tsunami from South America are plotted at respective positions in the lower half of the figure. The North Pacific tsunamis obviously pose the dominant threat to the Kahuku Point area. Since there are none of the natural protective conditions, such as offshore reefs, the tsunami inundation along the coastline is relatively uniform. Probably the variation is not more than 20 percent from the average along the coast.

THE RISK OF TSUNAMIS AT THE KAHUKU POINT AREA

Introduction

The historical tsunami data are listed in Table 2.1 by size of the tsunami run-up. Only those data directly or indirectly attributable to noninstrumental observations are included. These data span a time period of 131 years--from 1837 to the present. Presumably, tsunamis due to earthquakes are distributed in time according to Poisson's Law:

$$q = 1 - \exp(-ND)$$

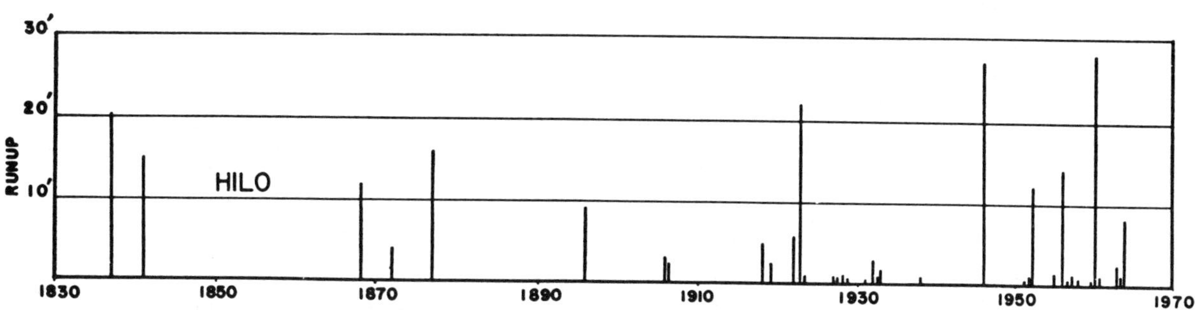

Fig. 1.1. Chronological tsunami data.

where q is the chance of a given height being exceeded in D years, D is a duration in years that a risk is accepted, and N is the frequency of occurrence of a given height per year.

Furthermore, each tsunamigenic earthquake is assumed to be independent of any other tsunamigenic earthquake. This is not true, deterministically, but the statistical dependence is small enough to justify the assumption.

Based on the historical tsunami data and these two explicit assumptions, a relationship of tsunami run-up to frequency of occurrence can be derived by standard statistical procedures. (See Figure 2.1.) The data points are given and the straight line faired through them. This graphical relationship is used in the following manner.

A. For an acceptable frequency of occurrence during any year interval, say 0.01, enter on the mantissa (the horizontal scale) at 0.01 and intersect the solid line. The corresponding tsunami run-up of 25 feet is found as the ordinate of the intersection.

B. For an acceptable frequency of occurrence, M, over all of D years, divide M by D to find the mantissa entry, N. As an example, for an acceptable frequency of occurrence of 0.01 over 10 years, enter at 0.001. This gives a corresponding height of 46 feet. This height is much larger because the same frequency of occurrence was demanded over a longer time.

In practice, an engineer knows the risk level which is acceptable and the duration for which the risk must be assumed. To convert the data given in Figure 2.1 to a more usable form, the method of Wemelsfelder (1961) is helpful.

Figure 2.2 shows the graphical relationship existing in the Kahuku Point area between tsunami run-up and duration of risk assumption, as a function of risk. Only three levels of risk are shown; others can be derived from Figure 2.1 using the given equation and a table of Napierian logarithms.

To use Figure 2.2, enter on the mantissa at the duration of the risk to be assumed, D. As an example, 40 years is taken. Proceed vertically to the level of risk considered acceptable over the total duration. If this is 5 percent, i.e., one in twenty, then the corresponding ordinate, 43 feet, gives the tsunami design height. This may seem unusually high, but remember that we are demanding that there be less than a one in twenty chance of this height being exceeded at any moment during the entire 40 years.

The risk considered acceptable may be higher; one in ten. Then for a duration, D, of 40 years, the corresponding tsunami design height is only 37 feet. Thus there is less than a one in ten chance of a tsunami inundation greater than this height at any moment throughout the 40 years.

The design tsunami run-up height might appropriately be the height above high tide of the second floor.

The graphical form of the situation shown in Figure 2.2 makes clear that, "Any height tsunami run-up may occur at any time." In situations involving extremes of a geophysical phenomenon, risk cannot be avoided.

The purpose of this report is to quantify this risk-taking in an objective manner.

Table 1.2. Probable Historical Tsunami Runups at Kahuku.

Year	Month	Source	Probable Runup in Feet
1837	Nov	Chile	7
1841	May	Kamchatka	15
1868	Aug	Peru-Chile	4
1872	Aug	?	4
1877	May	Chile	5
1896	June	Japan Is.	9
1906	Jan	Columbia	1
1906	Aug	Chile	1
1918	Sep	Kuril Is.	5
1919	Apr	Tonga Is.	1
1922	Nov	Chile	2
1923	Feb	Kamchatka	22
1923	Apr	Kamchatka	1
1927	Nov	California	1
1927	Dec	Kamchatka	0.5
1928	June	Mexico	0.6
1929	March	Aleutian Is.	0.5
1931	Oct	Solomon Is.	0.3
1932	June	Mexico	1
1932	June	Mexico	0.1
1932	June	Mexico	0.1
1933	Mar	Japan	1.5
1938	Nov	Alaska	1
1946	Apr	Aleutian Is.	27
1952	Mar	Japan	1
1952	Nov	Kamchatka	10
1956	March	Kamchatka	1.2
1957	March	Aleutian Is.	23
1958	July	Alaska	0.25
1958	Nov	Kuril Is.	1.2
1959	May	Kamchatka	0.5
1960	May	Chile	0.1
1960	May	Chile	9
1960	Nov	Peru	0.1
1963	Oct	Kuril Is.	2.5
1963	Oct	Kuril Is.	0.5
1964	March	Alaska	8

PROBABLE SEQUENCE OF EVENTS AT KAHUKU POINT FOR AN ALEUTIAN TSUNAMI

The effects of tsunamis in Hawaii have been recorded for more than a century and a half. Compilations have been made by Jaggar (1931), Powers (1946), and Pararas-Carayannis (1969). These reports allow an estimation of the events following an earthquake causing a tsunami. The order of the events, timing, and possible severity can also be estimated.

Because the Kahuku Point area has received maximum tsunami inundation from North Pacific tsunamis, a source in the Aleutian Islands will be used. The following description is hypothetical--being a synthesis of several tsunami histories and the current policies of Environmental Science Services Administration and Civil Defense of Hawaii.

Table 2.1

Rank	Runup Height (Feet, MSL)	Year	Source
1	27	1946	Aleutians
2	23	1957	Aleutians
3	22	1923	Kamchatka
4	15	1841	Kamchatka
5	10	1952	Kamchatka
6	9	1896	Japan
7	9	1960	Chile
8	8	1964	Alaska
9	7	1837	Chile
10	5	1877	Chile
11	5	1918	Kuril Is.
12	4	1868	Peru-Chile
13	4	1872	?
14	2.5	1963	Kuril Is.
15	2	1922	Chile

Table 3.1: Probable Sequence of Events.

TIME	EVENT	REMARKS
T	Earthquake	A fault ruptures at a depth of 20 miles in the Earth. The point above the rupture is at 53°N, 167°W, near Dutch Harbor. The rupture is 200 miles long and 10 miles deep. About 10^{23} ergs are released.
T plus 5 Minutes	Seismic Waves Trigger Alarm at Ewa Beach	The personnel of the U.S.C.G.S. view the visible seismic recordings. A preliminary estimate of location and size of the earthquake is made.
T plus 10 Minutes	Telegrams Sent	The U.S.C.G.S. requests data from seismological observatories around the Pacific Ocean.
T plus 20 Minutes	Telegrams Received	Data from California and Japan confirm the previous estimates of location and size.
T plus 30 Minutes	Telegrams Sent	Request water wave data from Adak, Alaska.
T plus 40 Minutes	Watch Issued	Civil Defense, Police, etc. alerted.
T plus 2 Hours	Telegrams Received	Water-wave height of 50 feet reported at Adak, Alaska.
T plus 2 Hours 30 Minutes	Warning Issued	Sirens sounded. Evacuation begins. Police patrol.
T plus 4 Hours	Midway Island reports 5 foot wave	Data about Midway relayed to public by radio.
T plus 4 Hours 24 Minutes	Kauai reports 35 foot wave	Data about Kauai relayed to public by radio.
T plus 4 Hours 30 Minutes	Kahuku reports start of 25 foot wave	Data reported to public by radio.
T plus 7 Hours	Tsunami "All Clear" issued	"All Clear" announced for Hawaii by Civil Defense.

The first wave is like a very rapid change in tide. Instead of changing two feet in twelve hours, the sea level changes 25 feet in ten minutes. A "wall-of-water", called a bore, occurs where there is concentration of wave energy by funnelling, as in bays, or by convergence, as on points. The tsunami waves at Kahuku Point closely resemble tidal bores. There is a steep front rolling in over relatively quiet water. Behind the steep front, the crest of such a wave is broad and flat, the wave velocity is about twenty miles per hour--much greater than longshore currents and faster than a man can run. There is a loud roaring and hissing. Huge coral heads, several feet in diameter, are ripped from the ocean bottom and carried landwards. Blocks of the reef, weighing tons each are torn up and moved shoreward. (Comparable design force should be a dynamic loading of 1000 pounds per square foot over submerged portions of any structure.)

The line of surf, which depends on the water depth, moves inland to approximately the same water depth. To a first approximation, waves break in water equal to their height, so the big outside surf will move to the area along the berm. The berm area will be subject to direct surf action, meaning the waves are beginning to collapse directly above the berm, for each period of ten to twenty minutes that the berm is covered deeply.

The sequence of events is repeated for several hours, during ten waves of varying crest heights. The maximum crest height happens to occur on the third crest.

Re-entry of the inundated area shows the following damage:

Wooden frame buildings that were at low elevations, on sandy foundations, and/or not well tied to the foundations are destroyed or floated away.

Reinforced concrete structures having adequate footings did not suffer noticeable structural damage.

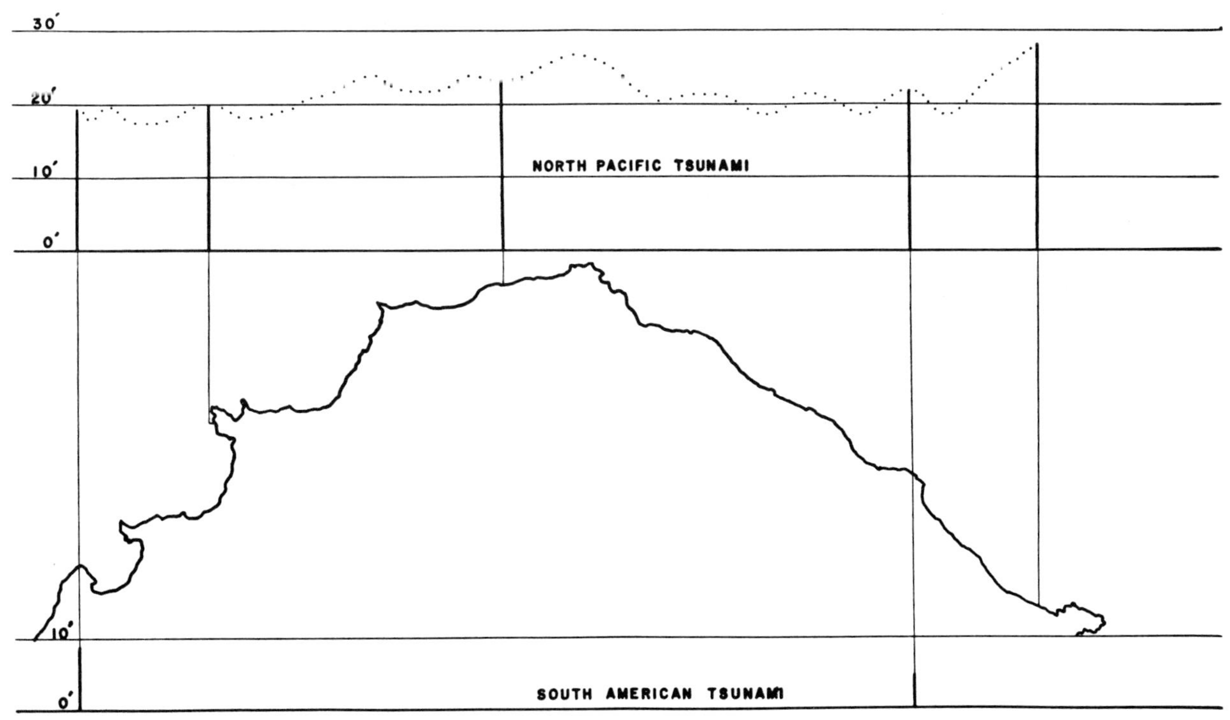

Fig. 1.2. Geographical tsunami data.

Objects having buoyancy or sand foundations, such as wharfs, railroad tracks, or houses were floated around or undermined.

Wave withdrawal eroded tens of feet of dune sand (measured laterally) from the beach.

Some damage to structures was attributable to objects such as cars, logs, etc. being brought into contact with the structures.

The net effect of the hypothetical tsunami on Hawaii is 50 persons died, 100 persons injured, and 20-million dollars property damage. Forty-five of the deaths and ninety of the injuries are to people in the tsunami evacuation zones and to people previously informed of the tsunami Fishermen and campers in remote coastlines comprise the other 5 killed and 10 injured.

All of the property damage occurred in the evacuation zones. Over 5 million dollars of the damage was in Hilo.

EVALUATION OF USING ARMOR ROCK OR SLOPING WALL AT KAHUKU POINT

A breakwater is designed to dissipate wave energy, reflect wave energy, or shift wave energy from some frequencies to others (Wiegel, 1964).

Breakwaters have been studied intensively, because billions of dollars have been spent building and maintaining such protective structures. Because so much of the energy in a wave is near the surface, submerged breakwaters are of relatively little use unless they extend almost to the surface and the waves are of short wave length compared to the depth. This is known both theoretically and observationally (Johnson, Fuchs, and Morison 1951). For both tsunamis and wind waves, an emergent breakwater is found desirable, with overtopping only occurring for a small percentage of the waves, say 15 percent. Overtopping can cause severe erosion on the lee of the breakwater, so elaborate expensive designs are required.

Fortunately, a recent study for protection of Hilo from tsunamis by a breakwater has been completed (Palmer, Mulvihill, and Funasaki, 1967).

The findings may be summarized by acknowledging that a breakwater could be constructed that would protect Hilo. This breakwater was to be built by the U. S. Army Corps of Engineers and to cost about 30 million dollars. Only about 5 million dollars was to be funded locally.

These conditions were evaluated by a local citizen's committee and an advisory committee. The results were that the esthetic consideration of having an attractive vista for the forty years was deemed more important than having physical protection for the few hours that a design tsunami might occur (only one chance in ten anyway).

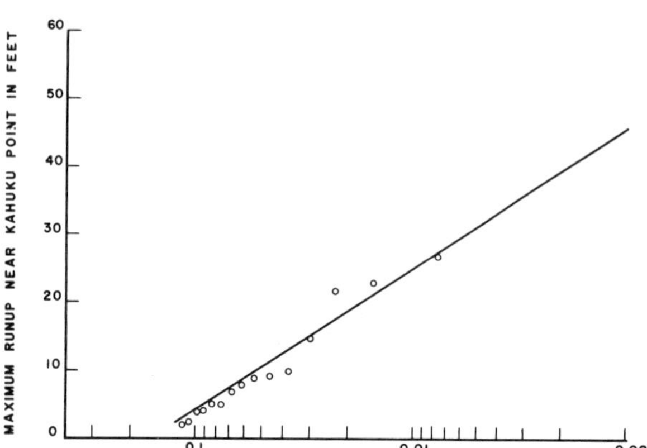

Fig. 2.1. Frequency of occurrence (N) per year.

The situation at Kahuku Point seems to be directly analogous to the Hilo situation. The major difference is that there seems little likelihood of the U. S. Army Corps of Engineers contributing 75 percent of the cost. The Kahuku Point terrain allows an easy opportunity to evaluate the esthetics of a dike or breakwater.

To achieve a "Before-the-breakwater effect": simply stand at the Kuilima site marker. Look out to sea and along the two coasts--to the northwest and to the southeast. The swell rises mystically up, crashes with a roar, and rolls shoreward hissing and fuming. The shear power, about a third of a horsepower per foot of coastline, awes any human and dismays a coastal engineer.

To achieve an "After-the-breakwater effect": move back to the site of the parking area (where a possible increment as a second tower has been noted). Now the point and, to some extent, the dunes serve as a breakwater or seawall. The fascinating surf is hidden from view. One feels hemmed in and somewhat oppressed. To return to the point is refreshing and exhilarating.

"De res gustibus, non disputandum est." There is no point in discussing matters of taste. Conduct your own experiment in esthetics at Kahuku Point. I believe you too will conclude that the esthetics of the natural environment far outweigh the benefits to be gained by any breakwater, armor, or sloping wall.

PREVENTATIVE STRUCTURAL DESIGN FACTORS FOR KAHUKU POINT

Introduction

There are appropriate actions which can be taken to decrease the extent of damage should a tsunami occur. One possibility, that of constructing a protective shell, such as a breakwater or seawall, has been discussed and rejected on esthetic grounds.

Observation of damage in past tsunamis has evolved some general "rules-of-thumb." (See Macdonald, Shepard, and Cox, 1947.) An obvious rule is, "Build up off the ground." This rule, in practice, seems far more effective than would be

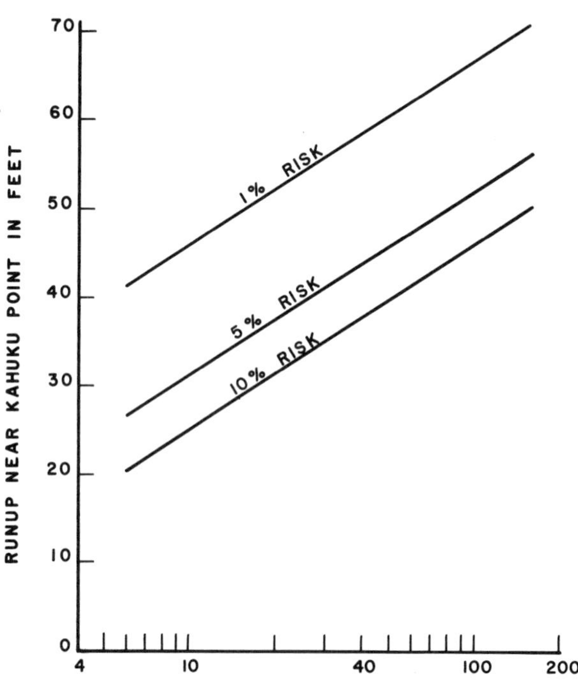

Fig. 2.2. Duration of risk (D) years.

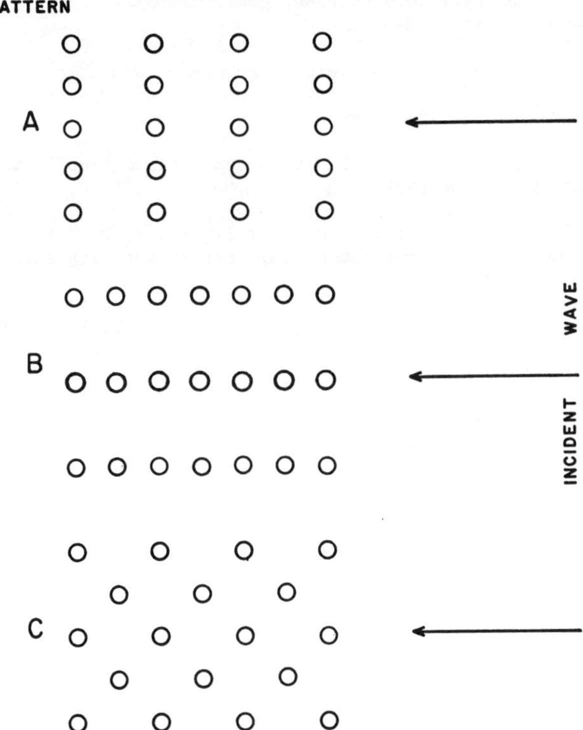

Fig. 5.1. Patterns of pilings studied for wave interactions.

theoretically predictable. This was the basis for suggesting in a previous report that the design tsunami height be taken as the height of the ground floor.

Orientation of Shear Walls or Pilaster Patterns

The interaction of a wave and a structure is dependent on the orientation of the structure with respect to the angle of approach of the wave. This interaction has been studied for rows of piles by Costello (1952). These are vertical circular cylinders, as might be used for the first floor of any structure.

Three orientations were studied (Figure 5.1). Pattern B had the least interaction, that is, transmitted the maximum energy. Therefore, the pilasters of the structure should be oriented like Pattern B with respect to the incident waves. A corollary of this recommendation is that shear walls, considered equivalent to a solid row of pilings, should be oriented in the direction of the incident waves. Use of this guideline requires that the direction of the incident waves be known.

Direction of the Incident Tsunami Waves

The propagation of energy in shoaling water, such as tsunamis running from the open ocean up onto Kahuku Point, has been intensively investigated. Graphical methods for constructing a diagram to show the rays or wavefronts were reported by Johnson, O'Brien, and Isaacs (1948). This is a ray-tracing procedure, so energy flux estimates are qualitative and justifiable only if the bathymetry has been spatially filtered to remove those spatial frequencies with wavelengths short compared to the wavelength of interest. In simpler terms, for tsunami waves, the topography of the ocean bottom must be severely smoothed.

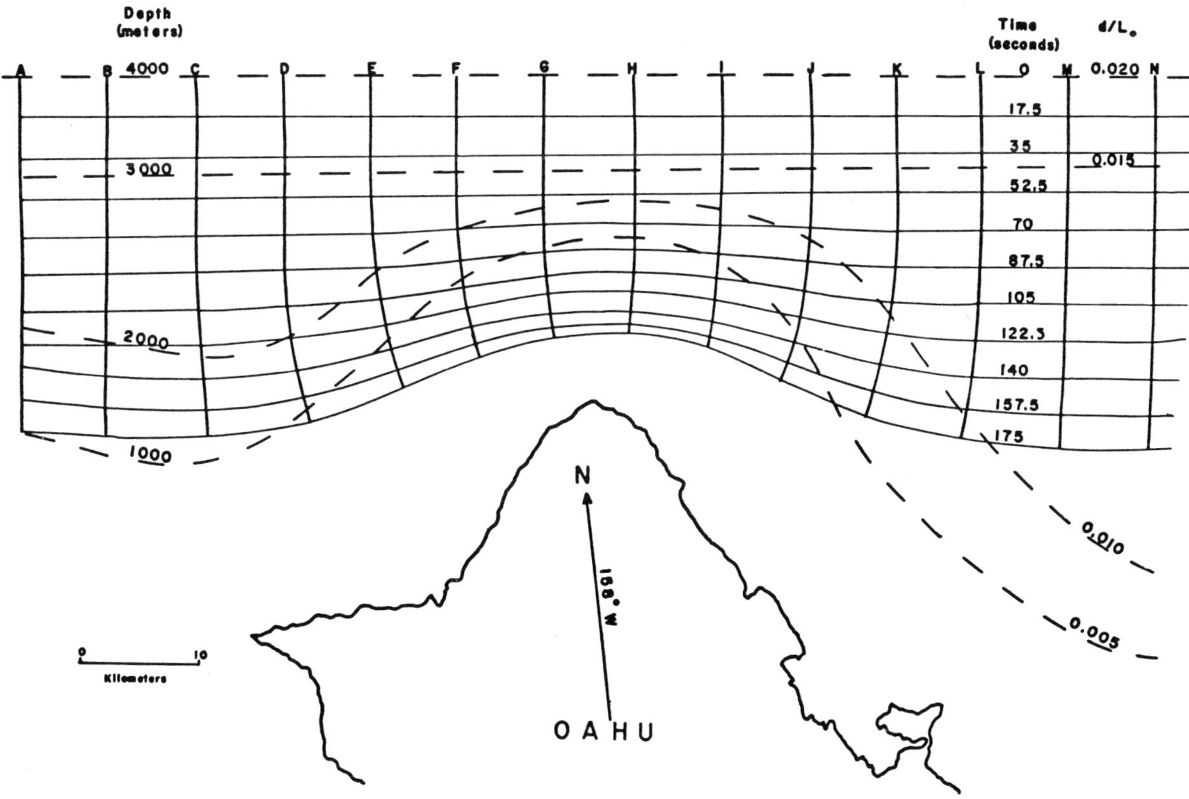

Fig. 5.2. Wavefronts and rays for tsunami propagating onto northern Oahu. A deep-water wavelength of 200 kilometers is given. Contours in 1000-meter intervals.

A diagram showing typical tsunami propagation onto northern Oahu is given in Figure 5.2. Oahu is the heavy solid outline at bottom center: the generalized bathymetry north of Oahu is shown by the dashed lines. The ratio of depth, D, to deep-water wavelength (200 kilometers) is given. The tsunami wave front is denoted for various times by light-weight solid lines and the rays by heavy-weight solid lines lettered A through N.

Note that the northerly "bumps" in the 1000 m and 2000 m contours cause the focusing on Kahuku Point. The spacing between the rays is approximately inversely proportional to the energy in the wave. (The closer the rays, the greater the energy, and the higher the waves.) The waves arrive directly off the point.

Similar refraction diagrams show that tsunamis generated from Kamchatka, the Aleutians, or elsewhere in Alaska, strike Kahuku Point from almost due north. Therefore, structures should be designed for waves incident from 10° east of north.

Design of the Columns

The columns in the first floor of the building will become pilings when a tsunami occurs. The drag coefficient of a cylinder is well understood. Note, in the Kahuku Point environment, the wave crests will come in over the point, but the ebb flow will be dominantly out the lows to either side of the point. Consequently, the drag of the columns can be notably reduced by giving them a streamline shape--like a fish, for example. (This modification would not be appropriate at most sites which have to survive a strong ebb in current, as well as a strong rising current.)

ACKNOWLEDGMENT

This study was initiated on behalf of the funding organization, Del E. Webb Inc. Permission for publication has graciously been granted by Mr. Fred Kuntz.

REFERENCES CITED

Costello, R. D. 1952. Damping of water waves by vertical circular cylinders. *Trans. Amer. Geophys. Union*, Vol. 33, pp. 513-519.

Cox, D. C. 1964. *Tsunami Height-Frequency Relationship at Hilo*. Hawaii Institute of Geophysics, University of Hawaii, unpublished manuscript, November.

Jagger, T. A., Jr. 1931. Hawaiian damage from tidal waves. *Volcano Letter*, 321:1-3.

Johnson, J. W., O'Brien, M. P., and Isaacs, J. D. 1948. Graphical construction of wave reflection diagrams. *U. S. Navy Hydro. Office Pub. No. 605.*

Johnson, J. W., Fuchs, R. A., and Morison, J. R. 1951. The damping action of submerged breakwaters. *Trans. Amer. Geophys. Union*, Vol. 32, pp. 704-718.

Macdonald, G. A., Shepard, F. P., and Cox, D. C. 1947. The tsunami of April 1, 1946, in the Hawaiian Islands. *Pacific Science*, Vol. 1, No. 1, pp. 21-37.

Palmer, R. Q., Mulvihill, M. E., and Funasaki, G. T. 1967. Study of proposed barrier plans for the protection of the city of Hilo and Hilo Harbor, Hawaii. *U. S. Army Corps of Engineers, Technical Report No. 1*, Nov. 1967, 76 pages plus plates and appendices.

Pararas-Carayannis, G. 1969. *Catalog of Tsunamis in the Hawaiian Islands.* World Data Center A-Tsunami 69-2.

Powers, H. A. 1946. The tidal wave of April 1, 1946. *Volcano Letter*, 491:1-3.

Wemelsfelder, P. J. 1961. On the use of frequency curves of storm floods. *Proc. Seventh Conf. Coastal Eng.*, Berkeley, Calif. The Engineering Foundation Council on Wave Research, pp. 617-632.

Wiegel, R. L. 1964. *Oceanographical Engineering.* Prentice-Hall, Inc., Englewood Cliffs, N. J. 532 pp.

Geological Factors in Rapid Excavation

Edited by Howard Pincus

with contributions by
Vinton Bacon, M. Friedman,
Richard Hamburger, George E. Heim,
Takeshi Iwasaki, Stanley A. Kling,
Dennis Lachel, John Logan,
Leonard A. Obert, Joseph M. Pugliese,
Carl H. Roach, George M. Sowers,
and James R. Swaisgood

Prepared for the
Engineering Geology Division of
The Geological Society of America

ENGINEERING GEOLOGY CASE HISTORY NO. 9

The printing of this volume has been made possible
through the bequest of
Richard Alexander Fullerton Penrose, Jr.

Library of Congress Catalog Card Number 72-82616
I.S.B.N. 0-8137-4009-6

Published by
THE GEOLOGICAL SOCIETY OF AMERICA, INC.
3300 Penrose Place
Boulder, Colorado 80301

Printed in the United States of America

Contents

Preface

Burgeoning interest in recent years in the technology of rapid excavation is manifest in a spate of titles of symposia, lectures, published articles, and technical reports on this subject.

The symposium, "Geological Factors in Rapid Excavation," presented as part of the Engineering Geology program of the 1970 meetings of The Geological Society of America, addressed itself to the roles of both geology and geologists in rapid excavation. Symposium speakers and attendants included many who were not geologists and, accordingly, the discourse was neither parochial nor self-congratulatory.

This volume is the vehicle for presenting the published versions of most of the papers presented at the symposium.

Some of the authors (Obert, Lachel, Heim and others, and Roach) call our attention to needs; for example, for refinement of measurements of "rock quality," for training engineering geologists to provide reliable evaluations on excavation problems, for training both engineers and geologists to relate soil and rock properties to stratigraphy, for reliably predicting rock properties ahead of the working face, and for geological data that will support a total systems approach to rapid excavation.

Other authors (Hamburger, Pugliese) show how geological information is used or should be used in particular excavation procedures or rock types. Elsewhere (Pugliese, Sowers, Logan and others) the volume presents approaches to rock fragmentation that rest primarily on field, laboratory, or theoretical work. The variety of analyses presented here, even in such a small number of papers, suggests that the armamentarium for attacking our technical problems may be considerably more formidable than many of us had realized. For example, the papers on extension fracture (Sowers) and sliding friction (Logan and others) lay the groundwork for far-reaching practical developments.

The opening paper by Obert was presented midway through the symposium as the featured address at the annual Engineering Geology Luncheon. Its style and content, however, qualify it as a highly appropriate technical introduction to the volume.

The closing note by Heim, Swaisgood, and Bacon, who presided as session co-chairmen and who concluded the symposium with a panel and floor discussion, succinctly presents their views and their reactions to some of the discussions. Especially noteworthy is the recommendation that prospective contractors certify that they have, in fact, reviewed the basic geological and geophysical data collected for the project: the acquisition of such data must become more than a contractual formality.

This collection of papers does not dispose of major obstacles to achieving routinely efficient "rapid excavation." (In fact, "efficient" seems to emerge here as a more important and inclusive adjective than "rapid.") However, the papers do seem to point the way to greater efficiency by both operational and scientific routes. If this volume should cast a little light along such a pathway, it will have served its purpose.

Howard J. Pincus
Department of Geological Sciences
University of Wisconsin
Milwaukee, Wisconsin 53201

Rapid Excavation and the Role of Engineering Geology

LEONARD OBERT
Science Advisor–Mining Research, Department of the Interior, U.S. Bureau of Mines

There is ample testimony that the demands for underground excavation will increase significantly in the future. The National Academy of Sciences Report on Rapid Excavation (1968) projects an expenditure for underground excavation of $69 billion during the next two decades. The Organization for Economic Cooperation and Development (OECD) reports on tunneling (1970) estimate that the world-wide demands for tunneling will double during the next 10 years. Because of these increasing demands, there is a special incentive for improving all phases of underground excavation techniques. The phase of this problem to which I will address myself is, What will be the role of engineering geology and the engineering geologist in improving the underground excavation process? In discussing this role I will presume that the engineering geologist is familiar with, and can utilize, those rock mechanics and geophysical techniques that have been developed for assessing rock quality.

In the last few years there have been a number of conferences and symposia on tunneling and rapid excavation. If the proceedings of these meetings are reviewed it would appear that the greatest concern and indicated need for research have been in developing improved methods for fragmenting rock and in transporting of muck and ground support materials—that is, in the development of improved equipment. In fact, mining systems are envisioned in which not only the mining machine will operate continuously, but fragmented rock will be transported from the face and the materials for ground support will be transported to the face in a continuous operation. Under the proper conditions such a system would undoubtedly increase the rate of excavation. In the OECD report on hard rock tunneling, of the seven major components of tunneling technology that, if improved, would accelerate the excavation process, rock disintegration had the highest averaged priority, and geology and hydrology the lowest.

However, in making on-site investigations or in reviewing the case histories of both tunneling and mining operations, one finds that it is the exception when delays, often of long duration, are not experienced owing to either a lack of information or a misjudgment of geological conditions that were encountered in the excavation process. Ground support requirements are underestimated, requiring the ordering, fabrication, shipment, and installation of additional or heavier sets; mining machines or mining methods are employed that are not adaptable to ground conditions; the face is advanced into unexpected bad ground or bad hydrological conditions. The delays that result are always costly, and the resulting loss of time can defeat any attempt to improve the rate of excavation.

To better understand this problem, consider a typical tunneling operation, although most of the problems experienced in tunneling are also experienced in shaft sinking or in developing a mine. Also, consider first the problems that develop during exploration, that is, before underground access is possible. A route is selected for the tunnel that meets anticipated needs. Usually, in making this selection, consideration is given to the projection of the surface geology to the underground route. However, once a route has been selected this phase has no bearing on the rate of excavation. The next step is to make a detailed geological study of the route. This is usually done by diamond drilling exploratory holes from surface along the proposed route. A drilling log is kept in which the penetration rate, bit wear, drilling water and mud loss, core recovery, and grouting details are recorded. Also, a geologist usually identifies the formations and rock type. In many instances, especially in developing a mine, this is the only information related to the rock quality that is obtained from exploration. If what appears to be a ground-water problem is encountered, drill stem tests may be performed to determine the pressure and flow rates. The techniques for ascertaining hydrological conditions are reasonably satisfactory.

Sometimes a more detailed examination of exploratory cores is made. In addition to identifying the rock type and formation, other mechanical characteristics of the rock are recorded, including the degree of alteration in intact rock, the decomposition products on fracture or parting surfaces, the degree of induration of decomposition products, the degree of recementation or bonding across fracture or parting surfaces, the degree of anisotropy as determined from either a visual or microscopic examination of the rock fabric, and evidences of discing if present. Discing is the breaking up of cores into relatively thin wafers—it is an indicator of adverse ground conditions. Note that these observations are qualitative; no numbers can be assigned to these factors that can be used to quantitatively prescribe a "rock quality." Two factors indicative of rock quality can be quantified: the fracture or parting spacing, and the rock quality designation (RQD), which is a modified procedure for indicating fracture spacing.

One might ask which of these factors are important in prescribing a "rock quality." In some tunneling operations, the principal judgments have been made on the basis of core recovery with consideration being given to the fracture spacing. These factors are especially important in near-surface excavations, especially those above the water table, where the degree of alteration on joint surfaces is extreme and the bond across joints virtually nil. However, below the water table and especially at depth, the degree of recementation or bonding across fracture surfaces becomes increasingly important. For example, consider an unsupported slab 5 ft thick over a mine opening. This slab will be supported against gravity if the average tensile strength across the parting separating it from the overlying rock is 5 psi. It is probable that an intact core cannot be obtained across a joint if the tensile strength is 5 psi or less. Or, consider this example. An unsupported mined area 20 ft × 20 ft will generally cave freely in a rock with an average joint spacing of 4 in., if the strength across the joints is nil. In contrast, spans from 50 to 75 ft have remained stable over long periods of time in rock with an average joint spacing of 4 in. when there is some degree of bond across the joints. At present, a Bureau of Mines investigation is under way to study the alteration and decomposition products that form on joint surfaces, and to determine which of these products tend to indurate and recement.

Core examinations can be supplemented by making borehole examinations, including inhole and crosshole seismic velocity and resistivity logs. Radiometric logs give information on the stratigraphy and jointing, and borehole photography, both optical and TV, can be used to determine joint spacing, joint orientation, and factors affecting core loss.

Finally, a variety of mechanical property tests can be performed to measure the mechanical properties of intact specimens taken from cores, including the unconfined compressive strength, the triaxial compressive strength, the coefficient of friction (intact rock), the coefficient of friction (fracture surfaces), the shear strength (Coulomb or Mohr envelopes), elastic or deformation moduli, seismic velocities, and indentation. The indentation test is performed by some equipment manufacturers to determine the performance of the cutting head on moles. It should be noted that the laboratory tests, with the exception of that for determining coefficient of friction on fracture surfaces, give the mechanical properties of the intact rock, not the rock mass. The seismic velocity ratio (or ratio squared) may be a measure of the mechanical characteristics or rock quality of the rock mass. The velocity of the in situ rock is compared with that of the intact rock (or the same quantity squared). It is presumed that if the rock is fractured or otherwise altered, the seismic velocity in the intact mass will be lower than that of the intact rock and, hence, a measure of the rock quality. Also, the absorption of seismic energy as it is transmitted through both intact and fractured rock is being investigated as a means of evaluating rock quality.

Of the various observations and tests listed in the preceding paragraphs (which is not a complete list of tests that could be performed), a part may have very little bearing on the excavation properties of rock. Also, it would be the exception if all of these tests were performed on any actual project. However, the generalization can be made that usually too few rather than too many observations and tests are performed. This is probably a very poor way to economize on the over-all cost of a project. It is also not the way to improve excavation rates.

Before underground access is possible, then, this is all the information that is available for decision making. In tunneling, the excavation method must be decided on: should conventional mining with or without a shield, or full-face or benching be employed, or should mining with moles be used. The type of ground support has to be preselected. This may range from bare rock to the use of heavy steel sets, with rock bolting, rock bolting with shotcrete, and so forth, as possible alternatives. An error in judgment at this point may completely defeat any attempt to improve excavation rates.

In developing a mine, decisions must be made at the conclusion of exploration because of both the timing and capital investment involved. A mining method must be selected on the basis of core studies. This will generally establish the production rate, which will then determine the size of shaft and the plant equipment and capacity. Also, major pieces of mining equipment, such as continuous mining machines and the conveyor system, are ordered at this time.

Thus, important decisions must be made on the basis of what is only a very small sample of the material that will form the underground structure, and one that is not too amenable to quantitative evaluation.

A rather serious problem often arises in evaluating this geological information. Often the engineering geologist is not sufficiently acquainted with mining methods and ground support procedures to translate the geological information into a specification for the various aspects of the excavation process. To about the same degree, the mining or excavation engineer usually cannot make a good interpretation of the geological data. Moreover, communication between geological and excavation engineers often is not satisfactory and, as a consequence, the best use of the geological information is not obtained. This problem is noted in the OECD reports. Also, because geological data are mostly subjective and, from project to project, highly variable and subject to errors in judgment, the problem of prescribing a mining system is subject to the same degree of error. If, for example, a mining system is specified that includes the transportation and installation of rock bolts as a means of ground support, and after initiation of mining it is found that steel sets are required, the loss of time involved in ordering, fabricating, and redesigning the system to handle sets would offset any efforts to accelerate excavation.

During the excavation phases of a tunneling, shaft sinking, or mine development operation—that is, after underground access is possible—another problem of geological nature is often experienced. Areas of bad ground or bad ground-water conditions are often excavated into without sufficient warning to take precautionary measures. In extreme instances combinations of bad ground and bad water conditions have buried mining equipment and shut down projects, sometimes for periods of months. This problem is of such importance that the International Society for Rock Mechanics committee considering needed research has listed it among the seven most important rock mechanics problems. In shaft sinking it is often found expedient to diamond drill several hundred feet ahead of the shaft bottom to ascertain geological and ground-water conditions. The Habbeger mole is designed to permit drilling through the central shaft of the cutter ahead so the conditions can be determined in advance of the working face. However, diamond drilling is a relatively slow procedure, and this method of determining geological conditions does not lend itself to expediting excavation. Usually, it would be sufficient if ground conditions could be determined not more than one or two diameters in advance of the working face. Despite the importance of this problem, very little research has been reported on investigations aimed at resolving it. Resistivity and refraction techniques might be employed to ascertain conditions within a limited distance in advance of the face. These are methods that would require only a minimum time for observation and would not seriously interrupt mining operation. It has also been suggested that infrared thermometers capable of detecting temperature changes of a small fraction of a degree might be sensitive to temperature changes on the face caused by the presence or percolation of ground water in shear zones and water channels ahead of the face.

The foregoing state of the art discussion suggests a number of problem areas related to excavation technology, the solution of which are subjects for engineering geology. Certainly, research is first in the order of priority. Unless a better evaluation, and especially quantification, of rock quality can be made during exploration, "surprises" that are so often encountered during excavation will remain a factor that limits the excavation rate. Moreover, this quantified rock quality must be related to both the selection of a mining system, and ground support requirements scaled to the dimensions of the opening.

During the excavation process the need for developing a method of determining the rock quality and hydrological conditions ahead of the working face is self-evident. If adverse conditions are anticipated they must be translated into requirements for remedial measures, such as grouting, prespiling, heavier sets, and the like.

Of the various investigations that can be performed during excavation, the point was made that, in general, too little rather than too much effort is expended to determine ground conditions. This raises the questions: In relation to the over-all cost of a project, what should be the optimum expenditure for geological studies? Can this problem be considered a subsystem amenable to an operational analysis? Such a study might make expenditures for geological investigations easier to justify.

One factor that would help in relating geological evaluations to operating conditions, both in regard to ground support and minability, would be a better documentation of tunneling and mining projects, especially those in which difficulties of geological nature are encountered. Such reports should not only describe the operating problems, but should include all core and borehole data and test results, together with whatever prognosis was made of anticipated problems. The International Committee on Large Dams (ICOLD) has found it profitable to document dam failure, that is, any problems requiring remedial action. A better documentation of similar problems in mining, tunneling, and other public works would provide a basis for analyses such as the operational analysis suggested in the preceding paragraph.

Finally, the training of engineering geologists is a problem that should be considered. If the NAS and OECD projections for future excavation demands are accepted, and if engineering geology becomes an even more important factor in excavation technology, it follows that there should be a corresponding increase in the demand for engineering geologists. Are sufficient numbers of engineering geologists being trained to meet this demand, and is their training directed toward the evaluation of excavation problems?

REFERENCES CITED

National Academy of Sciences, 1968, Rapid excavation, significance—needs—opportunities: Natl. Acad. Sci. Pub. 1690.

OECD Advisory Conference on Tunnelling, 1970, Report on tunnelling demand 1960–1980: Organization for Economic Cooperation and Development, Washington, D.C., p. 164.

—— 1970, Report on research and development related to tunnelling: Organization for Economic Cooperation and Development, Washington, D.C., p. 107.

—— 1970, Report on hard rock tunnelling—Rapporteur, T. E. Howard—United States: Organization for Economic Cooperation and Development, Washington, D.C., p. 73.

—— 1970, Report on soft-ground tunnelling–Rapporteur, J. Deschamps–France: Organization for Economic Cooperation and Development, Washington, D.C., p. 129.

—— 1970, Report on cut-and-cover construction–Rapporteur, J. Schmidbauer–Germany: Organization for Economic Cooperation and Development, Washington, D.C., p. 61.

—— 1970, Report on immersed tunnel construction–Rapporteur, H. C. Wentink–Netherlands: Organization for Economic Cooperation and Development, Washington, D.C., p. 42.

MANUSCRIPT RECEIVED BY THE SOCIETY APRIL 5, 1972

Engineering Geologist's Role in Hard Rock Tunnel Machine Selection

DENNIS J. LACHEL
U.S. Army Engineer Division, Missouri River, Corps of Engineers, P.O. Box 103, Downtown Station, Omaha, Nebraska 68101

ABSTRACT

The engineering geologist must accept the responsibility of providing planning leadership and applicable explorations for contractors and manufacturers to evaluate properly the possible use of a tunneling machine on a project. Explorations and laboratory and field tests must be conducted to determine in advance those characteristics which will affect the efficiency and economics of a machine-driven tunnel. To accomplish this will require more detailed data than that normally developed for conventionally driven tunnels. The geologic factors which influence tunneling machines, current laboratory and in situ borability tests, laboratory test programs, and the future of geologic explorations are discussed in light of the engineering geologist's role.

INTRODUCTION

The age of mechanization has caught up with the tunneling field. The day of the tunneling machine is here and the prospect of high-speed, high-quality tunnel boring is upon us. The question is, are we as engineering geologists ready to handle it? It is now up to us to put these tools to work in the most efficient and economical way. Accomplishing this will present a real challenge and impose a greater responsibility on the engineering geologist. Adequate geological information and interpretation are of the utmost importance in the design and application of a boring machine.

Until recently, the drill-and-blast method of conventional tunneling was forgiving and flexible if the geologic interpretation was wrong and could adapt itself to poor rock conditions. In the case of the machine-driven tunnel, if the geologic information is lacking or the interpretation faulty, the machine selected may be poorly suited to the site—or, even worse, the machine might be buried and the project delayed for months.

Tunneling machines, however, do have certain advantages over conventional tunneling, and some of these advantages are of particular interest to geologists.

1. With a machine, overbreak can usually be held below 5 percent, while 20 percent overbreak is not uncommon in conventional tunneling practice.
2. Using tunneling machines, the adjacent or surrounding rock is relatively undisturbed and a minimum of new stresses is introduced.
3. Many tunneling machines produce a circular cross section; this is usually the strongest geometrical configuration, gives excellent stress distribution, and hence requires less support.
4. Minimum disturbance of the tunnel rock also reduces the amount of water inflow and seepage.

Many of these items contribute significantly to the safety of the tunnel. Naturally, a major concern of geologists working in tunnels is the tunnel stability.

Other than the fact that most machines have to be designed and built for different diameters and geologic conditions, the disadvantages of machine-driven tunnels are outside the realm of geology. To be familiar with all aspects of machine tunneling, the geologist should also be aware of these disadvantages.

MODERN TUNNELING MACHINES

The engineering and design of the "mole" has made tremendous strides since the first modern rock tunneling machines arrived at Oahe Dam in 1954 (Underwood, 1965). The modern machine system is very much like the tunneling machine systems which were designed and used at the Hoosic tunnel and English Channel tunnels in the latter half of the 1800s. Basically, only the power source and the cutting tools are different.

Let us consider the different types of tunneling machines available. The first division that can be made is between the soft ground and the hard rock tunnelers.

The soft-ground tunnelers are typically equipped with shields and in some cases air pressure or pressurized mud. They are used in driving tunnels in soils and loose rock and are usually equipped with rotating or oscillating arms with picks. The shield provides protection to the personnel prior to the emplacement of support. The supports are usually set in place from the rear of the shield with lifting gear provided as an integral part of the machine. The machine then pushes itself forward by pressing on the last lining segment installed. Shield tunnelers up to 33 ft in diameter have been used and have advanced more than 650 ft per week.

In this paper we are mainly concerned with the hard rock tunneler, although the term itself is somewhat of a misnomer, since few *very* hard rocks (over 30,000 psi) have been machine tunneled successfully.

These tunnelers can be divided into two types (Proctor, 1970): (1) tunnel machines which work the full face at one time; and (2) machines that have cutterheads smaller than the tunnel diameter and work the face in sweeping movements.

A further division can be made by the type cutters used on the cutterhead: (1) roller cutters that crush the rock as they roll across it and attack the compressive strength of the rock; (2) disc cutters having a knife-edge rim followed by a roller wedge that breaks the rock to the kerf made by the disc cutter, gouging and shearing the rock; and (3) milling type cutters that fail the rock by undercutting it or wedging it and breaking it in tension.

All of the cutters used for rock tunneling employ carbide or hardened steel inserts along the periphery or, as in the case of roller bits, use carbide studs.

ENGINEERING GEOLOGY AND TUNNELING MACHINES

We are now on the threshold of continuous tunneling in almost every type of rock, and geologists must develop *their* skills to catch up with this potential. The first step toward this end is complete familiarity with the machine mining field. Equipped with this knowledge, a geologist or engineer may properly assess a geologic problem in light of the present technology. Only when the geologist knows what the machines are capable of accomplishing can he apply this knowledge to the project at hand.

Once a geologist has acquired this knowledge through experience or, since this is a relatively new field, through diligent research of the available literature on the subject, he is ready to view a project with the tunnel machine in mind. The state of the art of machine rock tunneling is such that all but the shortest tunnels should be considered with a mole in mind. Already there are a number of machines built and available for lease that can be altered to drill even short tunnels economically.

It is well to remember that there are some proposed tunnels where the requirement to minimize vibrations, as in urban areas, indicates machine mining as the only alternative. With increasing concern for the environment, and at the same time increasing needs for water supplies, rapid transportation, and sewers in the urban areas, a much higher percentage of engineering geologists will be faced with the problem of evaluating the use of a tunneling machine on a project.

In any case, a qualified engineering geologist should be included in the initial planning of a newly proposed tunnel. With his knowledge of the machine tunneling field, its advantages, disadvantages, and limitations, as well as a basic knowledge of the site geology, he can make useful contributions in establishing the alignment, depth, and turning radius requirements for the proposed tunnel. The result of his contributions may represent a substantial cost saving to the project.

EXPLORATIONS

Tunneling exploration, be it for conventional or machine tunnels, must begin at the same place. Indeed, since in most cases the option to use a tunneling machine is left to the contractor, all new tunnel explorations must begin with the thought that the use of a tunnel machine is a possibility. Although the proper exploration for a machine-driven tunnel will be more extensive and thus cost more than a conventionally driven one, the results will be applicable to either. This may be a blessing in disguise as there is usually a big difference between the predicted geologic conditions and those encountered in the tunnel.

Office studies are the first step in tunnel exploration. These would include literature and map studies of all data available in the area of the proposed project. Past experience in mines or projects in the general area can offer extensive information about the expected conditions. Also, data from all water wells drilled in the area may be extremely useful.

The next step in the program would be a surface geologic reconnaissance. This should encompass an area 4 or 5 mi on either side of the proposed tunnel alignment so that a "feel" can be made for the regional geology and its effects on the project site.

Detailed surface mapping of the tunnel line would be the next step. Many engineering geologists prefer to accomplish this field mapping on aerial photographs, then project this information to a topographic base. This map should show faults, shear zones, joint patterns, and types of overburden, as well as the rock types and areas of contact as they can be developed from surface exposures or previous work in the area. The emphasis should be on the rock defects and the engineering properties, rather than the classic geologic description of the rock.

In addition to the geologic map, geologic sections should be drawn and extended to the proposed tunnel line. From these sections possible trouble areas can be recognized and further explorations recommended.

Often a thick mantle of overburden covers the rock surface of the exploration area. Although geophysical surveys should properly be conducted with drill holes for correlation, with overburden present a shallow refraction seismic survey prior to drilling could be of assistance in laying out a meaningful drill program. The accuracy of the geophysical data may not be high, but an experienced, careful interpreter, familiar with the site, can provide enough information to help the geologist locate bore borings economically and in the areas most productive of relevant information. After the borings are made, the refraction data can be reinterpreted to supplement and fill in the data between the drill holes.

The consideration of a machine-driven tunnel on a project will usually require more drill holes than explorations for a conventionally driven tunnel. The high cost of mobilizing a tunneling machine to a project more than justifies the need of a detailed drilling program. In addition to samples of each of the rock types expected to be encountered in the tunnel, oriented core would also be extremely helpful in the test program to evaluate the rock. The strength of the rock, either normal or parallel to structural weaknesses, may be appreciably different, which could have a significant effect on the rate of advance. Additionally, exploratory borings should be made to determine the character of the discontinuities that will be encountered by the machine. However, it is not necessary for these borings to intercept the tunnel line. With forethought, careful planning, and a good knowledge of the structural geology of a site, shallow borings can usually be located to sample the range of expected geologic conditions, and this information can be projected to tunnel grade.

Boreholes are expensive and should be used to their full potential. In addition to supplying core to make petrographic studies, the hole may be used for providing information on the occurrence of ground water and gases, and as an indicator of permeability with drawdown and pressure tests. The use of downhole in situ stress measurements, geophysical probes, and downhole television and borehole cameras should be considered. Following initiation of construction, many of these same holes can be used to measure the effect of the tunnel on the surrounding rock mass. Ground-water measurements should be continued. This will monitor the effect of the tunnel acting as a drain, which may lower the water table and result in subsidence. Consideration should also be given to installing rock deflection gauges, such as borehole extensometers, in front of the approaching tunnel.

Numerous other geologic engineering tools are available for defining the character of the rock. Among these are geophysical surveys, joint stereo projection maps, RQD-type (Deere, 1970) systems, and in situ rock mechanics tests.

In the past, little emphasis has been placed on the joint frequency and the inherent structural weaknesses of a rock mass as they relate to machine-driven tunnels. Just as the occurrence of jointing can have an effect on the ripability of a surface outcrop, jointing can play an important role in the determination of the applicability of a tunnel machine on a project. A uniaxial compression test on unjointed rock may indicate that the rock is too strong to be economically bored, but if the joints are close together, the rock could be easily excavated with certain types of machines and bits. In one case, rock with a uniaxial compressive strength in excess of 50,000 psi was easily mined by machine because the rock broke out on existing joints.

The orientation of the joint systems is another consideration which should be closely evaluated from the standpoint of both stability and ease of excavation. Those joints parallel to the tunnel alignment may present stability problems, but at the same time result in easier excavation. This is because the rock is normally stronger perpendicular to the joints than parallel to them. Also, driving parallel to the joints would give the teeth on the cutterhead something to bite into.

The stratification of the rock and other sources of anisotropy can have a significant effect on the rate of excavation with a tunneling machine. Experience has shown that rocks with varying layers of soft and hard material can be more difficult to drill than a uniformly hard rock.

The engineering geologist must consider and evaluate all of these factors and make recommendations as to tunnel alignment and support requirements.

Thus far we have been discussing the effects of the rock on the machine, but the effects of the machine on the rock must also be evaluated by the engineering geologist. It has been reported that the shale (San Jose Formation) on the Navajo Irrigation Project failed by rebound due to the compressive forces imposed on it by the cutterhead (Bennett, 1967). In raveling ground, a full-face type machine could be preferable to cutterheads smaller than the tunnel diameter, because the former would act as a bulkhead and prevent the raveling from starting (Proctor, 1970). Another effect would be the "sinking" of the machines into a soft invert formation. All of these are geologic considerations unique to machine tunneling.

BORABILITY TESTING

The physical characteristic of the rock which has received the most emphasis is the quality which is known under a number of terms, including borability, drillability, and machineability. What they are all trying to find out is, can the rock be economically driven by a tunnel machine? The borability is actually composed of several physical properties, among which are the rock's strength, hardness, abrasiveness, percentage of quartz, and its shape and the rock fabric relative to the size and arrangement of its constituent minerals.

Until recently, the test most commonly used as the main parameter of borability has been the uniaxial compressive strength. It is common to pick up an engineering or construction trade journal and see where a particular machine is driving a tunnel in so many thousand pressures per square inch of rock.

This use of only the compressive strength of a rock to determine if and how fast a machine will break that rock is not a reliable one. The machine manufacturers are the ones who have used this test, and they are also the ones who have realized its shortcoming and are trying to develop more reliable methods. The design and selection of bits, torque and thrust to be used, and the rate of advance are all critical considerations which should be determined before a machine goes underground. Having bits either too strong or too weak, for instance, may each cause problems. In the too strong case the driving rate may be higher than the designed muck removal system can accommodate. In the too weak case the progress might be uneconomically slow. Both cases can cause delays and loss of efficiency.

Reliable tests will also assist the designer, estimator, and engineer to determine bit wear, type of bits to be used, and the cost of bits for a project.

The Mohs hardness test has been used to grade the rock according to its resistance to scratching. It is an empirical scale by which the hardness is determined as compared with a set of standard hardnesses. The disadvantage in using this test lies in the fact that it was designed to measure the hardness of a single mineral, not of a rock made up of a number of minerals. The scratch made on the sample will not necessarily be made in the mineral which wears away the bit.

The Schmidt-type hammer has been used to measure a certain type of hardness–namely, elastic modulus and unconfined compressive strength. This test is based on the rebound principle and a particular type of hammer has been developed for testing rock (Deere, 1970). The hammer used for concrete will fracture many types of softer rock.

Aside from the compressive strength, the most common test currently being used is the indentation test (Fig. 1). Several of the machine manufacturers use this test in a similar form. An oriented sample of the rock is encased in a confining cylinder and impressed with a full-scale rock cutter button for an assigned number of cycles. The resulting permanent deformation is used as a boring index or penetration indicator (Handewith, 1970).

The Department of Mining Engineering of the University of Newcastle upon Tyne has done considerable work on determining the machineability and abrasiveness of rock. The results are equated into a machineability index (Fowell, 1970). In one of the tests they have developed, a rock sample is sawed with a high-speed, workshop reciprocating hacksaw machine. The blades are weighed before and after the cutting test and the results are equated in light of the volume of rock removed and the loss of weight of the blades. This test is still undergoing appraisal.

The University at Newcastle has also been working in the areas of measurement of energy required to cut and chip a volume of rock and in designing better tools and materials for the cutterheads.

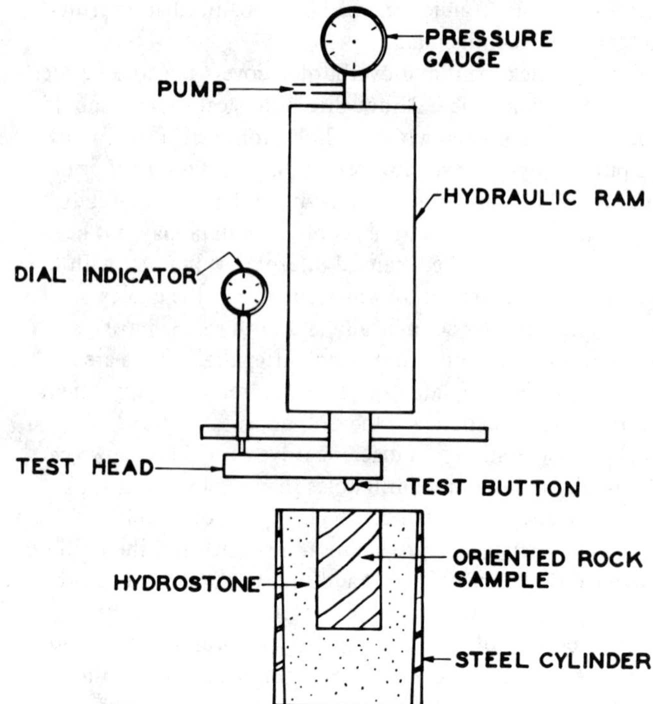

Figure 1. Schematic of indentation test apparatus–not to scale (compliments of Lawrence Mfg. Co., subsidiary of Ingersoll-Rand).

Ross and Hustrulid (1970) of the Colorado School of Mines have adapted a milling machine to accommodate a disc-type cutter and have tested it with various rock types. By instrumenting the machine, they can measure the forces needed to cut the rock. These tests are being conducted on rock from a bored tunnel and it is hoped that correlations can be made between the laboratory and field values of thrust, torque, specific energy, and the cutting coefficient. They will expand these tests to duplicate more nearly the prototype case.

Deere (1970) and Cook (1970), in their presentations at the Tunnel and Shaft Conference at the University of Minnesota in 1968, both described laboratory tests that can be related to the borability of machines.

Deere's test is concerned with the abrasive quality of the rock and is conducted on NX core samples. The abrasive qualities are of great concern and, as he points out, although a limestone and a quartzite could be in "the same strength category, the abrasive resistance of the quartzite is 15 times that of the limestone" (Deere, 1970).

Research conducted by the U.S. Bureau of Mines in the area of drillability has contributed significantly to the tunnel machine field. Reports on its test programs are available.

IN SITU TESTING

Perhaps the most relevant area in which testing should be accomplished is in the in situ rock that the tunnel will be

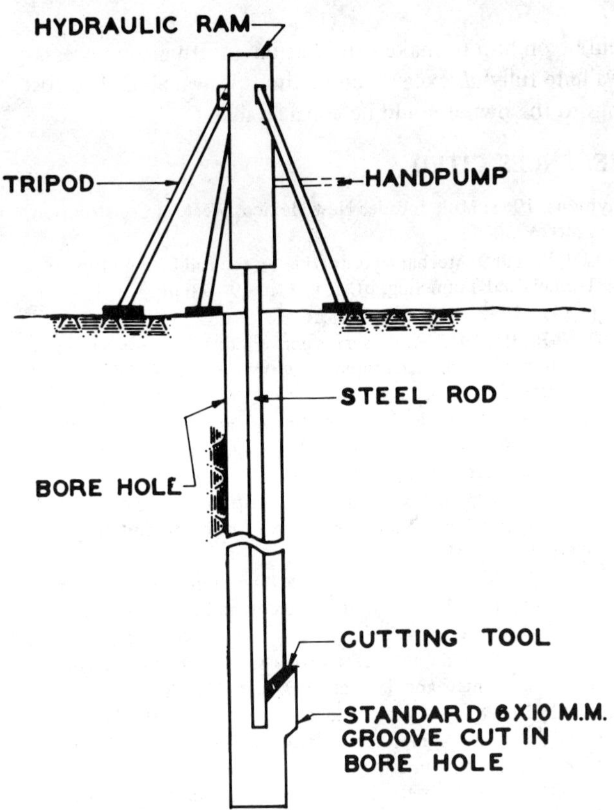

Figure 2. Schematic of in situ borability test–not to scale (University of Newcastle upon Tyne, *Tunnels and Tunneling*, September 1969).

driven through. A number of companies and organizations are looking into possibilities in this field.

Another University at Newcastle project, for instance, uses a core hole drilled into the rock and a small cutting tool attached to the end of a steel rod (Fig. 2). The tool is forced against the side of the hole and withdrawn with a hydraulic ram. The tool cuts a standard groove in the wall, and the force needed to accomplish this is related to the specific energy needed to cut the rock. The abrasiveness of the rock can also be assessed through the loss of weight of the tool (Roxborough, 1969).

One of the European tunneling machine companies is currently developing a downhole test which uses an actual machine bit. The bit is equipped with strain gauges and has the advantage of measuring the resistance to cutting on a full-sized prototype. This helps to lessen the scaling inaccuracies that normally accompany this type of test.

It can thus be seen that the question of borability is one which is receiving considerable attention, as well it should. A standardized borability test would probably be of assistance to the respective manufacturers but, since the different types of machines would require different types of tests for their particular method of attack on the rock, getting them to agree to one test would be a problem.

The high costs of exploration often limit the type of program or the number of exploratory holes we would like to have. With a machine tunnel it may mean concern over a particular shear zone or other unstable conditions. We may have already decided to tunnel through that section using conventional methods, but the question is, where is it? In any case, the answer to the problem is exploration ahead of the tunneling machine.

It would, in most cases, be impractical to move the machine back and out of the way to do these explorations. The engineering geologist may thus have an influence on the type of machine selected if he believes that these explorations are justified, since provisions can be made for making explorations through the machine while it is at the face. This is not a new idea. The Robbins machine used at the Azotea tunnel in 1965 had provisions for a gas analyzer mounted on the machine, connected to a 40-ft probe on the cutterhead (Anonymous, 1965). The Habegger (Atlas Copco) tunneler has the ability to drill an exploratory hole with a diamond drill mounted behind the cutterhead (Barendsen, 1969).

Perhaps the most promising prospect for in-tunnel explorations lies in the use of geophysical methods. At the 1968 Tunnel and Shaft Conference referred to above, Underwood (1970) proposed the use of a continuous exploratory drill hole 100 ft or so ahead of the face with a geophysical package as part of the drill string. Currently, one of the leading tunnel machine manufacturers is using this type of exploration with success. This system will allow the contractor to detect adverse ground conditions and large water inflows sufficiently in advance of the heading to adapt his methods to cope with the circumstances.

This type of system, however, should be used to complement and not to replace good, well-planned, and well-executed pre-excavation explorations. Our technology has not yet advanced to the point that one machine can handle all types of ground, because a really versatile machine will not perform as well as a machine designed to function under the specific geologic conditions which it will encounter.

LABORATORY TESTING AND REPORTS

Finally, the question is, which explorations and tests should a geologist perform or have performed to assist the contractor and the manufacturer in assessing the rock on his project? The answer is as many tests as he thinks are relevant to problems he anticipates on his project. Most certainly there will be limits on the funds available for testing. The geologist must evaluate those areas which could significantly affect the ability of the tunneling machine to accomplish its task and be sure that those areas are fully explored. No one will know the rock as well as he does. If a tunneling machine is selected for a project and cannot cut the rock, the geologist must certainly share some of the responsibility. The U.S. Bureau of Reclamation is supplying its contractors with test data as follows: the modulus of elasticity, Poissons ratio, compressive strength, absorption, specific gravity, and porosity of each of the rock types which will be encountered on a project (Hall, 1970). To this should be added tensile strength, unit weight, complete petrographic analysis including percent of mineral types for each rock, density, and sonic velocity. The completed report must include a geologic section along the tunnel line showing the probable engineering geologic conditions such as: information on the attitude and extent of faults, jointing, joint-filling material, schistosity, bedding, the presence of heavy or running ground, ground-water in-flow rates, rock temperature, the possible occurrence of noxious or explosive gas, an evaluation of the occurrence of tectonic stresses, and finally, the geologist's best recommendations on the amount and type of support that will be required. The geologist should be prepared to supply truly representative samples of the rock to the prospective bidders. Questionnaires from the manufacturers will occupy some of his time and should be completed as fully as possible.

CONCLUSIONS

The engineering geologist can make a useful contribution in all phases of a proposed tunneling project. With the recent advances made in the machine-tunneling field, the engineering geologist must assume increased responsibility in developing plans, explorations, and laboratory test programs that are applicable to this method of excavation.

Tunneling machines must be considered early in the planning stage. Explorations must be laid out and executed in a manner designed to gather information which can be used with any tunneling method. Laboratory testing must be complete and the data thoroughly evaluated. Reports must be concise, emphasizing the engineering properties and the geologist's best estimate on how the ground to be excavated will behave during tunneling.

It can well be argued that the owner's geologist should not recommend the method of mining to be used, but the burden should be on him to make sure that the information necessary to evaluate fully *all* excavation methods is available. The cost savings to the owner could be considerable.

REFERENCES CITED

Anonymous, 1965, Mole invades New Mexico: Western Construction, v. 40, p. 50–54.

Barendsen, P., 1969, Mechanized drifting by the full-face method: Tunnels and Tunneling, pt. 1, v. 1, p. 89–93; pt. 2, v. 1, p. 141–144.

Bennett, N. B., III, 1967, Mole versus conventional: A comparison of two tunnel driving techniques: Highway Res. Rec. 185, Highway Res. Board, p. 1–8.

Bruce, W. E., and Morell, R. J., 1969, Principles of rock cutting applied to mechanical boring machines: II Symp. Rapid Excavation Proc., October 1969, p. 3-1–3-43.

Cook, N.G.W., 1970, Analysis of hard-rock cuttability for machines, *in* Yardley, D. H., ed., Rapid excavation–problems and progress: AIME Jour., p. 39–52.

Deere, D. U., 1970, Indexing rock for machine tunneling, *in* Yardley, D. H., ed., Rapid excavation–problems and progress: AIME Jour., p. 32–38.

Fowell, R. J., 1970, A simple method for assessing the machineability of rocks: Tunnels and Tunneling, v. 2, p. 251–253.

Hall, C. E., 1970, The problems of tunnel exploration: 7th Ann. Eng. Geol. Soils Eng. Symp., Idaho Dept. Highways, Univ. Idaho, Idaho State Univ., p. 3–8.

Handewith, H. J., 1970, Predicting the economic success of continuous tunneling in hard rock: Canadian Mining and Metall. Bull., May, p. 595–599.

Proctor, R. T., 1970, Performances of tunnel boring machines: Assoc. Eng. Geologists Bull., v. VI, p. 107.

Ross, N., and Hustrulid, W., 1970, Use of a linear cutter to predict large diameter tunnel boring rates: Mines Mag., p. 10–11.

Roxborough, F. F., 1969, Rock cutting research for the design and operation of tunneling machines: Tunnels and Tunneling, v. 1, p. 125–128.

Underwood, L. B., 1965, Machine tunneling of Missouri River dams: Am. Soc. Civil Engineers Proc., Jour. Construction Div., v. 91, CO1, paper 4314.

—— 1970, Future needs in site study, *in* Yardley, D. H., ed., Rapid excavation–problems and progress: AIME Jour., p. 24–31.

MANUSCRIPT RECEIVED BY THE SOCIETY APRIL 5, 1972

Some Geologic Structural Influences in Quarrying Limestone and Dolomite

JOSEPH M. PUGLIESE
Twin Cities Mining Research Center, Bureau of Mines, Twin Cities, Minnesota 55111

ABSTRACT

The objective was to determine to what extent geology, mainly structure, influenced blasting practices and results in limestone and dolomite quarries, and to ascertain how geologic features could be utilized to advantage in quarrying. The rock in the quarries visited was mined by bench blasting with long cylindrical charges in vertical boreholes. The gross geologic features observed to influence blasting are: horizontal bedding without conspicuous jointing, horizontal bedding with conspicuous jointing, and folds, faults, unconformities, caves, and filled joints. As an example of how geology affects blasting, horizontal bedding with conspicuous jointing provided a condition where ammonium nitrate–fuel oil, AN-FO, misfired. The following reasons for the misfire are possible: (1) when poured into the borehole, the AN-FO was lost into open joints and bedding; (2) water migrating along joint and bedding planes intersecting boreholes desensitized the AN-FO because the blasting agent was not loaded into a plastic sleeve; (3) the undetonated explosive column in the borehole was cut off by rock shifting along bedding and joint planes. This shifting is caused by rock stresses produced by earlier detonation of nearby charges. A first step in planning a blast design is to identify the geologic features such as those mentioned above. The quarry officials attempt to adjust blasting techniques to minimize detrimental geologic effects and maximize beneficial geologic effects.

INTRODUCTION

Geologic structure and its influence on explosive blasting and fragmentation have been noted in previous studies. Atchison and Pugliese (1964), in their comparative studies of explosives in a tightly jointed limestone, found that repeated blasting in an area opened tight separations in rock. Their results reflected the influence of this joint opening. Belland (1966) concluded that geologic structural controls on rock fragmentation upon explosive blasting do exist, and explosive blast patterns should be designed to make optimum use of geologic controls. Davis (1953) determined that fragmentation of taconite at or near the surface is controlled or strongly influenced by joint and bedding planes. McMahon (1968) realized the fundamental problem in rock mechanics in accounting for the discontinuous and anisotropic nature of the jointed rock mass.

The investigations cited and others indicate that a field study should be made of geologic features important in explosive blasting and how these features affect quarrying by blasting. Because reports on this problem are few, scattered, and, in many instances, not complete in dealing with the subject, the Bureau of Mines has undertaken to study to what extent geologic factors, mainly structure, influence current blasting practices and results in limestone and dolomite quarries, and also to ascertain how geologic factors can be utilized to more advantage in quarrying. In collecting the following information, the author visited 30 quarries, studied the geologic structure in the quarries, observed the blasting practices and, after discussion with the quarry managers, foremen, and blasters, noted several geologic features consistently affecting blasting and mining and how adjustments were made to compensate for these geologic features. The relations inferred rest on empirical procedures.

In this report, explosive ingredients, properties, and field performance characteristics are included only when necessary. The reader is referred to Dick (1968) and Yancik (1969) if more extensive information is desired.

GEOLOGIC FEATURES OBSERVED TO INFLUENCE BLASTING

The following major geologic features were observed to influence explosive blasting and mining: horizontal bedding without conspicuous jointing, horizontal bedding with conspicuous jointing, and folds, faults, unconformities, caves, and filled joints. Some examples of how the above features affect blasting are presented here. The noted feature shown in each of the following figures is the major one for that particular photographed quarry face and surrounding area.

The word "explosive" is used as a collective term for both "blasting agent" and "explosive."

Horizontal Bedding without Conspicuous Jointing

The arrow in Figure 1 points to the cast of the 3-in. hole leading up to the smear of undetonated ammonium nitrate-fuel oil (AN-FO) and the rock shelf left in place because of no explosive detonation. The quarry operator used 3-in. holes loaded with free-running AN-FO and bottom primed with a stick of dynamite containing an electric blasting cap. The high (45- to 50-ft) dolomite bench is in the Prairie du Chien Formation with the top 10 to 15 ft thin-bedded, and the remainder massive.

The quarry operator's explanation for the lack of explosive detonation up the full length of charge was that the explosive column was only bottom primed. The propagation velocity of the rock compressional waves from explosive detonation of the bottom part of the charge was higher than the reaction velocity of the AN-FO charge in the small borehole. Because of this higher propagation velocity of the rock compressional waves, stress action could occur in the rock near the borehole top before the charge could detonate to that point. The stress action could shift the top rock slabs across the small diameter borehole and cut the explosive column before the column could fully detonate up the long length of borehole.

A possible solution to the above problem would be to multiple-prime the charge; that is, to have simultaneous initiation at several points along the charge column. With multiple priming, the charge could be completely detonated before the produced stress action in rock could cause rock slab shifting and chance of cutting the explosive before detonation.

If the bottom priming was necessary, the cut-off problem might be reduced or eliminated by using a charge of AN-FO with a higher detonation velocity. The detonation velocity would be such as to decrease the time difference between the arrival time of the rock compressional waves produced by detonation of the bottom portion of the charge and the arrival time of the explosive reaction front near the surface and top of the borehole. With this decrease in time differential, the top portion of the charge could be detonating before the top slabs could shift and cut the explosive column. A high detonation pressure primer for AN-FO should be used instead of a stick of dynamite.

Another reason for the undetonated charge column being cut could be that the slabs were shifted by stress wave action in rock produced by adjacent charges detonated earlier in the blasting sequence.

Horizontal Bedding with Conspicuous Jointing

Figure 2 shows the bottom section of a quarry face. The

Figure 1. Shifted horizontal rock shelf that cut explosive column in 3-in. borehole before complete charge detonation. Arrow points to cast of hole and smear of undetonated ammonium nitrate-fuel oil (AN-FO) and the rock shelf left in place because of no explosive detonation. Wabasha County, Minnesota.

Figure 2. Horizontal bedding and vertical joint intersecting 7.25-in. borehole producing condition for nondetonation of AN-FO. White arrow points to a vertical joint and a bedding plane intersecting the borehole. Black arrow points to the visible detonating cord and blasting cap leads. Crawford County, Ohio.

rock is the massive Columbus Limestone with thick distinct horizontal bedding, three vertical joint sets, and one inclined joint set. Note the cast of a 7.25-in.-diameter borehole near the base of the backface after a production round has been detonated. The bulk AN-FO did not detonate in this hole; observe the visible detonating cord and blasting cap leads (black arrow). Any one or all of the following reasons for this misfire are possible: (1) the AN-FO was lost into open joints and bedding upon loading (the white arrow points to a vertical joint and a bedding plane intersecting the borehole); (2) water migrating along bedding and joints came into the loaded borehole and desensitized the AN-FO because the blasting agent was not loaded into a plastic sleeve; (3) the detonating cord and explosive column could have been cut by rock shifting along bedding and joint planes because of earlier detonation of nearby charges.

The following suggestions are offered to minimize the chance of misfire:

1. When AN-FO is used under wet conditions, there must be sufficient water protection. Protection can be obtained through the use of a factory-produced cartridged product or a polyethylene borehole liner.

2. If a cartridge product or borehole liner is used, the AN-FO prills, or spherical pellets, will not flow into joints and bedding.

3. Subdrilling below the predetermined quarry floor depth will allow some of the mud to settle below the quarry floor depth. Before the holes are loaded with explosives, the hole depths should be measured. If the mud has built up to a great extent, the mud should be pumped or blown out or the holes redrilled.

4. The charges of AN-FO should be detonated as soon as possible. If economically feasible, slurry or gelatin should be used under very wet conditions.

Folds

In Figure 3, the rock mined from the 30-ft bench is folded Niagaran Dolomite. The quarry operator is attempting to mine horizontally, as indicated by the black arrow's direction. The small white arrow on top of the problem toe at the bench bottom points in the direction of dip. The top of the toe is a major bedding plane. The toe was left with light bottom loading of explosive in the boreholes and no subdrilling below the predetermined horizontal quarry floor. Upon being blasted, the rock broke at the bedding plane, leaving the toe. If the toe is not removed before another quarry round is attempted, the toe will block rock displacement from the subsequent production blast.

The quarry operator believes subdrilling of boreholes up to 10 ft and heavy bottom loading with powerful explosive slurry are necessary to overcome the fold's influence and to break the rock to the predetermined quarry floor level. Others contend that subdrilling this deeply is wasted drilling.

Faults

Viewed from left to right, Figure 4 shows the lower dolomite zone of the Tymochtee Formation (LT), the fault zone with breccia and gouge (FZ), and then the upper dolomite of the Tymochtee Formation (UT). The lower portion of the Tymochtee is used for roadstone, concrete aggregate, and railroad ballast. The upper portion of the Tymochtee is used only for road base aggregate and filler. The upper Tymochtee contains argillaceous laminae that are fairly tight and do not break open upon crushing, adding to the poor quality of rock for roadstone. The early operators had not located the fault and were confused by this abrupt lateral change in rock grade. They mined through the lower Tymochtee Dolomite, through the fault, and into the upper Tymochtee. Since the upper Tymochtee is not satisfactory for roadstone and concrete aggregate, the principal intended products, the early operators went bankrupt.

By exploratory core drilling and rock sampling, these quarry operators could have located the fault and could have blasted parallel to or away from it in the lower Tymochtee. They could have developed a market for products obtained from the upper Tymochtee Formation. The present owner core drills (exploratory) and takes toe and face samples well ahead of mining so that he can plan his blasting. He has developed a market for products obtained from the lower and upper dolomite of the Tymochtee Formation. In actual quarrying, he has mined through the fault zone so that he need blast in the faulted area once or only a few times before being in the solid rock again.

Figure 3. Dipping rock causing problem toe where quarry operator attempted to mine horizontally. White arrow on top of problem toe points in direction of rock dip. Black arrow points in horizontal direction. Racine County, Wisconsin.

Figure 4. Fault causing lateral change in rock grade. The fault zone with breccia and gouge (FZ) is between the lower dolomite zone of the Tymochtee Formation (LT) and the upper dolomite of the same formation (UT). Adams County, Ohio.

Figure 5. A 40-ft bench composed of the massive Peebles Dolomite overlain unconformably by the bedded Greenfield Dolomite. The arrow points to the unconformity. Adams County, Ohio.

Unconformities

Figure 5 depicts a 40-ft abandoned quarry face composed of the massive Peebles Dolomite overlain unconformably by the bedded Greenfield Dolomite. The arrow points to the unconformity. The early owners went bankrupt because they blasted the formations together in this one bench. Their principal products were roadstone and concrete aggregate. The Greenfield provided a good roadstone and concrete aggregate, but the Peebles did not. Mixing the rock from the two very different formations gave a low-grade product for roadstone and concrete aggregate use.

Figure 6 shows the present profitable operation in which the two formations are being blasted as separate benches. The new quarry operator has developed a market for products derived from both formations. The 25-ft bench in the background is composed of the Greenfield Dolomite. The top of the Peebles Dolomite is the unconformity (arrows). The operator realizes the unconformity is a surface of weakness and can cause loss of explosives and lack of explosion detonation confinement when blasting in the Greenfield. When he blasts in the Greenfield Dolomite, he does not subdrill and load explosives below the unconformity because the rock breaks to the unconformity anyway.

Caves and Filled Joints

Figure 7 is a view, looking from the west, of a quarry face composed of limestone in the Galena Formation. The lower three-fourths of the 75-ft bench is massive, and the top one-fourth is thin-bedded limestone and shale. Observe the 10-ft-high cave behind the pinnacle of rock.

Figure 8 is a view of the same quarry as seen in Figure 7, but looking from the east. The cave seen in this figure is the

Figure 6. The two different dolomite formations of Figure 5 being quarried as two separate benches. The top of the Peebles Dolomite is the unconformity (arrows). Adams County, Ohio.

same as shown in Figure 7. The cave extends completely around behind the pinnacle of rock. The black arrows point to filled joints (joints filled with clay and a few small open channels). Some tight joints are also present (white arrows). A major bedding plane is at the base of the face.

The following problems were encountered:

1. When the quarry operator blasted in the thin limestone and shale layers where caves are present, the rock broke at its weakest point into the cave and further rock breakage was reduced.

Figure 7. A 10-ft-high cave seen in a limestone quarry face (of the Galena Formation). Fillmore County, Minnesota.

Figure 8. Another view of the face seen in Figure 7 showing the 10-ft-high cave, filled joints (black arrows), and tight joints (white arrows). Fillmore County, Minnesota.

2. Large amounts of explosives were lost into caves when a production round was loaded.

3. When the pinnacle of rock was separated from the main rock body by the cave and a filled joint, and the main rock body blasted, this pinnacle could shift forward intact upon the basal bedding plane.

4. The filled joints posed more threat to loss of explosives or reduced explosive detonation confinement than did tight joints.

The quarry operator and the author propose the following possible reasons for the above problems and suggestions for improvement:

1. When holes are drilled near caves and loaded with explosives, stress-action and expanding gases from explosive detonation break the rock at its weakest point into the caves. The reduced borehole pressure and lessened expanding gas action from explosive detonation limit further rock breakage. When possible, drilling holes for charge loading at a distance from known caves has helped lessen this problem.

2. When a hole intersects a cave, explosives loaded into the hole can be lost into the cave. Drilling into caves can be avoided when a consistent cave pattern is present and has been determined from previous drill holes. The following procedure has provided satisfactory blasting results when a hole is drilled through a cave and continued into the rock below. The downline, or length of detonating cord slightly longer than the borehole depth, is lowered into the hole with the primer, or explosive-detonation initiator, attached to the downline bottom. The bottom of the hole is loaded with explosives until the top of the charge column is near the bottom of the cave. The hole is filled with sand or drill cuttings, or stemmed, until the cave bottom is reached. The hole is plugged at the cave top, loaded with explosives, and then stemmed. The downline should have primers spaced up the hole to assure complete detonation of the charge.

3. If the large pinnacle contains no explosive, the pinnacle will not be fragmented and may be displaced intact by the movement of fragmented rock and expansion of gases upon blasting in the main rock body. If the basal bedding plane is not well defined, no displacement of the rock pinnacle is possible. Holes should be drilled into the pinnacle of rock and also into the solid rock behind the pinnacle. The pinnacle would be blasted first, with the solid rock behind the pinnacle blasted in delayed sequence. This delayed sequence in blasting would allow the broken rock from the pinnacle to be displaced enough to provide room for displacement of broken rock from the main rock body.

4. The filled joints in this quarry cause more loss of explosive and explosive detonation confinement than do the tight joints. Drill rods and bits jam easily in the filled joints. Because the filled joints contain large amounts of soft clay and some small open channels, explosives can be lost into the channels, and the clay provides less explosive detonation confinement than does the solid rock bounding the tight joint. Filled joints are surveyed and drill holes are spaced so as to avoid as much as possible having holes in a production round intersect the filled joints.

CONCLUSIONS

As indicated by the above few examples, geologic structure—such as horizontal bedding with and without conspicuous jointing, folds, faults, unconformities, caves, and filled joints—influences quarrying limestone and dolomite. A first step in planning a blast is to identify geologic features such as these.

Blasting techniques should be adjusted to minimize detrimental geologic effects and maximize beneficial geologic effects.

The influence of petrofabrics as well as of macroscopic geologic features on quarrying by explosive blasting or other energy source should be investigated. Environmental considerations should be taken into account. The research should include other rock types that may contain other geologic features that affect rock fragmentation.

ACKNOWLEDGMENTS

The author wishes to thank David E. Fogelson, supervisory geophysicist at the Twin Cities Mining Research Center, Bureau of Mines, U.S. Department of Interior, for his suggestions and guidance in the early phase of this study. The author also thanks Richard A. Dick, mining engineer at this Center, for his helpful suggestions in interpreting the blast data. The author wishes to acknowledge the generous cooperation of those quarry operators who participated in this study by permitting access to their property and by providing much of the data that made this report possible. The information reported here was derived from quarries located in Minnesota (Fillmore and Wabasha Counties), Wisconsin (Racine County), and Ohio (Adams and Crawford Counties).

REFERENCES CITED

Atchison, Thomas C., and Pugliese, Joseph M., 1964, Comparative studies of explosives in limestone: U.S. Bur. Mines Rept. Inv. 6395, 25 p.

Belland, J. M., 1966, Structure as a control in rock fragmentation, Carol Lake iron ore deposits: Canadian Mining and Metall. Bull., v. 59, no. 647, p. 323–328.

Davis, Vernon C., 1953, Taconite fragmentation: U.S. Bur. Mines Rept. Inv. 4918, 34 p.

Dick, Richard A., 1968, Factors in selecting and applying commercial explosives and blasting agents: U.S. Bur. Mines Inf. Circ. 8405, 30 p.

McMahon, Barry K., 1968, Indices related to the mechanical properties of jointed rock: Status of practical rock mechanics: 9th Symp. Rock Mechanics Proc., Colorado School of Mines, Golden, Colo., April 17–19, 1967; New York, Am. Inst. Mining, Metall., and Petroleum Engineers, Inc., p. 117–128.

Yancik, J. J., 1969, AN-FO's explosive properties and field performance characteristics: St. Louis, Mo., Monsanto Blasting Products AN-FO Manual, Monsanto Co., 37 p.

MANUSCRIPT RECEIVED BY THE SOCIETY APRIL 5, 1972

Geologic Factors in Rapid Excavation with Nuclear Explosives

RICHARD HAMBURGER
U.S. Atomic Energy Commission, Washington, D.C. 20545

ABSTRACT

Many of the essential elements required to design safe nuclear excavation projects are geologic. Excavation dimensions depend on such rock properties as compressibility, porosity, water content, and strength. Estimates of ground motion at locations of natural or manmade structures depend on knowing the geology of the explosion site, the transmission path, and the structure site. The hydrologic environment must be understood in order to avoid unwanted effects. Although computational techniques have been developed to design nuclear excavations, the results of such calculations are greatly dependent on how well the geologic data provided represent the actual conditions. Better techniques to determine rock properties are required.

INTRODUCTION

Explosives have been used for many years for excavation purposes in the construction, mining, and agricultural industries. However, with the advent of nuclear explosives, the possibility of using explosives in the excavation of massive quantities of rock became, for the first time, an economic possibility, as was discussed at the First Plowshare Symposium in 1957.

I shall trace some of the history of the development of nuclear excavation theory, note the important geologic factors which have been identified to date in the application of this theory, and discuss how the geologist fits into the total picture. This, then, is a brief sketch of where we are in the understanding of the important geologic factors involved in the engineering design of excavation by nuclear explosions.

I shall not discuss the specifics of the geology involved or the details as to how geologic measurements and investigations are applied. These are readily available in the literature.

The term "excavation" is used in this paper in the same sense that it was used by the National Research Council's Committee on Rapid Excavation in 1968; that is, excavation means the making of usable space on or within the earth's crust. Excavation, in the sense of creating craters or ditches (Figs. 1, 2, and 3), is one example of making such usable space. The standing cavity (Fig. 4) formed by a nuclear explosion in salt also fits this definition, as does the chimney (Fig. 5) formed as the result of nuclear detonations in other types of rock.

In discussing the geologic factors important to nuclear excavation, the word "geology" is used here in its general sense to include all aspects of the science. Hence, I am also using the term to include the subfields of rock mechanics, seismology, and hydrology.

BACKGROUND

Most discussions on excavation are *process* oriented. Discussions on tunneling, drilling, and cut and cover, for instance, focus on how to improve the process; that is, how to excavate a tunnel faster, or larger, or both. Although nuclear explosives share many of the characteristics of conventional explosives, they bring a new dimension to our thinking because of the large energy contained in a small package and relatively low cost per unit of energy (Toman, 1969). Accordingly, one must consider nuclear explosives as a *new tool*—not just a larger stick of dynamite. The energy output of this new tool is of such a magnitude that, to get the optimum benefit, the whole excavation process, not just the explosive process, must be re-examined.

The most profitable approach, in fact, is to re-examine the purpose for which the excavation is being made. To phrase this another way, "What is the objective?" Accordingly, let us consider nuclear excavation in terms that are *objective* oriented. Some examples will clarify this point of view.

It has been demonstrated that productivity of a well is increased by increasing the effective well diameter (Coffer and others, 1964). The practical limit to which this increase can be accomplished when excavating with a drill is measured in inches to a few feet, plus the added effective well bore which can be made by fracturing with chemical explosives or with water pressure. However, the same objective, the increase of the effective well diameter, can be raised to tens to hundreds of feet by using nuclear explosives. The chimney (Fig. 6) and its adjacent cracks are the effective well bore.

In like manner, a series of adjacent explosively created chimneys might be used instead of a tunnel for water conveyance and storage, with the bonus of eliminating evaporative losses of surface storage. Where the geology is appropriate, the objective-oriented approach to excavation calls our attention to opportunities for creating underground space for waste management or for the storage of fuel (Project Ketch, 1967).

There is a reason for my urging focus on objectives. As the National Research Council's Committee on Rapid Excavation stated in its report:

> Two of the major challenges now facing the United States are urbanization and conservation of resources. Both challenges are affected to some degree by the nature, cost, and manner of application of available earth-moving or tunneling technology. Both challenges can be met in part by fuller exploitation of available earth-moving and tunneling technology. A very important step toward meeting these challenges will be the ability of technologists to bridge the gap between what is technically possible today in earth-moving and tunneling and what can become possible tomorrow.

If one takes the position that the objective is paramount and that the process chosen is the one that best accomplishes the goal, then the nuclear explosive is a new tool in the engineer's bag. As just illustrated, it may be used for more effective conservation and use of our natural resources, permit better use of the surface and subsurface, and help us in environmental control. Geologists, with their knowledge of the earth, can bring invaluable insight to the successful achievement of these important objectives.

The importance of this objective-oriented viewpoint cannot be overemphasized. To increase the living standard, all engineering technologies, which either reduce costs or make available resources not otherwise available, need to be devel-

Figure 1. Sedan nuclear explosive point charge crater, Nevada test site. Formed by detonating a 100-kiloton nuclear explosive buried at a depth of 635 ft in alluvium. Crater dimensions about 325 ft deep, and about 1,200 ft wide. Note D-8 caterpillar bulldozer in bottom of crater.

oped. Since there is a finite usable surface to the earth, technologies which permit us to use the depth dimension of the crust can also add to the quality of life on this planet. Nuclear excavation can contribute to the attainment of these objectives.

DEVELOPMENT OF THEORIES AND UNDERSTANDING OF NUCLEAR EXPLOSION TECHNOLOGY

In most excavation work, there are three major steps which are normally accomplished by three distinct operations. The first essential step is to break the rock; the second is to move the rock; and the third is to deposit it outside the area being excavated. In throw-out explosion excavation all three stages are accomplished as one integrated operation. In nuclear excavation, where the rock is not thrown out, the first step still is breaking the rock. The second step is that of moving the broken rock to create the useful spaces. The third step, removing the broken rock from the excavation, may not be needed. This movement of the broken rock may result in a subsidence crater, a chimney of broken and bulked rock with the attendant interstitial void spaces, or even a free-standing cavity. This also is an integrated operation.

No technology arrives on the scene fully developed. A technology is developed through iterative processes, observations are made that suggest certain empirical relationships, and a theory is developed to describe the relationship. Then new raw data are collected, against which this theory is tested, and if the theory does not account for all of the phenomena observed, new relationships are sought and the theory is either modified or redefined. Nuclear explosive excavation has, like other technologies, gone through such developmental stages and has arrived at a state where computational techniques, based on theoretical considerations, are sufficiently developed that some engineering applications could presently be designed. Although chimneys and craters are different end results of the excavation process, many of the geologic factors which must be considered are the same, and much of the analysis is identical.

Historically, however, analysis of explosive cratering mechanisms came first (Cherry, 1966; Cherry and Petersen, 1970). Early analyses of data collected from high-explosive

Figure 2. Buggy nuclear explosive row crater, Nevada test site. Formed by simultaneously detonating five 1.1-kiloton explosives, 150 ft apart, buried at a depth of 135 ft in basalt. Crater dimensions about 850 ft long, 250 ft wide, and 60 ft deep. Note pickup truck in lower right part of picture.

experiments were made and, based solely on dimensional analyses, it was postulated that the radius and depth of a crater would vary as the cube root of the explosive yield (Lampson, 1946; Fig. 7). As more experience was gained with higher yields, both high explosive and nuclear, it was shown that scaling to the 3.4 root provided a better fit to the data (Nordyke, 1962, 1964). All scaling laws assume that similitude exists, but in actual practice similitude does not occur for all parameters (Chabai, 1965).

It was early recognized that if the experimental data were divided according to rock types, a much better fit resulted. Nordyke (1964), for instance, separated the available experimental data between two rock types, desert alluvium and basalt. A later analysis separated the empirical data from high explosive (HE) and nuclear explosive (NE) craters for alluvium, shale, and dry, hard rock, scaled to the 3.4th root (Toman, 1969; Fig. 8). Information derived from the analysis of contained nuclear explosions in several rock types (tuff, alluvium, rock salt, granite) indicated that, within ±20 percent, the size of the initial cavity formed correlated with the water content of the rock (Boardman and others, 1964). In like manner, Higgins and Butkovich (1967) concluded that the cavity radius could be accurately predicted, provided the water content, the rock density, and the explosion energy were known.

By this stage of the development, several iterations, from experimental data to theory and back again, had been accomplished. It became obvious that to predict properly the size of the excavations in rocks with varying properties, either a fairly large number of empirically derived scaling curves would be required, or computation would have to be developed to have more general applicability.

Cherry developed such a computational procedure with the associated numerical solution for explosion-produced craters

Figure 3. High-explosive row excavation, Nevada test site. Explosion crater formed by simultaneously detonating five high-explosive charges in alluvium. Note that most of the rock is thrown out to the sides with very little throwout at the ends of excavation. Similar though less pronounced suppression of throwout at the ends of the excavation occurred on the Buggy Crater (see Fig. 2).

(Cherry, 1966). Later work applied similar numerical solutions to other explosion type excavations (Cherry and Petersen, 1970). From a geologist's point of view, the importance of Cherry's work is that it starts with first principles of physics and requires for its proper solution a definition of the important geologic parameters and the ability to measure these parameters. With appropriate geologic input this approach should be universally applicable.

Cherry uses a numerical description of stress wave propagation and a model for material in brittle failure. He starts with a feedback loop for the stress wave, as indicated in Figure 9. He divides the rock into discrete zones, in each of which the feedback loop for the stress wave may be calculated through time. The geologic input to this technique is in the equation of state. Rock properties, so far identified in the order of importance, are water content, shear strength, porosity, and compressibility, both in the original state and after fracture (Terhune and others, 1970). Measurements are made to determine these properties in the laboratory from drill hole samples, and also by in-hole logging techniques. The computational technique developed by Cherry permits consideration of varying geology to the extent that it is known. Similar numerical solutions have been developed by others (Moreau, 1970; Cameron and Scorgie, 1970). These calculations also depend on geologic data for their solution.

It is apparent from the above discussion that the geologist's ability to interpret drill hole data and to properly select samples for laboratory testing is of paramount importance. In addition, there are lateral as well as vertical variations within rocks, and for very large explosions these lateral variations might be important. The geologist's ability to provide a three-dimensional description of the rocks to be excavated, as well as his ability to recognize the limitations of his predictive capability so the need for additional measurements may be recognized, are requisite for use of this computational technique as an engineering design tool. Thus, a major contribution that geologists could bring to the entire field of rapid excava-

Figure 4. Gnome cavity, near Carlsbad, New Mexico. Underground excavation formed by detonating a 3.1-kiloton nuclear explosive in bedded salt at a depth of about 1,200 ft. Cavity dimensions: diameter 160 ft, height 75 ft. Note man standing near center (Rawson, 1966).

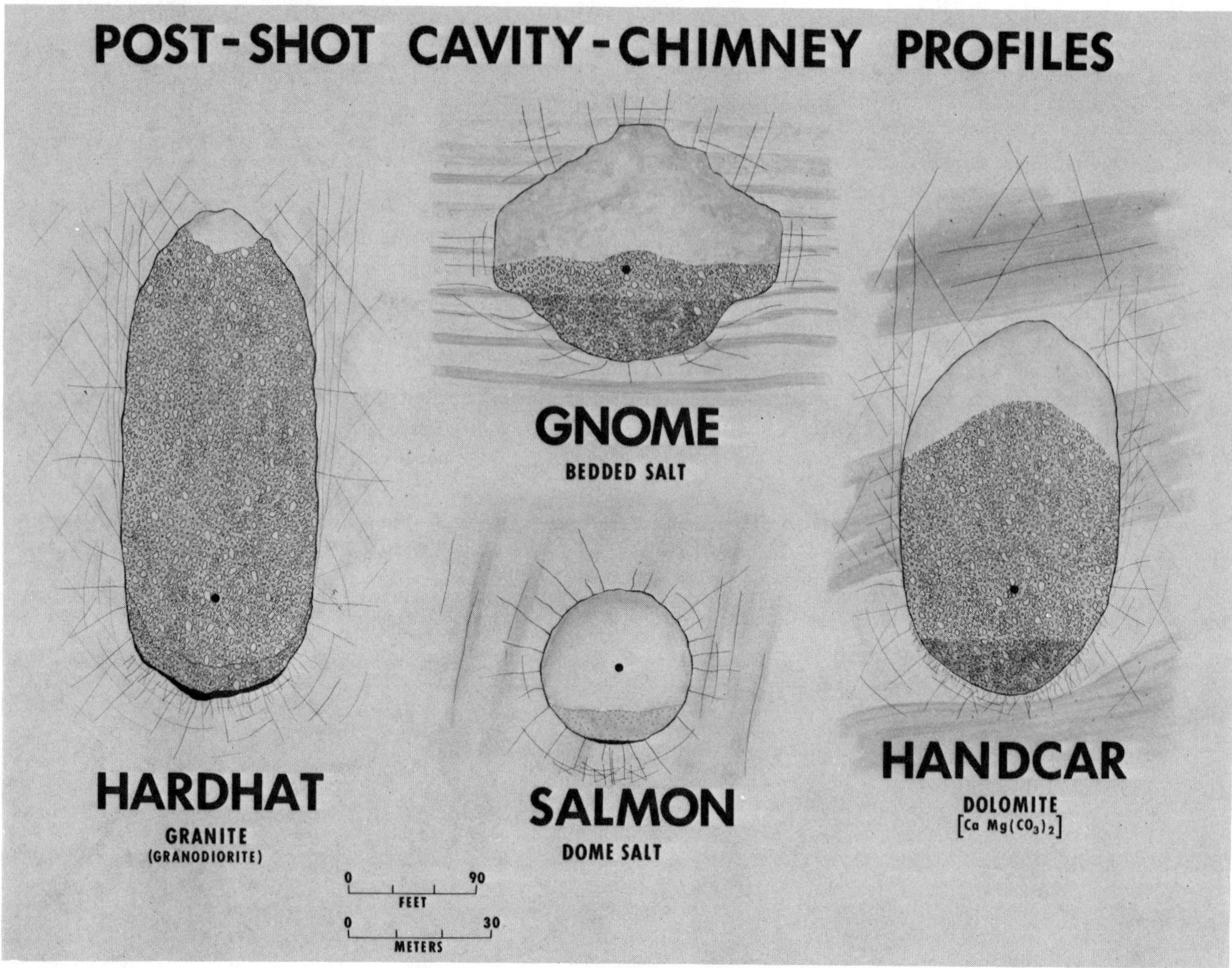

Figure 5. Size and shape of cavities and chimneys formed by nuclear explosives. Hardhat and Handcar are at the Nevada test site; Gnome is near Carlsbad, New Mexico; and Salmon is near Hattiesburg, Mississippi (Rawson, 1966).

tion would be standardized, quantitative descriptions of rock properties that are now often described in a qualitative manner.

In addition, as the technology advances the geologist must maintain close contact with the theoretician so as to be able to identify additional geologic parameters of importance in the calculation.

Geologists can also bring their knowledge to bear directly on some of the outstanding problems. Some requiring short-range solutions are: (1) measurement of adiabatic release paths from high pressure (100 kb to 1 mb) shocked states from various rock types, including both the wet and dry state; (2) laboratory measurements of stable and unstable failure (fracture propagation), including studies of the effect of strain rate.

Important longer range problems are: (1) development of a model for the equation of state of rocks from their mineral composition, density, and water content; (2) understanding the healing of fractured materials under different conditions of temperature, pressure, and water content.

Equipment to measure in situ rock properties is needed. Such equipment would include the following: (1) a logging type tool for measurement of static rigidity modulus; (2) a tool for determining the in situ strain in a drill hole at depth and/or recovering a core or sample in which the state of compression and strain has been preserved (Higgins, 1970, oral commun.).

OTHER GEOLOGIC QUESTIONS

Geology and geologists are involved in other aspects of nuclear excavation besides the ones mentioned. When an explosion occurs, not all of the energy is used for excavation—some small part of it leaves the immediate area as a seismic wave. The computational techniques mentioned before, particularly the displacement history of particles in the elastic region, provide information on the seismic coupling. In addition, there is a growing body of empirical information based upon both cratering and contained explosions on the efficiency and spectrum of the coupling of the energy as it

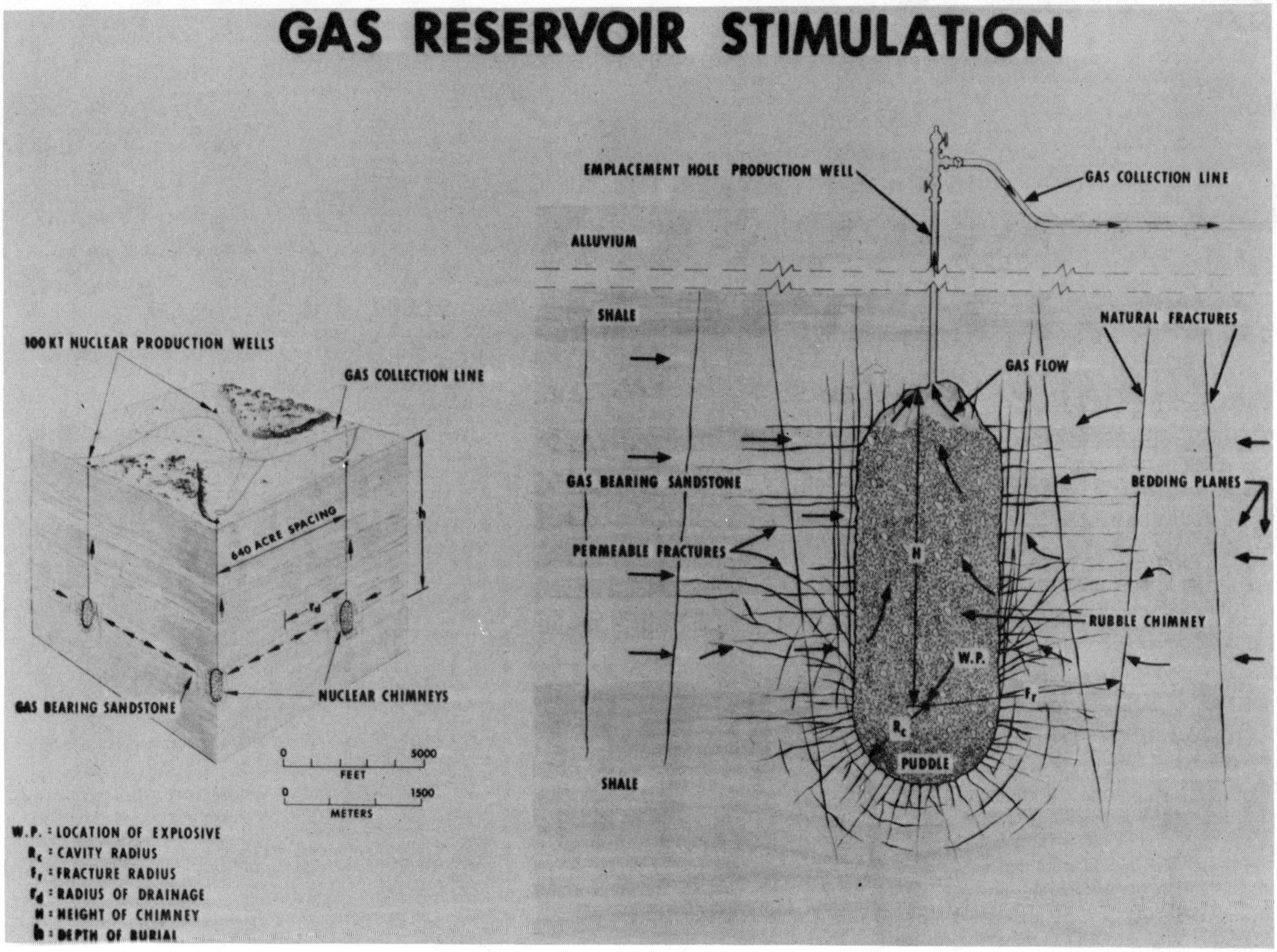

Figure 6. Chimney formed by nuclear explosion. Acts as well bore with an effective diameter of more than 100 ft (Rawson, 1966).

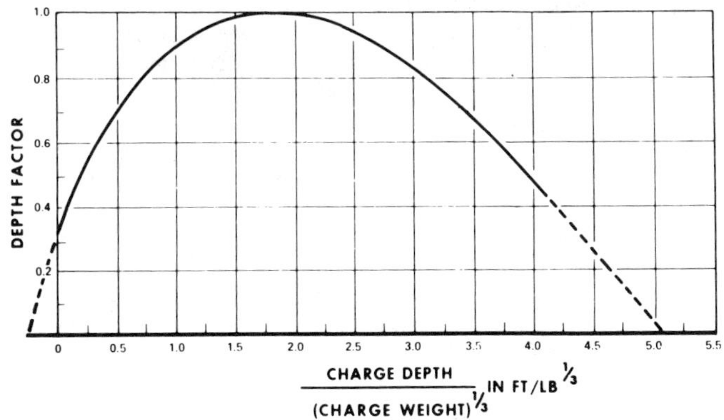

Figure 7. Depth factor for cratering by explosives (after Lampson, 1964).

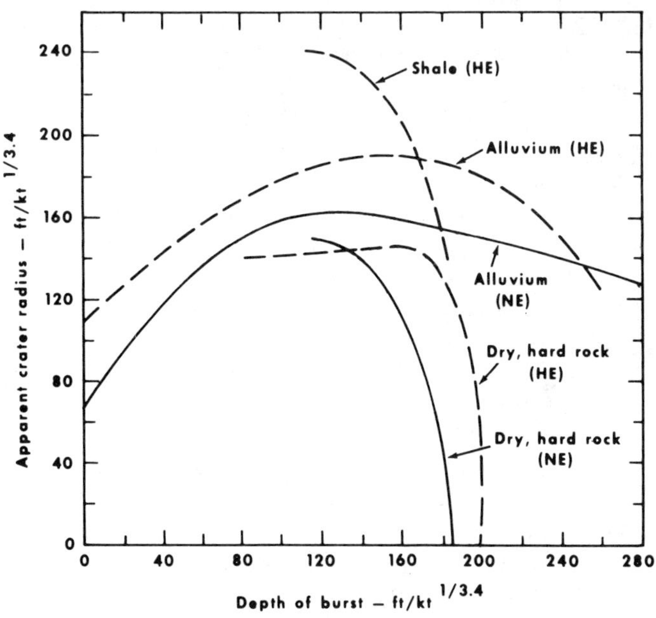

Figure 8. Cratering data, scaled apparent crater radius versus scaled depth of burst. Based on 3.4th root scaling. High-explosive data (HE). Nuclear explosive data (NE). (After Toman, 1969.)

varies with scaled depth (Hays and others, 1970). These data, collected at the Nevada Test Site, other areas in Nevada, Alaska, New Mexico, Colorado, and Mississippi, have been analyzed so that there is a good understanding of how the energy couples with the rock and what important parameters need to be considered when the energy reaches the location of structures. This body of information is certainly sufficient to

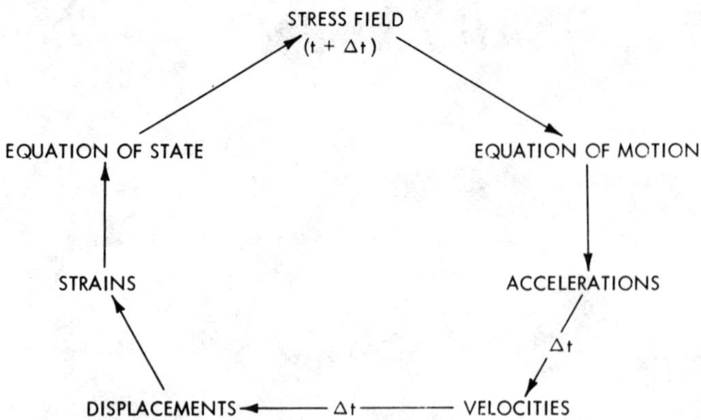

Figure 9. Cycle of interactions treated in calculating stress wave propagation (Cherry, 1966).

design safe nuclear excavation projects, especially if one allows, as we do, a large margin for the uncertainty of our information.

There are three major elements in this picture. They are: (1) the geology in the immediate vicinity of the explosion, (2) the geology along the transmission path of the signal, and (3) the geology at the point of interest. As indicated above, the geologic factors involved in the first and last of these two parts are reasonably well identified. The geology along the transmission path contains the largest uncertainties (Loux, 1970). Here, then, is another field of geological research of great importance. If geologists could translate the inferred geology along the transmission path into the quantitative effect on the seismic signal, and thus significantly reduce this uncertainty, the tolerances presently used in our safety programs might be significantly reduced without in any way degrading safety.

The hydrologic environment, both surface and subsurface, in which a nuclear excavation is made must also be investigated. The operation could then be conducted in such a manner as to ensure that potential radiation exposures through this and other pathways are kept as far below approved guides as is practicable. Fortunately, development of nuclear explosives and our knowledge of their interaction with the hydrologic environment have progressed to such an extent that, in fact, this is a very minor problem. Nevertheless, the geologist's talents will always be required to give assurance that in a specific instance there are no unusual circumstances which have not been considered.

Slope stability is another geologic factor of interest, and in designing cuts made with nuclear explosives it is an important consideration. For some rock types, for example, those which are strong, it appears that slopes of more than 1,000 ft will be stable (as described in IOCS Memo. NCG-34, 1970), although in some instances the geologic structure may play an important role, not only in stability of the slopes but in the shape of the excavation (Talobre, 1957). On the other hand, rocks such as clay shales are notoriously unstable and subject to slope failures when disturbed by excavation of any kind. The stability of nearby natural slopes that are disturbed by the seismic wave is also an important safety consideration.

CONCLUSION

Rapid excavation with nuclear explosions has the potential of raising our standard of living and concurrently improving the quality of our environment. Since such excavation results from the interaction of the explosion with the earth, geology provides an important input to the design of each project. Geologic factors needed to assure that the design of such applications are safe have been identified. Other factors, which will refine our understanding of nuclear excavation phenomena and permit designs with closer tolerances, are still to be identified. Tools are needed to measure and quantify geologic parameters more effectively. The geologist has played and will continue to play an important role in achieving these objectives.

REFERENCES CITED

Boardman, C. R., Rabb, D. D., and McArthur, R. D., 1964, Response of four rock media to contained nuclear explosions: Jour. Geophys. Research, v. 69, p. 3457.

Cameron, I. G., and Scorgie, G. C., 1970, Theoretical model of the early phases of an underground explosion: Peaceful Nuclear Explosions Panel Proc., Vienna, March 2–6, p. 331.

Chabai, A. J., 1965, On scaling dimensions of craters produced by buried explosives: Jour. Geophys. Research, v. 70, p. 5075.

Cherry, J. T., 1966, Computer calculations of explosion-produced craters: Internat. Jour. Rock Mechanics and Mining Sci., v. 4, p. 1.

Cherry, J. T., and Petersen, F. L., 1970, Numerical stimulation of stress wave propagation from underground nuclear explosions: Engineering with Nuclear Explosions Symp., Am. Nuclear Soc. CONF-700101, v. 1, p. 142.

Coffer, H. F., Bray, B. G., and Knutson, C. F., 1964, Applications of nuclear explosives to increase effective well diameters: U.S. Atomic Energy Comm. Rept. TID-7695, Third Plowshare Symp. Proc., p. 269.

Hays, W. W., Mueller, R. A., and Spikes, C. T., Jr., 1970, Analysis of seismic data from cratering and contained events: Environmental Research Corp. Rept. PNE 522.

Higgins, G. H., and Butkovich, T. R., 1967, Effect of water content, yield, medium, and depth of burst on cavity radii: Lawrence Radiation Lab. Rept. UCRL-50203, Livermore, California Univ.

Lampson, C. W., 1946, Explosions in earth, effects of impact and explosion: Office of Scientific Research Report, v. 1, pt. 2, chap. 3.

Loux, P. C., 1970, Seismic motions from Project Rulison: Engineering with Nuclear Explosives Symp., Am. Nuclear Soc. Rept. CONF-700101, v. 2, p. 1070.

Moreau, E., 1970, Methodes numeriques et usage industriel de l'explosif nucleaire: Peaceful Nuclear Explosions Panel Proc., Vienna, March 2–6, p. 327.

Nordyke, M. D., 1961, On cratering—brief history, analysis, and theory of cratering: Lawrence Radiation Lab. Rept. UCRL-6578, Livermore, California Univ.

—— 1962, An analysis of cratering data from desert alluvium: Jour. Geophys. Research, v. 67, p. 1965.

Nordyke, M. D., 1964, Cratering experience with chemical and nuclear explosives: U.S. Atomic Energy Comm., Third Plowshare Symp. Proc., TID-7695, p. 51.

Project Ketch, 1967, A feasibility study on creating natural gas storage with nuclear explosions: Columbia Gas Systems; U.S. Atomic Energy Comm.; U.S. Bur. Mines; and Lawrence Radiation Lab. Rept. PNE-1200.

Rawson, Donald E., 1966, Industrial applications of contained nuclear explosions: Lawrence Radiation Lab. Rept. UCRL-14756, Livermore, California Univ.

Talobre, J., 1957, La méchanique des roches: Dunod, Paris.

Terhune, R. W., Stubbs, T. F., and Cherry, J. T., 1970, Nuclear cratering from a digital computer: Peaceful Nuclear Explosions Panel Proc., Vienna, March 2–6, p. 419.

Toman, J., 1969, Summary of results of cratering experiments: HEW Public Health Service Rept., Public Health Aspects of Peaceful Uses of Nuclear Explosives Symp., p. 48.

MANUSCRIPT RECEIVED BY THE SOCIETY APRIL 5, 1972

Theory of Spacing of Extension Fracture

GEORGE M. SOWERS
Center for Tectonophysics, Texas A & M University, College Station, Texas 77843

ABSTRACT

A new theory of spacing of extension fracture in flattened, or compressed layered rocks, provides a basis for predicting some fracture patterns, and the theory is confirmed by model studies.

Fracture spacing may be controlled by elastic stress concentrations. In compressed layered rocks the initial uniform stress field may become unstable and periodic stress concentrations may develop. This instability is caused by differences in mechanical properties of the layered sequence, causing incremental interfacial stresses as deformation continues. A critical stress is required to initiate the instability. If critical stress exceeds the rock strength, then fracture occurs according to older theory, but loads of equal magnitude can cause elastic instability in bodies of other shapes.

By using strain energy methods a quantitative theory is developed. Work done in lengthening a stiff layer is computed and equated to the elastic strain energy of deformation. This energy is computed for the change in shape associated with the necking of the stiff layer and is then combined with the deformation energy computed for the loading on the restraining medium. When external work and strain energy (as computed above) are equal, the necked shape is favored over the straight form, which is the earliest stable shape in the deformation.

Various combinations of interfacial forces are used to obtain a measure of the effect of friction and welding at layer interfaces. Fracture spacing computed for these cases agrees well with experimental values. This theory may be applied in devising patterns of rock bolting, and perhaps in setting charges in advancing rock faces in rapid excavations.

INTRODUCTION

The instability of layered media conforming to the restrictions of classical elasticity is used to develop an explanation for the spacing of some fractures. The most general development of the theory of instability of interest to our study was given by Biot (1965), who offered an extended development of the mechanics of buckling. Biot investigated the effects of finite deformation for isotropic and anisotropic elastic and viscoelastic materials. For the most part, his theoretical results have not been tested experimentally, partly because the unusual moduli given in his stress-strain relationships have not been measured. Anisotropy is either an intrinsic property or it is induced by prestress. In the latter case the moduli are functions of the external load and, therefore, in all but rather brittle deformations, are subject to the conditions of the experiment. A materials testing program of this complexity should not be made without first showing need, which we feel must appear in a simpler study based on the theory of elasticity. Nevertheless, we will review aspects of Biot's work because of the insight it provides into the problem of instability and because it probably will be used in later work. In addition, some of the important generalizations made by Biot have not been widely circulated among structural geologists.

A theory of unstable elastic deformation of stretched layered rocks is developed to explain the occurrence of some pinch and swell structures and to account for spacing between some regularly spaced natural fractures. Biot applied his theory to buckling problems, but with minor changes it can be adapted to pinching or necking instabilities. Similar theoretical results can be obtained in a different way by using structural mechanics and strain energy methods. A practical method for computing approximate values of critical stress and fracture spacing is developed.

This theoretical study may be important in engineering geology. In all underground excavations practical use of the inherent anisotropic weakness of rocks is used. Application of this theory adds another way to make effective use of these weaknesses. Rapid excavation can be accomplished more effectively if use is made of a fracture, both existing and incipient. Knowledge of the conditions causing extension fracture in layered media can be used effectively in planning rapid excavations, and it then becomes possible to predict just where new fractures can be expected as a result of the change in stress caused by the excavation. This predictive capability may be useful in designing support for the excavation, because the theory predicts places where "heavy ground" will be caused by the fractures due to the opening. In the past it was necessary to judge this on the basis of existing fracture. In other words, this theory increases the factors to be considered in all aspects concerning the stability of the excavation.

Objectives

This study was undertaken to: (1) provide a better mechanical explanation for the distribution of certain fractures in heterogeneously layered rocks so deformed that a maximum axis of extension lies in the plane of bedding; (2) develop approximate theory that will permit the computation of critical loads for the initiation of instability and fracture of rocks in engineering activities such as tunneling; (3) provide an approximate basis for the computation of loads assumed to be the cause of some geologic fractures, where natural load conditions are as in (1), above; (4) make experimental tests of the theory by photoelastic modeling methods; (5) map fractures in suitable geologic structures and to apply these principles to provide an explanation for the distributions of the natural fractures; and (6) to provide a wider basis for the use of existing and incipient fractures that might be used in connection with the engineering of underground excavations. If the proposed theory is correct, the size, configuration, and spacing of some fractures related to the cavity are predicted. Furthermore, the approximate critical loading for the instability may be calculated.

PREVIOUS WORK

The theory of fracture spacing, not widely studied, has been developed largely by geologists seeking an explanation for the regular arrays of fractures they have mapped. My theory of fracture spacing is concerned with the elastic stability of layered media undergoing longitudinal extension and, specifically, with periodic stress concentrations caused by this instability which serve to localize fractures. There is a body of knowledge from physics that is pertinent to this problem which should be reviewed briefly. Direct application of this knowledge in geologic work is risky for several reasons. Rocks are anisotropic, and are often layered media that have been deformed, usually under high confining pressure over long periods of time and under loads that are complex at any instant, and will sometimes change drastically during the course of deformation. Moreover, sedimentary rocks may be regarded as plates that extend laterally to infinity, rather than small isolated test pieces. Nevertheless, the fundamental concepts which define stability are valid and useful when they are appropriately modified for geologic applications.

Results from Tensile Tests

While it is common knowledge that longitudinally loaded layers become unstable and buckle under compression, it may seem unexpected that loss of stability also occurs in extension. However, a familiar example is necking of rods and plates in ductile flow, which is related to the development of a geometric instability as the shape of the structure changes during stretching.

In tensile tests, rods and plates are drawn out longitudinally by forces applied at the clamped ends, or by confining pressure applied laterally to the sides in a pressure vessel while allowing the ends to extend freely. As load is increased, the bar first deforms elastically, but then, at the elastic limit, there are several alternatives: (1) the material ruptures on a brittle extension fracture; (2) the structure may draw out into a neck and then fracture on either ductile fractures or brittle fractures or some combination of both. The neck may occur without a preceding change in the geometry of the stress trajectories, or it may occur only as a result of instability of the elastic field. Finally, the influence of viscous behavior on this geometric instability must be considered.

Bridgman (1952) has discussed the deformations of rods and plates in high-pressure extension tests. He noted the following stages in refining his measurement of the "tensile strength" of rods. At first, strength was defined by dividing the maximum force of the force-displacement curve by the original cross sectional area of the test piece. Any necking was ignored. Later, a better approximation takes into account the change in cross sectional area, so that "true stress" is determined for "natural strains." The definition of strength becomes arbitrary and might be taken as equivalent to a state of unstable equilibrium.

Bridgman has shown that while the *applied* load declines after the onset of necking, the *true stress* continues to increase in the neck as its cross sectional area diminishes. The tensile strength defined as the quotient of applied load over true stress is always higher, especially if pronounced strain-hardening occurs. In this approximation it is assumed that only an axial stress component exists, which varies in magnitude only along the axis.

However, his observation of fractures in necked specimens convinced Bridgman that a third refinement, a complete stress analysis of the necked-down area, was necessary. Often, brittle extension fracture occurs first on the axis at the center of the neck, and this implies a nonuniform stress distribution not previously accounted for. Bridgman noted that the outer shape of the neck seemed to resemble an "infolding." Perhaps he meant that a symmetric instability occurs, but in any case he implies that necking is not the same as notching or the stable growth of a thickness variation. Bridgman did not consider the exact localization of the neck, nor did he consider the stability of the elastic stress field, but he speaks of the formation of the necked-down area as an instability.

Bridgman's solution for the plastic deformation is constructed from the equilibrium equations, an experimentally determined set of stress-strain relations, and the average Mises yield condition. Unfortunately, there is no way to measure the precise local variations in stress-strain relations that occur in the necked region as the test progresses. Bridgman does not analyze the mechanism by which the stress field changes from simple tension to the complex field. He defines the stress field as an axial stress consisting of an axial tension and a superposed hydrostatic load that varies along the radius of the neck. This radial variation of the axial load is related to the radius of curvature of the neck, and the maximum value of σ_2, the principal stress parallel to the axis, is at the center of the neck; the minimum value is at the surface, according to Bridgman. Even though the stress varies across the section, the plastic strain is constant. This axial stress distribution differs from that predicted by Neuber's (1933) elastic analysis of the different but related problem of elastic stresses in notched specimens. Stresses so obtained invariably give the larger values for σ_2 on the surface. Had Bridgman chosen a more appropriate elastic solution for rods not already notched in the ground state, but rather with infolded surfaces (Timoshenko and Goodier, 1970, p. 55, 61) superposed on the applied tension, he would have found a stress distribution that still qualitatively resembled the one he had obtained. He did not answer the question of the relation between elastic stress distributions and plastic necking.

In summary, Bridgman has shown that in rods stretched to the point of unstable plastic necking, the true axial stress continues to rise even after the maximum load is applied. His analysis of the stress field in the vicinity of the neck is not given in a way that can be used to investigate the stability of the rod. From the experimental results, he gives a final stress field which differs drastically from the initial one. Moreover, Bridgman's solution applies to short test cylinders; the stability of the cylinder as a whole was not studied. We are left with the problem of whether an instability in the stress field caused the neck to form or if it developed slowly in response to gradual stable change.

Nadai (1950, p. 70–88) has agreed in part with Bridgman. He noted, however, that as soon as necking starts, the test piece ceases to stretch uniformly, and the stress field becomes nonuniform. A principal stress axis is no longer uniformly parallel to the axis of the rod. These views, which may have been drawn from the work of Davidenkov and Spiridonova, emphasize the instability of the stress field. The stress variation is correlated with the flow from point to point. As flow increases and becomes finite, the stresses and strain become nonlinear.

Bridgman (1952) also obtained empirical formulas for the axial stresses in plates subjected to extension. These results illustrate the influence of the shape of a member of its stability. The longitudinal stress acting perpendicular to the cross section at the neck is equal to the tensile stress at the surface plus a term related to the radius of curvature of the neck (just as in rods). Consistent results from repeated experiments were obtained when the plate was brittle, but when ductile flow occurred, fracture patterns were not predictable, even empirically.

McClintock and Argon (1966) have given the reasons for this inconsistency. The ductile neck does not and cannot form at 90° to the axis of the plate. It forms such that the long direction of the neck makes an angle of 53° with the axis of tension on the flat surface, which is just the direction taken by Luders lines. Thus, the mechanism of necking in flat plates is "exactly the mechanism of Luders line formation." The neck does not form across the plate because of the restraint imposed by the condition of plane strain. If the neck were to form exactly across the long axis, it would require that the tensile stress be greater by $\sqrt{3/2}$ than the stress computed from empirical results, an impossible condition.

An exact solution for the distribution of plastic stresses and strains in the necked rods or plate has been (and is) difficult to obtain. Hill (1950, p. 273) accepted the empirical approximation obtained by Bridgman. So did McClintock and Argon (1966, p. 325), although they referred to several works that questioned the validity of Bridgman's results. In any event, in all these accounts it is assumed that the elastic stress field remains stable prior to the onset of the plastic deformation.

Panovko and Gubanova (1965) have studied the elastic stability of rods in extension. They have shown that the tensile load-displacement curve rises in some cases monotonically to a peak load, the critical load, after which it slowly declines to near-zero values for large displacements. At the critical load, the rod develops a geometric instability. (The rod is assumed to have "infinite" strength or to be infinitely deformable.) For large strains the incremental stress in the axial direction changes inversely with the square of the cross sectional area. The stress changes because the applied load increases while the area on which it acts decreases. The load, P, is

$$P = \frac{E \cdot A_o}{1 + \frac{\delta\ell}{\ell_o}} \cdot \log_\eta \left(1 + \frac{\delta\ell}{\ell_o}\right),$$

where A_o is original cross sectional area, ℓ_o is original length, $\delta\ell$ is increase in length, and E is Young's modulus. For stationary values, that is, when $dP/d\ell = 0$, the critical load is EA_o/e, ($e = 2.71$). This nonlinear elastic response applies to the deformation of the whole body, and it occurs even in rubbery materials. The same qualitative phenomena are supposed to occur for nonlinear deformation, $\sigma = \sigma(\epsilon)$, as well.

When the load is applied in steps of increasing constant force, there is a sudden loss of stability at the critical load, but when loading consists of steps of constant displacement, every point of the force-displacement curve is successively attained with no sudden loss of stability. Under these conditions the cylindrical shape of the rod loses its stability as soon as the maximum point on the P-δ curve is reached.

Consider a slight variation in original cross section where the critical load will be

$$P_{cr} = \frac{E(A_0 - \Delta A_0)}{e}$$

The rod becomes unstable at this point first, and this causes a local increase in length. Elsewhere, stress and force both decrease, while in the unstable region the stress may actually increase if the cross section shrinks quickly. Any small departure from the initial shape of the rod is enough to produce a large amount of necking-down in a small part of the rod.

Because of the formal similarity between viscous and elastic deformation, one might suppose that geometric instability could occur in stretched viscous rods. McClintock and Argon (1966) have noted from experiments that the tendency toward necking is less in materials that creep, and that necking does not occur at all in substances of linear viscosity. The *ratio* of initial variations in rod thickness remains constant even for large extensions, as in stretched stringers of honey in which no neck develops. The difference is that no critical load or peak on the force-displacement curve occurs in the linear viscous media. Fletcher (1970, personal commun.) has shown by similar arguments that this sort of geometric instability does not occur in layered viscous media either, but this is not to say that viscous rods cannot develop other types of instability, as will be shown later.

Drucker (1968), in the encyclopedic *Treatise on Fracture*, reviewed the mechanisms of brittle fracture. He emphasized the separate roles of necking and of notching in localizing fracture. According to Drucker's review, the occurrence and propagation of any single fracture is a complex process dependent on the nature of the material, residual stress, previous deformation history and other environmental factors, as well as deforming load. Necking instability is dependent on geometric properties of the structure, as well as the above factors. Drucker attributes necking to the well known and generally accepted mechanism of hole formation and coalescence in brittle deformation, and to gliding or shearing off in ductile deformation. He emphasizes the continued need for a continuum mechanics solution to the problem of fracture initiation and states that the elastic response controls local strain to a considerable extent, and also controls the relative crack opening and inelastic zones in brittle fracture in both notched and necked specimens. Finally, he notes the importance of instability, either plastic or elastic, or both, in localizing fracture. Usually necking is associated with plastic deformation and, as shown by Cohen and Vukcevich (1969), does not occur in metals where elongations are less than 20 percent.

In summary, the details of the physical process of necking are still not determined. The macroscopic process is known to be complex and, although generalized computations predicting the deformation of a tensile specimen can be made, exact solutions for specific problems are not yet available. In the case of brittle fracture the controlling role of elastic deformation is emphasized by Drucker (1968). Panovko and Gubanova (1965) suggest that necking may be initiated by an elastic instability. Where deformation has involved large flow over large parts of a structure, necking and fracture are controlled by inelastic behavior.

Finally, the word "stability" has many meanings. Fung (1965) points out that the loss of uniqueness of solution characterizes many practical stability problems. This can be due to loss of positive definiteness of the strain energy function, or to changes in the equations of equilibrium introduced by finite deformation. The first possibility arises when the *material* becomes unstable as yielding occurs. It is relevant to the *plastic* necking of rods. The second possibility may arise in several ways, according to Fung. Most important is finite deformation. Problems in elastic buckling or necking can be understood when we realize the basic changes in the equations of equilibrium introduced by finite deformation. We will show this in connection with Biot's (1965) theory of incremental deformation as used in developing an explanation for fracture spacing. Multiple solutions (nonlinearity) arise because strains and rotations depend on the stresses. The second possibility gives rise to instabilities which are dependent on the geometry of the deformation or structure. When necking or buckling occurs it may be due to the shape (that is, dimensions) of the structure and not involve material instability.

Drucker (1968) illustrates the stability concept by referring to the stress-strain curve or the load deflection relationship. Fung's first possibility, listed above, may be referred to the "microscale," that is, to stresses in local regions of a larger body, while the latter reference is made to changes in the stress field of an entire structure. Stability is achieved when either curve continues to rise during deformation.

Geologic Studies

The mechanics of extended layered media have been of concern to structural geologists who are interested in boudinage and in the spacing of extension fractures. These problems are very different from that of the tensile test on a single rod or plate. The size and shape of each geologic structural element favors instability. Layers may extend virtually to infinity in the plane of layering, which is also the plane of extension. The ratio of length to thickness is large. As deformation proceeds, loads appear at interfaces between layers of different mechanical properties. These forces induced during deformation may, when superposed on external loads, cause necking or buckling instabilities. Stress and strain fields in brittle layers, which are embedded in softer ones, may become unstable.

In my study, the extension of layered sedimentary rocks by distributed compressive load acting perpendicular to bedding is an essential condition. The end loads may be greater than, less than, or equal to this uniformly distributed load. These end loads are difficult to specify on an unfractured bed because it is essentially a semi-infinite plate. After the bed has been broken and fragmented, stress components may be transmitted by interfacial friction. Finding a realistic way to specify interfacial forces has been an important objective because they may ultimately determine fracture and boudin spacings.

Griggs and Handin (1960) extended concentric rock cylinders under high pressure to simulate boudinage. In these tests the inner core was relatively more brittle (dolomite, for example) than the enclosing rock (limestone, for example). The ends of the composite specimen were restrained so that *total* extension of both cylinders was the same, although the longitudinal strain varied widely in different parts of the specimen. The inner core was ruptured in one or more places, and individual pieces were engulfed by the influx of the ductile matrix into the cracks. Sometimes the inner core fractured into several more or less equant blocks (Fig. 1) along either shear (ductile) or extension (brittle) fractures that usually terminated against the outer cylinder without crossing it. Fractures often opened, and the ductile material penetrated the openings. As the tests were done on dry rocks, fracture formation could not involve any hydraulic mechanisms. When over-all deformation was more nearly brittle, the shape of the longitudinal cross section was nearly constant. Even after large flow, the outer walls were drawn in along a smooth curve despite the large inflow between boudins. Even where deformation was brittle, more or less rectangular blocks of uniform length were formed, within which permanent deformation was negligible and lengthening took place by separations of these blocks.

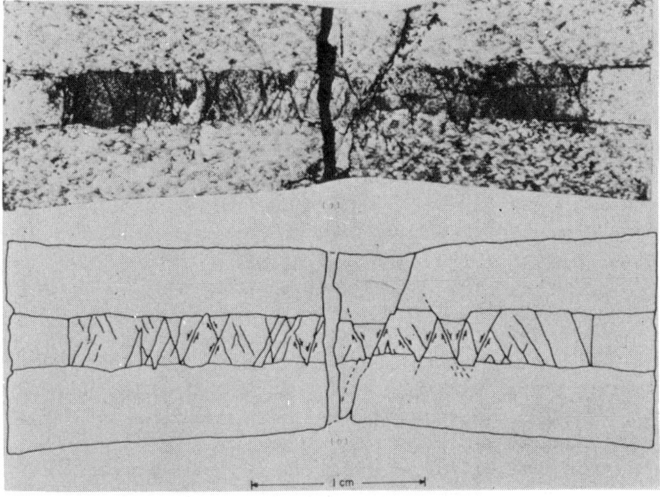

Figure 1. Triaxial extension test, outer core limestone, inner core dolomite (after Griggs and Handin, 1960).

These experiments closely simulate some, but not all, geologic conditions. There are also over-all lengthening, differential internal displacement, and complex internal stress conditions. Apparently, fracture blocks and boudins are formed by a similar arrangement of forces, but in materials of widely different ductility. The exact analysis of the two geologic features must differ. Where the deformation is essentially brittle (less than 2 percent permanent strain), elastic analysis should adequately describe the phenomenon.

Nonetheless, the close relations between the formation of fractures and boudins require consideration of all significant work on both problems. In the late 1930s, Goguel (1946) first applied mechanical principles to the problem of boudinage. From geological observation, he decided that the plastic flow of weaker members of a geologic sequence controlled the deformation of the whole sequence, including the markedly stronger layers. He used Prandtl's solution for the plastic flow between two appressed rigid platens to obtain the displacement and flow velocity fields that might develop in the plastic layers. He postulated that strong and brittle layers were first fractured into large blocks by regional longitudinal extension, and the ductile material could then flow into the voids. The large blocks of adjacent layers were supposed to assume the role of the rigid platens of Prandtl's model that were displaced toward one another in a direction normal to bedding with uniform velocity and displacement, and that the strong layer was acted on by tensile forces at the interfaces. These interfacial forces were supposed to be caused by viscous flow of the weak material, and were computed from well-known formulas for viscous drag. By computing strain energy Goguel was able to approximate the lengths of boudins.

In the late 1940s, Russian geologists became interested in the problem of fracture spacing. Bogdanov (1947), Kirillova (1949), and Chertkova (1950), in association with Beloussov (1952), measured fractures in the field, made model studies, and proposed a qualitative physical explanation for fracture spacing which is widely held today. Their field measurements show that fracture spacing in brittle beds of alternating brittle and ductile layers depends linearly on the brittle-layer thickness. Some fractures are wholly contained in the brittle layers, and all fractures studied are normal to bedding, although in some places layers are separated into boudins. Models were built of petrolatum and "gun grease" which was "shot" into or between compressed layers. Compression varied in different parts of the model and allowed the less "viscous" layers to flow from places of high to low pressure. Viscous flow in the ductile layers produced local tensile stresses of sufficient magnitude to break the brittle layers. Shorter boudins occurred where the compression across the layer was the larger (by amounts which were proportional to the reduction in thickness of the more viscous layer).

In later papers, Beloussov (1964) stated that tensile fractures formed by a "plastic" process of necking in the less ductile layers. He correctly assumes that the more ductile material will have variable flow velocities, as shown in Figure 2. However, his qualitative statement on the development of interfacial tensile stresses is erroneous for truly plastic materials. Frictional force is said to be greatest where flow velocity is highest, but solid friction is largely independent of velocity, according to Amonton's Law and as recently reaffirmed by Bowden and Tabor (1954). This point is important because it is necessary to the argument that variable tangential stresses occur along the interfaces, and hence to Beloussov's explanation of fracture spacing. Furthermore, it seems probable that plastic, rather than viscous, flow has occurred in the geologic model he described. If deformation is indeed plastic, the interfacial stresses can only reach the maximum set by the yield value, equal to the friction on the interface which is independent of flow velocity. Apparently this problem has troubled Beloussov, for sometimes in geologic papers he refers to "plastic flow" (1964) and at other times to "viscous flow" (Beloussov and Gzovsky, 1967) in reference to the same case.

Ramberg (1955), Gzovsky (1960), and Voight (1965) have all given quantitative solutions for the lengths of fracture blocks in a viscous regime. Ramberg (1955) made models of clay, putty, and cheese, which were compressed perpendicular to the layering. He created boudins which strikingly resemble natural ones. In his qualitative discussion of the physics of his experiments, he identifies the ductile flow as *plastic* and proposes a scheme (Fig. 3) that correctly describes the formation of boudins in relatively ductile materials. He assumes an initial fracture followed by extension, which pulls apart blocks of harder layers of finite length and provides sinks for viscous infilling. Ramberg supposes that forces of variable intensity are generated on the surfaces of the hard layers. He attributes them to *viscous* drag, which he relates to variation in flow velocity.

In the scheme he used for computation, he assumed an initial tension and applied uniform compression across entire interfacial surfaces. Viscous interlayers may flow out the unrestricted ends of the model or into internal sinks created

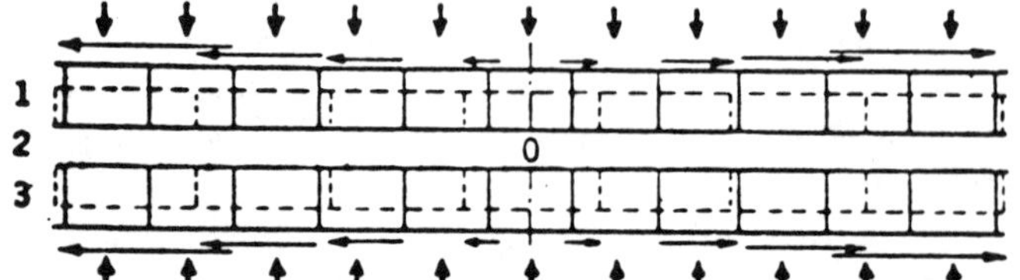

Figure 2. Flow velocity distribution. Arrows show distribution of components of viscous flow velocity, which are proportional to drag forces on interfaces. Solid rectangles are deformed to dotted ones (after Beloussov, 1964).

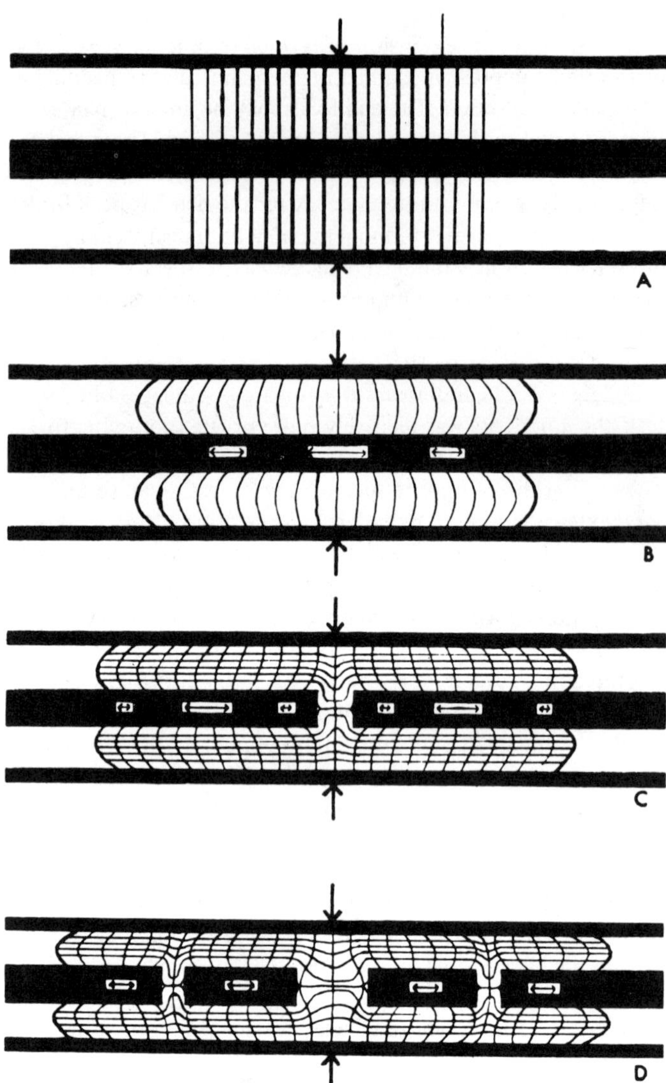

Figure 3. Successive steps during formation of boudinage structures. A. State prior to compression. The black layer in the middle is the competent body on either side of which incompetent bodies exist (vertically lined regions). Compressive stress is transmitted by means of two stiff lined regions. Compressive stress is transmitted by means of two stiff sheets (black). B. Compression has started, the plastic flowage in the incompetent layers is indicated by the distortion of the originally vertical lines. Arrows in the competent black layer indicate tensile stress. C. A more advanced step than B; competent layer is now ruptured in the middle where tension was greatest. The network in the incompetent layer indicates pattern of flowage. This network is not a further evolution of the deformed network in B, but rather the evolution of an imaginary rectangular network not shown in B. Horizontal arrows show tensile stresses in the competent layer. D. A stage more advanced than C; competent layer has ruptured at two new places. The pattern of the network indicates plastic flowage in the incompetent layers during evolution from C to D. Again, the deformed network in D is supposed to have developed from a rectangular imaginary network not shown in C.

by separation of brittle layers. Rapid flow of low-viscosity layers exerts lateral forces on the interfacial contacts and causes viscous drag whose magnitude depends on the thicknesses and viscosities of the less viscous layers. Viscous flow from between appressed rigid platens has been solved exactly for Newtonian fluids of relatively small viscosity (Nadai, 1950, p. 349–352), but not for slow viscous flow as in rocks. In his computed solution, flow is symmetrical from the center of the low-viscosity layers at the rigid platens that move toward one another at uniform velocity. The velocity of flow is greatest at the center of the outer ends of the viscous layers. Shearing stress on the interfaces increases linearly from the center to the ends. Integration of these stresses would provide the tensile force that could cause additional cracks to form in the brittle layer. As these layers are embedded in low-viscosity material which does not fracture, fracture of brittle layers gives a temporary stress drop. Stress within brittle fragments depends on interfacial conditions but does not drop to zero. Fracture of thicker layers requires proportionally larger forces. The maximum block length is determined by the thickness of the interlayer assuming surface forces remain the same as before.

Not all extension fractures can be explained by this mechanism. In the experiments of Griggs and Handin (1960), little or no viscous flow occurred, yet fractures and boudins were formed. The high-viscosity ductile rocks were deformed at such high rates that the forces generated by such viscous flows would not be possible. Probably, in Ramberg's clay experiments, and certainly in similar experiments done in my laboratory, the ductile flow is plastic rather than viscous, and stress distributions cannot be analyzed by the laws of viscous flow. Ductile flow is plastic because areas of plastic flow are separated from elastic fields by discontinuity in the displacement field. Moreover, in the specimens of Griggs and Handin, neither the amount of flow nor the flow field matches that required by viscous flow. There are discontinuities in the displacement field, and even the outer shape of the deformed cylinder near necked areas fits plastic-flow patterns. Finally, Ramberg's assumption that the "platens" remain straight and rigid up to the moment of necking is not always valid in natural rock deformation.

Some approximation is needed if calculation is to be made at all. Adoption of the simple viscous model leads to large errors in computations of viscous drag forces. If natural flow is indeed viscous, it is very slow and the coefficient of viscosity is very large. Therefore, the drag forces computed by Ramberg become unrealistic for thick viscous layers.

If one accepts these uncertainties, quantitative relations developed by Ramberg can be used. The rate of deformation,

$$\frac{\delta z}{\delta t} = \frac{Z_i^3}{\ell^2 6\eta} (P_o - P_\ell),$$

is related to the cube of the thickness, Z_i, of the incompetent bed and the inverse of the square of the length, ℓ, of the block of rigid material, times the viscosity, η. It is directly related to the difference in pressure, P_o, at the center of the block, and that, P_ℓ, at the end.

This pressure gradient is related to the thickness ratio,

$$\left(\frac{Z_i}{Z_c}\right) S_s > (P_o - P_\ell) > 1/2 \left(\frac{Z_c}{Z_i}\right) S_s,$$

and the strength of the "competent" layer, S_s. Z_c is the thickness of the competent layers.

The length of each block is related to the drag force, and is a function of flow velocity, which varies in several ways. At a given point the velocity depends on the applied load rate, on the thickness of the ductile layer, and on the location relative to the sink. Flow rate can obviously change during deformation. Uniform block length suggests that stresses must have attained a uniform maximum value at all points. In Ramberg's theoretical development, cracks supposedly form in sequential stages and increase in number at each step in a geometric progression as the number of local sinks increases.

In this development no regard is given to the possibility that blocks can become unstable as structural units so that viscous flow could become nonuniform before fracture.

Gzovsky (1960) modified Ramberg's formula and writes for the length, L, of the block,

$$L \approx K \frac{S_s + (\sigma_{iz})_o}{|\text{grad } P|} \frac{Z_c}{Z_i},$$

where $(\sigma_{iz})_o$ is the "value of the reactive compressive strength along the more viscous layers," grad P is the "gradient of the pressure normal to layering," and K is an empirical constant to be determined in models. Block dimensions are proportional to the change in the thickness of the more viscous layer.

Price (1966) has supposed that fracture frequency is related to stored elastic strain energy, and he has applied this concept to the development of fractures in sandstones and coal beds. As these rocks were unloaded by uplift and erosion, stored energy was released and used to form new surfaces. Other things being constant, the tensile surface load required to reach fracture strength depends on the amount of new surface area to be created. Obviously, this depends in turn on the thickness of the layer so that surface forces must be integrated over longer blocks in thick beds.

Hobbs (1967) gives a more elaborate theory of fracture spacing that applies to elastically deformed layers. His theory is based on the explanation of Cox, whom he quotes, for the deformations of fibrous aggregates. Hobbs assumes that rocks are elastic layers held together by both cohesive forces and the weight of the overburden. He supposes that tensile stresses are caused by uplift accompanied by lateral strain relief. As lateral stress increases in a layer, a joint forms at its weakest point. The effect of the fracture is to increase the stress in neighboring beds, and then, because of differences in elastic properties, different displacements occur in adjacent rock layers, and shear stresses in each layer differ. A relation between the gradient of the tensile stress and the differential displacement is first developed. Finally an expression for the tensile stress, P, is obtained,

$$P = E\,t\,e\,(1 + A \sinh Dx + B \cosh Dx),$$

where

$$D = (\frac{2C}{Et}).$$

In these equations Young's modulus, E, layer thickness, t, and strain, e, in the layer are proportional to P; and C, A, and B are coefficients related to shear stress and relative displacements and are determined empirically.

For fracturing subsequent to the initial fracture, Hobbs finds the block length, ℓ_2, to be

$$\ell_2 = t\,(\frac{E}{G_n})^{1/2} \cosh^{-1} (\frac{e_2}{e_2 - e_1}),$$

where G_n is the shear strength in the neighboring beds, and e_1, e_2 are the longitudinal strains in the softer and harder layers, respectively. Spacing depends directly on bedding thickness, and the block length is smaller as the strain e_2 increases in the neighboring beds. This explanation may account for fracture spacing in rocks if an instability cannot develop.

Secor (1968) suggests that many extension fractures are caused by the mechanism of hydraulic intrusion. The volume of pore fluid available, and the rate at which it can be delivered to various parts of a porous bed from a less permeable reservoir, might then determine fracture spacing.

In summary, the previous work explains some features of fracture of the spacing of extension fractures or boudins in extended, layered rocks; but in some cases, especially where the permanent deformations are very small in both "competent" and "incompetent" beds, older theories which require large viscous flow fail. We will show by studying the stability of layered media undergoing extension that an alternate theory can explain fracture spacing without requiring large viscous flow.

THEORY OF FRACTURE SPACING

Introduction

In this section a theory is developed for fracture spacing in stretched layered rocks. A basic assumption is that fractures start at sites of high stress difference in a nonuniform stress field. Such points are located by developing equations for the static stress field by well-known methods. In geology, the choice of stress-strain equations is always difficult and at best can only be approximated. We have chosen to define the problem as an elastic one for two reasons: (1) because the particular field examples studied involve very little ductile flow, and (2) because it is more convenient. In sequences where all layers are ductile, the theory can be used directly by applying the correspondence principle, if deformation is visco-elastic. Possibly the solutions can be used as a guide to the geometry of plastically deformed sequences, as suggested by Drucker (1968).

Description of the way instabilities arise can be best appreciated by using the incremental-deformation theory of Biot (1965). His theory provides a simple way to understand the role of initial tension and anisotropy in the physics of the deformation of layers. An alternative derivation of the stress fields associated with the necking instability is developed by using strain energy methods. This derivation was developed independently of Biot's theory before the relationship between the two methods was realized. Later, calculations will be made for a much reduced and simplified case by the methods of structural mechanics.

The older theories of Ramberg (1955) and Beloussov (1964), and even those who have taken instability into account in various ways, differ from mine in several fundamentals. For the first time in developing an explanation of fracture spacing, a variation is made in the shape of the entire "brittle" member (as opposed to the variation of a diameter of a rod). Instability is introduced by means of a sinusoidal perturbation of the interfacial forces. Such perturbations may result in various unstable configurations similar to those which Biot (1965) has described in buckling. An internal instability can occur where the shape of "brittle" layer changes, but the outer margins of thick embedding ductile layers do not change. Under other conditions, the outer layers may change into shapes that are more or less concentric to the inner ones (Fig. 4). Biot points out still another possibility, which is that the interfaces become unstable and wrinkle into folds that die out exponentially away from the interface.

Theory of Internal Instability

Biot (1965, p. 182–259) has discussed the compressive buckling of layered media. The following general theory of tensional instability is substantially the same, although the

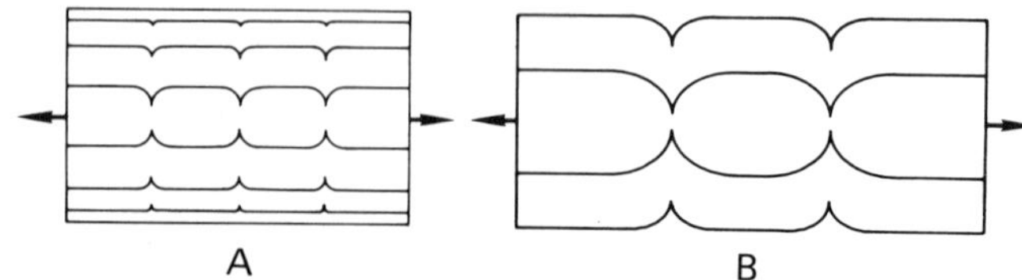

Figure 4. A. Internal instability of layered media. Under action of external tension outer margins of rectangular domain are deformed to produce longer, narrower rectangular body shown. Inside the body the central stiff layer has deformed into a pinch and swell structure because of the growth of a sinusoidal perturbation. B. Necking instability of entire domain. Top and bottom soft layer as well as inner stiff layers are deformed.

difference in sign of the axial prestress alters the effect of moduli in some important relationships which determine stability. Internal instability occurs in either anisotropic or layered media; moreover, in the latter case, layers must be alternately elastically hard and soft but not of too large contrast.

In addition, if layers are thin relative to their length the layered material may be considered a continuous anisotropic material in approximate theory. The basic theory is greatly simplified by using this assumption, but as shown by Biot (1965, p. 216), it is not a necessary one. It is assumed here that the material has orthotropic or orthorhombic symmetry: one axis is normal to "bedding," and the others lie in the bedding plane.

We consider a case of plane deformation in which the ends of the layers are subjected to uniform tensile stress. In theory, at least, this is equivalent to uniform pressure perpendicular to the layers, after superposition of isotropic compression. The upper and lower surfaces are confined between effectively rigid walls but in perfectly free contact with them (Fig. 4). The reference, load, and symmetry axes are taken as coaxial. Biot further simplifies the problem by assuming homogeneity and incompressibility of the entire domain. Individual layers may be either isotropic or orthotropic. Instability develops if the material is isotropic to prestress, but it must be anisotropic to incremental stresses. Otherwise the behavior would be that of a stable isotropic material.

Under these conditions internal instability will occur for a certain range of values of the prestress. These can be found by investigating the stability of a system of equations of incremental elasticity. The possibility of unstable solutions arises because of the prestress and the anisotropic properties of the domain. In the following equations, s_{ij} are the incremental components of stress, and T is the tensile prestress. For this particular case we adapt from Biot (1965, p. 192) the equilibrium equations,

$$\frac{\partial s_{11}}{\partial x} + \frac{\partial s_{12}}{\partial y} + \frac{T\partial\omega}{\partial y} = 0,$$

$$\frac{\partial s_{12}}{\partial x} + \frac{\partial s_{22}}{\partial y} + \frac{T\partial\omega}{\partial x} = 0, \qquad (1)$$

$$T = S_{11} - S_{22}.$$

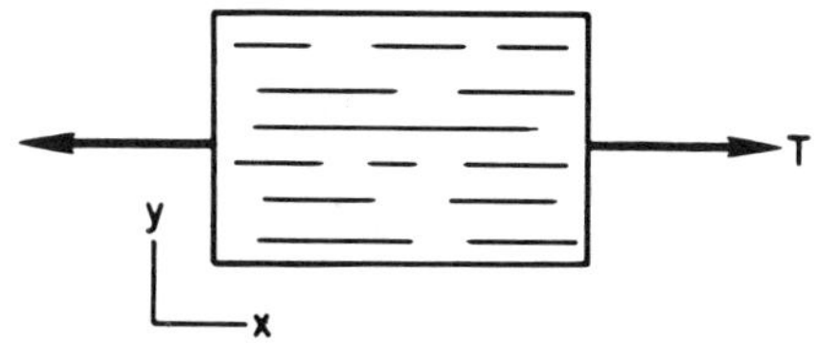

Figure 5. Laminated medium under tension, T: reference axes, x and y, are also axes of orthotropic symmetry.
$T = S_{22} - S_{11}$
T = applied tension; S_{ij} prestress components.

These equations show that initial stress or prestress is a tension acting in the x direction. The angle of rigid rotation, ω, is the amount of rotation which occurred during deformation from ground state to the state of critical loading. These equations differ from the usual equations of infinitesimal analysis in the following ways: (1) the s_{ij} are components of the *incremental stress*, and (2) the terms $T\frac{\partial\omega}{\partial y}$ and $T\frac{\partial\omega}{\partial x}$ are related to the prestress S_{ij}. Perturbations and incremental stresses are treated by infinitesimal theory while the prestresses use finite deformation theory (Fig. 5).

The two different sets of stress components arise from the fact that as deformation proceeds from initial to final state, the body changes shape as do its initial reference axes, as surely as if they were attached at the outset. Stresses are defined in terms of load conditions related to the original shape and axes. Strains are measured after deformation but referred to orthogonal axes. So long as only linear terms are kept in, say, a Taylor series expansion of the deformation about a point, the initial and final axes can be taken as identical without serious discrepancy as, for example, with incremental stresses. But for large deformations, both stress and strain must be referred to the same axes. Biot considers a special finite deformation which allows large rigid rotations but neglects second-order terms in pure deformations. By this simplification he reduces a difficult problem to manageable proportions for certain types of deformations. In order that these equations can be used, there must be some way to measure the actual rigid body rotation of the coordinate axes and it must be shown that the prestrains are small relative to rotation.

Besides the equilibrium equation (1), we need stress-strain relations which take a somewhat different form for these problems of the finite deformation of anisotropic material. They are

$$s_{11} - s = 2Ne_{xx}$$

$$s_{22} - s = 2Ne_{yy} \qquad (2)$$

$$s_{12} - 2Qe_{xy}$$

where s is mean incremental stress, and N and Q are elastic moduli for anisotropic materials. In addition, N is a composite modulus correctly proportioned to represent the discrete properties of the separate layers as a uniform continuum. N and Q will include a factor relating to prestress, if it induces anisotropy. This is because anisotropy may be: (1) an intrinsic physical property, (2) induced in isotropic materials by prestress, or (3) a combination of both. In practice, this effect appears in the moduli of the stress-strain relations, which are composed of two parts: one dependent on intrinsic anisotropy and the other a function of the prestress.

The condition of incompressibility

$$e_{xx} + e_{yy} = 0 \qquad (3)$$

completes the system of elastic equations except for the boundary conditions. We consider boundary conditions which are derivable from a potential ϕ, then by well-established analytical principles (that is,

$$-\frac{\partial u}{\partial x} = +\frac{\partial v}{\partial y};$$

then let

$$\phi = f(x,y);\ -\frac{\partial \phi}{\partial y} = u;\ \frac{\partial \phi}{\partial x} = v, \text{ and } \frac{\partial^2 \phi}{\partial x \partial y} = \frac{\partial^2 \phi}{\partial x \partial y}$$

as required for)

$$u = -\frac{\partial \phi}{\partial y},\quad v = \frac{\partial \phi}{\partial x}, \tag{4}$$

is a condition imposed by (3), and the potential function, ϕ, so defined can be an Airy function. By expressing the strain in terms of ϕ and using the stress-strain equations (2) and then the equilibrium equations (1), and by using these to eliminate s, we obtain

$$(Q + \frac{T}{2}) \frac{\partial^4 \phi}{\partial x^4} + 2(2N-Q) \frac{\partial^4 \phi}{\partial^2 \partial y^2} + (Q-\frac{T}{2}) \frac{\partial^4 \phi}{\partial y^4} = 0. \tag{5}$$

This equation can be solved for ϕ to obtain a complete elastic solution.

The form of this fourth-order partial differential equation and its coefficients is such that both stable and unstable solutions are possible. Unstable solutions are tested for by seeking hyperbolic solutions. This is done by substituting for ϕ, another function θ $(x - \xi y)$. (See Kreyszig, 1963, p. 535–539.) Elliptic and hyperbolic equations are separated by forming an equation from the coefficients of an equation such as (5). If the product of coefficients of the first and third terms of equation (5), minus the square of the coefficient of the second term, is greater than zero, the equation is elliptic. It is hyperbolic if this quantity is less than zero. Elliptic solutions are stable, so a unique solution occurs for each set of well-posed boundary values, but hyperbolic solutions are periodic sets instead, and this is a condition found in one kind of instability (Fung, 1965).

As Biot notes, the characteristic equation then becomes:

$$\xi^4 + 2m\,\xi^2 + k^2 = 0, \tag{6}$$

where

$$m = \frac{2N - Q}{Q - T/2},$$

and

$$k^2 = \frac{Q + T/2}{Q - T/2}.$$

The roots of this equation are:

$$\xi^2 = -m \pm \sqrt{m^2 - k^2}. \tag{7}$$

Biot then identifies the parameter, ξ, and he finds it to be the ratio of the wave length in the x direction to that in the y direction of a sinusoidal solution. If equation (6) has real roots, then hyperbolic solutions exist. Real roots of ξ occur when (Case 1)

$$m > 0,\ k^2 < 0;$$

or equivalently,

$$T > 2Q,\ Q > 2N,$$

and (Case 2)

$$m < 0,\ m^2 > k^2 > 0$$

or

$$T < 2Q,\ Q > 2N \text{ and } |2N-Q| > |Q-T/2|.$$

Otherwise, equation (5) is elliptic and gives only stable solutions. Case 1 occurs only if the brittle higher modulus layer is much thinner than the embedding medium; otherwise the applied tension would exceed the strength of material before the onset of instability. The second case above may occur under more widespread conditions and give rise to what might be called shear instability. This case is probably related to the development of ductile fracture in necked rods.

We have shown by slight modification of Biot's theory of buckling that unstable solutions do occur in a stretched anisotropic domain when tension along layers exceeds a certain critical value. This explains the development of high stress concentrations crucial to the development of this theory of fracture spacing. For this condition to occur in nature the material must have the strength to support the required critical applied tension; otherwise unstable stress conditions will never come into existence before fracture occurs.

Using these methods a closed-form solution can be obtained by assuming different functions, ϕ, in terms of x and ξy. Substitution of these values in equation (5) leads to direct evaluation of all elastic quantities. Sometimes the assumed function is suggested by the geometry of deformation or the physics of a particular problem.

Biot (1965, p. 185) has shown that *surface* instability, another mechanism of instability, may occur in problems of the sort under consideration. He has used a particular solution of the form $\phi = \frac{f(y)}{\ell^2} \sin \ell x$ in equation (5), which reduced it to the ordinary equation

$$\frac{\partial^4 f}{\partial y^4} - 2m \frac{\partial^2 f}{\partial y^2} + k^2 f = 0. \tag{7a}$$

This equation has solutions of the type $f = \exp^{\beta \ell y}$, where β is the solution of the characteristic equation,

$$\beta^4 - 2m\,\beta^2 + k^2 = 0. \tag{7b}$$

If β were equal to $i\xi$, equation (7) would become equation (5) except for the sign of the second term.

With $f = c_i e^{\beta i \ell y}$, and for the condition $f = 0$ when $y = {}^{-\infty}$ unstable solutions occur if (Case 1)

$$T < 2Q \text{ or } m > 0$$

and

$$Q < 2N \text{ or } k^2 > 0,$$

and also (Case 2)

$$T > 2Q \text{ or } m < 0,$$

and

$$m^2 < k^2 < 0 \text{ or } Q < 2N; |2N-Q| < |Q-T/2|.$$

These are exactly the conditions excluded from *internal* instability, and accordingly extend the range of loading conditions under which instabilities can form. One infers from the above conditions on coefficients that Case 1 should occur easily and frequently in nature. If surface instabilities form easily, they will not affect thick layers because the stress dies out exponentially away from the interface. Surface instability is favored by low ratios of N:Q or in deformations where large normal strains occur relative to shear strain.

Still another type of instability occurs under deformation conditions under discussion and it is called layer instability.

Had we chosen for our solution the function

$$f = C_1 \cosh \beta_1 \ell y + C_2 \cosh \beta_2 \ell y \sin \ell x, \qquad (7c)$$

where β is a solution for the characteristic equation (7b), and where boundary conditions are set at top and bottom of the layer, then the entire layer would become unstable as Biot (1965) has pointed out in the corresponding case for compressive loading. Layer instability occurs for the same conditions of T, Q, and N as for surface instability. In this case the solution does not contain the exponential decay term in y, and the entire layered sequence becomes unstable, the rectangular outer form changing to the pinch and swell shape of boudinage.

In summary, from general considerations of the incremental deformation of anisotropic media in extension, several types of geometric elastic instability have been defined as generalizations of those recognized by Biot (1965). These are internal, interfacial, and layer instability. Each of them has periodic stress concentrations associated with it which might serve to localize fractures. In both layer and internal instability high stress concentrations may occur either on the surface or along the axis, depending on whether or not layer interfaces are free to slip. Periodic stress concentrations are formed only at the interface in interfacial instabilities.

For these theoretical solutions to be tested or applied to real deformations, material coefficients must be evaluated in the range of stress of each particular application. As yet N, Q, and other moduli have not been measured for rocks under tension. In the geologic examples of brittle fracture presently being considered there is no record of either elastic stress field or even the displacements that may have occurred prior to fracture. These difficulties limit the use of these theoretical methods and preclude the likelihood of obtaining a closed-form solution in this way; hence we turn to energy methods.

Strain Energy Methods

Biot has shown in detail (1965, p. 73-82, 135-159, 203-204) that the direct methods are fully equivalent to a strain-energy formulation of the problem. This is an important step because the original formulation of spacing theory was made and all computations were done by using classical strain energy methods. Biot shows the more general strain energy methods of incremental deformations. By using them, the connection between the direct methods already outlined and the approximate methods of the calculations becomes clear. Biot writes

$$\Delta V = 2Me^2_{xx} + 2Le^2_{xy} + T/2 \left(\frac{\partial v}{\partial x}\right)^2 \qquad (8)$$

for the total incremental strain energy per unit area, ΔV, within the domain. In this equation, T is the tension $S_{11}-S_{22}$, and M and L are moduli. M equals N–T/4, and L, the slide modulus, is Q–T/2 in this equation. The equation states that the incremental strain energy (related to incremental stresses) and the work done by the tension in lengthening the body equal the total incremental strain energy. The total incremental strain energy potential, P, for the entire body is obtained by integrating ΔV over the entire area (which is of unit thickness). The general equation (8) for potential energy expresses the instability condition and is equivalent to the equations of equilibrium (1). Therefore, solutions of (8) provide the stress field needed to locate the high stress concentrations believed to localize fracture.

Biot (1965, p. 137) writes the following equation for the incremental strain energy density in the development of equation (8):

$$\Delta V = 1/2\, t_{11}e_{xx} + 1/2\, t_{22}e_{yy} + t'_{12}e_{xy} + 1/2\, T \left(\frac{\partial v}{\partial x}\right)^2. \qquad (9)$$

This equation differs from the corresponding equation of infinitesimal theory in that it involves terms related to the prestress. For an orthotropic medium, the corresponding equation for ΔV is, using stress-strain relations (1),

$$\Delta V = 1/2\, C_{11}e^2_{11} + 1/2\, C_{22}e^2_{yy} + C_{12}e_{xx}e_{yy} + 2Le^2_{xy} + 1/2\, T \left(\frac{\partial v}{\partial x}\right)^2.$$

The t_{ij}, an alternate form of s_{ij}, are incremental stresses related to *undeformed* area. The t_{ij} must be used in strain energy equations, and Biot (1965, p. 61) develops equations which define t_{ij} in terms of s_{ij} and $S_{ij}e_{ij}$. By using these and equations (2) and (3) he derives

$$t_{11} = S + (2N - T)\, e_{xx}$$
$$t_{22} = S + 2Ne_{yy} \qquad (10)$$
$$t_{12} = 2Le_{xy}.$$

Substitution of equations (10) in equation (9) gives (8). In this derivation Biot used nonlinear equilibrium equations (Fung, 1965) relating to incremental, not total strain energy.

Accordingly, solutions of these equations are for the instability and not for the initial stable conditions described by classical direct methods of small deformations.

Biot has shown that in incremental analysis the variation of the total strain energy potential, δV, is

$$\delta V = -\int\int\int_\tau \partial A_{ij}/\partial x_j \, \delta u_i d\tau + \int\int_s A_{ij} n_j \, \delta u_i ds, \qquad (11)$$

where

$$A_{ij} = t_{ij} + S_{ij} + S_{kj}\omega_{ik} + 1/2\, S_{kj}e_{ki} - 1/2\, S_{ki}e_{kj}.$$

The virtual work done by all external forces on the boundary and the body forces, X_i, also equal δV.

$$\delta V = \int\int\int_\tau X_i(\xi_\ell)\,\delta u_i \;\rho d\tau + \int\int_s f_i\,\delta u_i \, ds. \qquad (11a)$$

Remembering that for any equilibrium condition the strain must be stationary, we seek the solution by combining equations (11) and (11a):

$$\int\int\int_\tau \left[\frac{\partial A_{ij}}{\partial x_j} + \rho x_i(\xi_\ell)\right] \delta u d\tau + \int\int_s (f_i - A_{ij}n_j)\,\delta u_i \, ds = 0. \qquad (12)$$

As this equation applies to arbitrary variations, then it follows that

$$\frac{\partial A_{ij}}{\partial x_j} + \rho\, X_i(\xi_\ell) = 0; \;\; f_i = A_{ij}\, n_i, \qquad (13)$$

which are exactly the equilibrium equations and the boundary condition, respectively, as derived by the usual methods of stress analysis of finite deformation as restricted by Biot. By following Biot's method the equivalence of the equilibrium equations and the variational relation are established.

This also can be written for the incremental strain energy potential when the body force is derived from a potential:

$$\int\int\int_\tau \delta\,(\Delta V + \rho\,\Delta V)\, d\tau = \int\int_s \Delta f_i\,\delta u_i\, ds$$

and

$$\delta\Psi_\tau = \int\int_s \Delta f_i\,\delta u_i\, ds$$

and

$$\Psi_\tau = \int\int\int_\tau (\Delta V + \rho\,\Delta V)\, d\tau.$$

ΔV is the incremental body forces potential, and Ψ_τ is the total incremental potential.

In the special case of internal instability, Biot shows

$$\Psi_\tau = \int\int\int_\tau \Delta V \;\, d\tau, \qquad (14)$$

thereby, using T for the stress difference and integrating equation (8) over the area, we obtain

$$\rho = \int_{-h/2}^{h/2} dy \int_{-h/2}^{h/2} \Delta V\, dx.$$

Using sinusoidal displacement fields in ΔV, following Biot (1965, p. 204), we derive

$$\rho = 1/8\, h_x h_y\, C^2 \ell^4\, [L\,\xi^4 + 2\,(2M-L)\,\xi^2 + L + T]. \qquad (15)$$

Letting $\rho = 0$ we obtain the characteristic equation for internal instability

$$L\xi^4 + 2\,(2M-L)\,\xi^2 + L + T = 0,$$

which is identically

$$\xi^4 + 2m\,\xi^2 + k^2 = 0.$$

This identifies the variational principle with the internal instability condition. It has also been shown that it is equivalent to an equilibrium condition.

Biot has used moduli of anisotropic materials subjected to incremental deformations. In developing fracture spacing theory we have used exactly parallel strain energy methods but have greatly simplified the procedure. For practical reasons already mentioned, it is necessary to specialize and treat the material as isotropic layered materials not under prestress. We return to the classical strain energy methods. We have used Biot's general methods to display a very general theoretical basis for a fracture spacing theory. To test this hypothesis in both the laboratory and the field, approximate methods have to be used.

By using conventional strain energy methods, I have derived the following fracture spacing theory independent of the more exact theory of Biot. The well-known principle of virtual work (Timoshenko and Goodier, 1970, p. 244–254) is

$$\Delta V - \Delta W = 0. \qquad (16)$$

To use this principle we must compute ΔV, the *incremental* strain energy stored in the system, and we must compute ΔW, the work done by the external forces for the *incremental* deformation. Equation (16) shows these quantities must be equal. We use this equation to find the critical tension, T, for elastic instability.

To compute the location of zones of high shearing stress that may localize fracture, three separate terms in equation (16) must be found. If one simplifies the problem to a 3-layer sandwich in which the stiffer layer is the embedded member in a rectangular domain (Fig. 6), one can describe the required terms relatively easily. The first term is the strain energy associated with the deformation of the stiffer layer, the second is the strain energy of the outer layers induced by their restraining effect on the stiffer layer, and the third is the work done by the external load or the prestress, T, in extending the stiffer layer.

Neglecting the effect of the surrounding medium for the time being, one can compute the strain energy associated with deformations caused by assuming sinusoidally distributed loads on the surface of the stiffer layer (Fig. 7). This loading

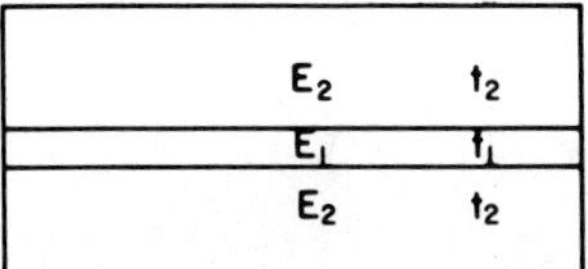

Figure 6. Layered domain for which theory has been developed. $E_1 > E_2$ $t_1 \ll t_2$. The central layer is the stiffer one.

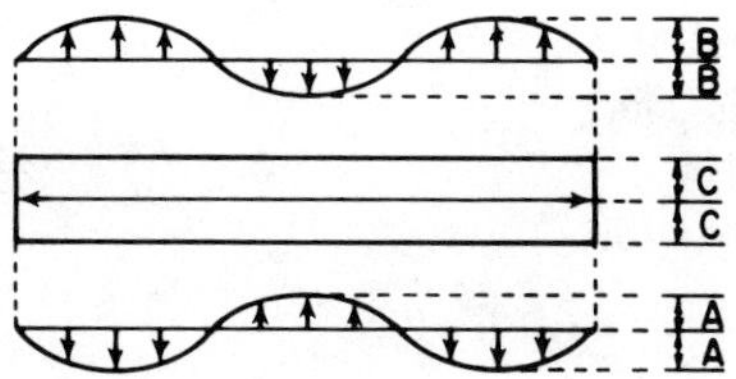

Figure 7. Load distribution on stiffer layer.

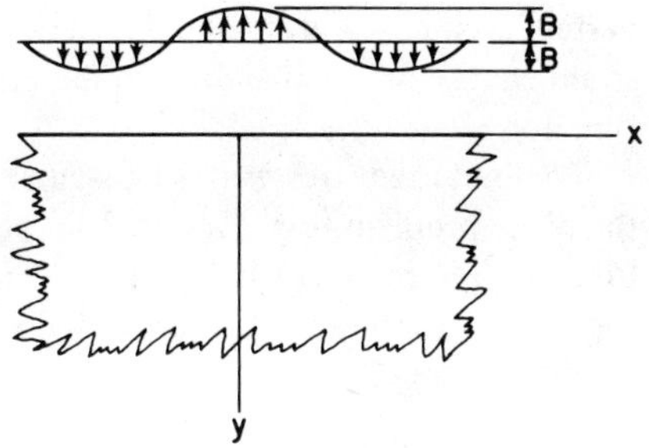

Figure 8. Load distribution on lower embedding layer, E_2, t_2.

is assumed because of the boudinage or necking in stretched layered rocks observed in the field experiments (Griggs and Handin, 1960). The strain energy of the deformation of the restraining material is computed by regarding the medium as a semi-infinite half-space also loaded sinusoidally (Fig. 8). Later each case will be discussed fully.

This procedure is approximately equivalent to the one described in connection with Biot's theory. The strain energy which is computed is associated with an assumed sinusoidal perturbation, and the work done by external forces is related to shortening caused by the perturbation. The procedure is an incremental analysis in the sense that the deformations which are associated with the simple lengthening of the sandwich are not evaluated nor is the work done in this lengthening computed. Integration of equation (1) produces the same sort of solution, but in that case Biot refers to an incremental deformation of anisotropic material which is prestressed. In each procedure the deformation of only the central bar need be considered, as the constraining effect of the embedding material is included in the energy principle. If the same elastic moduli are used in Biot's equation and in equation (16) the computations become nearly identical.

The sinusoidally distributed loads are assumed above in connection with the occurrence of a sinusoidal perturbation on the shape of the brittle layer. They exist only if this internal perturbation can occur in equilibrium with externally applied simple stretching or flattening loads. At the outset of loading the stable configuration must consist of straight layers. Therefore, the assumption of the sinusoidal form implies a departure from a stable equilibrium that can exist in equilibrium with higher values of applied load only if its strain energy is lower than that of the straight form. As the load is applied, first the straight form is stable, then after passing a critical load the sinusoidal form is stable. This can happen only if the strain energy of the sinusoidal form increases nonlinearly with increasing deformation. By assuming that the sinusoidal form can exist, and by using equation (16), we can find the critical load at which it will be the preferred form. If this load is less than the strength of the material then necking occurs. We can compare the strain energy of the sinusoidal form and the straight form at loads slightly above the critical load, and we find that the strain energy of the necked form is the smaller.

There is no guarantee that the shape we assumed is the one that actually occurs. By using methods of variational calculus it would be possible to find the exact function that minimizes the strain energy and describes the shape the straight bar assumes at the critical load. I could not solve this problem, and the approximate solution used must give somewhat higher strain energy than the real value. I have followed the general procedure recommended by Timoshenko (1936) for this eventuality.

That an instability should occur can be anticipated from consideration of the fact that the elastic properties of adjacent layers do not match. For small externally applied tensions or compressions small values of uniform shear and normal stress occur at the interface between the layers. Higher external loads generate larger interfacial stresses so that is seems likely that small perturbations of the interface might generate large local bending moments which cause buckling in compression and necking in extension.

Since the exact form of the deformation cannot be known beforehand, the critical tension to be computed will be larger than the true value. This is not a serious problem unless the deformation is large or unless the exact shape of the instability is required. The fracture-spacing problem does not depend on these factors. Finite-amplitude buckling problems are, on the other hand, difficult to solve by these methods (Chapple, 1968). The strain-energy method has the distinct advantage of displaying the physics of the process and oftentimes can be used to provide useful relations among physical variables. For these reasons I have chosen to develop the fracture-spacing argument by these methods.

Let us return to the problem of computing the strain energy in the stiffer layer. The second order part of the incremental strain energy for an element of the beam is given by Timoshenko and Goodier (1970) as

$$\Delta V = \frac{1}{2E}(\sigma_x^2 + \sigma_y^2) - \left(\frac{\mu}{E}\sigma_x\sigma_y + \frac{1}{2G}\tau_{xy}^2\right), \tag{17}$$

and the total incremental strain energy for the domain as a whole is then

$$V = \int_0^{\ell}\int_0^{c}\left\{\frac{1}{2}E(\sigma_x^2 + \sigma_y^2) - \frac{\mu}{E}\sigma_x\sigma_y + \frac{1}{2G}\tau_{xy}^2\right\} dx\,dy. \tag{18}$$

In the computation given below the complementary strain energy is used for convenience because forces are specified and stresses are obtained directly. Strains are not computed. As perfect elasticity and small deformations are assumed, the strain-energy function and the complementary strain-energy function are identical.

To obtain the stresses it is convenient to use the solution to the plane problem first given by Ribiere, according to Filonenko-Borodich (1965, p. 171). The beam is loaded on top and bottom surfaces sinusoidally so that the boundary conditions are (Case 1):

when

$$y = +c; \tau_{xy} = 0; \ \sigma_y = -B \sin \alpha x;$$
$$y = -c; \tau_{xy} = 0; \ \sigma_y = -A \sin \alpha x.$$

The maximum amplitudes of the applied forces are A and B, and $\alpha = \frac{n\pi}{\ell}$, where n is an integer equal to the number of half waves, and ℓ is the length of the beam. The ends of the beam are free of normal stress, and the integral of the small shear stress taken along the ends is also zero.

This problem is then solved by using the Airy stress function in the biharmonic equation. The assumed function, ϕ, is

$$\phi = \sin \alpha x \,(c_1 \cosh \alpha y + c_2 \sinh \alpha y + c_3\, y \cosh \alpha y + c_4\, y \sinh \alpha y). \quad (19)$$

In this case, the stresses are

$$\begin{aligned} \sigma_x &= \sin \alpha x \,(c_1\, \alpha^2 \cosh \alpha y + c_2\, \alpha^2 \sinh \alpha y \\ &\quad + c_3\, \alpha\, (2 \sinh \alpha y + \alpha y \cosh \alpha y) \\ &\quad + c_4\, \alpha\, (2 \cosh \alpha y + \alpha y \sinh \alpha y)), \\ \sigma_y &= -\alpha^2 \sin \alpha x \,(c_1 \cosh \alpha y + c_2 \sinh \alpha y \\ &\quad + c_3\, y \cosh \alpha y + c_4\, y \sinh \alpha y), \\ \tau_{xy} &= -\alpha \cos \alpha x \,(c_1\, \alpha \sinh \alpha y + c_2\, \alpha \cosh \alpha y \\ &\quad + c_3\, (\cosh \alpha y + \alpha y \sinh \alpha y) + c_4\, (\sinh \alpha y \\ &\quad + \alpha y \cosh \alpha y); \end{aligned} \quad (20)$$

where α is $\frac{n\pi}{\ell}$, and ℓ is the length of the beam.

The constants can be evaluated as

$$\begin{aligned} c_1 &= \frac{A + B}{\alpha^2} \ \frac{(\sinh \alpha c + \alpha c \cosh \alpha c)}{(\sinh (2\, \alpha c) + 2\, \alpha c)}, \\ c_2 &= \frac{A - B}{\alpha^2} \ \frac{(\cosh \alpha c + \alpha c \sinh \alpha c)}{(\sinh (2\, \alpha c) - 2\, \alpha c)}, \\ c_3 &= \frac{A - B}{\alpha} \ \frac{\cosh \alpha c}{(\sinh 2\, \alpha c - 2\, \alpha c)}, \\ c_4 &= \frac{A + B}{\alpha} \ \frac{\sinh \alpha c}{(\sinh 2\, \alpha c + 2\, \alpha c)}. \end{aligned} \quad (21)$$

This solution is associated with a pinch-and-swell type of deformation of a beam of thickness 2c and length ℓ.

Sinusoidally distributed shearing stresses on the top and bottom of the layer provide another loading condition (Case 2) of interest. The boundary conditions are:

at

$$y = c, \tau_{xy} = B \cos \alpha x; \sigma_y = 0,$$
$$y = -c, \tau_{xy} = A \cos \alpha x; \sigma_y = 0.$$

The stresses are the same as for Case 1, but the constants become

$$\begin{aligned} c_1 &= \frac{A - B}{2\alpha} \ \frac{c \sinh \alpha c}{\alpha c + \sinh \alpha c \cosh \alpha c}, \\ c_2 &= \frac{-(A + B)}{2\alpha} \ \frac{c \cosh \alpha c}{\alpha c - \sinh \alpha c \cosh \alpha c}, \\ c_3 &= \frac{A + B}{2\alpha} \ \frac{\sinh \alpha c}{\alpha c - \sinh \alpha c \cosh \alpha c}, \\ c_4 &= \frac{A - B}{2\alpha} \ \frac{\cosh \alpha c}{\alpha c + \sinh \alpha c \cosh \alpha c}. \end{aligned} \quad (22)$$

Equation (18) was integrated by using expressions (20) for the stresses to obtain algebraic expressions for strain-energy terms that might be used to display various relations between physical variables. Unfortunately, the integrated expressions are so lengthy that the desired manipulations cannot be made, so the integrations were then programmed for numerical solution on an IBM 360/65 computer. The program finally adopted made use of hand integrations to give greater flexibility in output than could be obtained by using a Gaussian integration technique. Computer solutions finally checked the tedious hand integrations. Chapple (1971, written commun.) pointed out that a better approximate solution could have been obtained by adjusting these solutions so that all deformation would be toward the center of the layer rather than both toward and away from it.

One obtains the strain energy for the stiffer layer deformed by forces acting on its top and bottom that are sinusoidally distributed. There are normal forces (Case 1) or shear forces (Case 2). These solutions may be superposed to give the strain energy contained in the stiffer layer deformed by both shear and normal forces acting simultaneously.

The strain energy stored in the restraining medium can be computed by using an elastic solution obtained by Biot (1937, p. A1-A6). He regarded the restraining medium as a semi-infinite half-space, edge loaded by sinusoidally distributed forces. Biot found the stress function for this case to be

$$\phi = \frac{B}{\alpha^2} \sin \alpha\, x\, e^{-\alpha y}\, (1 + \alpha y). \quad (23)$$

When $\sigma_x = \sigma_y = \tau_{xy} = 0$ as y becomes infinite, and $\sigma_y = B \sin \alpha y$, and $\tau_{xy} = 0$ at $y = 0$, the stress components are (Case 3)

$$\sigma_x = B\, (\alpha\, y - 1)\, e^{-\alpha y} \sin \alpha x,$$
$$\sigma_y = -B\, (\alpha\, y + 1)\, e^{-\alpha y} \sin \alpha x,$$
$$\tau_{xy} = -B\, \alpha\, y\, e^{-\alpha y} \cos \alpha x.$$

These stress components are substituted into equation (18) and integrated to give the strain energy of the restraining medium under the action of normal surface forces only. In this case the elastic moduli used in equation (18) are those of the restraining media.

The same stress function is used to evaluate the strain energy in the restraining media due to shear forces distributed sinusoidally on the plate. Here (Case 4) the boundary conditions are

$$\sigma_x = \sigma_y = \tau_{xy} = 0 \quad \text{when y is infinite,}$$

and

$$\tau_{xy} = B \cos \alpha x; \; \sigma_y = 0 \text{ on } y = 0.$$

Then

$$\sigma_x = B(\alpha y - 2)\, e^{-\alpha y} \sin \alpha x\,,$$

$$\sigma_y = -B \alpha y\, e^{-\alpha y} \sin \alpha x\,,$$

$$\tau_{xy} = B(1 - \alpha y)\, e^{-\alpha y} \cos \alpha x\,.$$

These components are substituted into equation (18) and the strain energy is computed.

This completes the set of strain-energy terms that are associated with incremental forces either normal or shear acting on the upper and lower surfaces of a stiff layer embedded in a softer material. The strain-energy terms computed for all cases can be added as boundary conditions required to provide the total incremental strain energy of the deformed domain.

Finally, the work done by the external forces is calculated. This is done by finding the change in length of the beam caused by the sinusoidal surface loading. The change in length $\Delta \ell$ is multiplied by the tension, T, to give work, W:

$$W = T\,\Delta \ell. \tag{24}$$

There are various ways to compute $\Delta \ell$. Had we been able to determine the exact perturbation function, it would furnish a way to compute $\Delta \ell$ exactly. As this was not possible we assumed a function which provides a good approximate value for the strain energy but which does not have an associated

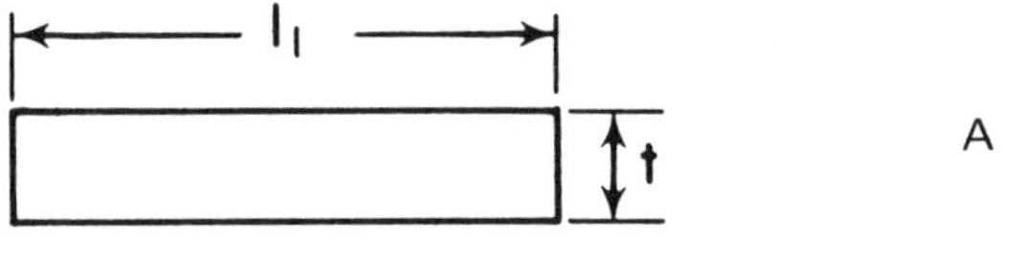

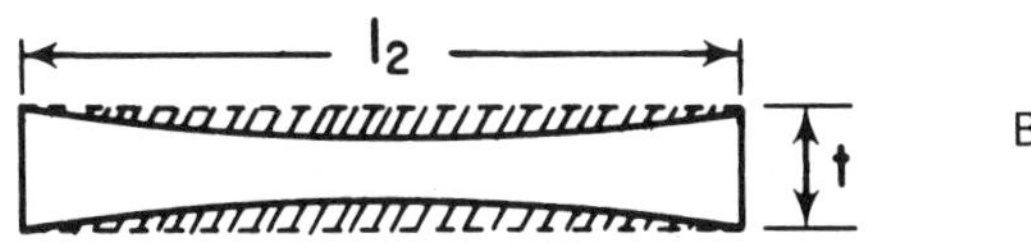

Figure 9. A. Rectangular form of stiffer beam of unit thickness. B. Necked form of stiffer beam; shaded areas show material which must be subtracted from $V = tl_2$ so that volume is equal to $V = tl_1$.

change in length. To compute the external work we must make another assumption. We must assume a function similar to the first which can give a measure of shortening. One way to do this is to assume that the surface forces cause a sinusoidal deflection β' of amplitude $B/|E|$, where E is the Young's modulus of the stiffer beam. It is also assumed that the total volume is constant (Fig. 9).

The volume for a plate of unit width before the alteration in shape is

$$V_1 = \ell_1 t_1\,,$$

where ℓ_1 is the length prior to necking. After necking the volume is given by

$$V_2 = \ell_2 t_2 - 2 \int_0^{\ell_2} (\beta' + \beta' \sin \frac{n\pi}{\ell_2} x)\, dx, \tag{25}$$

where ℓ_2 is the altered length, and t_2 is the thickness after deformation. The integral is the area under the curve of the sinusoidal deflection if the original surface of the layer is tangent to the outside of the swells. This is multiplied by 2 as there are segments on top and bottom of the layer.

As volume is constant, $V_1 = V_2$, so that

$$\ell_1 t_1 = \ell_2 t_2 - 2 \int_0^{\ell} (\beta' + \beta' \sin \frac{n\pi}{\ell} x)\, dx.$$

For a whole number of waves we find after integration

$$\ell_1 t_1 = \ell_2 t_2 - 2\beta' t_2\,.$$

We must consider that t_2 contracts due to simple thinning during the necking process. Then

$$t_2 \approx t_1 - \frac{T}{2E}\,,$$

and

$$\Delta \ell = \frac{2\beta' \ell_2}{t_1} + \frac{T \ell_2}{E t_1}\,. \tag{26}$$

The work done by the external load is then

$$W = (2\beta' \ell_2 + \frac{T \ell_2}{E}) \frac{T}{t_1}\,. \tag{27}$$

An alternative approximation is to use the difference in length between the chord and the arc in the sinusoidal function used to describe the strain energy. This is the procedure used in the case of symmetric buckling.

We can assume the deflection curve is

$$y = \beta' (1 - \cos \frac{n\pi x}{\ell}).$$

The change in length, $\Delta \ell$, is then

$$\Delta \ell \approx \frac{1}{2} \int_0^{\ell} (\frac{dy}{dx})^2\, dx\,, \tag{28}$$

by using methods of elementary calculus and assuming small rotations associated with the perturbation.

Evaluating (28) we obtain

$$\Delta \ell = \frac{\beta'^2 n^2 \pi^2}{2\ell}\,,$$

and the external work, W, is

$$W = \frac{\beta'^2 n^2 \pi^2}{2\ell} T. \qquad (29)$$

We assume that the change in length along the axis due to necking is approximately the same as the stretching of the surface as it goes to the deflected form. We neglect the stretching due to simple extension during the deformation. This assumption, although not very accurate, is convenient for it gives in equation (29) the work as a function of the square of the amplitude β. Using (29) to calculate the work means that all terms in the equation of virtual work depend on the square of the amplitude; therefore the equation (16) becomes independent of the deflection and a critical T can be found. If one neglects the second order terms in equation (27) then the external work is linearly dependent on β' and a critical load seems indeterminate or nonexistent; that is, the pinched shape seems stable. As the change in length due to the perturbation is well described by equation (27) we need to look more carefully at this problem.

If the external load is constant, as in this case in the incremental deformation, it often happens that ΔW is a linear functional of the incremental displacement vector. As Langhaar (1962) explained, in this case ΔW vanishes identically and the stability criterion that we have been using reduces to the condition that ΔV is positive definite. Therefore, at the critical load, ΔV becomes positive semidefinite; that is, its minimum value is zero. Moreover, this value is attained at some nonzero increment of the displacements. The Trefftz criterion is used to find the critical value in this case. We set $\Delta V = Q$ where Q is a quadratic expression. Q has a nontrivial minimum and by setting the variation of Q equal to zero we find certain nonzero values of the displacement.

The variation is carried out by applying certain arbitary values to the displacement. In this case, we vary the wave number of the sinusoidal perturbation and seek the least value of the tension for each perturbation. The solution is obtained very much as if this were an eigenvalue problem obtained from differential equations. Its validity rests on the fact that the equations for Q and ΔV are homogeneous linear equations with homogeneous boundary conditions. This procedure and its derivation from the general methods of the calculus of variations is discussed at length by Langhaar (1962, p. 206–211).

EXPERIMENTAL STUDIES

Fracture spacing intervals predicted by theory have been verified by experimental work done on photoelastic gel models. Theory is tested by a three-layer sandwich model flattened in plane strain between appressed "rigid" platens. The center layer is thinner and elastically harder than the embedding layers. Examples of both free and welded interfacial contacts were studied. In these models lateral ends were free of constraint. Visual monitoring of the photoelastic stress field during testing confirmed the assumption that these simplified models embody the significant aspects of the physics and furnish an adequate check of the theory.

The laboratory tests were designed to give a quantitative measure of the fracture spacing and the critical load for instability. Spacing as a function of thickness and length of layer and its elastic modulus was studied.

In the foregoing section, theory was developed for extension of layered media in which the external load was a tension applied parallel to the layering. In the models, because we found no way to apply tension to the ends of the weak model materials used, extension was achieved by flattening, that is, by loading the model with a compression applied normal to the layering.

Nevertheless, this load arrangement does test the theory. The external load enters into the Biot equations (1) where the prestress S_{ij}, called T, was defined as $S_{11} - S_{22}$. Taking tension as positive we assigned a value of T to S_{11} and zero to S_{22}. By superposing a hydrostatic compression just equal to $|T|$, S_{11} is reduced to zero and S_{22} becomes equal to -T and the load conditions for prestress remain unchanged in the systems of equations Biot used to derive the instability.

In application of strain energy theory the problem is more complicated. It makes little difference in the perturbation whether load is applied to the ends or along the lateral sides since either system can cause discontinuity to develop along the interface in one or the other of the stress or strain fields. Normal loads applied to the upper and lower surface can generate shearing and normal stresses at the interference between elastic media of different properties. Any sinusoidal deflection of the interface will experience similar possibilities for instability as if tension were applied at the ends. The difference between the two types of loading conditions is a superposed hydrostatic load which does not greatly affect the instability.

Two other types of model tests were made: (1) multiple layers of alternating clay and putty were compressed to produce fractures and boudins as Ramberg (1955) did, and (2) rock interlayers were stretched in high-pressure testing machines. The latter models have been technical failures to date because of failure of their jackets.

Experimental Procedures and Equipment

Experiments were performed under a 15-in. clear field Chapman research polariscope. As the instrument is of the split bench type, it has been equipped with a selsyn motor drive which rotates analyzer and polarizer simultaneously so that they can be completely separated by the rather large model and strain frame (Fig. 10).

The strain frame consists of a large aluminum U-frame 40 mm thick and covered front and rear by ¼-in. plate glass. The frame holds the gels in plane strain and at the same time allows light to pass through the model. The frame is rigid relative to the loads generated in the model. Small forces

Figure 10. Strain frame and polariscope showing model set up. Frame is about 90 cm by 50 cm.

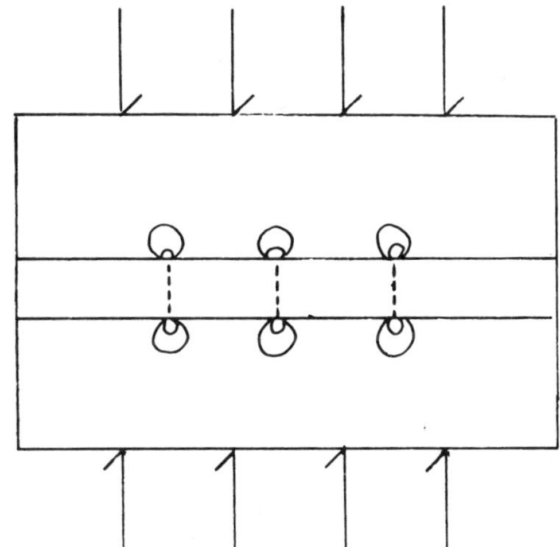

Figure 11. Early indication of instability. Small concentric circles are isochromatic fringes that appear in the gelatin adjacent to agar layer. These indicators of high stress concentration appear before the fractures occur. Fractures will appear along dashed lines.

ranging up to 9×10^6 dynes are applied to the top of the model by a small Blackhawk ram. The reaction bar at the base of the model is rigidly fixed to the U-frame (Fig. 11). The strain frame can accept models as large as 38 cm X 75 cm X 4 cm. The large frame is carried on a tram which follows a track embedded in the laboratory floor; moreover, the model can be raised or lowered as well as moved from side to side so that the entire model can be put into the central part of the field and at the same time kept perpendicular to the optical bench.

Special equipment is required for the measurement of the elastic properties of weak gels. No commercially available equipment was found for measuring such fragile elastic gels whose Young's modulus ranged from 0.15×10^6 dynes/cm^2 to 2.3×10^6 dynes/cm^2. For example, the standard vane tester for obtaining the shear modulus of plastic clays does not work on brittle or rubbery materials because it causes an extremely nonuniform stress field and the materials rupture in tension. We made a small tester in which the twist of a cylinder of gel could be measured to obtain the shear modulus of the rubbery gelatin gels. Stress concentrations near pins embedded in the ends of agar rods caused local fracture of this brittle gel and the device was useless for the agar. Because of these local irregularities this tester always gives an apparent modulus which is considerably less than the true value. A difficult problem connected with testing weak gels is devising a way to grip the specimen without creating a large local stress concentration and premature local failure. One modulus can be obtained in isotropic materials by measuring the deformation of free standing prisms of gels and computing the modulus from the displacements. Success depends on having an independent value of one other modulus. Further work is indicated if very exact values for stresses measured in models are required. Free standing prisms deformed under their own weight were successfully used to measure Young's modulus of both gels.

Special laboratory built equipment is required in casting weak gels. We use plate glass rectangles which are taped and puttied to form watertight moulds. Glass is treated with wax and Dow Corning mould release compound. Very large models are cast directly onto one of the glass sides of the strain frame; then the entire frame is placed on a special turntable for safe handling of the fragile model. Very weak (2 percent) gels cannot be touched and must be supported at all times. Silicone oils are used between glass walls and gelatin to minimize both adhesive and friction between these materials. Nonetheless, adhesion causes unresolved problems with the deformation of these low modulus models.

By standardizing preparation procedures gels are made which have nearly homogeneous and consistent properties. Homogeneity and optical properties were greatly improved by homogenizing gel and water in a blender before heating; then the mixture is heated for 2 hrs at 90°C to improve dispersion of colloidal particles. Gelatin gels can be made more brittle and with higher elastic modulus by varying procedures. As gelatins age, it is important to use gels that have aged for equal periods. For example, gels that have set for only about an hour have very much lower elastic limit and, surprisingly, fail by shear fracture. Fracture is ductile and response appears to be ideally plastic.

Gelatin gels used were usually composed of 16 percent gelatin, 16 percent glycerin, and 68 percent distilled water by weight. Some weaker gels of 8 percent or 12 percent gelatin were made. The standard gelatin gel has a Poisson's ratio of about 0.5 and a Young's modulus of about 2×10^6 dynes/cm^2. The fringe value of these gels (the stress difference which causes one full wavelength of retardation) is about 10 dynes/cm^2. The agar gel was varied from about 1.4 percent to 2.9

percent. Young's modulus for a 2 percent agar gel was about 3×10^6 dynes/cm^2. The ratio for the modulus of hard to soft layers was about 3:1 but was varied to as much as 6:1.

Models were loaded so that a displacement was applied normal to the bedding at two rates. The fast loading rate caused a normal strain of about 15 percent in less than 5 min. The upper bar was displaced about 2 cm during this time. The same displacement required about 30 min at the slower rates. Displacement was applied intermittently by a small Blackhawk jack. Pressure in the slower loadings was somewhat lower. The difference in results between the two loading methods was caused by the slow unloading of frictional restraints on the front and rear glass walls resulting in two truly different loading rates. Time-dependent material behavior is not significant in experiments. Rapid loading permitted a metastable stress field to develop which was relieved by shattering of the model.

Load and total deformation are measured in experiments. According to theory, to compute strain energy from the test we must measure dimensions, elastic moduli, applied loads, and displacements. From this information a spacing can be computed and compared with the experimentally observed spacing. It is not necessary to measure the magnitudes of stress components at various points in the model, although this capability exists in photoelastic models.

In photoelastic work the pattern of stress distribution in a plane model is visible because model materials are optical stress transducers. That is, in the polariscope we can see optical patterns caused by changes in the index of refraction which are directly proportional to changes in stress. As Frocht (1948) discusses, we observe color fringes called isochromatics. These are proportional to the differences in principal stresses in a two-dimensional model. By making a simple calibration test we measure the proportionality constant that enables us to measure the amount of birefringence change and correlate it with the magnitude of the stress difference at a point. We also measure dark lines, called isoclinics, which are the loci of points where the principal stresses are equally inclined and coincident with the vibration directions of the polarizer and analyzer. The lines are black because the principal axes of stress and accordingly the principal optic planes, as well, are so oriented that extinction occurs.

In models made from weak gels, experimental errors caused by friction are appreciably large. They cannot be measured because they occur on front and rear glass walls and are of opposite signs, so their optical effects are self-cancelling in this setup. (Frictional forces are dependent on the normal force acting across the surface, all of which vary in the nonhomogeneous deformation field existing in these models.) Even in experiments with higher modulus materials the actual stress magnitudes must be obtained from graphical integration along a stress trajectory (Frocht, 1948). The critical places in these experiments are loci of stress concentrations, which are very small areas of high stress gradients where the exact pattern of stress trajectories cannot be accurately measured, a common problem in photoelastic analysis. However, the pattern of isoclinics and isochromatics reveals these areas without equivocation. Once located, these points may be compared with the location of actual fractures, thereby establishing the desired correlation between stress concentration and fracture spacing.

Description of Model Deformation

The geometric properties of the stress field show directly the physical instability, so a rather detailed description of the development of the photoelastic stress patterns follows. At low stress levels the entire field takes on a nearly uniform color interference showing homogeneous distribution of the stress difference over the model. Broad and very diffuse isoclinics appear when the vibration directions of polarizer and analyzer parallel the load axis, which shows that a principal stress axis is parallel to the load axis. This axis is also in the direction of the long side of the rectangular model. Principal stress axes are uniformly subparallel in all parts of the model, except for randomly distributed local stress concentrations of small magnitude caused by flaws. These effects die out within three or four times the average diameter of the imperfection and are of no consequence at any time during the experiment. As the regional stresses become larger, the effect of the local irregularity becomes a smaller percentage of the total, and these are stable imperfections.

When elastic instability occurs, at a critical stress the stress pattern becomes nonuniform. At first the local changes are small and affect an area in the gelatin of about 1 cm radius (Fig. 12). Agar gel is much less transparent and much less

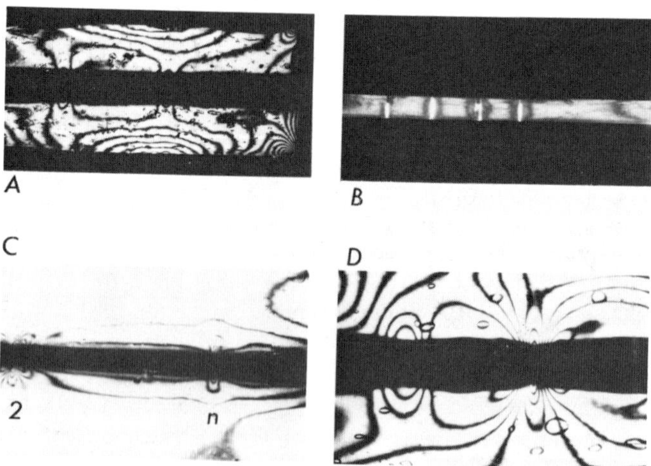

Figure 12. Photoelastic patterns. A. Expt III isochromatics in gelatin gel prior to failure. Agar layer covered. B. Expt III fracture in agar layer and weak isochromatic pattern associated with fracture. Pattern appears before fracture, gelatin layers covered. C. Expt X pattern in gelatin shows two stages. At left (2), fracture exists in covered agar layer; at right (n), no visible fracture. D. Expt XI isochromatic pattern in gelatin in which necking has occurred. Flaws at the surface and in either layer between necks are stable. Necking is a perturbation of the *entire* brittle rod, not unstable growth of randomly located thin places.

optically sensitive than gelatin gel, so that these small changes are very hard to detect in it. Also, initial stresses associated with the fabrication of gel layers paralleling layer contact obscure these small stresses. As load is increased the agar gel develops periodic changes in its internal pattern that correlate directly with fractures which form almost immediately afterward. Generally, patterns such as this occur whether the interfacial surface is welded or not, but in the latter case only one central locus is formed, which intensifies as load is continued. Even though the precursor change appears in the gelatin, the crack often starts in the center of the agar beam (Fig. 12). Fractures in the agar beam and large stress concentrations in the gelatin layers coincide with the small precursor pattern of Figure 12. The entire stress field changes in the more rigid agar layer, which indicates the pervasive nature of the elastic instability. This illustrates the advantage of the photoelastic model, that is, the instability is made visible as it occurs photoelastically. One can observe the nearly simultaneous formation of a set of stress concentrations and see that they form before fracture starts. Additional patterns develop after the first set of fractures forms and as the load is increased.

If contacts are welded, stress concentrations are larger and more closely spaced than in those with either unwelded or freely lubricated interfaces. In some of the latter type, instabilities did not seem to develop and failure occurred on a single fracture in the middle of the model.

Instability appears only when thickness, Young's modulus, and specimen length are such that tensile stresses exceeding the critical value can occur prior to fracture. For example, if agar layers exceed 3 cm, fractures do not form in models where layer interfaces are unwelded until loads exceed the crushing strength of the entire sandwich (length of model is 260 cm).

We made more than 100 photoelastic models of which 63 were successful, that is, were not in some way mechanically destroyed. A tabular summary of results is given (Table 1) in which the number of fractures is related to variations in layer thickness, interfacial conditions, and relative moduli. All of the results apply to models 25 cm long and of total thickness of all layers of 10 to 15 cm.

In most models several modes of fracture occur. Other fracture modes form beside the predicted ones related to elastic instabilities. Some are made when loads are increased beyond first-formed fractures. These fractures seem to form without disrupting the stress field geometry until fracture migrates into the surrounding gelatin gel. This second fracture set consists of short small fractures and probably forms as the layer is crushed by compressive stress that exceeds the strength of material.

In experiments where large loads have crushed the middle layer into small fragments of 5 to 15 mm on a side, they are formed by cataclastic or granular flow into a pinch and swell

TABLE 1. EXPERIMENTAL RESULTS, GEL MODELS

Thickness		Young's modulus					
Layer	Medium X 2	Layer (X 10⁶ dynes)	Medium (X 10⁶ dynes)	Average number of fractures	Spacing interval	Number of tests	Applied load
A. Welded contacts							
40.8 mm	121.8 mm slow	0.42	0.21	2.1	76 mm	2	1.7×10^6 dynes
25.4 mm	125.7 mm slow	0.42–0.28	0.25–0.21	3.8	55 mm	16	2.2×10^6 dynes
20.6 mm	128.5 mm slow	0.42	0.25	4.2	50 mm	6	2.4×10^6 dynes
13.0 mm	135.0 mm slow	0.42	0.25	7.1	38 mm	6	2.9×10^6 dynes
25.4 mm	125.7 mm fast loading	0.36–0.42	0.21–0.25	15.5	40 mm	11	1.9×10^6 dynes
20.6 mm	128.5 mm fast loading	0.32–0.36	0.21	5.1	51 mm	12	
B. Frictional contacts							
21 mm	128 mm	0.4	0.2	5.0	50 mm	1	Fractures appear in second stage; 2 then 3 fractures
25 mm	125 mm	0.4	0.2	1.0		2	Second test; rapidly overloaded layer shattered
C. Free interfaces–very low friction							
13 mm	135 mm	0.4	0.2	0		3	Water lubricant
7 mm	142 mm	0.4	0.2	0		2	
60 mm	120 mm	0.4	0.2	1.0		1	
25 mm	120 mm	0.4	0.2	100s			Foam lubricant, shatter
20 mm	120 mm	0.4	0.2	0-100s		1	Shattered

structure (boudinage) in which the number and location of necks are the same as would be predicted by the instability theory. This occurs after the initial elastic instability has formed and disappeared. It is noteworthy that these small agar fragments are very mobile and easily "flow" into pre-existing loads in the surrounding gel.

A third mode of fracturing exists in a few of the gel experiments and nearly all putty experiments. Very small, very closely spaced fractures form on the interface, usually in the outer few millimeters of the more brittle layer. Fractures are usually less than 1 cm in length and are separated by only 2 to 3 mm. These tension fractures may be arranged on an echelon pattern along a line which is parallel to a possible shear trajectory. In a gel model, small fractures of this sort formed along the line of potential shear fracture well within the layer. Fractures close to the surface are probably controlled by interfacial instability, a possibility that will be investigated at a later date.

To summarize, tensile fractures are produced in experiments having approximately the same location as predicted; some form as predicted by earlier theory; and some cannot be explained yet. In experiments, fractures formed at smaller loads occur as predicted, even though the external load is applied along rather than across the layers. Load levels must be lower, so fewer fractures form and the simpler conditions predicted by theory are not overprinted by crushing. In nature, one would expect theory to be more easily applied in areas that were not crushed; that is, where large vertical movement does not occur.

Because of the multiple modes of fracture the spacing at the end of an experiment is variable. First-formed fractures are rarely perfectly spaced and vary from predicted spacing in individual experiments by a considerable amount.

In experiments with strong brittle layered materials, such as epoxy resins, the fracture patterns are not related to anisotropy in any evident way. This is because strength, modulus, and size cannot be combined to produce unstable elastic deformation at laboratory scale for these materials. One material must be able to flow while the other material fractures. Flowage may be by easy large elastic strain or by equally easy ductile flow.

Of the 63 photoelastic tests tabularized most frequently, 53 were made on models with welded contacts (Table 1). Fast deformation produced more fractures at larger loads in about half of the tests. Thickness was varied from 13 to 41 mm and correlates directly with the increase in spacing and applied load. Tests with free interfaces or with low frictional contact have few or no fractures, which dramatically shows the influence of interfacial shear on the formation of the instability.

Clay-putty models were made to indicate the influence of material property. They differ drastically from gelatin models in that the materials have much higher Young's moduli (by 1 to 2 orders of magnitude), larger contrast between layers, and, relative to Young's moduli, very low yield strength. In Table 2, results of 10 tests show different fracture spacing to thickness relationships.

Fracture spacing is about the same in clay layers of about one-tenth the thickness of gel layers in corresponding experiments.

For computations of spacing we need measured values of Young's modulus for these materials. None could be made with any certainty. The shear modulus measured in the torsion device was 2.8×10^6 dynes/cm^2. Soil mechanics handbooks give values roughly 100 times larger. Very small yield values of clay and putty (which make these substances "typically plastic") require special unavailable equipment for precise measurement of elastic properties.

COMPUTATION OF FRACTURE SPACING

Computed fracture spacings are to be compared with observed values obtained in experiment and in outcrop. Computations were made using strain energy methods discussed earlier. Computed spacings of areas of high stress concentration (that is, nodal points) are to be correlated with observed values. Such nodal points are determined by the dimensions of the rectangular layer, by its mechanical properties and those of adjacent layers, and finally on the interfacial stresses. A specific solution is obtained by assuming a sinusoidal perturbation field of arbitrarily assigned amplitude. The physical parameters of the problem are obtained from measurements of laboratory models or field materials. The critical load is obtained by plotting tension against wave number. The minimum tension for a specified layer thickness is the critical load and determines the model spacing or computed fracture spacing.

To compute the incremental strain energy, ΔV, due to the perturbation, Hooke's Law was used to convert strains to stresses. The interfacial perturbation was taken as a small disturbance of the boundary stresses rather than displacements, and the strain energy was computed from the stresses. In reality we use the principle of least work, or Castigliano's theorem (Timoshenko and Goodier, 1970, p. 257), to obtain the energy. By fixing some small definite variation of boundary forces we limit the problem to infinitesimal deformation which permits solution

TABLE 2. EXPERIMENTAL RESULTS, CLAY MODELS

Clay layer length		Number of fractures		Fracture interval	Number of tests
Initial	Final	Major	Secondary		
A. Clay layers–3 mm thick, clay plasticine embedded in putty					
203 mm	254 mm	4.1	20-25/in.	65 mm	3
280 mm	240 mm	4.0	20-25/in.	70 mm	1
175 mm	240 mm	3.0	20-25/in.	70 mm	3
B. Clay layers–3 mm thick					
203 mm	245 mm	2.0		90 mm	3

by classical methods. Such a solution gives the energy of a possible variation from the stable rectilinear stress field that exists below the critical stress. To obtain the total strain energy we must include the energy due to stretching below the critical load; and, the strain energy due to stretching during the incremental deformation. As the problem is no longer infinitesimal we must take care of second-order terms relating to the prestress (Timoshenko and Goodier, 1970, p. 252; Biot, 1965). In our computation we have neglected terms arising from the products of stress components due to stretching, and stress components due to necking instability, thereby omitting finite deformation terms that do not influence the stability criterion.

In computations, the amplitude of the perturbation is readjusted for each variation of layer thickness; we use the force required to neck down the bar by 1 percent at maximum deflection.

Interfacial forces differ, depending on the nature of the interfacial contact. If the layers are welded, particle displacements on either side of the contact are equal. On the other hand, at free contact slip may occur and stresses need not be continuous across the interface. Slip accompanied by friction results in intermediate conditions between the end member cases. These cases differ in degree but if interfacial shear stresses are zero the deformation differs considerably and the instability is of different form.

The actual perturbation which occurs is probably not the simple sinusoidal function that was assumed. The form of the photoelastic stress pattern suggests that, at least in models, the perturbation has a more cuspate form with sharp local stress concentration of high intensity. Nevertheless, the local concentrations are periodic, so their spacing may still be predicted from the smooth sine curve. If the difference between the arbitrary displacement field and the true one is not too large, then the difference in strain energy is probably not large. The computed strain energy is too large, and thus the computed critical tension is also too large because of this difference.

The restraint of the surrounding medium is accounted for by computing the strain energy stored in it by the sinusoidal perturbation of its stable configuration and adding it to the equation for total incremental strain energy. This is computed separately, as described in a previous section. The amplitude of the applied disturbance is adjusted so that equilibrium of normal stresses across the interface between the layers is preserved.

Work done by external loads is computed so that the strain energy equation can be solved for the applied tension. To evaluate the change in length during the perturbation we neglect the terms discussed earlier that describe the thinning due to pure stretching that occurs during the necking process. This simplification eliminates the possibility of computing the critical stress exactly.

By using purely geometric methods we have obtained an evaluation of the lengthening which, in the first approximation, depends on the first power of the amplitude of the assumed perturbation. By neglecting the higher order terms in T, we construct a solution of the principle of least work (equation 16) in which the applied tension is a linear function of the amplitude of the perturbations. The critical tension is obtained by using the Trefftz criterion. It is to be emphasized that the solution gives a good approximation for fracture spacing. Moreover, the strain energy of an incremental pure stretch of the straight form at a given applied tension equal to that determined above for necking is much larger than that of the necked form. Thus, the straight form is assuredly not stable. In other words, additional external work would have to be done to obtain the same amount of shortening by simple stretching as we obtained by the necking process.

By setting the incremental strain energy equal to the work done by the external forces in a given set of physical parameters (number of nodes, amplitude of perturbation), we obtain a value of the applied tension for the stable-necked form in the slightly postnecked state.

The following measurements were used in the computations. The models' length was taken as 28 cm, the width of the stiff beam was varied from 1.2 to 4.1 cm, and that of the embedding medium was about 5.5 cm. Poisson's ratio of about 0.5 was used for both gel layers. Young's moduli of 0.28 to 0.42 $\times 10^6$ dynes/cm^2 were found for the agar layer, while they are 0.2 to 0.25 $\times 10^6$ dynes/cm^2 for the embedding gelatin gel. A convenient deflection is assumed to be 0.25 mm for the perturbation, which is too small to be measured when it first forms.

We plot T(n) to find the number of nodes which will occur at the lowest possible T, which is called the critical tension, T_{cr}, for a given amplitude of 0.25 mm (Fig. 13). By assigning an arbitrary amplitude to the incremental displacement we give a value to Q in the Trefftz criterion which is varied by changing (n). The (n) dependence is not very sensitive to amplitude variation and remains about the same over changes in B of three orders. In models, fractures are spaced at distances about equal to internodal distances so that the number of nodes and the number of fractures in a given length are about equal. Tensions are computed for each of several nodes of necking for layers, using first one, then another thickness as a parameter. For each thickness, minimum tension corresponds to a different (n) or fracture spacing. The minimum value is the critical tension. For below-critical tensions the rectangular shape is stable; for larger values the sinusoidal form with (n) nodes is formed. Thus, the curve for each thickness represents the boundary for the stability of the rectangular form at that thickness. If an additional constraint prevents formation of the instability at the minimum value (for example, friction against the glass walls of the strain frame), two possible nodes exist for one tension.

The physical nature of the instability is brought out by plotting tension against only normal surface forces (Fig. 14) and against only shear forces (Fig. 15). In the former graphs, the number of nodes increases nearly linearly with tension, but

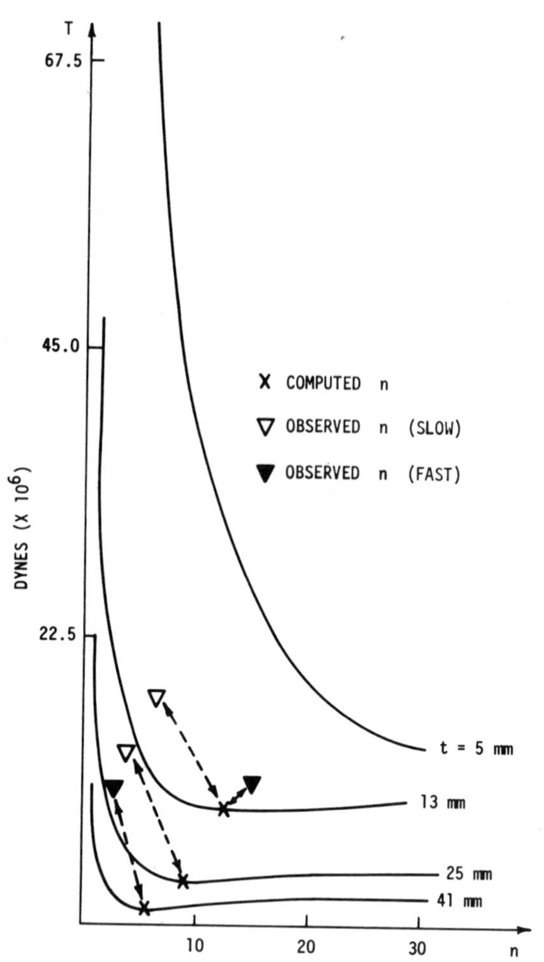

Figure 13. Graph of tension versus number of nodes (fracture sites) at various layer thickness. Critical tension and fracture spacing indicated for observed values in gelatin-agar model (gelatin E = 0.42 X 10^6 dynes/ cm^2; agar E = 0.21 X 10^6 dynes/cm^2), length 28 cm. Note linear arrangement of three critical tensions from slow tests at three-layer thickness. Slope differs from slope of three corresponding computed values. Combined action shear and normal forces on interface produces minima (critical values).

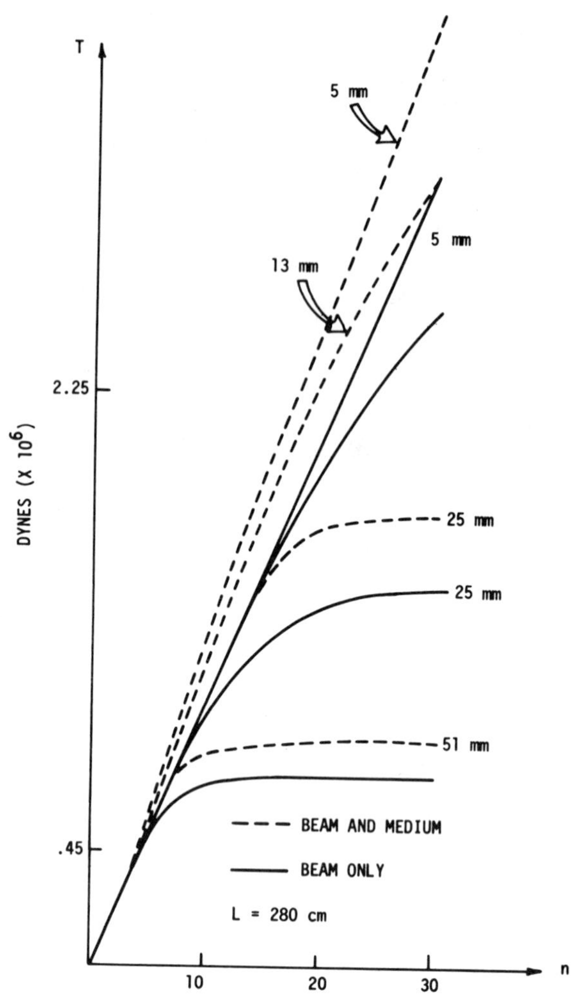

Figure 14. Graph of tension versus number of nodes at various thickness for only normal forces on interfacial surface; otherwise as in Figure 13.

in the latter ones the nodes are inversely related to the tension. As normal and shear forces act on the interface, the critical value is obtained from the superposition of the two forces. In the absence of shear the first node (n = 1) forms for very small value of T, and no higher node will ever appear unless constraints are applied; that is, the graphs do not imply as tension increases that first one node appears, then two, and so on.

These results show generally that nodal points (fractures) are more frequent in thin layers than thick ones, other things being equal. This relationship is not linear in models. In computed tensions shown here, critical tensions in thin layers are higher than in thick. Because these computed values are sensitive to the amplitude of the force function, in other computations we find higher tensions in thicker layers. An exact evaluation of critical tension would require more exact equations and the measurement of additional properties which were not made at this time. Inasmuch as the experiments themselves are crude there is little need to refine the computation at this time. The computed values and the measured values are shown in Figure 16. Although offset from one another, the slope is parallel and differences are not large. The inverse load relation shown by computation also occurs in the models.

Another feature of the instability is shown in Figure 16 by considering the stiff layer only. These graphs are nearly the same as those for the whole three-layer sandwich over most of the range. These are similar because the deformation of the embedding layer contributes to the solution for only large values of (n). The influences of variation in the contrast in elastic properties between the layers will be studied later, but first we will compare the computed and measured results. In Figure 15 the critical value and fracture spacing are shown for 5 mm, 13 mm, 25 mm, and 41 mm. The small differences shown are within the range of experimental error and the approximations used in computation. For thick layers the spacing compares

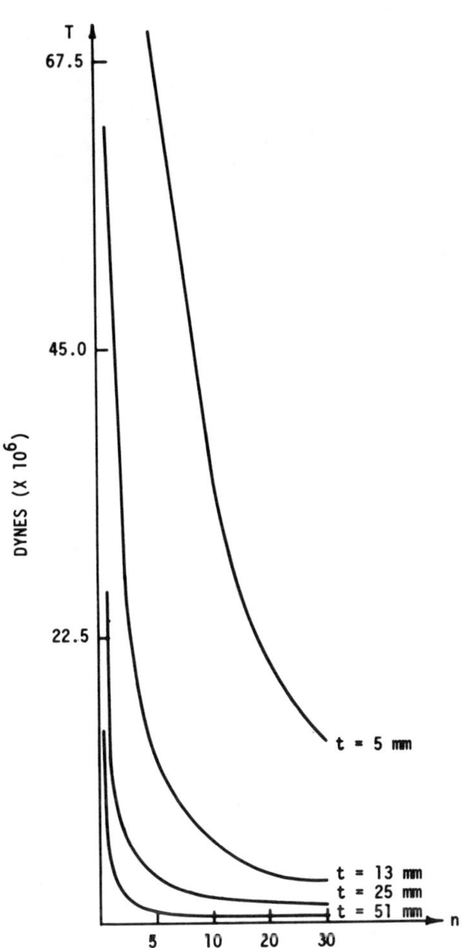

Figure 15. Graph of tension versus number of nodes at various thickness, for only shear forces act on interfacial surface. Other conditions as in Figure 13.

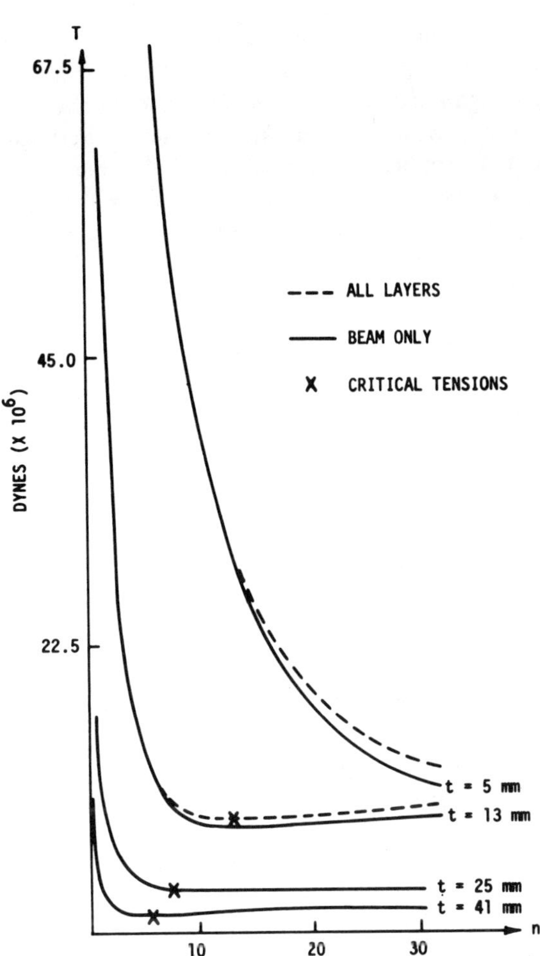

Figure 16. Comparison of critical values computed for three-layer model with those obtained for the stiff layer alone. Conditions as in Figure 13.

very well with experiment but the computed critical value is somewhat low. The inverse prevails in 25-cm layers deformed rapidly. These results are regarded as a confirmation of the general physical validity of the theory.

The effect of interfacial slip on the instability is shown in Figure 16. As already noted, shear stress seems to be required for the instability to form. The single centrally located stress concentration remains stable as load increases because no added constraint allows high-order nodes to form. A graph of the tension against thickness for constant (n) (Fig. 17) for both welded and frictionless contacts makes this evident. The tension on free interfaces is nearly uniform for all (n), which contrasts sharply with the tension on welded interfaces. Again, experimental results verify this result of computation. In experiments where interfaces are free to slip, shear stresses were made very small (indeterminate) by lubricating interfacial surfaces with a very weak solid. Either one fracture occurs at the center of the beam, or none forms. An exception occurs in high-speed loading when many fractures form, showing the possibility of a dependence on high wave numbers in shattering.

The effect of variation of thickness of the embedding layers was computed. As the stress dies out exponentially away from the interface in the embedding medium, effects of thickness are most important in thin layers. For example, the critical tension for a 5-mm layer sandwiched between two 5-mm layers is about half the value of that for two 50-mm layers.

It would indeed be surprising if fracture spacing in clay layers could be computed from the elastic strain energy methods. As the elastic range is very small in these plastic substances, it would seem that little influence would be exerted on events that occur at large strain. It was suggested earlier by Panovko and Gubanova (1965) that elastic instability might determine the site of necks in ductile materials. Therefore, it is of interest to try the computation. Because of uncertainty in Young's modulus, calculations were made for the same deformation on materials of 4.9, 49, and 490 $\times 10^6$ dynes/cm^2 moduli. The assumed value of 4.9×10^6 dynes/cm^2 provides results which approach experimental ones (Fig. 18) but, as expected, this theory does not seem to describe fracture spacing in clay.

The effect of large variation of elastic modulus is found to be nearly linear as shown by the above computations. While holding other variables constant the critical load is increased by an almost exact order of magnitude by increasing the modulus one order. Spacing is not sensitive to these changes in modulus, although fractures are more closely spaced in harder materials, other things being equal. The dependence of critical stress and fracture spacing on the ratio of moduli of the two materials has also been studied. Fractures are more widely spaced for large contrast in material. The spacing for constant ratio is less variable than is the apparent critical load.

Discussion of Results

The elastic instability mechanism explains many features of natural and experimental fracture spacing. High stress concentrations caused by compression normal to layering or extension parallel to it are caused by the instability, as can be seen in photoelastic experiments. At increased loads, fracture occurs at or near these sites and often many or all fractures form nearly simultaneously, as expected according to theory. On the other hand, stable mechanisms such as proposed by Beloussov (1952) explain the development of single fractures on some occasions. Many fractures occur in sequence; first one at the center, then others to either side, as described by Ramberg (1955), which is taken as a physical evidence of the stable mechanism.

In all theoretical and experimental work, attention has been directed toward fractures that form in hard brittle layers embedded in soft ones. My theory assumes elastic response only but supposes that large strains occur in the low modulus layer. The critical stress for fracture of the hard layer will often be larger than the strength of the embedding medium, which then flows permanently. In geology, weak interlayers are often unfractured or less fractured than "competent" beds, suggesting ductile flow of the embedding layer under moderate confining pressure. Otherwise, fractures would form throughout the whole mass with the greatest number in weak beds.

In natural deformation the regional stress or the prestress may not be composed of a single compression or tension component, nor is it likely to be uniformly distributed. For example, extension fractures in folds are uniformly spaced and perpendicular to layering, as has been shown by Harris and others (1960), and Stearns (1968), and in the laboratory by Friedman and Logan (1970, oral commun.). Reformulation of the computation of the total potential energy can take into account differences in regional loading so theory can be applied more generally. In folding applications the influence of interfacial slip and the change of curvature will influence fracture spacing. This problem will be investigated in the future in connection with folding studies.

The existence of a quantitative relation between critical stress for instability and fracture spacing may become useful geologically by indicating the magnitude of deforming stress components at the time of instability. If fracture spacing can be simply related to elastic instability the ability to compute stress components at the time of instability is implicit in existing theory. Natural fracture spacing is probably influenced by

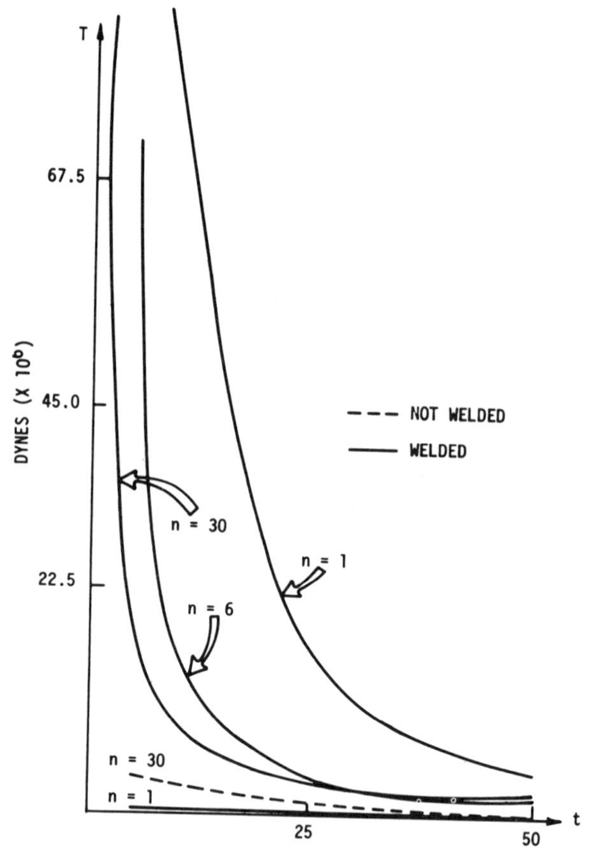

Figure 17. Graph of tension versus thickness at various constant number of nodes (fracture sites). No critical tensions occur in unwelded case.

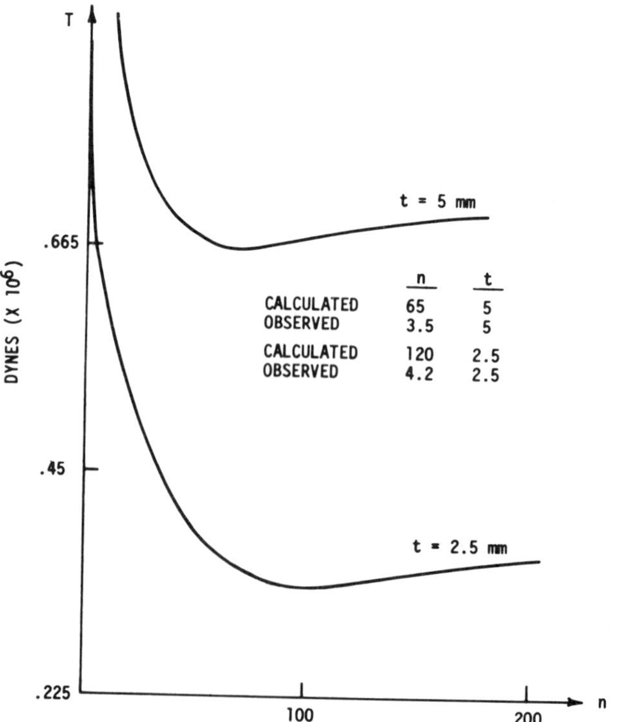

Figure 18. Graph of tension versus number of nodes for clay-putty model. Length 25.0 cm. Assumed values: clay E = 4.9 X 10^6 dynes/ cm^2; putty E assumed to be negligible.

time-dependent responses in rocks. Biot has shown how this theory can be modified to apply to a wide range of viscoelastic responses. The correspondence principle applies as long as displacements are not large, so that formal solutions from elastic theory can be converted to viscoelastic solutions by using Biot's operators in place of elastic moduli and changing strain to strain rate. The theory bids to map the magnitude of stress at one instant of geologic deformation.

Another use for theory suggests itself in conjunction with shock that may create metastable situations. The suggestion here is that very large normal stresses cause instability, with closely spaced nodes causing extension fracture swarms to occur before shear response can occur. Possibly this can provide a method for computing impact stresses or explosive loads needed for efficient rock fragmentation. This theory might be used in comminution problems encountered in peaceful applications of nuclear explosions. Thus, in very rapid deformations fractures may well transect all beds and anisotropy may be of little or no importance.

Natural fractures often pass through all beds. In experiments, as deformation is carried beyond the formation of the first set of fractures, early fractures enter the gelatin and a second sequence of short fractures forms, which are confined to the brittle layer. The throughgoing fractures mark the spacing of the original instability, and the second set fits the computed value for a new shorter layer. These fractures form where stress concentrations occur. They form only after the strain has increased and the thick embedding layer has fractured. Some throughgoing fractures are localized by stress concentrations formed earlier. These fractures persist and are often filled with intrusions of fragmented brittle agar which flowed cataclastically into the void. (The fractures were not wedged open as in hydraulic fracturing.)

In experiments it is interesting that initial imperfections and irregularities are stable. Small stress risers, such as air bubbles, are not usually loci of fractures. Small thickness variations do not grow and become unstable or localize fracture until the loads are increased to levels in excess of the critical load. This suggests that fracture phenomena in layered media are physically related to the behavior of the entire body, and points to the fundamental difference between layer instability and the deformation of a notched layer. In nature, fractures may be localized by local flaws but these are features of the same size as layer thickness itself.

Fracture spacing varies by as much as 100 percent in nature and experiment. Other than the above-mentioned influence of flaws, little is seen in experiment to account for these departures from prediction. Possibly stresses are lowered to values below fracture strength in some spaces between fractures after some fracturing occurs. Interfacial shear stresses are not always uniformly distributed. In some experiments an obvious correlation between presence and absence of fracture occurred where voids opened along layer interfaces. Further work is indicated to solve this problem.

Entire sequences occur in which fractures are very closely spaced and where they pass through all layers. This is often seen in the Rockies where large differential vertical uplift has occurred. In regions of large vertical uplift, possibly the superincumbent sedimentary load acted as a restraint against the ram effect of the uplifted basement block, allowing sediments to be crushed. In localities where the top of the basement is uniform except in down-dropped blocks, fracture spacing seems to conform to the prediction of instability theory. In these areas that were pulled apart or stretched parallel to the earth's surface, one would not expect vertical stresses to be as large, and crushing might not occur.

Fracture spacing and critical loads are dependent on contrasting rock properties. Theory predicts that the spacing of fractures in sandstone embedded in shale differs from the expected spacing of limestone embedded in the same shale, even when thickness is constant because one layer is unable to fracture; that is, that it is ductile, or elastically very soft.

Rapid Excavation

Application of these concepts in engineering geology and particularly in rapid excavation may be most useful. Theory and experiment show the great importance of interfacial shear stresses in producing fracture. Layered sequences which have been deformed at depth have a zero or ground state of high stress. Unloading places a severe stress on rocks that have such high residual stresses. This probably causes fracture, as Price (1959, 1966) has discussed. To stabilize a tunnel opening in layered prestressed rocks, it might be suggested that every attempt be made to minimize interfacial stress. This might be done by fixing stiffeners or bolts into the ductile layer (for example, the *shale* sequence). The idea would be to increase its strength and elastic modulus artificially so that it more nearly matched that of embedded brittle rocks. If bolts are laced through both rock types, they will not alleviate the problem caused by shear stresses in materials of contrasting properties. An experimental study in stabilization of heavy ground in a suitable area should be initiated to test this idea.

Another application of this concept in design of tunnel equipment and its use may be appropriate. Advantage of the natural ease of fracture can be taken in drilling the tunnel. First map the direction of fracture and measure the directions of residual stress as usual; then theory may be used to infer directions of expected natural fracture (possible places where spontaneous fracture and rock burst might occur). Measurement of fracture spacing might be taken as an indication of localities of high stress and instability that would otherwise be overlooked. More specifically, a study of spacing might indicate sites of expected fracture. Equipment might be made to take advantage of this geologic knowledge and greatly speed operations. Moreover, planning can be greatly improved when it is known which natural fractures might be expected to open spontaneously, and where along an advancing tunnel this might happen. Obviously, operations differ when these directions are parallel to or at close angle to the tunnel direction and when they are normal to it. That is, if a direction of fracture instability can be predicted to occur in a direction nearly parallel to the tunnel direction (or if it is found to occur), it might be more efficient to advance the tunnel along one long wall and then pull the other wall toward the tunnel axis. If, on the other hand, spontaneous fracture direction is nearly normal to the tunnel direction, the tunnel should advance across the entire face and then the tunnel would not have to be widened further. This suggestion needs experimental study and would be applicable in large bore tunnels where it would be convenient to use multiple front-working faces.

If these suggestions are geologically sound, and if it can be shown by experiment that it is economically worthwhile to take advantage of them, then it is evident that these ideas must be considered in the design of any tunnel boring machinery to be used in rapid excavation. Much greater flexibility in the operation of boring equipment is needed than is presently available. Application of this theory would require that the boring machine be capable of advancing the tunnel by making a thin slot or wedge—a long working face in nearly the same direction as the tunnel itself in some places. Elsewhere, as geologic conditions change, the machine must somehow telescope so that the advancing face is normal to the tunnel direction.

CONCLUSIONS

An explanation for the distribution of some fractures in layered rocks has been developed. This theory is based on the classical methods for the stability of beams for which deformation is finite. It was developed independently from the more general and exact methods given by Biot (1965). From the theory, the critical load for the initiation of an elastic instability has been formulated as a function of layer geometry and elastic properties. This provides a basis for estimating loads that cause some regularly spaced geologic fractures in layered sediments. It is possible to predict the spacing of certain types of fractures. Theory has been tested and confirmed by somewhat crude laboratory models and has been applied to natural fractures with an approximately correct result. The elastic instability theory does describe spacing of fractures that form in brittle rock under loads at or near the earth's surface. It is natural to look for suitable practical applications in the field of rock mechanics. Accordingly, it is suggested that it is possible to compute the size, configuration, and spacing of some fractures that may be encountered in rapid excavation practice. Moreover, this theoretical explanation can be used to describe geometric properties of some geologic fractures that formed under deeper-seated conditions by reformulating the equations for viscoelastic deformation.

ACKNOWLEDGMENTS

Those who know John Handin are aware of the importance of his contributions to any work done by his colleagues and it is always a pleasure to acknowledge his help. Critical discussions with D. W. Stearns were significant in establishing geologic perspective of the work. Critical review of the manuscript by William Chapple, Helmer Odé, and Ian Dalziel resulted in very significant improvements. Discussion with R. Fletcher clarified ideas concerning geometric instability. David Lasch, Lyle Baie, Jack Gallagher, Dan Heinze, and especially W. Z. Savage have helped extensively in laboratory work done at Texas A&M University. Mrs. Cherie Hale wrote the program used to compute the results presented. Finally, my thanks go to Dr. Howard Pincus, for all his many services at the symposium, and as editor of this volume.

Financial support was provided in part by National Science Foundation Grant GA1227 and also by ARPA DACA73-68-C0004.

APPENDIX I. LIST OF SYMBOLS

Symbol	Meaning
A, B, β'	Amplitude of sinusoidal perturbation
c, h	Thickness of beam
E	Young's modulus of elasticity
e_{xx}, e_{yy}, e_{xy}	Infinitesimal strain components of incremental strains
G	Shear modulus of elasticity
ℓ, ℓ_1	Original length of element
ℓ_2	Deformed length of element
$\delta\ell$	Change in length of structural element
m, k^2	Coefficients in characteristic equation for stability
L, M	Biot's elastic moduli (refer to the deformation of entire structural element)
N, Q	Two of Biot's elastic moduli for orthotropic materials; may include prestress
n	Number of nodes or half waves
P	Load or external load
s_{ij}, t_{ij}	Incremental stress tensor; t_{ij} refers to undeformed areas
s	Mean incremental stress
S_{ij}	Prestress tensor
T	Applied tension
T_{cr}	Critical tension
u, v	Components of displacement in x,y directions
V	Strain energy
ΔV	Incremental strain energy
W	External work
ΔW	Incremental work
X_i	Body forces
x,y	Reference axes
α	$\alpha = \frac{n\pi}{\ell}$
β	Parameter in characteristic equation
θ, ϕ	Functions of x,y
μ	Poisson's ratio
ξ	Parametric ratio of wavelengths in x and y directions
ρ	Density
ψ	Total incremental potential energy
ω	Angle of rigid rotation
Z	Bedding thickness

REFERENCES CITED

Beloussov, V. V., 1952, Spacing of fracture in rocks: Akad. Nauk. SSSR Trad. Geofiz. Inst. 17, p. 144.

—— 1964, Basic problems in geotectonics: New York, McGraw-Hill, p. 809.

Beloussov, V. V., and Gzovsky, M. V., 1967, Experimental tectonics, physics and chemistry of the earth: London, Pergamon Press, v. 6, p. 409–497.

Biot, M. A., 1937, Bending of an infinite beam on an elastic foundation: ASME Appl. Mech. Div., Trans., p. A1–A6.

—— 1965, Mechanics of incremental deformation: New York, John Wiley & Sons, Inc., 519 p.

Bogdanov, A. A., 1947, The intensity of cleavage as related to bedding thickness: Soviet Geology, v. 16, p. 147.

Bowden, F. P., and Tabor, D., 1954, The friction and lubrication of solids: Oxford, England, Oxford Univ. Press, p. 427.

Bridgman, P., 1952, Large plastic flow and fracture: New York, McGraw-Hill, 375 p.

Chapple, W. M., 1968, Finite amplitude folding: Geol. Soc. America Bull., v. 79, p. 47–68.

Chertkova, E. I., 1950, Some results of the modeling of tectonic faults: Akad. Nauk SSSR Izv. Ser. Geol. 14, issue 5, p. 42–44.

Cohen, M., and Vukcevich, M. R., 1969, Statistical treatment of cleavage in iron, *in* Argon, ed., Physics of strength and plasticity: Cambridge, Mass., M.I.T. Press, 398 p.

Drucker, D. C., 1968, Macroscopic fundamentals in brittle fracture, an advanced treatise, *in* Liebowitz, H., ed., Treatise on fracture: New York, Academic Press, v. 1, p. 474–532.

Filonenko-Borodich, M., 1965, Theory of elasticity: New York, Dover, p. 378.

Frocht, M. M., 1948, Photoelasticity: New York, John Wiley & Sons, Inc., p. 916.

Fung, Y. C., 1965, Foundations of solid mechanics: Englewood Cliffs, N. J., Prentice-Hall, 529 p.

Goguel, J., 1946, Introduction á l'étude mechanique des déformations de l'écorce terrestre (2d ed.): Paris, France Service Carte Geol. Mem., p. 182–197.

Griggs, D. T., and Handin, J. W., 1960, Observations on rock fracture, *in* Rock deformation: Geol. Soc. America Mem. 79, p. 347–364.

Gzovsky, M. V., 1960, Model testing of tectonic processes: Tectonophysics, p. 315–344.

Harris, S. F., Taylor, G. L., and Walper, J. J., 1960, Relation of deformational fractures in sedimentary rocks to regional and local structure: Am. Assoc. Petroleum Geologists Bull., v. 44, p. 1853–1873.

Hill, R., 1950, Plasticity: Oxford, England, Oxford Univ. Press, 554 p.

Hobbs, D. W., 1967, Formation of tension joints: Geol. Mag., v. 104, p. 550–556.

Kirillova, I. V., 1949, Problems of the mechanics of folding: Akad. Nauk SSSR Trad. Geofiz. Inst. 6, p. 27 (in Russian).

Kreyszig, E., 1963, Advanced engineering mathematics: New York, John Wiley & Sons, Inc., p. 856.

Langhaar, Henry L., 1962, Energy methods in applied mechanics: New York, John Wiley & Sons, Inc., 350 p.

McClintock, F. A., and Argon, A. S., 1966, Mechanical behavior of materials: Reading, Mass., Addison-Wesley, 770 p.

Nadai, A., 1950, Theory of flow and fracture of solids: New York, McGraw-Hill, v. 2, 3, 1269 p.

Neuber, H., 1933, Elastic solution for notch problems in extended bodies: Math u mech., v. 13, p. 439–442 (in German).

Panovko, Y. G., and Gubanova, I. I., 1965, Stability and oscillations of elastic systems: New York, Pergamon Press, p. 245.

Price, N. J., 1959, Mechanics of jointing: Geol. Mag., v. 96, p. 149–167.

—— 1966, Fault and joint development in brittle rock: London, Pergamon Press, p. 176.

Ramberg, H., 1955, Natural and experimental boudinage: Jour. Geology, v. 63, p. 512–526.

Secor, D. T., Jr., 1968, Mechanics of natural extension fracture, *in* Baer, A. J., and Norris, D. K., eds., Conference on tectonics: Canada Geol. Survey Paper 68-52, p. 3–48.

Stearns, D. W., 1968, Fracture on a mechanism of flow, *in* Baer, A. J., and Norris, D. K., eds., Conference on tectonics: Canada Geol. Survey Paper 68-52, p. 79–95.

Timoshenko, S. P., 1936, Theory of elastic stability: New York, McGraw-Hill, 495 p.

Timoshenko, S. P., and Goodier, J. N., 1970, Theory of elasticity (3d ed.): New York, McGraw-Hill, 525 p.

Voight, Barry, 1965, Plane flow of viscous matrix with an interned layer compressed between long rectangular parallel rigid plates: A geologic application and a potential viscosimeter in distorted rocks (abs.): Geol. Soc. America Abs. for 1965, Spec. Paper 87, p. 176–177.

MANUSCRIPT RECEIVED BY THE SOCIETY APRIL 5, 1972

Experimental Investigation of Sliding Friction in Multilithologic Specimens

JOHN M. LOGAN, TAKESHI IWASAKI, and M. FRIEDMAN
Center for Tectonophysics, Texas A&M University, College Station, Texas 77843
STANLEY A. KLING
Cities Service Oil Company, Exploration and Production Research, Box 50408, Tulsa, Oklahoma 74150

ABSTRACT

To evaluate the properties of rock masses, particularly the influence of friction across surfaces of mechanical discontinuity, sliding friction studies of triaxial compression have been made of all combinations of Tennessee Sandstone, Solenhofen Limestone, and Blair Dolomite. Right-circular cylinders, each with a polished saw-cut at 45° to the load axis, were deformed at confining pressures to 1.4 kb at room temperature, and at a constant rate of shortening of 10^{-4}/sec.

For intact samples, dolomite has the highest ultimate strength and limestone the lowest, while for precut monolithologic samples, limestone has the highest and dolomite the lowest. Where sliding takes place by brittle behavior along the surface, the coefficient of sliding friction, μ, is 0.5 to 0.6 and is independent of confining pressure. Where ductile behavior accompanies sliding, μ decreases with increasing confining pressure and has a mean value of about 0.7. In dilithologic specimens, μ lies between those of the corresponding monolithologic samples. The rock with the more ductile behavior along the sliding surface controls the behavior of the composite specimen. For brittle sliding, μ is maximum after 0.1 to 0.2 cm of sliding at 350 bars confining pressure; for ductile sliding, μ reaches a maximum at 0.05 cm of displacement. This suggests that ductile behavior along the sliding surface increases the area of contact. Optical and scanning electron microscopy support this conclusion. At all pressures investigated, monolithologic dolomite shortened by stable sliding, sandstone by stable sliding followed by stick-slip, and limestone by stable sliding and faulting. Dilithologic specimens showed stable sliding in all cases except for dolomite-sandstone at 350 bars where stick-slip followed stable sliding.

INTRODUCTION

Techniques of rapid-excavation cannot be utilized to their full potential until the mechanical properties of the rock mass are accurately known for all anticipated conditions. The economic implications of this knowledge are obvious. An understanding of the mechanical properties of the rock mass in the area of mineral deposits may make marginal or low-grade bodies economically feasible, especially in the face of rising mining costs. More explicit predictions of the mechanical response during drilling operations would allow greater freedom in the choice of excavation directions, shaft dimensions, pillar size, and optimum drilling rates. Removal problems can be more accurately predicted. Potential subsidence problems associated with urban transit systems and mining operations may be more intelligently investigated. Secondary effects, such as transmission of shock waves or release of stored energy, can be considered before the fact. "Overkill" techniques, empirical determinations, and engineering guesses of the bulk behavior of the rocks in their in situ condition are inefficient, costly, time consuming and, frequently, misleading.

Extensive research is in progress to develop rapid excavation and removal techniques (Howard, 1969), but the technological advances cannot be fully exploited until the rock mass can be completely characterized in terms of lithology, fracture arrays, faults, and bedding surfaces, and until the mechanical response under expected boundary conditions can be predicted. Although more than 35 yrs of laboratory study of the mechanical properties of intact, homogeneous, and isotropic specimens have contributed to our understanding of the deformations of these materials, it is still very tenuous to predict the behavior of these same materials when they contain fractures, faults, layers of other material, or other planes of potential mechanical discontinuity. Fractures are the most pervasive fabric element that may potentially create a discontinuity during the mechanical response of the rock mass. Thus, one of the first steps in assessing the behavior of the rock mass is to determine the role of friction as it acts along fractures or similar surfaces.

The effect of friction on the behavior of rocks can be assessed in a variety of ways. (1) Information gathered on the frictional properties of metals may be applied in part to rocks (for example, Carlisle, 1965, p. 277–285). However, even as a first approximation, many of these ideas, such as the independence of friction on the surface area, do not seem to be applicable to fracture-surface geometries found in rocks. (2) Laboratory and in situ measurements can be made of the bulk behavior of masses of rocks which contain arrays of fractures (Pratt and others, 1970; Jaeger, 1970). These techniques provide specific values of strength under the applied conditions, but the results are not directly applicable to other rock types, fracture arrays, or other variables. (3) Model studies can be conducted on materials simulating rocks. This approach, while promising, has limits imposed by the problems of scaling the rock properties and processes. (4) Direct laboratory experiments can be done on real rocks to determine the nature of the frictional processes.

In this case, one starts with the simplest rock characteristics – that is, homogeneous, isotropic material containing a single saw cut or fracture, deformed under controlled laboratory conditions – to evaluate one variable at a time. This approach generally ignores the influence of surface geometry upon the frictional processes. Despite this assumption, this method of research has been chosen because it is believed that controlled experiments coupled with observations of the frictional surface and the adjacent rock might provide insights into the fundamental nature of friction in the deformation of rocks. If generalizations on the frictional processes can be obtained, then one may try to integrate these with the effects of the surface geometry and the existing data on the behavior of intact rocks to predict the mechanical response of the rock mass without the need to test all possible combinations of rock types and fracture arrays.

STATE OF KNOWLEDGE

Laboratory studies to this time have provided only tentative statements on the role of friction in rocks. Displacements induced along either saw cuts or fracture surfaces have been by stable sliding, stick-slip movement, or uncontrolled sliding (Handin, 1969; Byerlee, 1967, 1970; Jaeger, 1959; Jaeger and Cook, 1969, p. 58–63; Brace, 1970; and others).

Jaeger (1959, p. 154) studied the behavior of a quartz porphyry containing a lightly ground, saw-cut surface inclined at 25° to the load axis in triaxial compression. At confining pressures from 200 to 1,000 bars, and at a temperature of 25°C, an initial period of stable sliding was observed to be followed by violent stick-slip. Byerlee (1967) investigated the behavior of both ground saw cuts and natural fractures in Westerly Granite at confining pressures to 11 kb. The saw cuts were oriented at 45° to the load axis, and fractures predeveloped at 30° to the load axis. In both cases, and at all confining pressures, the sliding was characterized by stick-slip. Byerlee concluded that sliding in the granite under these conditions involved brittle fracturing of the asperities along the sliding surface. Brace (1970) extended the work on Westerly Granite to temperatures of 200° to 500°C at confining pressures from 2 to 5 kb. At high temperatures and low pressures, for lightly ground saw cuts, stable sliding was the mode of displacement, while at low temperatures and high pressure, stick-slip was the principal mode. In a series of experiments at constant pressure, stick-slip was replaced by stable sliding as the temperature was increased. Varying the rate of shortening from 10^{-4}/sec to 10^{-6}/sec had no effect.

Byerlee and Brace (1968) reported results for fractured and lightly ground, saw-cut surfaces of three gabbros and two dunites, two limestones, a marble, a tuff, and a granodiorite. These surfaces were inclined at approximately 30° to the load axis and deformed at room temperature and confining pressure from 0.5 to 5 kb. At low confining pressures, stable sliding occurred, while at high pressures, stick-slip was the sliding mode in two of the gabbros, one dunite, and the granodiorite. At intermediate pressures, both initial stable sliding followed by stick-slip and the reverse occurred.

Coulson investigated ten rocks in direct shear at normal stresses to about 85 bars (1970). Included were basalt, dolomite, two limestones, two granites, two sandstones, a gneiss, and a siltstone. The saw-cut surfaces were prepared by sand blasting or lapping with either 80 or 600 grit and investigated under wet and dry conditions. Stick-slip always occurred in rocks containing a high percentage of quartz – the gneiss, both granites, and both sandstones – except where large amounts of clay were also present, as in the siltstone. The basalt also showed stick-slip. The carbonate rocks did not show stick-slip under any conditions. In most cases, stick-slip followed stable sliding; only on the surfaces ground with 600 grit and under the highest normal stress did stick-slip take place at the onset of sliding. The presence of water seemed to inhibit stick-slip except for surfaces finished with 600 grit, or for sandstones where there was no discernible effect.

The coefficient of sliding friction, μ, has been found to decrease with increasing normal stress for limestone, dolomite, and sandstone by Handin (1969), for a marble, trachyte, and sandstone by Jaeger and Cook (1969, p. 60), and for a serpentinized peridotite by Raleigh and Paterson (1965). In studies on Westerly Granite, Byerlee (1967) reported that for ground surfaces, μ reached a maximum after approximately 0.1 cm displacement along the sliding surface, and decreased to a nearly constant value after approximately 0.5 cm of sliding. The maximum value of μ was found to decrease as the normal stress increased. Maurer (1965) also reported that brittle materials, such as granite, showed a decrease of μ with increasing confining pressure. Coulson (1970) found that for more than 60 percent of his tests on essentially flat surfaces, the residual coefficient exceeded the initial coefficient of friction. In the remaining tests, the two values were either the same, or the residual was the lower value. He also found that the initial coefficient usually tended to decrease with increasing normal stress. Although μ also varies with surface roughness (Hoskins and others, 1968; Byerlee, 1967; Patton, 1966; Coulson, 1970), Coulson (1970, p. 150, 207) concluded that the frictional characteristics are controlled principally by the lithologic type, even with large contrasts of surface roughness.

EXPERIMENTAL PROCEDURE

Our study was initiated to determine the mechanisms whereby sliding takes place along a simulated fracture surface. We have studied monolithologic specimens with the same rock on both sides of the surface and, for the first time (as far as we know), dilithologic specimens, with different rocks on either side of the sliding surface. Triaxial compression tests were conducted on right-circular cylinders of dry Tennessee Sandstone, Blair Dolomite, and Solenhofen Limestone. Rock joints were simulated by saw cuts at 45° to the load axis, and ground with 600 grit. The surfaces were planar to ±0.001 cm and were matched together to ±0.002 cm. Confining pressures were in the range of 350 to 1,400 bars. All tests were done at room temperature and a constant rate of fractional shortening of 10^{-4}/sec. The specimens, 1.3 cm in diameter and approximately 3.5 cm long, were encased in an annealed copper jacket with a wall thickness of 0.02 cm (Fig. 1).

For comparison purposes, intact specimens were also tested under the same conditions. All experiments were conducted in an apparatus similar to that described by Heard (1963). The internal force gage provides an accuracy of about ±2 percent. The load capacity limited the confining pressures to 1,000 bars for the limestone and 1,400 bars for the dolomite and sandstone. Once sliding was initiated, it was assumed that all the axial shortening was accomplished by displacement along the precut surface, except where new macrofracturing also occurred. This assumption was confirmed from observations of the specimens after deformation. As reproducibility of friction data has been difficult in the past, at least two experiments were done for each condition.

The surfaces of slip were studied microscopically in reflected light. Selected surfaces were also photographed with a scanning electron microscope. Some specimens were retained in their jackets and thin-sectioned parallel to the cylinder axis and the sliding direction. These sections were studied on a flat stage under transmitted light. An attempt was then made to relate observed deformational features to the experimental conditions.

The coefficient of sliding friction, μ, is defined as the resolved shear stress divided by the normal stress calculated for the sliding plane. All calculations were based upon values at the initiation of sliding.

Stable sliding is herein defined as displacement which is macroscopically without interruption. If the load is applied in a continuous manner, as in these experiments, the sliding should also be continuous. As there is considerable confusion in the use of "stick-slip" (for example, Jaeger and Cook, 1969, p. 63), the term is here used as sliding in which there are abrupt changes of force with displacement. Uncontrolled sliding is defined as displacement with an abrupt loss of load. Energy is released by the system without any further input and the specimen loses all of the load. The jacket may or may not be ruptured.

Figure 2 shows the force versus time for a dilithologic sandstone-dolomite specimen. Following an initially elastic portion of the curve, an abrupt change in slope occurs. This knee indicates the onset of stable sliding. Stable sliding is terminated when force oscillations begin, indicating stick-slip behavior. The test is terminated when uncontrolled sliding or a complete loss of load occurs.

EXPERIMENTAL RESULTS

Monolithologic Specimens

Monolithologic specimens of Tennessee Sandstone were tested at 700 bars confining pressure with the sliding surface inclined at 5° increments over the interval of 20° to 55° to the load axis to confirm that 45° was a representative angle for frictional studies for the rocks selected (Fig. 3). Sliding occurred in all specimens where the surfaces were inclined from 20° to 45°. When the angle between the saw cut and the load axis was increased to 50° and 55°, new fracturing and sliding both took place. At 60° only fracturing occurred. This agrees with the predictions of Handin (1969) for these rocks. Although there is a systematic variation in the ultimate strength with the angle of sliding, the variation is small, and 45° is regarded as representative. Fracturing occurred in the limestone and dolomite at 55° and 65°, respectively, under the same conditions.

Blair Dolomite was tested at confining pressures to 1,400 bars (Fig. 4A). Separate tests were stopped at various increments of shortening, and the surfaces were observed to determine where sliding started. At all confining pressures, initial sliding was indicated by the marked change of slope of the stress-shortening curve (Point A, Fig. 4A). Displacement along the surface was by stable sliding until uncontrolled sliding accompanied by audible "snap" terminated the test.

The slopes of stress-shortening curves are positive at all confining pressures, although the amount of shortening prior to uncontrolled sliding decreases with increasing pressure. The energy released during uncontrolled sliding is contributed in part by that stored in the specimen, but also by that stored in

Figure 1 (left). Specimen 1.3 cm in diameter jacketed to upper and lower pistons after test. Light lubrication between the piston face and upper spacer reduces misalignment as displacement progresses.

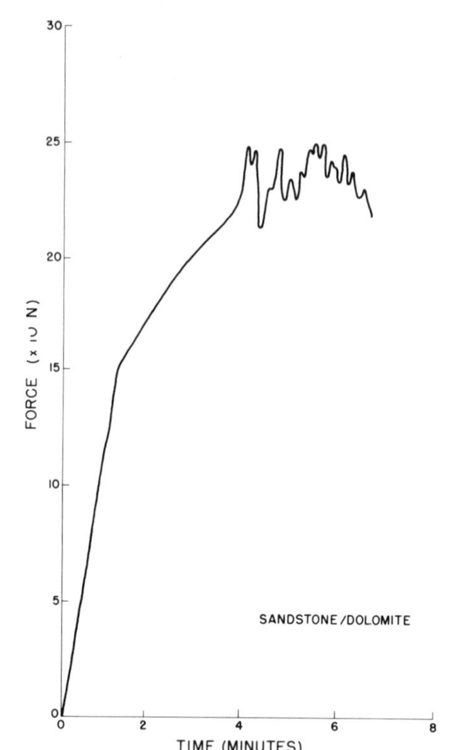

Figure 2 (middle). Differential axial load versus time for dilithologic Tennessee Sandstone–Blair Dolomite specimen deformed at 350 bars. The initial elastic portion of the curve is followed by stable sliding which is initiated at the abrupt change in slope. Stick-slip follows stable sliding but is terminated by uncontrolled sliding.

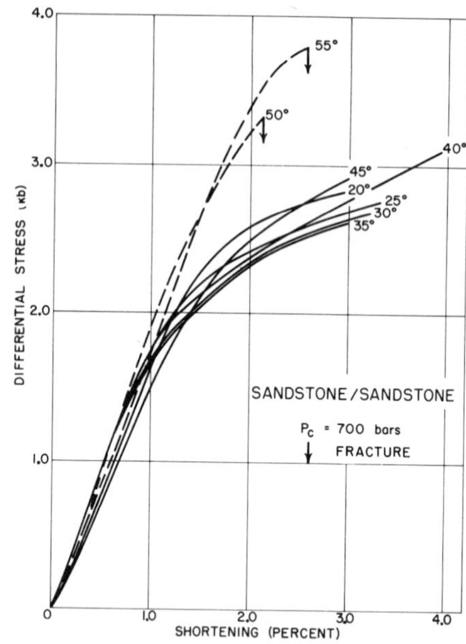

Figure 3 (right). Differential stress versus axial shortening for saw-cut Tennessee Sandstone deformed at 700 bars confining pressure. Shortening by displacement along the precut is shown by the solid curves. New faulting and sliding is shown by the dashed curves. The angles are measured between the precut and the load axis.

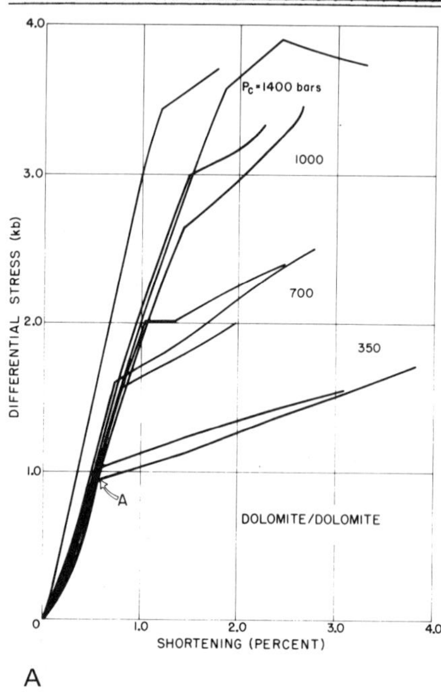

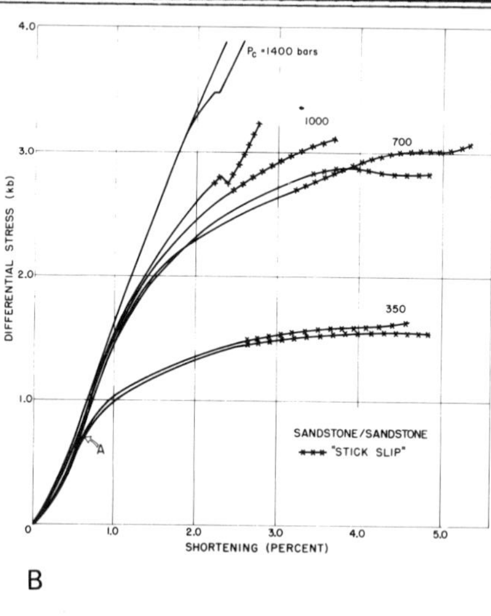

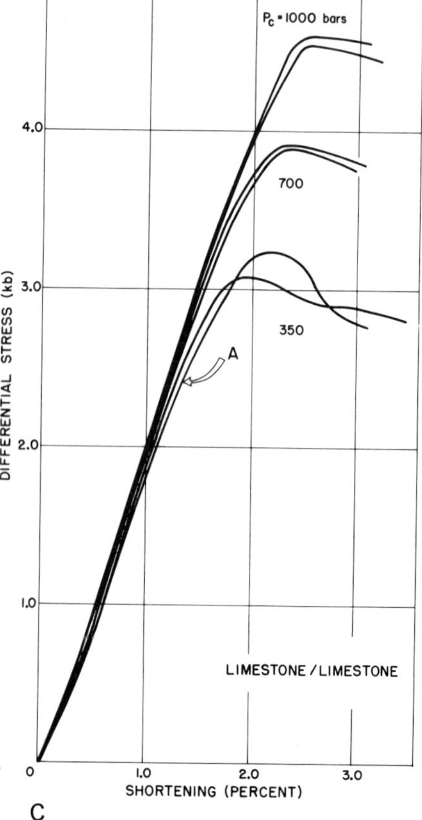

Figure 4. Differential stress versus axial shortening for specimens containing a precut at 45° to the load axis. Confining pressures shown in bars. Point A indicates the onset of sliding. A. Blair Dolomite. B. Tennessee Sandstone. C. Solenhofen Limestone.

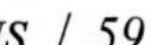

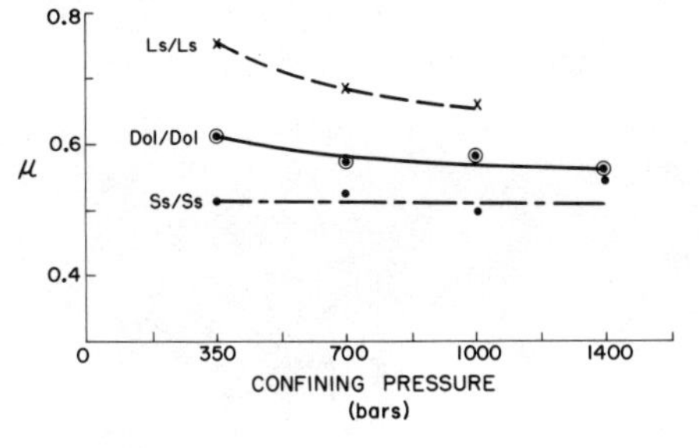

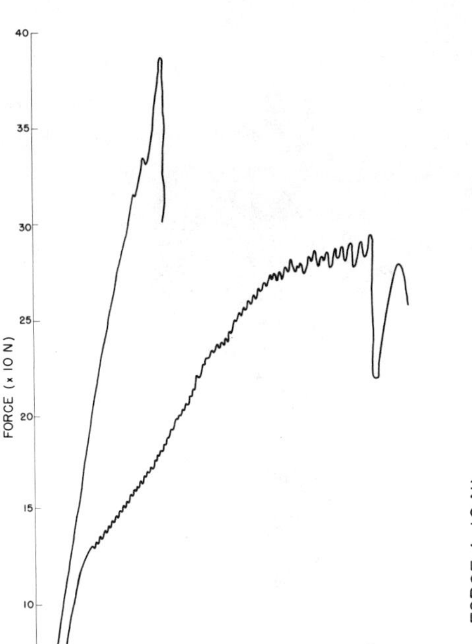

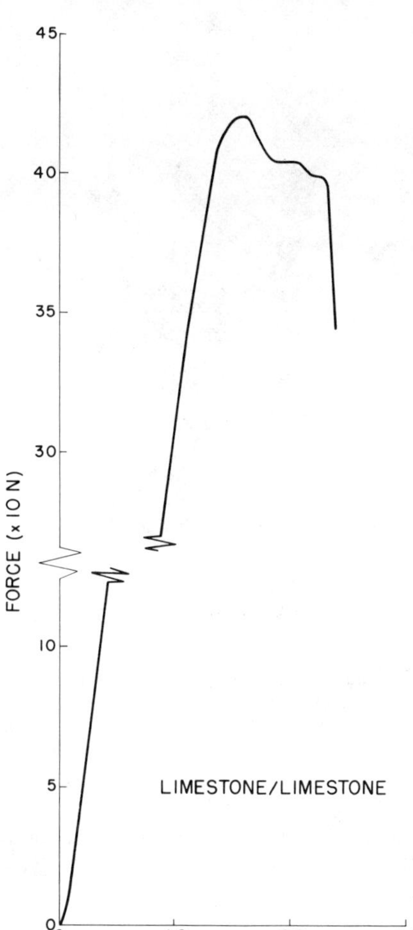

Figure 5 (left). The initial value of the coefficient of sliding friction, μ, versus confining pressure for Solenhofen Limestone, Blair Dolomite, and Tennessee Sandstone. Curves are visual approximations.

Figure 6 (middle). Differential axial load (newtons) versus time for tests on saw-cut Tennessee Sandstone. The lower curve is typical of 350 bars confining pressure; the upper is characteristic of 1,000 bars. At a constant rate of 10^{-4}/sec, the abscissa is equivalent to axial shortening of the specimen.

Figure 7 (right). Differential axial load (newtons) versus time for saw-cut Solenhofen Limestone. This curve is typical for confining pressures of 350 bars where stable sliding was followed by new faulting.

the loading column. At higher confining pressures, higher loads are necessary to sustain sliding; more energy is therefore stored in the loading column, and uncontrolled sliding is favored.

As calculated at the initiation of sliding, μ is approximately constant at all confining pressures investigated (Fig. 5).

The behavior of the Tennessee Sandstone is similar in some respects to that of the Blair Dolomite (Fig. 4B). Sliding initiates after an essentially linear elastic deformation (Point A, Fig. 4B) at all pressures to 1,400 bars. Permanent shortening of the specimen is accomplished by displacement along the cut surface. The initial sliding is always stable, but as displacement increased, stick-slip occurs. The experiments finally terminate with uncontrolled sliding.

Stick-slip in the sandstone is accompanied by periodic, although irregular, oscillations of the force (Fig. 6). Each load drop is accompanied by an audible "snap." Stick-slip occurs in the sandstone at pressures to 1,000 bars, although the number of load drops and the amount of displacement preceding the ultimate (peak) strength decrease with increasing pressure. At 1,400 bars, although stick-slip may have taken place before uncontrolled sliding terminated the test, the record is not definitive. Stick-slip follows stable sliding at all pressures, but always precedes the ultimate strength of the specimen (Fig. 4B). The amount of stable sliding decreases at the higher pressures until it becomes almost unrecognizable. The force drops are very small, but they increase with both displacement and force until uncontrolled sliding occurs. The irregularity of the oscillations also increases with displacement.

The initial coefficients of friction of the sandstone are the lowest of the three rocks investigated, about 0.5, and are almost independent of confining pressure (Fig. 5).

At 350 bars confining pressure, displacement in the Solenhofen Limestone is initiated in the region where the stress-shortening curves become nonlinear (point A, Fig. 4C). Stable sliding is followed by faulting of the rock immediately adjacent to the saw cut. The fractures form subparallel to the saw cut and in a zone on either side of it (Fig. 8B). The width of this zone is notable when compared with monolithologic dolomite (Fig. 8A). The ultimate strength is attained very early in the sliding history. After the ultimate strength is reached, displacements occur during which no further changes in load are recorded (Fig. 7). No audible noise accompanied the shortening.

A

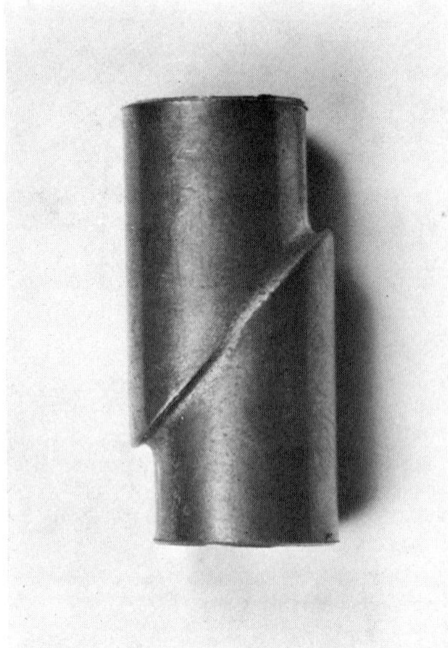

B

C

Figure 8. A. Photograph of saw-cut Blair Dolomite deformed at 350 bars confining pressure. No macroscopic fractures are associated with sliding along precut. B. Photograph of saw-cut Solenhofen Limestone deformed at 350 bars confining pressure. Fracture zone occurs subparallel to precut surface and intersecting it near the middle of the specimen. C. Photograph of Yule Marble deformed at 700 bars confining pressure. A small amount of stable sliding along the precut surface preceded formation of the fault. All specimens are 1.2 cm in diameter.

At 700 and 1,000 bars confining pressure, displacement occurs along the pre-existing surface and also along the newly formed fractures. As the confining pressure increases, the displacement decreases along the saw cut but increases along the new fractures. To confirm this behavior, some experiments were conducted on Yule Marble at 350 and 700 bars confining pressure. Very similar behavior was found. At 700 bars new fractures form at an angle characteristic of the intact material (Fig. 8C). The initial coefficient of sliding friction for limestone decreases with increasing confining pressure (Fig. 5), in contrast to those of the sandstone and dolomite. Its values are the highest of all the rocks investigated at all pressures.

Dilithologic Specimens

For the first time, the frictional properties of unlike rocks sliding by each other have been investigated in the laboratory. Of interest are (1) whether the character of the sliding is controlled by one lithologic type or is different from that of both individual rocks, and (2) how the strengths of the dilithologic specimens compare with those of the monolithologic ones.

The stress-shortening curves for the sliding of Tennessee Sandstone against Solenhofen Limestone are shown in Figure 9A. Initiation of sliding is indicated where the curves become nonlinear (point A). Stable sliding occurs at all pressures, in contrast to the faulting and sliding seen in the monolithologic limestone specimens or the stable sliding and stick-slip seen in the monolithologic sandstone. The ultimate strength and character of the stress-shortening curves are similar to those of limestone (Fig. 12), although the dilithologic curves are always intermediate. This apparent control by the limestone of the behavior of the composite specimen is further reflected in Figure 9B. Here the initial coefficient of friction is very close to that of the monolithologic limestone specimens and it decreases with increasing confining pressure. The parallelism of the dilithologic values with those of the limestone also is in striking contrast to the sandstone alone, which is independent of confining pressure.

The behavior of the dilithologic combination of Solenhofen Limestone and Blair Dolomite is shown in Figure 10. Only stable sliding occurs. The shapes of the curves and the ultimate strengths both indicate a very strong control by the limestone on the composite behavior (Fig. 12). The stress-shortening curves lie between those of the monolithologic specimens, ex-

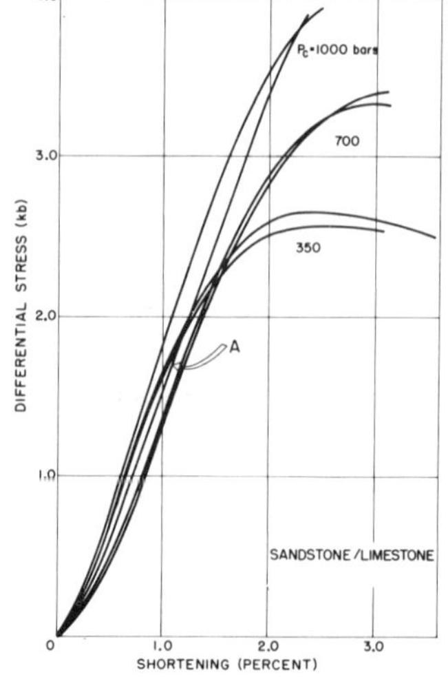

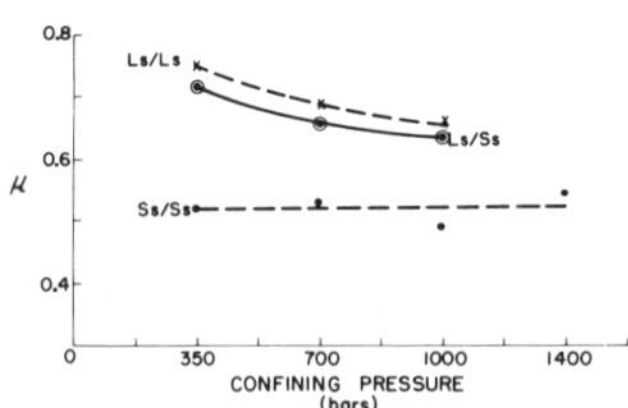

Figure 9. Left: Differential stress versus axial shortening for dilithologic Tennessee Sandstone–Solenhofen Limestone specimens. Confining pressures noted for each set of curves. Sliding was initiated at about Point A of the 350 bar curve. Right: Initial value of the coefficient of sliding friction versus confining pressure. Dilithologic values connected by solid curve; monolithologic counterparts by dashed lines.

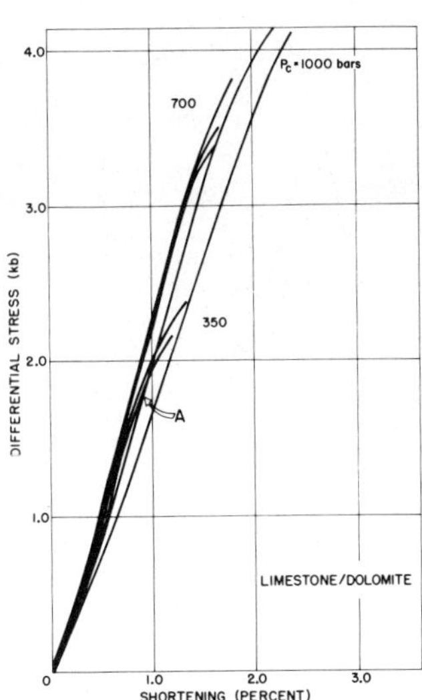

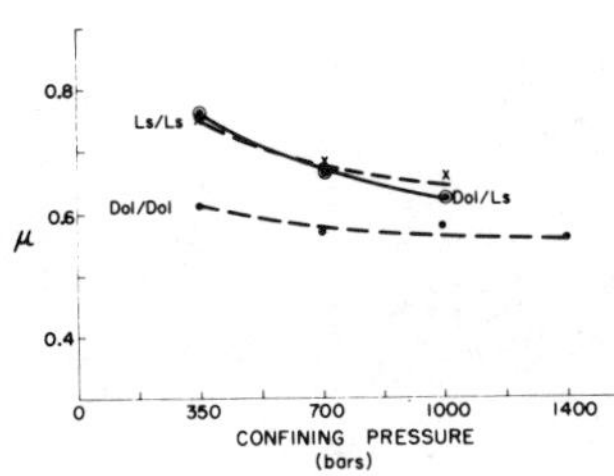

Figure 10. Left: Differential stress versus axial shortening for dilithologic Solenhofen Limestone–Blair Dolomite specimens. Confining pressures are indicated for each set of curves. Sliding was initiated at about Point A of the 350 bar curve. Right: Initial value of the coefficient of sliding friction versus confining pressure. Dilithologic values connected by solid curve; monolithologic counterparts by dashed lines.

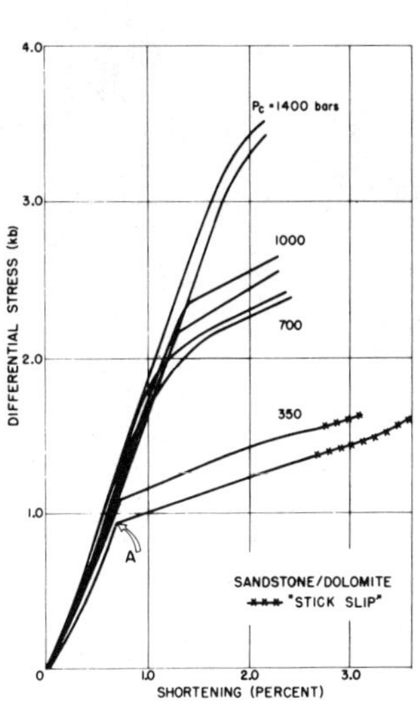

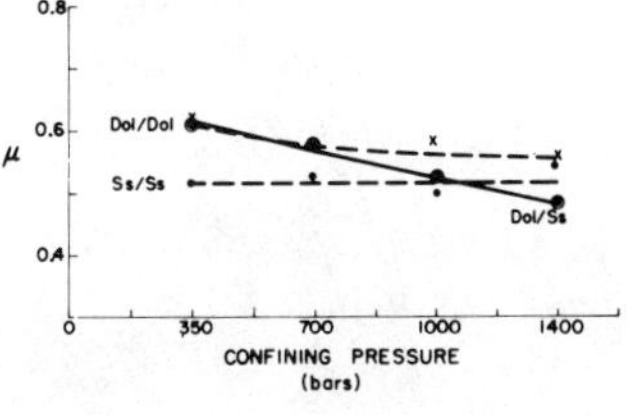

Figure 11. Left: Differential stress versus axial shortening for dilithologic Tennessee Sandstone–Blair Dolomite specimens. Confining pressures are indicated for each set of curves. Sliding was initiated at about Point A of the 350 bar curve. Right: Initial value of the coefficient of sliding friction versus confining pressure. Dilithologic values are connected by solid curve; monolithologic counterparts by dashed lines.

cept at 1,000 bars confining pressure. Here the slope of the curve is somewhat higher, but the reason for this is not known. The significant control of the limestone upon the frictional character of the specimens is further manifested in Figure 10B. At pressures to 700 bars, the coefficient of friction has values identical to those of the limestone alone.

In the sliding of Tennessee Sandstone against Blair Dolomite (Fig. 11A), stable sliding is followed by stick-slip at 350 bars confining pressure. The nature of the stick-slip is similar to that seen in sandstone alone, except that the ultimate strength occurs at about the initiation of stick-slip here, in contrast to stick-slip initiation before the ultimate strength in the sandstone. The load drops are accompanied by audible "snaps." At confining pressures of 700 bars and above, stable sliding occurs until the test terminates in uncontrolled sliding. The stick-slip is characteristic of the monolithologic sandstone, but not dolomite, at the lower confining pressures. However, at 1,000 bars, stick-slip is still characteristic of sandstone alone.

Figure 12 suggests that while the stress-shortening curves for dilithologic specimens of sandstone against dolomite lie between the respective monolithologic curves, no pervasive control by either is obvious. Even the long period of stable sliding preceding stick-slip is more characteristic of monolithologic dolomite than sandstone. Consideration of the values of μ and their variation with confining pressure helps to clarify the behavior (Fig. 11B). The initial coefficient of friction of the dilithologic specimens is identical to that of the dolomite alone at 350 bars confining pressure, and μ decreases linearly until 1,000 bars, where it is about the same as that of sandstone alone. At 1,400 bars μ is less. This change in behavior is also suggested by comparing the behaviors at 350 bars and 1,000 bars (Fig. 12). It is paradoxical that the coefficient of friction at the lower pressures is very close to that of dolomite alone, but the mode of sliding, stick-slip, is characteristic of the sandstone. The reverse is true at 1,000 bars confining pressure.

DEFORMATION MECHANISMS ACCOMPANYING SLIDING

Monolithologic Specimens

An understanding of the behavior of the material along the sliding surfaces is critical to the predictions of the role of friction in other untested rocks. Tests on intact, homogeneous specimens of the sandstone and dolomite show that these rocks are macroscopically brittle under all the conditions investigated here (Fig. 12). Failure is due to macroscopic fracture. Microfracturing is also present in both rocks, but is more prevalent in the sandstone because of the angular nature of the grain boundaries. It seems reasonable to suppose that sliding in the sandstone and dolomite alone, and even in the dilithologic sandstone-dolomite, is accompanied by brittle deformation of the rocks adjacent to the sliding surfaces. Within the brittle regime of granite, Byerlee (1967) has suggested that at low normal stress, sliding occurs by lifting over the asperities. At higher stresses, the asperities are broken through by microfracturing. This same behavior is indicated here. Thin-section observations of the surface show that the zone of sliding is very narrow, apparently a maximum of a few grain diameters. Figures 13 and 14 show scanning-electron photomicrographs of the sliding surfaces of sandstone and dolomite, respectively. Note the large amount of microfracturing in both rocks and the granular debris. Some grains have subrounded shapes, which suggests that they have been rolled along during the sliding. Much of the debris forms by microfracturing and rigid-body rotation, and seems to collect in depressions. This tends to reduce the surface relief as displacement progresses and to increase the actual area of contact, bringing it closer to the apparent area. This increase in contact area requires an increase in shear stress if displacement is to continue. This may explain the positive slope of the stress-shortening curves (Fig. 4). If brittle behavior is the dominant mode of deformation along

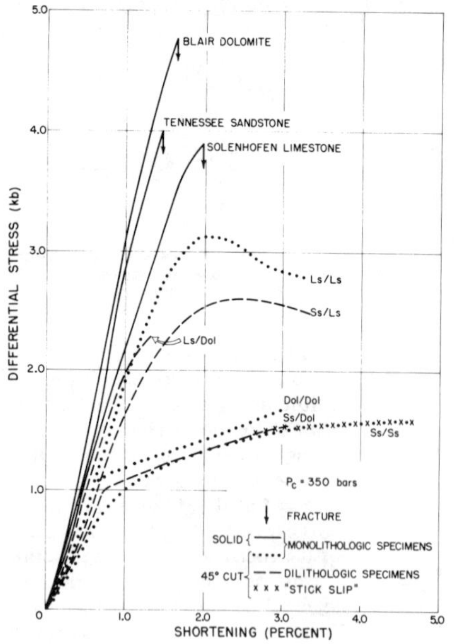

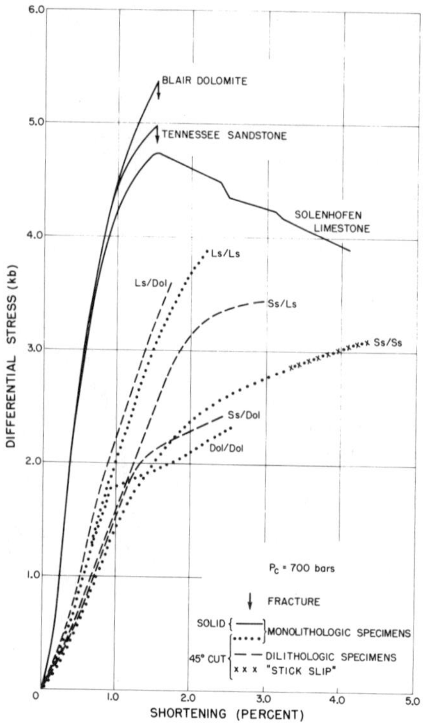

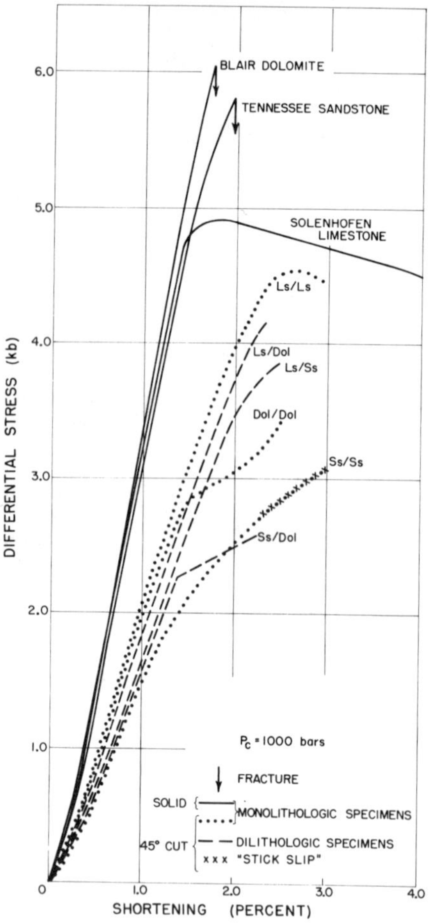

Figure 12. Differential stress versus axial shortening curves at three different confining pressures for intact specimens (solid curves), monolithologic saw-cut specimens (dotted curves), and dilithologic specimens (dashed curves). Curves are averages of two or more experiments.

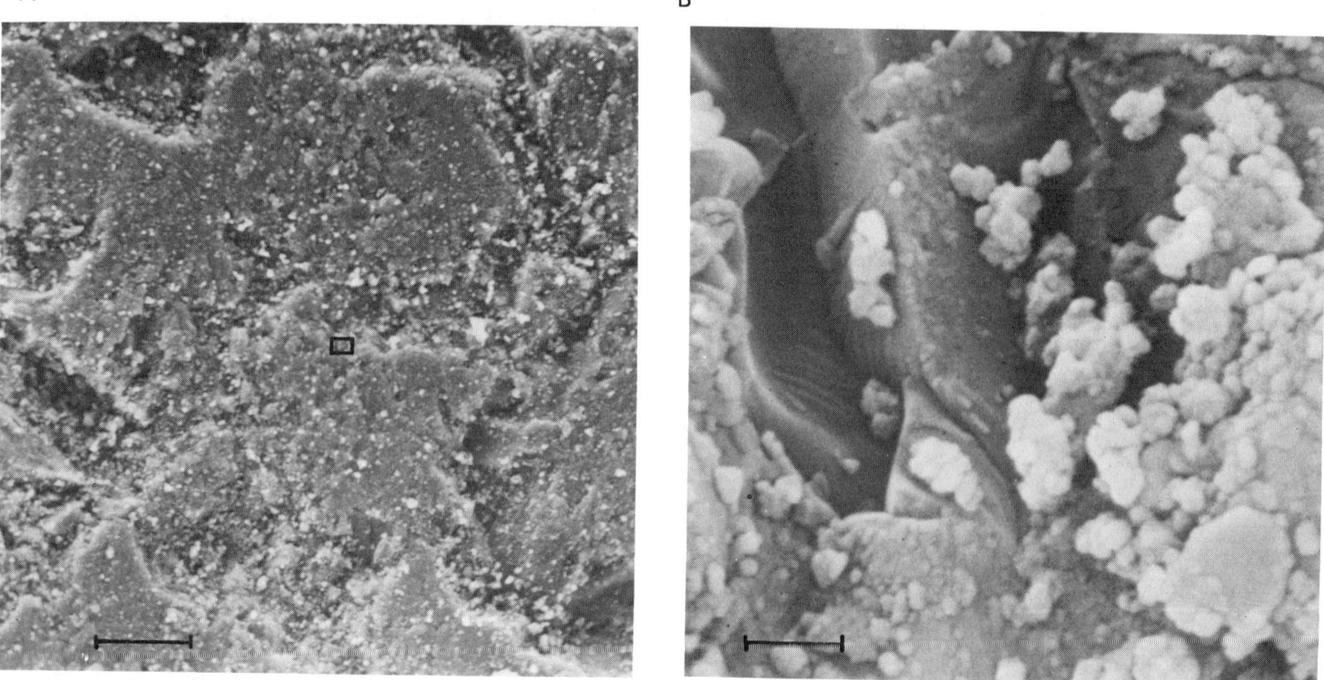

Figure 13. Scanning electron photomicrographs of Tennessee Sandstone sliding surface after deformation at 700 bars. Specimen is a dilithologic sandstone-dolomite. The sandstone moved toward the top of the page. A. 300 X magnification. Scale equal to 42×10^4 A. Negative areas occur along grain boundaries. Square indicates enlargement seen in B. B. 10,000 X magnification. Scale equal to 1.2×10^4 A. Conchoidal surfaces are probably microfractures. Débris is rounded.

these surfaces, the coefficient of sliding friction should be independent of confining pressure (Handin, 1969). This is true also of the coefficients for the initial sliding of the dolomite and sandstone (Fig. 5).

The behavior of Solenhofen Limestone is very different from that of Tennessee Sandstone and Blair Dolomite under our testing conditions. Displacement along the cut surface is accompanied by faulting at all confining pressures. The amount of sliding decreases, however, until at 1 kb, none is discernible. The fracture zone adjacent and subparallel to the cut developed at 350 bars (Fig. 8B), while at 700 bars the saw-cut limestones faulted at 30° to the load axis, which is exactly the angle in the intact rocks. The same behavior observed in Solenhofen Limestone is also seen in Yule Marble deformed under the same conditions (Fig. 8C).

The restricted development of the fracture zone adjacent to the precut at 350 bars suggests that strain induced by slip along the surface promotes fracturing of the limestone. The limestone is macroscopically brittle to confining pressures of approximately 1 kb, where the transition to ductile behavior occurs (Fig. 12). Within the brittle regime, macroscopic failure is by throughgoing shear fracture. However, the low critical resolved shear stress necessary to cause twinning in calcite (about 70 to 170 bars at 25°C; *see* Heard's Fig. 2, *in* Friedman, 1967), suggests that some twinning may be activated even here, and this has indeed been substantiated by Friedman (1963) in other limestones. It is here postulated that

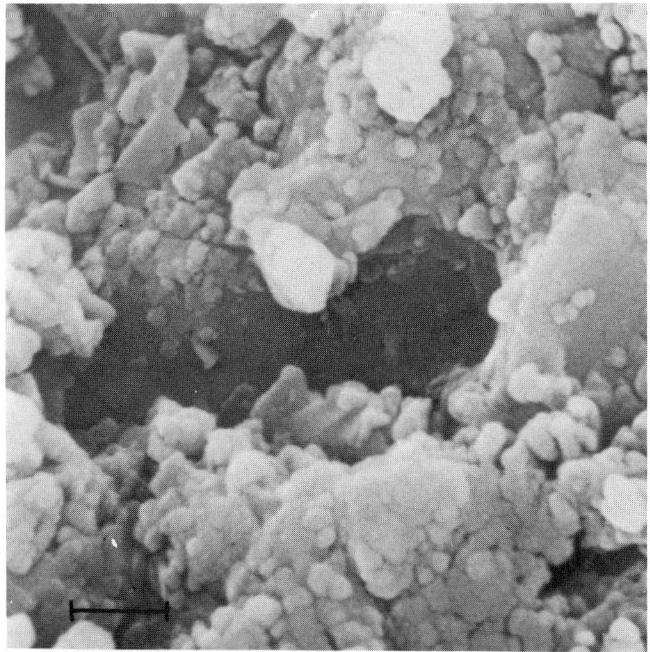

Figure 14. Scanning electron photomicrographs of Blair Dolomite surface, magnification 10,000 X. Scale equal to 1.2×10^4 A. This surface was in contact with the one shown in Figure 13. The displacement was toward the top of the page. The disaggregated debris is characterized by fractures, some angular, and some rounded fragments.

twinning of the limestone occurs at the asperities on the sliding surface. This localized microscopically ductile behavior would tend to remove the asperities and to fill in the less deformed areas between. The net result would be an increase of the area of contact until it approached that of the intact, homogeneous rock.

At 700 bars, faulting is favored over sliding, and the fault lies at an angle characteristic of the intact rock. Petrographic observations of the Solenhofen Limestone are precluded by its very fine grain size, but thin sections of Yule Marble have been studied in transmitted light. Only those grains immediately adjacent to the sliding surface and possessing a favorable orientation show evidence of twinning. Moreover, in twinned grains the twin gliding is restricted to that portion of the grain immediately adjacent to the sliding surface.

Further evidence in support of this hypothesis can be found by considering the displacement needed to reach the maximum value of μ during any single experiment (Table 1). This maximum value should correspond to the maximum interlocking of the asperities, that is, to the greatest area of contact along the sliding surface. If microscopic flow occurs in the limestone under stress concentrations produced at the asperities, one would predict that the maximum value of μ should be found after very small displacement; that is what is observed. Notice in Table 1 that the smallest displacements are seen in the limestone, while much larger ones occur in the sandstone and the dolomite, which are postulated to slide by brittle mechanisms. Also notice that this displacement becomes smaller as the confining pressure increases. If the sliding is by a ductile mechanism, we would expect μ to decrease with increasing confining pressure, and this is just what occurs in the limestone (Fig. 5) and also in the Yule Marble.

TABLE 1. SLIDING (CM) BEFORE MAXIMUM COEFFICIENT OF SLIDING FRICTION IS ATTAINED FOR MONOLITHOLOGIC SPECIMENS

Specimen	Confining pressure (bars)			
	350	700	1,000	1,400
Tennessee Sandstone	0.18	0.14	0.09	0.02
Blair Dolomite	0.13	0.07	0.04	0.03
Solenhofen Limestone	0.05	0.03	0.01	

The types of mechanisms that accompany sliding are reflected in the relative strengths of the precut specimens as compared to the intact ones. At all confining pressures, tests on the intact specimens show Blair Dolomite to be the strongest and Solenhofen Limestone to be the weakest. However, for the monolithologic specimens containing a saw cut, the limestone is always the strongest. The ultimate strengths of all saw-cut specimens increase with increasing confining pressure, but only that of the limestone approaches that of the intact rock. Any attempt to predict the behavior of limestone, marble, or other rock that is microscopically ductile under the expected conditions must take account of the possibility of increased area of contact, which alters the relative strengths of fractured rocks and their mechanical response.

Dilithologic Specimens

Dilithologic specimens have sliding characteristics that lie between their monolithologic counterparts. If either rock exercises control in the sandstone-dolomite specimens, the effect is difficult to detect. The stress-shortening curves and the variation of μ with confining pressure both suggest that dolomite exerts a slightly stronger influence at low pressures, while the sandstone may be more important at the higher pressures, but nothing significant is revealed by microscopic studies to confirm this speculation. The presence of stick-slip at 350 bars pressure is clearly characteristic of the sandstone, but even here the processes have been modified. The long-region of stable sliding is characteristic of dolomite, and the coincidence of the ultimate strength and the onset of stick-slip is not typical of sandstone. Testing under a wider range of confining pressures might help to answer this question.

The influence of the limestone in contact with dolomite or sandstone is very pronounced. The form of the stress-shortening curves, the ultimate strengths, the values of μ, and their variation with confining pressures, all indicate that the limestone is dominant. We have pointed out that the limestone behaves in a microscopically ductile manner wherever asperities exist, but the brittle response of the other rocks still limits the actual area of contact, so that here sliding rather than faulting is the mode of shortening. However, even here, the amount of sliding that precedes the maximum value of μ is noticeably lower in the limestone combinations than for dolomite-sandstone (Table 2). Notice that, as in the monolithologic specimens, the amount of sliding also decreases with increasing pressure.

Another possibly significant fact is that the limestone, because of its low ductility, is softer than the dolomite or sandstone so that when sliding does occur the harder rocks plow through the limestone and produce well-defined and continuous grooves (Figs. 15 and 16). Striations also appear on the dolomite and sandstone surfaces, but they are less distinct and intermittent. This behavior is consistent with that found in metals by Tabor (1945) and Gemant (1950).

TABLE 2. SLIDING (CM) BEFORE MAXIMUM COEFFICIENT OF SLIDING FRICTION IS ATTAINED FOR DILITHOLOGIC SPECIMENS

Specimen	Confining pressure (bars)			
	350	700	1,000	1,400
Tennessee Sandstone and Blair Dolomite	0.13	0.05	0.04	0.03
Tennessee Sandstone and Solenhofen Limestone	0.06	0.06	0.02	
Solenhofen Limestone and Blair Dolomite	0.02	0.01	>0.01	

TYPES OF SLIDING

The type of sliding – stable, stick-slip, or uncontrolled – is most important in defining the behavior of the rock mass. Our experiments have shown that when sliding occurs in mono-

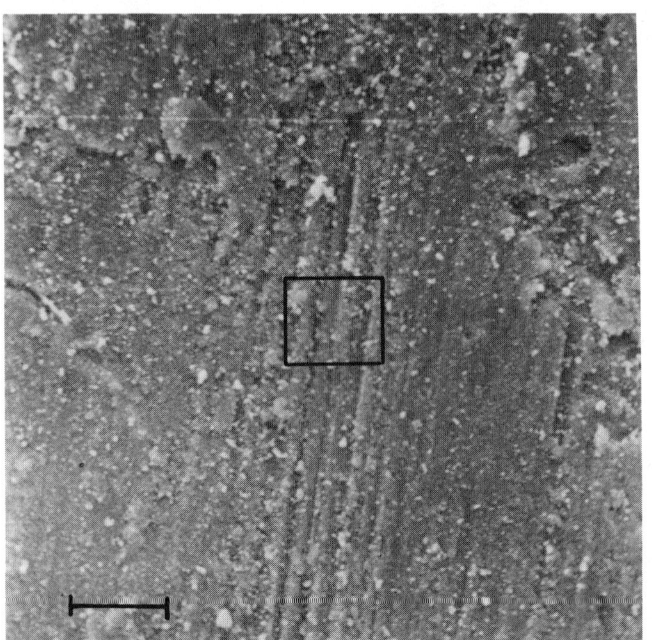

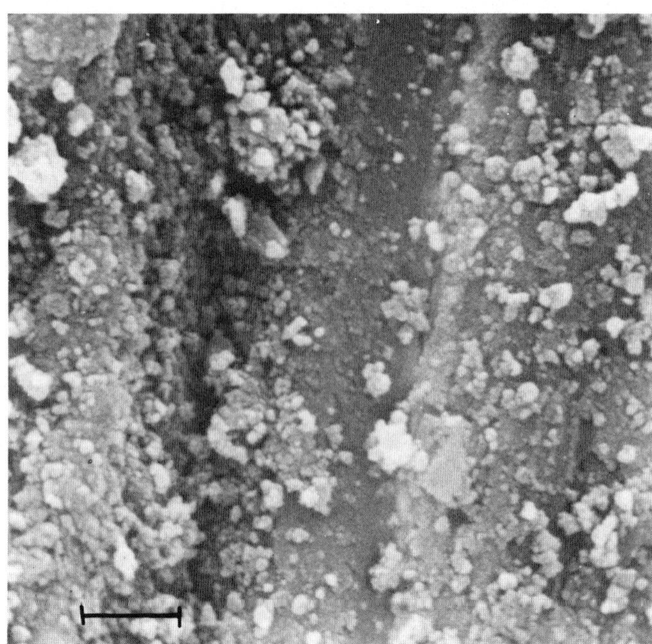

Figure 15. Scanning electron photomicrographs of Solenhofen Limestone surface. This surface was in contact with sandstone and deformed at 350 bars. Displacement of the limestone was toward the bottom of the page. Left: 300 × magnification. Scale equals 42 × 10^4 A. Striations or grooves left by plowing of sandstone show sliding direction but not sense. Right: 3,000 × in area outlined at left. Scale equals 4 × 10^4 A. Debris left on surface is both angular and rounded. Trace of groove is apparent.

lithologic specimens of limestone, dolomite, and marble, it is stable. This agrees with the results of Coulson (1970) and Byerlee and Brace (1968).

Monolithologic specimens of sandstone characteristically slip by stable sliding, followed by stick-slip. Stick-slip has been reported in rocks that contain a high percentage of quartz (Coulson, 1970; Byerlee and Brace, 1968; Byerlee, 1967; Jaeger, 1959; Rabinowicz, 1965). However, not all such rocks show stick-slip. Coulson (1970, p. 110) finds only stable sliding in a siltstone; apparently the clay minerals control its behavior. Preliminary tests on Coconino Sandstone, which has approximately 18 percent porosity, reveal only stable sliding. This same effect occurs in a highly porous tuff (Byerlee and Brace, 1968).

Stick-slip does seem to be characteristic of the sliding of monolithologic, quartz-rich specimens which have low porosity and do not contain a large amount of clay, and stick-slip is not confined to quartz-rich rocks, having been observed in some, but not all, dunites, gabbros (Byerlee and Brace, 1968), and a basalt (Coulson, 1970). Sliding of the sandstone in dilithologic specimens shows stick-slip only when the sandstone is in contact with the dolomite, and then only at the lowest confining pressure. Thus, modification of stick-slip in quartz-rich rocks is expected from contacts with other lithologic types and with variations in confining pressure.

The transition to uncontrolled sliding confirms the suggestion by Jaeger and Cook (1969, p. 62), Rabinowicz (1965), and Bombolakis and others (1970) that the load drop associated with stick-slip involves contributions from both the loading machine and the material itself. On the other hand, Byerlee and Brace (1968) conclude that apparatus stiffness does not affect stick-slip in their specimens of Westerly Granite. The uncontrolled sliding in our experiments probably occurs when the energy stored in the apparatus is a large part of the total stored energy, and this behavior is due more to experimental conditions than to material response. Thus, uncontrolled sliding is due merely to a load loss associated with stick-slip, but the overwhelming amount of energy released from the loading column ruptures the jacket and causes termination of the test.

It seems unlikely that stick-slip is due entirely to the response of the apparatus. If the energy stored in the loading column leads to stick-slip, then the higher the load, the more energy stored in the column, and the more likely is stick-slip. However, this is not what is observed. Limestone has much higher ultimate strength than does sandstone, but the former never shows stick-slip. Dilithologic sandstone-dolomite specimens show stick-slip at the lower confining pressures, and thus low differential loads, but only stable sliding occurs at the higher confining pressures and higher loads.

In all tests where stick-slip is recognized, the force drop is accompanied by an audible "snap," caused by the release of seismic energy. Neither stable sliding, sliding and faulting, nor movement along the composite zone in the carbonate rocks,

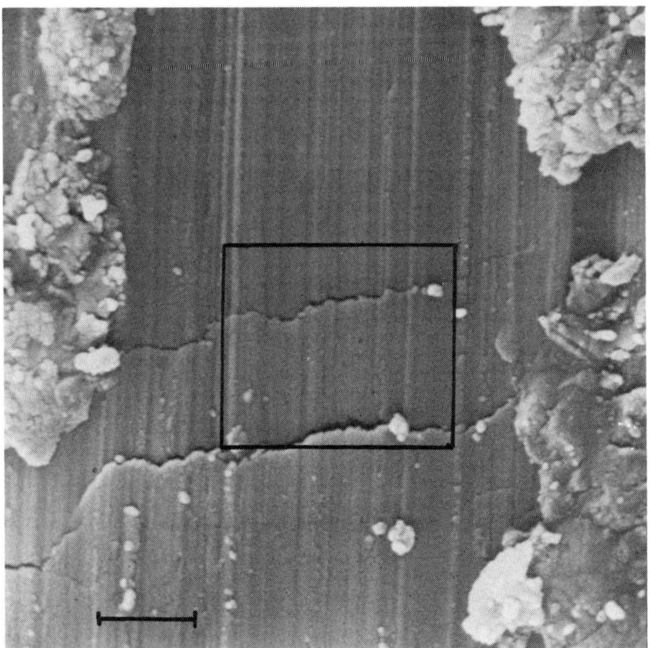

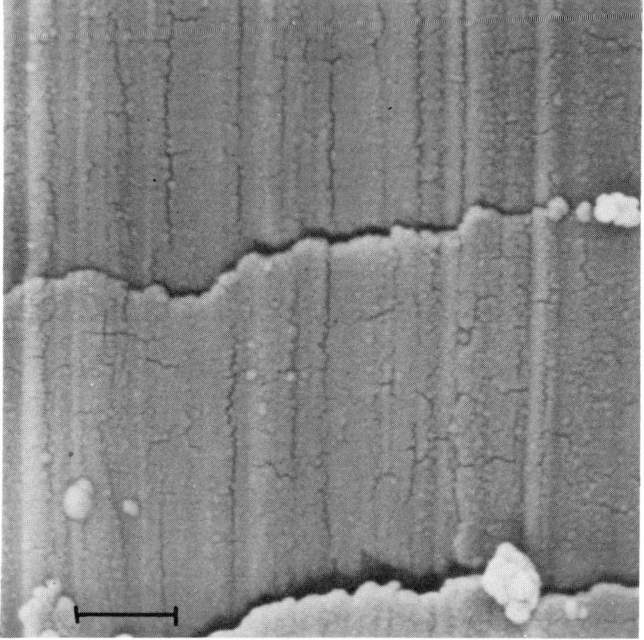

Figure 16. Scanning electron photomicrographs of Solenhofen Limestone surface after deformation. This surface was in contact with dolomite and deformed at 700 bars confining pressure. Displacement of the limestone was toward the top of the page. Left: 3,000 $\times$ magnification. Scale equals 4×10^4 A. Surface appears relatively smooth, striations are continuous, and "steps" occur at high angles to the direction of movement. Right: 10,000 $\times$ magnification of area outlined at left. Scale equals 1.2×10^4 A. Microfractures occur parallel and perpendicular to sliding direction, that is, the striations.

is associated with audible events. The paucity of data precludes any definitive correlation of stick-slip to seismic events, or stable sliding or sliding and faulting in carbonates to aseismic behavior. Scholz (1968) has investigated the seismic energy release during the loading history of intact Westerly Granite, which commonly shows stick-slip. It would be profitable to apply his line of investigation to frictional processes, particularly those in the carbonate rocks.

Our experimental data show clearly that the small oscillations of force and the accompanying releases of small amounts of seismic energy do not assure that larger releases of energy will not follow. It is also evident that neither stick-slip nor stable sliding precludes the increase of the ultimate strength of the rock mass as displacements take place along frictional surfaces.

We have pointed out that the relation of stable sliding to stick-slip is not always clear. In all experiments where we observed stick-slip, stable sliding preceded it. The amount of preceding stable sliding seems to decrease with increasing confining pressure, but further work must be done to confirm this.

CONCLUSIONS

More information is needed on the mechanisms of friction in rocks before precise predictions of the mechanical response of the rock mass can be made. This information, which is necessary for the optimum use of rapid-excavation techniques, may not be available until well after the technological feasibility of rapid-excavation is firmly established. Meanwhile, the following conclusions from our work may help to achieve a somewhat better understanding of the role of friction.

1. Sliding is accompanied by the brittle behavior of the grains along the sliding surfaces in rocks such as Tennessee Sandstone, and by ductile behavior in specimens such as Solenhofen Limestone.

2. Ductile deformation of the material along the sliding surface increases the actual area of contact, promotes new faulting in addition to sliding, and raises the ultimate strength of the precracked rock to nearly that of the intact material. The ultimate strength of intact Blair Dolomite is the highest and that of the Solenhofen Limestone is the lowest under the conditions investigated, but of the monolithologic precut specimens limestone is the strongest.

3. Where sliding is accompanied by brittle deformation of the grains along the surface, the coefficients of friction, as measured where sliding is initiated, are independent of confining pressure. Where ductile behavior predominates, μ decreases with increasing confining pressure.

4. The stress-shortening curves for dilithologic specimens reflect behaviors between those of their monolithologic counterparts. The limestone markedly controls the behaviors of both dolomite and sandstone. The behavior of sandstone-dolomite combinations is inconclusive.

5. Stable sliding, together with faulting, is characteristic of monolithologic limestone specimens. Stable sliding followed by stick-slip characterizes monolithologic sandstone specimens and occurs also in the dilithologic dolomite-sandstone specimens at 350 bars confining pressure. All other specimens show stable sliding. Stick-slip does not occur in carbonate rocks, but it is typical of rocks with high quartz content and low porosity. The amplitude of stick-slip oscillations of force increase with displacement. Stick-slip always precedes the peak strength. The early small force drops do not eliminate the much larger energy releases that occur later in the deformational history.

6. Specimens with brittle deformations along the sliding surfaces require more displacement before μ reaches a maximum value than does the limestone, which deforms ductilely. The total slip before the maximum value of μ is attained decreases with increasing confining pressure for all specimens.

Much additional careful work is necessary before the role of friction in rocks is fully understood. At the very least, predictions of the response of the rock mass based upon the mechanical properties of intact specimens are clearly inadequate. Rapid-excavation techniques should exploit the brittle deformations of the rocks along fracture surfaces, and the normal stresses along these surfaces should be as low as possible. Jointed rocks like limestone and marble, their intact properties to the contrary, may well be among the most difficult to excavate rapidly.

ACKNOWLEDGMENT

The authors are indebted to John Handin and James Byerlee for their comments and critical reading of this paper.

REFERENCES CITED

Bombolakis, E. G., Rizer, W. D., and Rainville, G. D., 1970, Study of closed crack behavior during preliminary stages of brittle fracture in uniaxial compression: Geol. Soc. America, Abs. with Programs (Ann. Mtg.), v. 2, no. 4, p. 269.

Brace, W. F., 1970, Some effects of high temperature on the frictional sliding of granite [abs.]: Am. Geophys. Union Trans., v. 51, p. 423.

Byerlee, J. D., 1967, Frictional characteristics of granite under high confining pressure: Jour. Geophys. Research, v. 72, p. 3639–3648.

—— 1970, Sliding characteristics of finely ground surfaces of gabbro and dunite: Geol. Soc. America, Abs. with Programs (Ann. Mtg.), v. 2, no. 4, p. 275.

Byerlee, J. D., and Brace, W. F., 1968, Stick-slip, stable sliding, and earthquakes – effect of rock type, pressure, strain rate, and stiffness: Jour. Geophys. Research, v. 73, p. 6031–6037.

Carlisle, D., 1965, Sliding friction and overthrust faulting: Jour. Geology, v. 73, p. 271–292.

Coulson, J. H., 1970, The effects of surface roughness on the shear strength of joints in rock: U.S. Army Corps of Engineers, Missouri River Division, Tech. Rept. MRD-2-70, 283 p.

Friedman, M., 1963, Petrofabric analysis of experimentally deformed calcite-cemented sandstones: Jour. Geology, v. 71, p. 12–37.

—— 1967, Description of rocks and rock masses with a view to their physical and mechanical behavior: First Cong. Internat. Soc. Rock Mechanics Proc., v. 3, p. 181–197.

Gemant, A., 1950, Frictional phenomena: Brooklyn, Chemical Publishing Co., Inc., 497 p.

Handin, J., 1969, On the Coloumb–Mohr failure criterion: Jour. Geophys. Research, v. 74, p. 5343–5348.

Heard, H. C., 1963, Effect of large changes in strain rate in the experimental deformation of Yule Marble: Jour. Geology, v. 71, p. 162–195.

Hoskins, E. R., Jaeger, J. C., and Rosengren, K. J., 1968, A medium scale direct friction experiment: Internat. Jour. Rock Mechanics and Mining Sci., v. 5, p. 143–154.

Howard, T. E., 1969, Development of a new rapid excavation technology – how certain is success?, *in* Cook, E., ed., The new dimension – rapid underground excavation: AIME, p. 33–39.

Jaeger, J. C., 1959, The frictional properties of joints in rocks: Geofis. Pura. Appl., v. 43, p. 148–158.

—— 1970, Behavior of closely jointed rock, *in* Somerton, W. H., ed., Eleventh symposium on rock mechanics: p. 57–68.

Jaeger, J. C., and Cook, N.G.W., 1969, Fundamentals of rock mechanics: London, Methuen and Co., Ltd., 513 p.

Maurer, W. C., 1965, Shear failure of rock under compression: Soc. Petroleum Engineers Jour., v. 5, p. 167–175.

Patton, F. D., 1966, Multiple modes of shear failure in rock: First Cong. Internat. Soc. Rock Mechanics Proc., v. 1, p. 509–513.

Pratt, H. R., Brown, W. S., and Brace, W. F., 1970, In situ determinations of strength properties in a quartz diorite rock mass, *in* Somerton, W. H., ed., Twelfth symposium on rock mechanics: p. 27–44.

Rabinowicz, E., 1965, Friction and wear of materials: New York, John Wiley & Sons, Inc., 302 p.

Raleigh, C. B., and Paterson, M. S., 1965, Experimental deformation of serpentine and its tectonic implications: Jour. Geophys. Research, v. 70, p. 3695–3985.

Scholz, C. H., 1968, Microfracturing and the inelastic deformation of rock in compression: Jour. Geophys. Research, v. 73, p. 1417–1432.

Tabor, D., 1945, The frictional properties of some white metal-bearing alloys: The role of the matrix and the hard particles: Jour. Appl. Physics, v. 16, p. 325–337.

MANUSCRIPT RECEIVED BY THE SOCIETY APRIL 5, 1972

Total Systems Approach to Rapid Excavation and Its Geological Requirements

CARL H. ROACH
Chief, Mine Systems Engineering Group, U.S. Bureau of Mines, Federal Center, Denver, Colorado 80225

ABSTRACT

A total systems approach to the development of an efficient rapid excavation system should consist of three phases: (1) *preliminary systems analysis* to identify critical blocks of technology in present excavation systems that will need significant advances before the desired rapid excavation systems can be developed; (2) *systems analysis of subsystems research*, initiated as a result of *preliminary systems analysis*, to understand clearly all subsystem or element interfaces, and to obtain research output that is as quantitative as possible and in a format suitable for design of a total rapid excavation system; and (3) *final system design, testing, evaluation,* and *demonstration* of "best" rapid excavation systems possible under the set of constraints present.

Even the crudest sensitivity analysis and technological forecasting studies reveal that our present inability to predict accurately critical geological or hydrologic conditions and engineering rock properties of ground ahead of an advancing excavation is one of the greatest deterrents to the successful development of an efficient rapid underground excavation system. Earth scientists must develop new and accurate means of detecting and quantitatively describing critical geological and hydrologic conditions during the pre-excavation, excavation, and immediate postexcavation phases of these operations. Geologists should also learn to characterize environments subject to excavation with probability functions for the occurrence of each type of geological hazard to be expected in the area so that maximum use can be made of mathematical models used in a stochastic mode.

INTRODUCTION

The purpose of this paper is to bring to the attention of the earth sciences community the increasing need for the United States to initiate a vigorous and continuing nationally oriented research and development program designed to improve greatly the "state of the art" in the underground rapid excavation technology. More specifically, an attempt is made to show that success or failure of this nationally oriented program may very well rest on the ability of earth scientists to solve some of the critical geological problems associated with the field of underground excavation.

The geological problems for which solutions are desperately needed constitute some of the most difficult problems in geologic prediction and delineation. Solutions, however, must be found if the nation is to develop the rapid underground excavation systems that are essential if we are going to use effectively the subsurface ("the third dimension") to solve some of the critical problems associated with the nation's ever-increasing urbanization, and if we are going to produce domestically, at a reasonable and acceptable "total cost to the public," the continually increasing amounts of minerals that will be needed by the United States in future years.

The scope of this report is limited to a consideration of the geological requirements of the total systems approach to the development of an underground rapid excavation system. The total systems approach by its very nature requires that a detailed study be made of all activities that make up the underground excavation process, the interrelationships among all of these activities, and the relationships between the total excavation system and its environment. A total systems study of the development of an underground rapid excavation system is much too comprehensive for the purpose of this paper.

NEEDS AND GOALS OF A NATIONAL PROGRAM ON RAPID EXCAVATION

The timely development of efficient underground rapid excavation systems can contribute greatly to the solution of at least four types of problems facing our nation today and that will become more critical with each passing year: (1) impact of accelerating urbanization; (2) conservation and development of mineral resources; (3) increasing demands for underground military facilities; and (4) preservation and enhancement of our environment.

The impact of increasing urbanization can perhaps be placed in perspective when it is realized that the present worldwide rate of urbanization is about four times the present rate of population growth. "Metropolitanization" and "megalopolitanization" as a way of living in the United States are certain to increase in the future (McNee, 1969). It has been estimated that in a few years, 90 percent of the population of the United States will be living and working in 10 percent of the land area. "It will benefit 90 percent of the population if in that 10 percent of the land area we can put power plants, sewage processing systems, transport conduits, garages, and other solid and liquid storage facilities underground economically" (Cook, 1969).

Our nation's continuing demands for minerals in the face of our rapid depletion of accessible and known ore reserves, more and more mineral imports, and our recognition that increasing urbanization requires us to not only preserve but to enhance our environment, all combine to emphasize the need for us to design drastically different underground mining systems in this country. "Mineral deposits not now available for exploitation because of excessive depth, discontinuity or low grade of mineralization, proximity to population centers and other restricting conditions can be included in future reserves if methods and costs of obtaining access can be improved" (Garfield, 1969). The successful development of an efficient underground rapid excavation system can contribute greatly to the solution of these and other problems related to the conservation and development of our mineral resources.

The military requirements for various types of underground facilities are continuing to increase as a result of the very rapid rate of advance in the technology of offensive and defensive weapons systems. Many of these needed underground facilities have specialized requirements in a time frame that necessitates more rapid underground excavation systems than those now in use. Some aspects of our national defense posture bear on the availability of more rapid methods of completing specialized underground facilities.

The surface environment would be preserved and enhanced by placing underground those facilities (such as power plants, mineral processing plants, sewage plants, storm sewer systems, storage facilities, parking garages, and utilities) that are normally placed above ground. Underground solid and liquid waste disposal would do much to improve our surface environment. Rapid and efficient underground excavation capability might make these and other types of environmental improvements possible.

A U.S. Bureau of Mines contract research study of the projection of applications and national benefits of a new rapid excavation technology revealed that the results of a vigorous program to advance the development of rapid excavation technology over the next 5 to 10 yrs will return benefits to the nation which will be measurable over the next 20 yrs in tens of billions of dollars (Lago and others, 1967).

TOTAL SYSTEMS APPROACH TO RAPID EXCAVATIONS

Systems engineering is the application of the *"scientific method"* to the study of large complex systems. The good systems engineer uses this methodology in his work because it disciplines him to ask the right questions at the right time, innovate at the right time, and tells him when to test those innovations. The method also tells him when he has solved his problem. The scientific method can be considered to consist of the following steps (Ackoff, 1968; Churchman and others, 1957; Hall, 1968):

1. Formulate the problem.
2. Construct a mathematical model of the system.
3. Derive a solution from the model.
4. Test the model and its solution.
5. Establish controls over the solution.
6. Implement the solution.

The scientific method can be illustrated by the way in which Bohr determined the physical nature and characteristics of the hydrogen atom (Morton, 1964). He first clearly defined his problem as the precise determination of the physical structure and characteristics of the hydrogen atom. Second, he constructed a planetary model of the hydrogen atom that was compatible with his hypothesis of the structure of the atom. Next, he made some predictions of what his physical model would do under various conditions. Following his predictions, he carefully designed and performed experiments with real atoms to see if they behaved the way his hypothetical model predicted. Finally, he compared the experimental results with his predicted model behavior. Although many predictions did agree with the model and with real nature, some conditions did not agree. Therefore, he re-examined his statement of the problem and modified his model in an attempt to satisfy all experimental conditions. This is the feedback principle that makes the scientific method the powerful error-correcting, problem-solving methodology that it is. He later found that the exclusion principle and certain quantum effects had to be added to his model to explain certain results of his experimental work. After he was able to satisfy all experimental results with his model, he then concluded that he had solved his problem and understood the fundamental physical structure and nature of the hydrogen atom.

The total systems approach employs the scientific method and is useful for solving large, complex, interdisciplinary scientific problems. It is based on the fundamental principle that each element, or subsystem, of a complex system has some interaction with or effect on every other element of the complete system. The total systems engineering approach therefore attempts to identify all of the significant subsystem interactions and to evaluate their combined impact on the performance of the whole system. The final objective of the total systems approach to problem solving is the design of an "optimum" or "best" system which is seldom, if ever, composed of a number of individually optimized subsystems or elements.

Although most scientists accept in principle the basic elements of the total systems approach, they rarely apply the principles in their own research. Most scientists try to reduce a complex problem down to a size that can be "easily solved by standard techniques," whereas the total systems approach requires that the problem be intentionally enlarged until all significantly interacting elements are identified and thoroughly understood.

Recommendations and guidelines for the development of an efficient rapid underground excavation capability to contribute to the national needs were recently specified by the National Research Council Committee on Rapid Excavation (National Academy of Sciences, 1968). This committee recommended a 10-yr, federally funded research program in the amount of $200 million (in constant 1964 dollars). The program goals were to achieve a 30 percent reduction in cost, and a 200 to 300 percent increase in the sustained rate of advance of underground excavation in soft-medium and hard rock by 1990. The committee stressed the need for a total systems engineering approach to the development of a rapid excavation capability and identified the following critical areas of research, given in approximate order of priority (National Academy of Sciences, 1968, p. 39):

1. Development of processes and equipment for boring tunnels and shafts in hard abrasive rock and for reducing the downtime experienced with current mining and tunneling machines in soft or medium rock masses.
2. Improvement of processes and equipment for removing muck and handling supplies and materials for compatibility with advanced mining and tunneling machines.
3. Development of geological-geophysical techniques for determining rock and ground water conditions qualitatively prior to excavation operations, and for probing immediately ahead of the working face during excavation operations to determine rock quality and ground water conditions more precisely.
4. Development of rock-mechanics techniques for measuring mechanical-electrical, and thermal properties and behavior of rocks in situ.
5. Improvement of rock-mechanics techniques for measuring subsurface stress field in deep boreholes and for determining stress distribution in walls of underground excavations.
6. Improvement of processes, materials, and equipment for temporarily and permanently supporting long tunnels and deep shafts in wet, broken rock masses.
7. Improvement of processes, materials, and equipment for producing temporary support, in a wide range of rock mass conditions, at speeds compatible with advanced mining and tunneling machines.
8. Development of rational methods for designing temporary and permanent support systems.
9. Development of standards, processes, and equipment for attaining adequate environmental quality and safety in long, continuously excavated tunnels and shafts.
10. Improvement of drilling and blasting techniques.

Recently, a comprehensive survey was made of the amount and type of tunneling research that was being conducted in 17 OECD countries (Organization for Economic Cooperation and Development, 1970). The set of reports covering the results of the OECD Advisory Conference on Tunneling held in Washington, D.C., June 22–26, 1970, are an excellent summary of current research on tunneling and give recommendations for needed research, the priorities of which are in essential agreement with the recommendations of the National Research Council Committee on Rapid Excavation. The goals and recommendations of the rapid excavation program proposed by the National Research Council are used in this report as the goals to be achieved by the total systems approach to design a more efficient rapid underground excavation system than exists today. The need to use a systems engineering approach to the design of an efficient rapid excavation system has also been stressed by Howard (1967a, 1967b, 1968, 1969).

The problem under consideration in this report can then be structured by three questions: Where are we now? Where do we want to go? How do we get there? The answer to the first question involves a detailed study of underground excavation systems in use today. The answer to the second question is contained in the specific objectives of the rapid excavation program recommended by the National Research Council and adopted for this report. The answer to the third question is obtainable by the successful application of the total systems engineering methodology that will be briefly described in the following paragraphs.

The total systems approach to the development of an efficient rapid excavation system should consist of three phases: (1) *preliminary systems analysis*, (2) *systems analysis of subsystems research*, and (3) *final system design, testing, evaluation, and demonstration.*

Preliminary Systems Analysis

The initial step in the development of an improved underground rapid excavation system should be to develop first an accurate life-cycle simulation model and use that model to make a systems analysis of present excavation systems to identify those critical blocks of technology that will need significant advances, or breakthroughs, before the desired or improved excavation systems can be perfected. The life-cycle simulation model should provide for the accurate and complete simulation of all activities from project inception through planning, design, development, construction, and completion of the excavation project being studied. Although an excavation system can be subdivided in a number of ways, I believe the life-cycle model should provide for accurate simulation of the following subsystems: (1) delineation, or engineering geology; (2) design of excavation system; (3) fragmentation; (4) materials handling; (5) ground control; (6) environmental control and safety; and (7) excavation completion. These subsystem categories characterize the excavation system in a continuous manner from the initial engineering geology investigations to determine feasibility of the excavation project, through the design stage, the excavation stage, and includes the excavation completion phase of the total excavation operation.

The life-cycle simulator can be used to perform sensitivity analysis (Hillier and Lieberman, 1969), which measures the effect in quantitative terms that a specific improvement of some component of the system has upon the output of the entire system. This type of analysis would be useful to identify those elements of present excavation systems that have greatest impact on the entire excavation system and that would be best candidates for significant research and development to achieve the improved excavation system desired.

Although sensitivity analysis will identify those activities of the excavation system that may appear to be most worthy of intensive research and development, this technique will not predict the probability that extensive research in a specific area will be successful and achieve its intended results. Therefore, some additional techniques will be needed to structure the priorities of research projects to be undertaken in an attempt to have the maximum probability of achieving the required improvements so that the specified excavation systems can be developed under the set of constraints imposed on the research and development program.

Technological forecasting is "the description or prediction of a foreseeable invention, specific scientific refinement, or likely scientific discovery that promises to serve some useful function" (Prehoda, 1967). It is the art of knowing which way to direct development so that the desired systems have the maximum probability of being developed within the constraints under which the research and development program is operating. Technological forecasting combined with sensitivity analysis represents a powerful approach to the problem of optimally allocating total resources (manpower, research facilities, money, and time) to specific areas of research for the purpose of achieving the specified underground rapid excavation system to be designed and developed.

The "laws" and methodology of technological forecasting are fairly well known. The state-of-the-art in most areas of technology tends to have a growth with time that follows a characteristic "S"-shaped curve, called a "Gompertz curve" in honor of the mathematician who discovered it (Fig. 1). The Gompertz curve predicts the general way that the performance achievement of a particular type of technology will increase with increasing time over the life span of that particular technology.

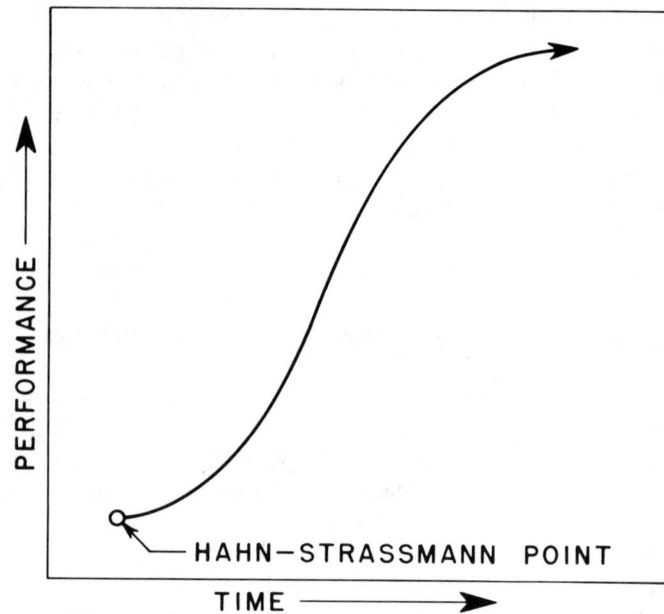

Figure 1. Gompertz "S"-shaped curve.

The life span of development of a particular technology starts with a technology discovery point which has been called the "Hahn-Strassmann point" in honor of Otto Hahn and Fritz Strassmann, who discovered uranium fission, thereby opening a completely new type of technology (Cole, 1965). It is very important to be able to recognize that a Hahn-Strassmann discovery point has been reached in a particular field of science or technology development when an attempt is being made to develop either a new or an improved technological system. After the Hahn-Strassmann point discovery has been achieved, the particular technology will undergo dynamic growth which follows the "S"-shaped curve, whereby performance increases slowly with time at first, then increases more rapidly with time, and finally levels off or becomes nearly horizontal, which is an indication that the technology has reached maturity and balance with its total environment. When a particular technology has truly reached maturity and balance with its environment, no significant increase in performance of this general field of technology is likely to be achieved until a new Hahn-Strassmann point in a related technological field is discovered, thereby allowing the "S"-shaped dynamic growth to occur again.

In large scientific research and development programs, one of the most valuable contributions of successful technological forecasting is the attainment of telescoped achievements (Fig. 2). Trend curve "B" is the rate of increase of performance with time that would be expected if decision makers do not rely on technological forecasting and a normal evolutionary pattern of technology development is followed. Trend curve "A" is the pathway that performance can take with time if

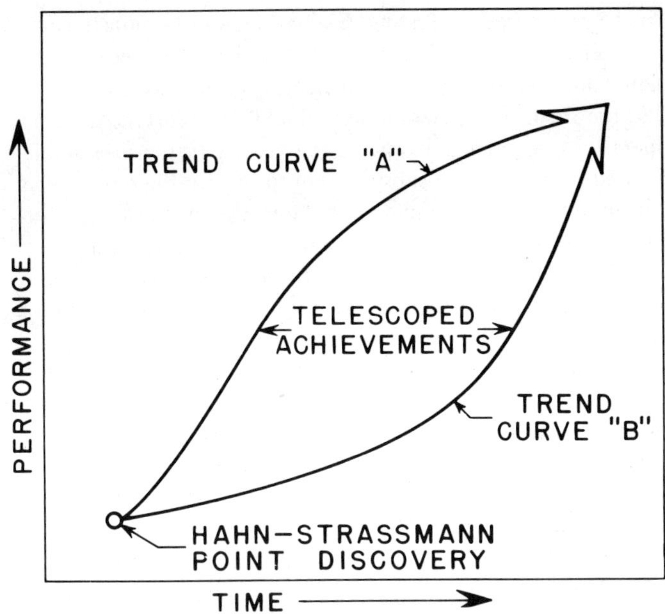

Figure 2. Telescoped performance gains achieved by full technological forecasting implementation.

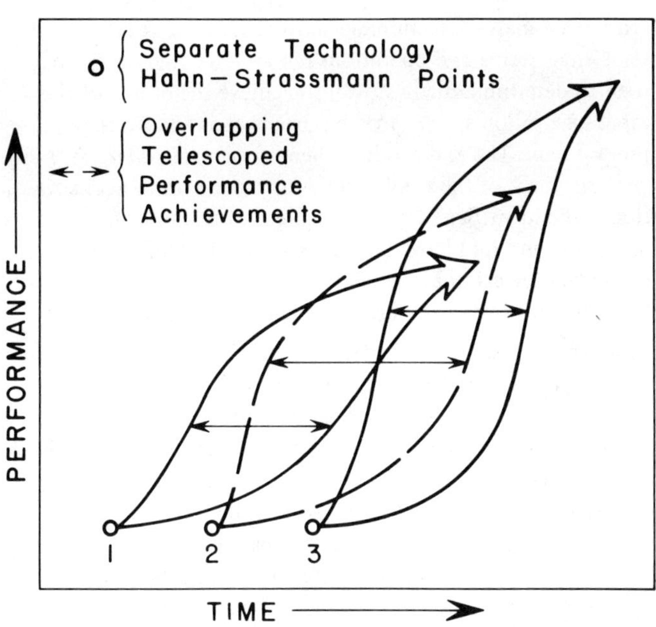

Figure 3. Overlapping telescoped performance gains achieved by full technological forecasting implementation.

decision makers make full use of technological forecasting combined with optimum application of total resources (manpower, funds, research facilities, and time). Good technological forecasting can therefore accomplish telescoped achievements by compressing the time scale and developing a desirable performance level in less time than by normal procedures not utilizing technological forecasting. Trend curves "A" and "B" combined form an envelope for the particular technological field represented on Figure 2.

It is important to realize that overlapping telescoped performance gains can be achieved also by full utilization of technological forecasting techniques (Fig. 3). For example, in the field of underground excavation, technology envelope "1" could represent the historical development of black powder as a means of achieving rock fragmentation. Technology envelope "2" could represent the historical development of a rock fragmentation method utilizing smokeless powder and other modern chemical explosives. Technology envelope "3" could represent the fragmenting of rocks by the use of underground nuclear explosions. These methods overlap each other, but the envelope common to all three technologies represents a part of another Gompertz curve that could be used to predict future potential developments in the general area of rock fragmentation as applied to the general field of underground excavation.

For purposes of the present discussion, it is perhaps worthwhile to consider how technological forecasting methods could be used to predict or determine realistic objectives for a research and development program that is aimed in the direction of improving presently used underground tunneling or excavation methods (Fig. 4). A study of the historical and projected

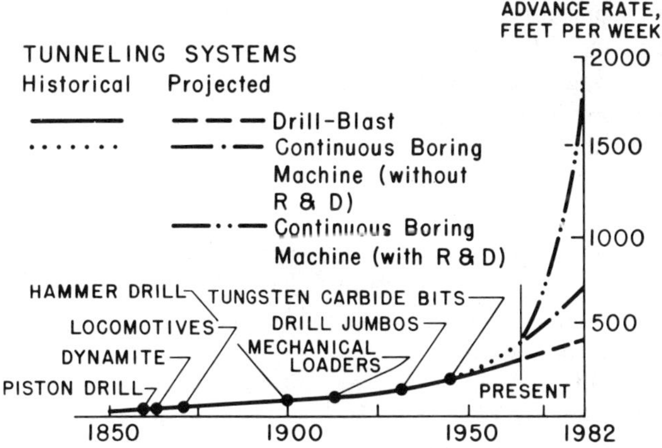

Figure 4. Historical and projected development of tunneling technology.

development of tunneling technology in this country focuses attention on several interesting possibilities. First, it indicates that the continuous boring machines developed over the past few years may, in fact, be a Hahn-Strassmann point discovery with respect to tunneling technology. If true, these machines may represent a point in technology from which we can expect fairly substantial future improvements in the state-of-the-art of tunneling or underground excavation technology (Howard, 1968). Figure 4 also indicates that great potential exists for effectively applying technological forecasting to obtain a substantial telescoped achievement (as previously discussed) in the average tunneling or excavation advance rate in feet per week!

Once an accurate life-cycle simulator of present-day underground excavation systems has been successfully developed, sensitivity analysis, technological forecasting, and other systems engineering and operations research techniques can be used to determine the best way to achieve the design of the rapid excavation system that has been set as an objective of the program. The end product then of the *preliminary systems analysis* phase of the total systems approach to rapid excavation is the identification of those areas of research that are most vital and that have maximum probability of producing the technological improvements necessary to develop successfully the improved excavation system required.

Systems Analysis of Subsystems Research

After the *preliminary systems analysis* has identified the research areas that are most needed, the research in each appropriate subsystem or element will be started so that the state-of-the-art can be improved to the point where the required system can be developed within the time frame, and other constraints specified for the research and development programs. These new subsystem research projects should be coordinated and followed closely by the group responsible for the total systems engineering.

The systems engineering group should monitor the subsystem's research closely to achieve a balance in the over-all development program and to assure progress along all needed lines, and at the same time make the best use of research and development manpower and other resources allocated to the program. The subsystem research projects should be coordinated by the systems group in such a way as to achieve an output from each research project that can be combined in a preplanned way with the output from all other research projects to form the well-balanced final system to be developed.

During the subsystem research phase of the program, the life-cycle simulation models should be updated to incorporate the positive results of the research effort as they become available and to understand clearly all subsystem or element interfaces as the new research progresses. Special emphasis should be given at this stage to the development of methods to insure that each new research project produces output that is as quantitative as possible and is in a suitable format to be integrated with the output of all other projects when the final system is to be integrated and designed. Development of methods of quantifying critical geological conditions and engineering rock properties in terms compatible with the excavation process being developed, and necessary for systems design and analysis, is one example of the type of systems engineering research study that should be conducted at this stage.

The main responsibility of the total systems engineering group during the *subsystems research stage* is to evaluate continuously all subsystems research for the purpose of determining when each research project has completed its assigned objectives, and to determine at what point the preliminary attempt should be made to design the improved rapid excavation system that incorporates the achievements of all subsystems research undertaken on the program.

Final System Design, Testing, Evaluation, and Demonstration

After the *subsystem research projects* have been completed and the total systems engineering group has concluded that the final total system design should be undertaken, various operations research and systems engineering techniques should be used to integrate all subsystem outputs into the design of an "optimum" or "best" rapid excavation system, and to develop an updated and accurate simulation model for the newly designed rapid excavation system. After the "best" system has been designed and sufficiently tested by simulation techniques, the final system designed should be used as a basis for a full scale field testing program of the total excavation system. Standard and newly developed operations research and systems engineering techniques should then be used to evaluate carefully results of the total system field testing program. Deficiencies in the total system detected during the field testing program should be overcome by further research and development so that the finally designed total rapid excavation system can be successfully demonstrated and made ready for industrial implementation.

GEOLOGICAL REQUIREMENTS OF A TOTAL SYSTEMS APPROACH TO RAPID EXCAVATION

"Geological conditions, more than any other factor, determine the degree of difficulty and the cost of excavating and supporting underground openings" (National Academy of Sciences, 1968). Sensitivity analysis and technological forecasting studies reveal that our present inability to predict accurately, on a real-time or near real-time basis, critical geological or hydrologic conditions ahead of an advancing excavation operation is one of the most critical deterrents to the successful development of an efficient underground rapid excavation system.

If geologists and other earth scientists are going to make their required contribution toward the development of an efficient rapid excavation system, they must make one or more fundamental breakthroughs ("Hahn-Strassmann point discoveries") in the science of predicting geological and hydrologic conditions ahead of an underground mining or excavation operation. Great advances must be made in the development of techniques for quickly and accurately obtaining the following types of delineation data that are essential for the efficient operation of an underground rapid excavation system: (1) physical and chemical properties of intact blocks of rock; (2) frequency, spacing, orientation, strength, and alteration characteristics of all geologic discontinuities; (3) in situ stress conditions; (4) hydrologic conditions to be encountered; and (5) manmade hazards that might be encountered by the rapid excavation operation. Most important of all, earth scientists must develop a fundamental understanding of exactly how the geological and hydrologic conditions in an excavation environment will affect the efficiency of the rapid excavation system being used. The quantitative functional relationships between environmental parameters and the excavation system being used in that environment must be accurately identified and made available in a time frame that is compatible with operating the excavation system under the most efficient conditions.

Three time frames are important when considering the geological requirements of the efficient underground rapid excavation system that we have specified as our objective: (1) pre-excavation, (2) during excavation, and (3) immediate post-excavation. The pre-excavation geological requirements are mainly related to the need for much improved techniques for accurately delineating those physical characteristics of the environment that will help determine the best site of the excavation project, and that will be needed for planning the operational characteristics of the excavation system to be used on the specific project. The requirements for geological data *during* the excavation activity are very demanding in some geological environments, and determine the degree to which the excavation process can be continuous and efficient. Great improvements, or breakthroughs, are needed to develop these geological prediction techniques needed for acquiring in real time the geological data essential for continuing the efficient uninterrupted operation of the underground excavation system. Improved prediction techniques are also badly needed for determining geological and hydrologic conditions in the immediate vicinity of the underground opening within a short time frame after a given volume of rock has been excavated.

The geological requirements of a truly efficient underground rapid excavation system must satisfy several constraints or conditions. First, the geological data must be obtained, interpreted, and made ready for planning within the three time frame constraints mentioned previously. Second, the geological data for each time frame requirement should be *accurate* and *quantitative* enough to be used in analytical models for engineering construction and planning purposes. These types of data are required for maintaining the most efficient operating conditions for the excavation system. If weak, water-saturated ground is to be encountered in the next hundred feet or so ahead of the excavation, this condition must be accurately and quantitatively predicted so that ground support materials can be obtained and brought into the underground workings in sufficient time to provide proper ground support and, therefore, maintain efficient continuous operations. Every possible effort should be made to develop geologic prediction techniques that are *accurate* enough to be used for planning and programming the excavation operation.

Where complete and specific information on geological and hydrologic conditions ahead of an excavation operation is not possible, all of the data should be presented as probability distribution functions to make maximum use of mathematical models used in a stochastic mode. A stochastic model has a probability structure and includes random variables whose values depend on a parameter such as "time." Geologists would greatly improve the efficiency of operation of underground rapid excavation systems if they would specify for each unit length of rock ahead of the excavation the probability of occurrence of each type of geological and hydrologic condition that is known to characterize the environment containing the proposed excavation project. These probability distribution functions would be quite helpful in predicting the method of operation of the excavation system that would most likely keep the system operating with highest efficiency and, therefore, with minimum downtime and associated high cost.

DEVELOPMENT OF IMPROVED PRE-EXCAVATION SITE DELINEATION TECHNIQUES

Several areas of research can be identified that should be pursued vigorously if the required pre-excavation site delineation techniques are to be properly developed. Improved geophysical techniques (capable of uniquely identifying certain geological conditions) are greatly needed for airborne and ground surface reconnaissance required during the pre-excavation planning and design stages of rapid excavation projects. Airborne remote sensing techniques that are recommended for investigation include the following: (1) multiband and color high-resolution photography; (2) infrared imagery and color photography; (3) side-looking radar; (4) gamma-ray spectroscopy; (5) aeromagnetic surveys; and (6) gas "sniffer" techniques. The objective of this investigation would be to evaluate and assemble an improved suite of airborne remote sensing techniques that can be applied quickly and at low cost to look for large-scale, serious geological or manmade hazards (such as major fault zones, weak geological formations, abandoned oil wells and buried cables, underground caves, abandoned mine workings, and the like), and other major rock conditions that might adversely affect the completion of the desired excavation project.

Ground surface geophysical techniques that are recommended for research and development include: (1) seismic refraction and shallow reflection surveys; (2) electrical resistivity techniques; (3) magnetometer studies; (4) gravity meter investigations; (5) gamma-ray spectroscopy measurements; (6) gas "sniffer" techniques; and (7) seismic or acoustical holographic imagery. The objective of this type of investigation is to develop improved surface geophysical techniques necessary for making intensive ground follow-up investigations of airborne geophysical anomalies that suggest the presence of serious geological or manmade problems at the site or along the proposed route of an underground excavation project.

Research to develop improved geophysical well-logging techniques for accurately delineating the critical physical characteristics of rock and soil at the site or along the route of a proposed underground excavation project is urgently needed. These techniques should be accurate enough to be used for engineering construction purposes. Techniques recommended for investigation include: (1) neutron attenuation and activation; (2) natural gamma-ray and gamma-gamma measurements;

(3) sonic pulse velocity and waveform evaluation; (4) electric and magnetic logs (resistivity, induction, spontaneous potential, induced polarization, and magnetic susceptibility); (5) oriented caliper and fracture detection logs, including the use of mechanical, photographic, television, and sonar techniques; (6) physical and chemical properties of core samples and rock outcrop samples needed to calibrate and interpret the above logging measurements; and (7) between-hole geophysical surveys. The objective of this research is to develop new or improved techniques for determining the in situ geological, physical, and hydrologic properties of the rock masses surrounding the excavation project that are of critical significance in planning and constructing the project by rapid excavation techniques.

Most important to the development of these improved pre-excavation site delineation techniques is, of course, a fundamental understanding of the quantitative effects that the total geological environment will have on the rapid excavation system to be used for accomplishing the specific project. The time frame for this type of research program calls for the conventional or normal time requirements for accumulating and interpreting all necessary data. Sufficient time is normally available for this type of study to utilize home-based computers for analyzing all data under office conditions. However, use of a field-based teletype in communication with a home-based computer might be very helpful in reducing costs of some types of expensive pre-excavation field work, such as exploration drilling and airborne remote sensing investigations.

DEVELOPMENT OF IMPROVED DELINEATION TECHNIQUES USED DURING EXCAVATION PROCESS

The development of new geophysical methods for looking ahead of an underground excavation while the excavation is in progress is urgently needed if truly efficient rapid excavation systems are ever going to be developed and used efficiently. Significant breakthroughs ("Hahn-Strassmann point discoveries") in the science of geological delineation will be needed to develop the required techniques. The following potential techniques are recommended for research to develop the required delineation techniques: (1) high-resolution seismic reflection techniques using focused source and receiver arrays; (2) high-resolution sonar techniques for accurately and three-dimensionally delineating reflecting-type geological discontinuities; (3) high-resolution electromagnetic reflection techniques; (4) surface or in-hole detection and analysis of acoustical energy generated by the continuous boring machine; (5) continuous-wave electromagnetic measurements of attenuation and wave tilt; (6) nuclear magnetic resonance techniques for detecting "free" water in rock pores and fractures, (7) electrical resistivity measurements made by focused electrode systems; and (8) seismic or acoustical holographic imaging techniques.

The previously mentioned delineation techniques do not exist at this time for continuous real-time monitoring of critical geological and hydrologic conditions ahead of an underground excavation operation. Real-time or near real-time computer analysis and interpretation of all delineation data will be required if the previously mentioned techniques are successfully developed and are to be used in the necessary time frame for making decisions concerning the most efficient operation of the underground excavation system. The real-time or near real-time requirements for delineation data acquisition and interpretation present earth scientists with one of their first requirements for on-line computers to make decisions regarding the geological data that are required to support an underground excavation operation!

Interesting possibilities exist for designing continuous real-time geophysical monitoring or detection systems that utilize systems composed of detectors placed in drill holes previously drilled in front of the advancing excavation, detectors along the ground surface over the excavation, and detectors located at various strategic locations in the underground excavation, in combination with energy sources either within the excavation itself or at some point in the surrounding environment (Fig. 5). Continuous monitoring of data received from these complex systems can be very quickly analyzed by on-line computers, and the progressive changes in interpretation with each iteration can hopefully be made more accurate as excavation advances toward some unknown hazardous feature.

Although the development of these new delineation methods needed for use during excavation is a severe problem and the probability of successful development for some techniques

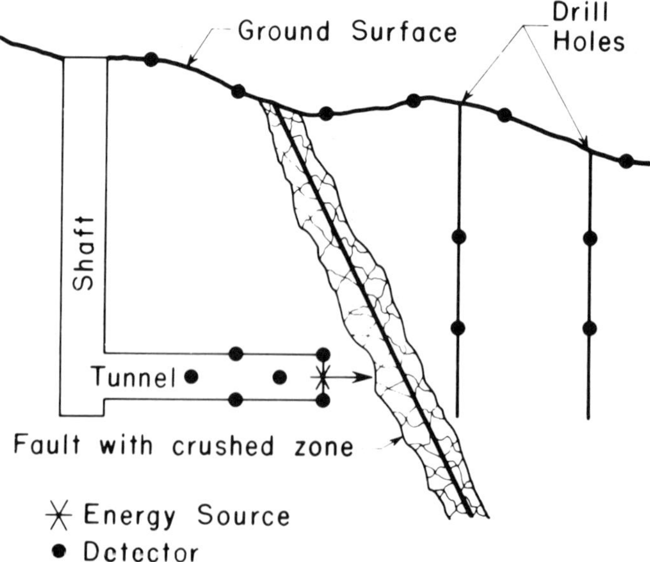

Figure 5. Continuous real-time geophysical monitoring system for delineating geological and manmade hazards ahead of a rapid excavation operation.

may be low, this need is extremely urgent and attempts *must* be made to develop the required techniques. If satisfactory site delineation methods capable of being used during the excavation activity cannot be developed, it is unlikely that rapid excavation systems having the sustained rate of advance and estimated cost reductions, as specified by the National Academy of Sciences, can be developed.

DEVELOPMENT OF IMPROVED DELINEATION TECHNIQUES FOR IMMEDIATE POST-EXCAVATION USE

Recent geophysical research (Scott and Carroll, 1967), in the Straight Creek tunnel pilot bore in central Colorado has shown that seismic velocity and electrical resistivity of rocks behind the disturbed layer surrounding the pilot bore can be used to predict accurately such economic and engineering parameters as: (1) time rate of construction and cost of construction per foot (Fig. 6); (2) rock quality; (3) set-spacing required; (4) percent lagging and blocking required for support; (5) type and amount of steel support required; (6) height of tension arch; and (7) stable vertical rock load.

It should be possible, in some cases, to modify the seismic and resistivity techniques used in the Straight Creek tunnel study to make possible the prediction of the same engineering and economic parameters immediately after a given volume of rock had been excavated by a rapid underground excavation system. If modified techniques of this type can be successfully developed, it should be possible to use the techniques to determine the optimal time for installing ground support and tunnel linings after the excavation opening has been made. Such delineation techniques are vitally needed to complete underground excavations to the required specifications and within a time frame that allows the total excavation system to operate most efficiently.

CONCLUSIONS

1. Earth scientists will play a key role in the development of improved rapid excavation systems that are urgently needed in the United States.

2. Earth scientists must develop the ability to delineate *accurately* and *quantitatively* those manmade, geological, and hydrologic conditions that adversely affect the operation of an underground rapid excavation system in a specific environment.

3. When it is impossible to describe *accurately* and specifically the delineation data needed, earth scientists should characterize environments subject to underground excavation with

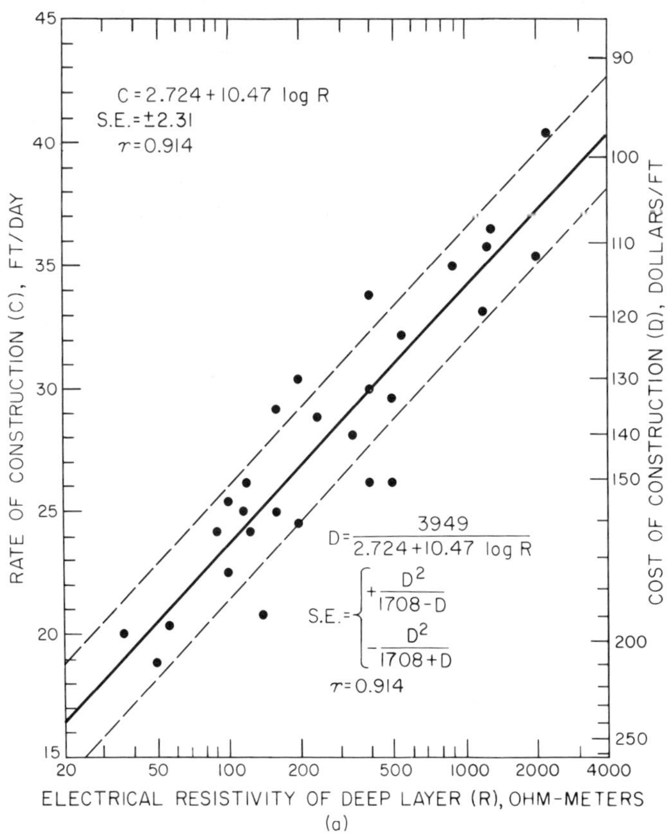

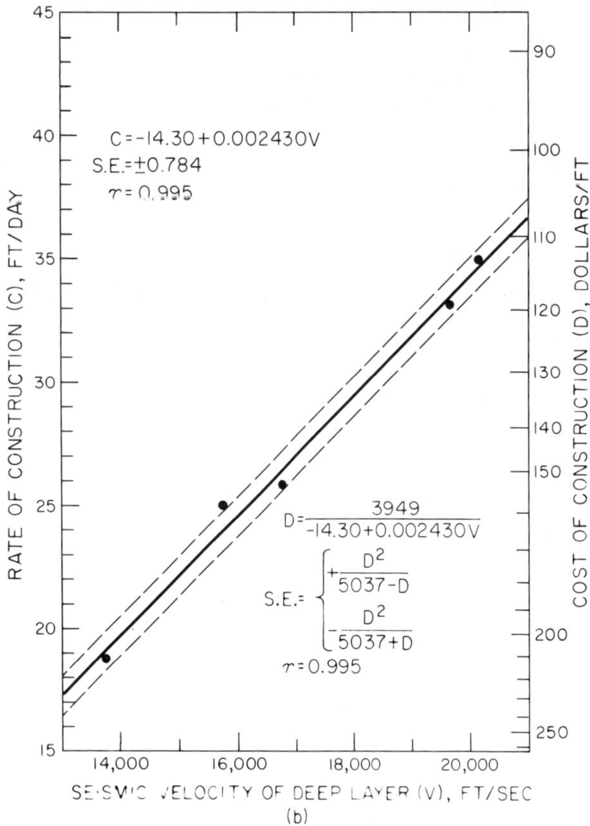

Figure 6. Straight Creek tunnel pilot bore rate of construction and cost per linear foot plotted against: A, electrical resistivity of deep layer, and B, seismic velocity of deep layer.

probability functions for the occurrence of each type of geological hazard to be expected in the area so that maximum use can be made of mathematical models used in a stochastic mode.

4. The essential delineation studies must be completed and ready to implement at a rate (or in a time frame) that is compatible with the rate of operation of all subsystems of the rapid excavation system.

5. Some time frame requirements for geological and hydrologic data are such that real-time monitoring systems and near real-time computer analysis systems will be needed to maintain the continuous and efficient operation of the underground rapid excavation systems.

6. Some of the geological prediction techniques that must be developed will require significant improvements of present techniques and, in some cases, breakthroughs in certain areas of the technology.

7. Failure to develop the delineation techniques capable of accurately and quantitatively delineating appropriate geological and hydrologic conditions in the required time frame will largely determine the highest sustained rate of excavation for a specific rapid excavation system.

ACKNOWLEDGMENTS

I am greatly indebted to Dr. T. B. Johnson and to C. O. Bunn for many stimulating discussions on the systems engineering approach to problem solving. Although I have blended their thoughts freely with my own, I alone am responsible for any inadequacies in the present report. James H. Scott has spent many hours discussing with me the areas of geophysics and geology that need extensive research if the desired rapid excavation systems are to be successfully developed, and I am pleased to acknowledge his unselfish help in this regard.

REFERENCES CITED

Ackoff, R. L., 1968, Scientific method – optimizing applied research decisions: New York, John Wiley & Sons, Inc.

Churchman, C. W., Ackoff, R. L., and Arnoff, E. L., 1957, Introduction to operations research: New York, John Wiley & Sons, Inc.

Cole, Dandridge M., 1965, Beyond tomorrow – the next 50 years in space: Amherst, Wis., Amherst Press, p. 22, 23.

Cook, Earl, 1969, The national interest in rapid excavation, *in* The new dimension, rapid underground excavation: Am. Inst. Mining, Metall., and Petroleum Eng. Ann. Mtg., Washington, D.C., p. 1–4.

Garfield, L. A., 1969, Rapid underground excavation in relation to conservation and to the development of mineral resources, *in* The new dimension, rapid underground excavation: Am. Inst. Mining, Metall., and Petroleum Eng. Ann. Mtg., Washington, D.C., p. 23–32.

Hall, Arthur D., 1968, A methodology for systems engineering: Princeton, N. J., Nostrand Company, Inc.

Hillier, F. S., and Lieberman, G. J., 1969, Introduction to operations research: San Francisco, Holden-Day, Inc.

Howard, Thomas E., 1967a, Underground mining – mine systems design: The next effort will focus on tunneling: Eng. and Mining Jour., v. 168, p. 158–164.

—— 1967b, Rapid excavation: Sci. American, v. 217, p. 74–85.

—— 1968, Outlook for the future in mining technology: Paper presented at II International Surface Mining Conference, Minneapolis, Minn.

—— 1969, Development of a new rapid underground excavation technology – how certain is success?, *in* The new dimension, rapid underground excavation: Am. Inst. Mining, Metall., and Petroleum Eng. Ann. Mtg., Washington, D.C., p. 33–39.

Lago, Armando, Williams, P. D., Nisselson, Harold, and Kushner, Harvey, 1967, Projection of applications and national benefits of a new rapid excavation technology: Natl. Tech. Inf. Service, U.S. Dept. Commerce, Pub. PB177-784, Springfield, Va.

McNee, R. B., 1969, Urbanization – new directions for both social and technological research, *in* The new dimension, rapid underground excavation: Am. Inst. Mining, Metall., and Petroleum Eng. Ann. Mtg., Washington, D.C., p. 5–13.

Morton, Jack A., 1964, From research to technology: Internat. Sci. and Technology, no. 29, p. 82–92.

National Academy of Sciences, 1968, Rapid excavation, significance–needs–opportunities: Natl. Acad. Sci. Pub. 1690.

National Academy of Engineering Committee on Rapid Excavation, 1968, Panel reports of the committee on rapid excavation: Natl. Tech. Inf. Service, U.S. Dept. Commerce, Pub. AD674-604, Springfield, Va.

OECD Advisory Conference on Tunnelling, 1970, Report on tunnelling demand, 1960–1980: Organization for Economic Cooperation and Development, Washington, D.C., p. 164.

—— 1970, Report on research and development related to tunneling: Organization for Economic Cooperation and Development, Washington, D.C., p. 107.

—— 1970, Report on hard rock tunnelling – Rapporteur, T. E. Howard, United States: Organization for Economic Cooperation and Development, Washington, D.C., p. 73.

—— 1970, Report on soft-ground tunnelling – Rapporteur, J. Deschamps, France: Organization for Economic Cooperation and Development, Washington, D.C., p. 129.

—— 1970, Report on cut-and-cover construction – Rapporteur, J. Schmidbauer, Germany: Organization for Economic Cooperation and Development, Washington, D.C., p. 61.

—— 1970, Report on immersed tunnel construction – Rapporteur, H. C. Wentink, Netherlands: Organization for Economic Cooperation and Development, Washington, D.C., p. 42.

Prehoda, Robert W., 1967, Designing the future – the role of technological forecasting: New York, Chilton Book Co.

Scott, James H., and Carroll, Roderick D., 1967, Surface and underground geophysical studies at Straight Creek tunnel site, Colorado, *in* Highway research record number 185 – tunneling, geophysical studies, and soil-culvert interactions: Natl. Research Council Pub. 1516, p. 20–35.

MANUSCRIPT RECEIVED BY THE SOCIETY APRIL 5, 1972

Summary of Comments by Panel Discussants

GEORGE E. HEIM
Dames & Moore, Consulting Engineers, 1550 N.W. Highway, Park Ridge, Illinois 60060
JAMES R. SWAISGOOD
Dames & Moore, Consulting Engineers, 100 South Union Boulevard, Denver, Colorado 80228
VINTON BACON
Department of Mechanics, College of Applied Science and Engineering, University of Wisconsin, Milwaukee, Wisconsin 53201

There are three points which we believe are worthy of consideration by those involved with engineering geology. The first point pertains to the basic education of the geologist and the civil engineer. It is rare to find a recent graduate in either of these fields of specialization who has more than a cursory knowledge of soils and soil deposits. Seldom can these new graduates identify soils and apply the correct geologic name, describe soils, determine the origin of a soil deposit, apply the basic geologic principles of structure and stratigraphy to determine the distribution of soil deposits, or appreciate the type of variations which may occur in different soil deposits.

This is rather an anomalous situation in eduction, because both the geologist and the engineer, in their first course in geology, are taught all of these items as they pertain to rocks, although throughout much of the United States, soil deposits are more easily observed than rock deposits. When the basic geologic principles are not applied to soil deposits, both the geologist and the soils engineer are greatly handicapped in their ability to predict horizontal and vertical changes in characteristics or in anticipating potential problems which are commonly associated with soils of similar origin. Perhaps this lack of ability to describe adequately soil deposits and to recognize the types of variations which may be expected is in part responsible for the numerous claims by contractors of changed conditions. We encourage educators to re-evaluate their course content and expand their emphasis on soil deposits in the classroom, in the laboratory, and in the field.

The second point pertains to basic research. It would be of great value to have soil and rock engineering properties established and related to the stratigraphic column for each state. Unless the various soil and rock properties are linked to the stratigraphy, there is no way to anticipate the range of variations within a stratigraphic unit. Knowledge of the general range of engineering properties of the various stratigraphic units would greatly aid in the initial phases of conceptual planning for a variety of engineering projects, and could stimulate more rapid growth and development than otherwise would have occurred.

The third point pertains to assuring that basic data are reviewed by contractors prior to construction. This point was made during the discussion periods which followed the sessions, and we readily concur. We recommend that all project data, soil or rock samples, and geophysical data be made easily available to contractors for inspection, and that the prospective contractors be required to sign an affidavit stating that they have seen and reviewed the presented data prior to being allowed to submit a bid.

Using the Chicago deep-tunnel scheme for the solution to the flooding and combined sewer overflow problems in the metropolitan area as an example, it was emphasized that there must be a much closer coordination with and between the geologist and the civil engineer. In the Chicago metro area, tunnels up to 17 ft in outside diameter are being driven with the mechanical mining machine (mole) in dolomite and limestone rock strata some 250 ft below the surface. Consideration is also being given to drilling tunnels and mining room-and-pillar storage areas in a lower stratum, some 800 ft below surface. The Southeastern Wisconsin Regional Planning Commission, which serves seven counties, including the Milwaukee metropolitan area, has adopted as its official plan to use deep tunnels in rock strata to solve the combined sewer overflow problem. The consulting engineers for the Boston Metropolitan

Sewerage District have recommended rock tunnels as the solution to the drainage problems of that area. Thus, it appears that much of the future utility construction will be deep underground, and most often in rock. Design engineers will have many structural and construction questions which the rock and soil mechanics specialists will be called upon to answer. An immediate closer relationship between the educator, the designer, and the contractor is a must, because it has not often existed in the past. The future underground projects in the drainage field alone will run into the billions of dollars, and elected officials will be demanding better coordination among the contributing disciplines and arts.

MANUSCRIPT RECEIVED BY THE SOCIETY APRIL 5, 1972

Index

Geologic Mapping for Environmental Purposes

Edited by H.F. Ferguson

with contributions by
Robert G. Font
Robert F. Legget
Christopher C. Mathewson
Samuel J. Meltz
Charles R. Meyers, Jr.
Hugh B. Montgomery

Prepared for the
Engineering Geology Division of
The Geological Society of America

ENGINEERING GEOLOGY CASE HISTORIES NO. 10

The printing of this volume has been made possible
through the bequest of Richard Alexander Fullerton Penrose, Jr.,
and the generous support of all contributors
to the publication program.

Library of Congress Catalog Card Number 73-90839
I.S.B.N. 0-8137-4010-X

Published by
THE GEOLOGICAL SOCIETY OF AMERICA, INC.
3300 Penrose Place
Boulder, Colorado 80301

*Printed in the United States of America
by Edwards Brothers, Inc., Ann Arbor, Michigan 48104.
Type composed by The Geological Society
of America, Boulder, Colorado 80301.*

Contents

Preface

The rapid increase in the development of land, the exploitation of minerals, and the related accelerated environmental impacts have caused an explosion of demand for information that can be used as a guide to land use decision-making.

"Environmental Geology Mapping" was the topic of an engineering geology symposium at the 1972 annual meeting of The Geological Society of America; natural resources planning and the roles and interrelations of geology and geologists, planning and planners were considered and discussed. This book presents the coverage of those subjects because of their continuing timeliness and the need for a reminder that we must provide data that are relevant and usable for interdisciplinary considerations in natural resources planning. The authors express their ideas on how to translate professionally the traditional, basic earth-science data into forms that are adaptable to interdisciplinary solutions of environmental problems. They unanimously state that this conversion of data has to result in a viable input for decision-making, and it must also stand the scrutiny of the real world; that is, it must receive public endorsement and support.

We conclude that current planning practices are inadequate in predicting the effect of specific technological projects or specific social activities or policies on the natural environment. Presently, large sums of money are being spent on environmental assessment and resource planning. Whether this expenditure of time and money is justified depends upon how effective the natural science and selected engineering data are in the land use decision-making.

The development of the land and all the ramifications that contribute to the environmental impact will continue whether or not good geonatural resource decisions are reached. The existing social-political system demands that the final product not only be completed as growth requires but also that all impact, good or bad, be carefully assessed and considered to determine the least-effect approach from among the alternatives. With the increasingly constrictive requirements of an environmental impact statement on most public works and regulated private utilities, it is imperative that future highways, housing, and other land and resource developments be planned and built with the quality of the environment considered and evaluated in an operating framework that guarantees the best possible developmental and environmental results.

Despite all this concern over the effective uses of natural resources data input, we re-emphasize that at the policy-making level (state and federal land use policy and planning assistance acts) there is need for much more knowledge and skills than the natural and physical sciences can contribute. The ability of geologists to fit into environmental planning, not only in their professional capacity but also in a policy-making capacity, is questioned. Hopefully, an enlightened geological educator and professional practice leadership will emerge and grow and eventually prove that there is a tremendous potential in using geology input to aid in solving our developmental-environmental problems.

HARRY F. FERGUSON
Department of the Army
Corps of Engineers
Pittsburgh, Pennsylvania 15222

What Kinds of Geologic Maps for What Purposes?

HUGH B. MONTGOMERY
Hugh B. Montgomery & Associates, 122 Mt. Pleasant Boulevard, R.D. 6, Irwin, Pennsylvania 15642

ABSTRACT

Geologic mapping is evolving from primarily a method of communication among earth scientists into a decision-making tool for multidisciplinary and multiprogrammatic activities. The natural philosophy for this change in geologic display is the desire for ecological harmony in an environment favorable for the survival of mankind. The pragmatic dictates for the change are public pressures for a cost-accounting approach to provide fiscal control over development strategies. Control is expected to provide more value to a wider range of benefits and satisfy the growing demand for improved quality of life as growth proceeds. Total land use planning, unfolding on a national scale, requires more than a local geologic data array; it must include area-wide approaches to land, water, air, people, wildlife, and amenity sheds as a total ecological logic. Short-term cleanup goals must integrate with responsibilities for natural amenities and resources conditions within the environments of future generations. Comprehensive planning processes are unfolding as a major means of reaching solutions for comprehensive problem analysis, understanding, and solution programming. Comprehensiveness raises special problems as to jurisdictional, organizational, budgetary, and disciplinary logic systems as guides to future planning and program activities. Geonatural planning, although important, must be integratable with other planning elements. These influences require reassessment of the suitability of traditional biological-chemical-geologic data criteria, banks, and displays to assist decisions about logic systems for planning, priority setting, programming, and project activities. The geologic profession has not yet established the best relativity of its traditions to the requirements of social and economic expansion.

MULTIDISCIPLINARY TRENDS

Whether there is such a field as environmental geology is currently being debated; therefore, whether there is any special identifiable geologic mapping technique for environmental purposes is equally undefined. There is broad acceptance among geologists that geologic information when contributed in an environmental analysis can be influential in arriving at valid conclusions and predictions; however, the use of geologic information for this purpose awaits some basic changes in how geologists are educated and how they practice as professionals—hopefully, as diversified professionals.

Let us first examine the presumption that the pursuit of environmental concerns is an identified legitimate activity. Should geologists, then, be part of that pursuit?

Over the past few sessions, Congress has expressed in legislation its desire that the allocation of funds for the solution of problems concerning solid waste, air, and water cover approaches that are area-wide in scope, comprehensive in concern, and coordinated in skilled analysis and solution.

In the 1972 session of the 92d Congress, both the Senate and the House spent considerable time considering bills that dealt with a national policy for land planning and use with parallel special consideration over the management of Federal land and resources.

The Senate concerns were embodied in "Title I—Findings, Policy, and Purpose" (S. 632) passed by a vote of 60 to 18. The following is a partial quote:

> Sec. 101. (a) The Congress hereby finds that there is a national interest in a more efficient system of land use planning and decision making . . . land use management decisions of wide public concern often are being made on the basis of expediency, tradition, short-term economic considerations, and other factors which too fre-

quently are unrelated or contradictory to the *real concerns* of a sound national land use policy.

(b) that a national land use policy is needed to develop a national awareness of, and ability to measure, the land use impacts *inherent* in most public and private programs and activities.

(c) that adequate data and information on land use and systematic methods of collection, classification, and utilization thereof are either lacking or not readily available . . . , and that a national land use policy must place a high priority on the procurement and dissemination of *useful* land use data.

(d) that a failure to conduct competent land use planning has, on occasion, resulted in delay, litigation, and cancellation of proposed significant development . . . without regard to *sound* environmental, economic, and social land use considerations.

(e) that . . . costly conflicts between the Federal agencies and between Federal, State, and local governments [exist], and thereby subsidize undesirable and costly patterns of development; and that a concerted effort is necessary to *coordinate* existing and future . . . public and private decision making in accordance with a national land use policy.

(f) that . . . the manner in which this responsibility [Federal, State, and local planning] is exercised has a tremendous influence upon the *utility, the value, and the future* [of all lands] and that the failure to plan or, in some cases, the existence of poor or ineffective planning . . . poses serious problems of broad national or regional concern and often results in irreparable damage to commonly owned assets of great national importance.

(g) that . . . a national land use policy ought to take into consideration the *needs and interests*, and invite the *participation*, of State and local governments and members of the public.

(h) that such [land use] decisions and programs should seek to provide the maximum freedom and opportunity, consistent with *sound and equitable* land use planning and management standards, for all citizens to live and conduct their activities in locations of *convenience and personal choice.* (Italics added; U.S., Congress, Senate, Committee on Interior and Insular Affairs, 1972a)

The Senate postponed any action on its proposed legislation concerning the management of national resources but stated its policy clearly in S. 2401. It is partially quoted as follows:

Sec. 3. Declaration of Policy.–(a) Congress hereby declares that the national resource lands are a vital national asset containing a wide variety of natural resource values and that the national interest will best be served by retaining the national resource lands in Federal ownership unless . . . the Secretary determines that disposal of particular tracts of national resource lands . . . is consistent with the purposes, terms, and conditions of this Act.

(b) Congress hereby directs that the Secretary shall manage the national resource lands under the principles of *multiple use and sustained yield* in a manner which will, using all practicable means and measures, protect the environmental quality of such lands to assure their continued value for present and future generations. (Italics added; U.S., Congress, Senate, Committee on Interior and Insular Affairs, 1972b)

The House of the 92d Congress combined its intentions for national and Federal conduct in land and resource management in proposed bill H.R. 7211. Although not passed in this session, its intentions were clearly stated in its "Findings, Goals, and Objectives." A partial quote follows:

Sec. 101. (a) The Congress finds that there is an urgent need for land use planning with respect to both the Federal public lands and the land in non-Federal ownership in order to promote and secure the proper allocation of resources and to provide for the protection of the environment.

(b) The Congress further finds and declares it to be in the national interest that–

(1) Congress establish a policy with respect to the Federal public lands;
(2) public land management agencies coordinate the management of lands under their jurisdiction with one another and with the States;
(3) the people . . . be brought into the planning process;
(4) State and local governments be assisted in planning;
(5) patterns of population distribution be influenced to make . . . environmental, cultural, and social amenities available to large numbers of people;
(6) there be encouraged the development of new communities, the revitalization of existing rural and urban communities, and economic diversification of all communities;
(7) the administration of Federal programs be coordinated;
(8) systematic methods for the exchange of land use, environmental, and ecological data be developed at all levels of government.

(c) decisions about the use of land significantly influence the quality of the environment. (U.S., Congress, House, Committee on Interior and Insular Affairs, 1972)

Both Congress and the Federal government's "fiscal watchdog," the General Accounting Office, agree that there are serious environmental concerns and deficiencies in approaches. The General Accounting Office has issued a series of reports indicating that the resource development programs of the nation have fallen short of their objectives, increased the social and economic costs, and failed to sustain an environment of quality.

Additional positions presented by the Water Resources Council, Secretary of the Army, Appalachian Regional Commission, and the Congress in its recently vetoed Rivers and Harbors Omnibus Bill have both formally and informally endorsed analytical procedures that include measures of environmental quality, social well-being, and regional development along with national economic development benefits as a measure of the worth of proposed water resources construction projects.

All of these expressions state that every technical, economic, and social facet of a proposed government act should have the benefit of the appropriate professional.

At this point I want to emphasize particularly the words italicized in the previous quotes. The astute geologist will note that their use implies considerations not only far beyond geology but beyond other natural and physical sciences. Are we fit or are we planning to be fit as a profession for these levels of national and state-wide determinations?

PROFESSIONAL ASPIRATIONS OF ENVIRONMENTAL GEOLOGY

Many geologists are cognizant of the roles they are capable of playing in the national economic development effort. However, I am not convinced that the profession recognizes its full potential or how it can best be expressed—the dynamics of using its influence to the betterment of the profession and the nation. The following quotes are illustrative:

The purpose of this study was to provide guidelines, based on geologic analysis, which may be used by planners to reduce costs resulting from adverse geologic conditions. (Reid and Montgomery, 1972)

The need for geological information in urban and regional planning is now widely recognized, but unfortunately there often exists a gap between what the geologist is able to translate into usable information and what the county or regional planning commissions are able to understand. (Easterbrook, 1972)

A knowledge of rock unit distribution and associated physical properties is requisite to the recognition of areas where adverse interaction between urbanization and natural systems occurs. (Garner, 1972)

There is, therefore, a need to present natural resource information that is easily understood and readily available. . . . These single resource characteristic maps can be variously combined to meet needs of specific planning problems. Maps can be combined according to local priorities, and combinations can be adapted to changing technologic, economic, statutory, and social conditions. (Pessl, 1972)

It is generally assumed that the conferring of a degree indicates that certain academic standards of excellence have been met. . . . With the awarding of a diploma, the new graduate becomes an "instant pro" endowed with all the wisdom necessary to cope with the avarice and greed of a cynical business world. (Stead, 1972)

The professional geologist of the future . . . will need more science, more mathematics, and above all he will need to know how to solve problems in a quantitative way. . . . Future geological education must aim to produce professionals, as well as scientists, who are well versed in geology and ready to apply their knowledge to the discovery of resources, to orderly development of resources, and to planned use of the environment. (Berg, 1972)

Geology should relate to interdisciplinary discussions; currently the biological and social scientists are largely unaware of the roles geology and geologists play in the development of ecology and human affairs. Individual geologists must assume the responsibility for development and maintaining communication with their communities. (McKee, 1972)

One of our professional duties is to let the general public know something about what geology is and what geologists do: After all, in one way or another society pays our bills. (Bates, 1972)

Geologists, however, until recently, have not seen as a high priority the necessity of showing the relevance of the earth sciences to the taxpayer. . . . Problems which geologists can solve better than any others are those relating to natural resources and environmental geology . . . geologists must . . . improve the public understanding of the relevance of the geosciences to the solution of problems of society through . . . use of all the communications media. (Skehan, 1972)

Population centers are mislocated. These people must be protected from their ignorance and circumstances. . . . Geologists must play a vital role in defining hazardous areas, in public decision-making, and in helping to educate the public. (Parizek, 1972)

Professions and public officials are realizing the need to increase effectiveness of joining forces. . . . What is required is expansion of the profession's role in urban affairs; willingness to work with policy, planning and operational personnel; increased influence upon the political process; and effective communications. (Zeizel, 1972)

Steps should be taken lest this new information breed indecision . . . or warp the planning process through a mis-weighting of our environmental factors . . . we should . . . plan for breadth and flexibility in research, teaching and professional work, so that geologists may participate in the many new forms of decision-making. (Twiss, 1972)

Beyond providing the physical setting for informed, effective decision-making, geologists can also serve effectively as leaders in the decision-making process. . . . The integration of processes in three dimensions, through time, is not customary in most other fields. . . . Thus, geologists have both substantive and methodological contributions to make in the increasingly important decisions affecting the environment and resources, decisions which ultimately determine the quality of life in America and the world. (Everett, 1972)

A field study course . . . offers several advantages over the traditional field experience . . . the relationship of geology to biology, history, or present (or past) environment can be observed—that between volcanism and the history of certain Indian groups, or the relation between topography and air pollution. (Thomas, 1972)

To me the deficiency appears to be the general persuasion of the profession that if it explains its wares and produces volumes of information, interpreted to end-use purposes where possible, the world will beat a path to its door to buy the new "mouse trap." This is far from the truth. The only way the profession can become a dynamic social force is to define the interprofessional and social relevance of geology and then practice as professionals free from geologic bias within those findings. I would like to note that the views of Zeizel, Twiss, Everett, and Thomas offer guidance for a course of action with far greater potential for the future of the geologic profession.

Two basic actions need to be taken: First, geologists must ask what kinds of current and future questions need geologic information to aid in finding answers and then formulate information for those answers on a problem priority basis.

Second, geologists must participate in the policy-making, budgeting, and planning processes necessary for comprehensive area-wide approaches. It is in this triple-headed arena that the resources to do the jobs needed by this country are determined.

PROBLEMS WITH DATA AND THEIR USE RELATIVE TO COMPREHENSIVE ENVIRONMENTAL ANALYSIS AND PLANNING

Some very fundamental concerns about data must be evaluated. Traditional state and Federal social and physical resource programs have collected data to quantify characteristics of the program area of concern in a form and manner that are designed to serve the traditional purposes, goals, and objectives of those programs. For example, the relative acidity of water (pH) is a commonly measured factor that conveys a certain relative degree of the condition of water which is useful for communication purposes. It is not a measurement that allows calculations of changes in acid contaminants to predict pollution level changes in response to contamination reduction or increase. On the other hand, the biological oxygen demand (BOD) is the type of measurement that does lend itself to such calculations and predictions, but is information that has not been collected generally.

The recent trends toward funding of air, land, and water projects with environmental implications and the statement of national policy in the National Environmental Policy Act of 1969 require that advance comprehensive planning be a precondition for the granting of funds. The data available must now also serve the purpose of making analyses, projections, and predictions for environmental decisions.

The environmental decisions must apply not only to environmental matters concerning public investments and decisions that are directly related to environmental impacts, such as a sewer facility, but must also deal with matters concerning the spending of public funds for cleaning the residuum of wastes and degradation inherited from the less sophisticated contamination controls that dominated the economic development of the past.

Figure 1 diagrams the data needs for discussion purposes. The diagram illustrates that there are two data collecting and analyzing processes that meet in the field where man's public and private investment programs actually occur physically in the growth process. The upward flow of data usually results in the establishment of a specific number or permissible range of values to which the environmental impact effect must conform or a legal violation is presumed, which may be followed by enforcement actions and penalties.

The downward flow of data may pass through a variety of quantitative and graphic forms such as numerical tables, graphs, and maps. The chief difference from the upward flow is the divergent paths taken through the development of the form for the use of the data. In other words, the data may vary with the form required to match the needs of the specific activity it is meant to serve. It is possible that the range, scope, and level of detail of the data and their presentation may be different for overall economic development decision-making than for environmental analysis and decision-making. It may be true that eventually all the data that might be taken for economic development would be gathered for environmental decision-making. The approach, however, would be modified in the following ways:

1. The type of data may vary and be tailored specifically to the type of environmental decision that must be made; for example, water quality for health facility rather than for water supply facility planning.

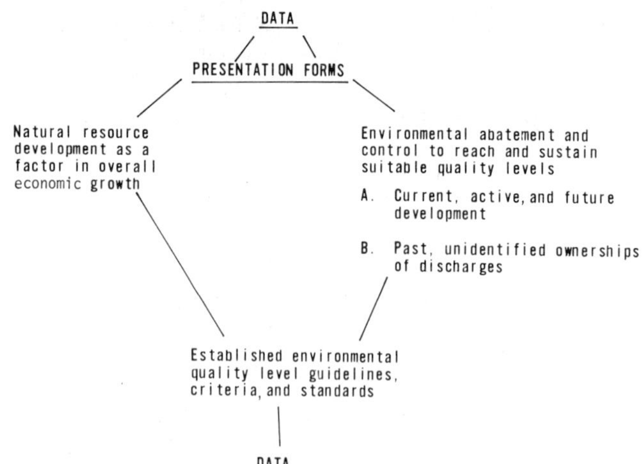

Figure 1. Data collection and analysis processes.

2. The level of detail would vary with the type of environmental problem; for example, the construction of a 100-acre shopping center as compared to a 10-acre apartment complex.

3. The areas where data would be gathered would be taken where the environmental impacts are first expected to be felt; for example, at or near population growth centers or corridors.

Once these basic data differences are confirmed, then the data should be formulated to serve the following tasks:

1. Permit identification of interrelations among environmental components (soil, rock, air, water, biota, and so forth).

2. Allow projections to identify future needs and their quantity and quality parameters, relative to other indicator information such as population growth.

3. Be translatable into planning the process for environmental achievements.

4. Define accurate implementation to accomplish the environmental plan.

5. Permit evaluation of the state of development of the data collecting, data interpretation, projecting, planning, implementing, and, ultimately, the actual effect or use of the environmental service or project to determine the degree of success or modification needed to obtain the desired environmental achievements.

6. Be easily interpreted into technology choices and financial variables so that alternative approaches can be technically, economically, and logically sound.

We recognize that data collecting is a costly and time-consuming effort. It is vitally important, therefore, that the organization of such an effort be carefully analyzed and precisely formulated to suit the design of the environmental administrative and management effort that is anticipated.

First, the following objectives should be considered and the relative degree of emphasis each is to receive at the early, advanced, and final stage of the management effort should be determined: (1) develop an overview of the total environment region-wide, (2) construct a planning framework, (3) identify a broad range of environmental problems, (4) develop a state-wide environmental monitoring system, (5) examine the impact of area-wide development policies and programs on environmental policy, (6) construct an operational and continuing analytical framework and system, and (7) examine the nature of existing and alternative policies and institutional arrangements.

Second, a set of procedural steps should be outlined to maintain a continuity and assurance that the selected objectives in the proper depth of detail will be reached. One approach might be:

1. Define the project objectives relative to which decisions are expected to be made on legislation, government reorganization, program and projects, and monitoring to meet existing quality standards, the establishment of quality standards, and determining the relevance of quality levels to user needs.

2. Define the environmental status as to existing legislation, government organization, and current data on land, water, and other resources in terms of their relevance to decision-making for reaching the carefully defined objectives.

3. Define data gathering as being specific to fulfill clear needs rather than being a total inventory. With these data, legislation could identify controls and programs, not just in terms of development policy but in development without quality loss; agency reorganization would be somewhat different using these data in addition to material previously provided to the executive branch; program design could be more relevant to environmental quality control within economic development; and budgets for the continued development of planning processes and environmental programs would be more meaningful.

4. Launch the planning process in data gathering, analysis, and program development that would have a clearer meaning to identified environmental policy and quality level goals in terms of user needs.

5. Complete information data analysis and identification of roles and recommendations quickly (within six months); planning analysis and alternatives development and statewide, regionwide, or nationwide goal setting should come second, although some conceptual thinking could evolve from the start. Next, the planning process could begin to formulate, and subplans could begin to crystallize toward an overall guide plan that should be constantly updated.

Being alert to the obstacles and opportunities that are presented by the need for, acquisition, array, and analysis of data in view of selected objectives can avoid loss of time, repetition of available or useless data, and provide substantial savings, and great assurance that a usable planning or budget product will be available for the environmental management and administration process.

SOME SPECIAL PROBLEMS PRESENTED BY A COMPREHENSIVE AREA-WIDE APPROACH TO ENVIRONMENTAL PLANNING AND ADMINISTRATION

Some special concerns about the comprehensive approach and the general nature of geologists must be considered.

The increase in the demands of Federal legislation to require that funds for the solution of problems concerning air, water, land, and waste be provided only when the project can be shown to be an integral part of a comprehensive area-wide environmental improvement plan places a burden on all of us to produce creative, interdisciplinary, multipollutant insight into integrated planning recommendations. This approach is not now developed to a high degree of sophistication. Consequently, the changes in approach, concepts, analysis, plans, implementation, and the very nature of projects themselves will be revolutionary over the next few years.

The complex matter of measuring and quantifying many of the problems, projecting their future levels, and integrating solutions at a level of design appropriate to the size of the problems and the desired results is a major hurdle to overcome; however, these physical obstacles may be easy to solve relative to some of the less tangible philosophical, economic, and social problems.

One of these less tangible aspects is the ability to conceive of the interweaving of the environmental quality systems about which we are not yet fully knowledgeable, or to perceive adequately those systems that may be understood but not yet widely known among the various environmental professions. The lack of experience in applying a variety of skills in a coordinated effort to approach a problem as to all of its physical, social, and economic facets is at present an impediment. Even universities are not well advanced in interdisciplinary team approaches to environmental problems.

In my opinion, the future is not especially promising in regard to turning bright young people into environmentalists who can teach, build, and administer in the way that the intricate environmental problems require. Since all of us who are moving into the environmental mainstream must work together, let us share some thoughts as a basis for future cooperation.

Interfaces

Most of us have heard of the water or hydrologic cycle. It is a concept that helps us to understand the chain of events and the path that a molecule of water takes as energy is applied and the molecule responds to its force. To a lesser degree, we understand similar systems expressed as crop rotation, food chains, birth, life, and death.

Young students are beginning to receive advanced training in the nature of the environment and the interrelations of plants, animals, and minerals, and the systems they comprise. The September 1970 *Scientific American* published a series of articles on the component parts of the environment that should be of interest to all who are concerned with learning about the environment. Too often we are presented with only pieces of knowledge about environmental subjects when we are asked to participate in environmental problem-solving. For example, if we are looking for information on the physical parts of the environment, we will find articles or ideas dealing with environmental compartments, such as soil, minerals, water, land, air, and amenities (that is, historical or archaeological treasures, scenic vistas, unique scientific or ecological niches). We often find ourselves confused about the commonalities and interrelations of the parts of the environment. Conversely, we are often faced with literature that separates a problem and so thoroughly covers a *specific* emphasis that we fail to obtain the overall perspective that is needed. For example, the concern and sometimes hysteria over the mine-drainage effect on an area may completely mask the fact that municipal, industrial, domestic, or agricultural polluters so contaminate the area that it would still be unusable even without the effects of mine drainage. It is important, therefore, to recognize that hardly anything happens or exists in the world that does not have an impact on at least one other event or thing. Every environmentalist, whether he is an administrator or a practitioner, should condition himself to look for these relations, to evaluate their relative importance, and to incorporate this knowledge into a comprehensively considered judgment.

TABLE 1. DIGEST OF ENVIRONMENTAL SYSTEMS

Interfaces	Examples
Atmosphere to outer space	Space junk to orbit
Outer space to land	Meteorites to Crater Diablo, Arizona
Outer space to atmosphere (air)	Cosmic rays to ozone
Atmosphere to astral body	Space station to moon
Air to air	Smoke stack to air shed
Air to land	Dust to ground
Air to water	Sulfur dioxide to hydrolized sulfuric acid (H_2SO_4)
Water to air	Vapor to cloud
Water to water	Fresh stream to salty ocean
Water to land	Rain to ground water
Land to water	Drainage to streams
Land to land	Sand beaches to dunes drifting inland
Land to air	Dust bowls to reddened skies
Biota to biota	Roast beef to your dinner
Biota to air	Transpiration of plants to air
Air to biota	Carbon dioxide to photosynthesis
Biota to land	Dinosaurs to fossils
Land to biota	Minerals in our milk
Biota to water	Seaweed oxygenation of water
Water to biota	Maple sap for pancakes

Table 1 is a digest of environmental systems with examples to serve as a check list for identifying systems, chain reactions, and linkages that may produce a final condition as the result of an initial judgment with a completely different primary purpose.

Interdisciplinary Reactions

Being a geologist, I often hear my colleagues complain about the failure of civil engineers, planners, or others to recognize the value of geology and to use geologists in their work and decision-making. I have heard other professionals make similar sounds. Certainly, some of the examples have been legitimate, and undoubtedly there will continue to be examples of unnecessary expenditures of time, money, and even health and life because of a lack of a particular professional input.

Several years ago, while working on the geologic engineering, land value, and multiple use aspects of several dam projects (Montgomery, 1963), I became humbly aware of the fact that the contribution of one man or even one profession is important, but only when combined with the input of many other men and professions; perhaps hundreds of men, dozens of professions, and decades of time are involved before a project becomes a reality.

Geologists as well as other professionals must recognize their obligation to contribute to the input process. Their contributions will vary in direct relation to the need for their particular knowledge. For example, a geologist may be intimately involved in the early investigation stages of a dam, but his role may diminish in importance when construction begins and the structure begins to rise. After construction is completed, the geologist's talents may be needed only sporadically during intermittent inspection. In other words, some professionals play an input or secondary intermittent role, whereas others play an integral or primary continuous role even though the interdependence of roles is vital to the success of any project.

An environmentalist must never lose sight of the need for a multidisciplinary team to attack the complexity of environmental problems. He must understand the range of disciplines needed for solving a particular problem, the sequencing of their inputs, and the repetitive follow-up. The multidisciplinary team approach to problem-solving makes it possible for the various disciplines to interact as they are developing their input so that all of the input is available to all team members to aid in solving all components of the project.

Table 2 is a digest of the interdisciplinary concerns and will serve as a check list for identifying the reservoir of talent, the selection process, and the development of the climate for the interdisciplinary team approach.

Although our society needs trained environmentalists,

TABLE 2. DIGEST OF INTERDISCIPLINARY CONCERNS

Needs	The number of environmental components: soil, minerals, water, land, air, amenities (that is, historical and archaeological treasure, scenic vistas, unique scientific or ecological niches), construction and administrative demands, demographic, and economic concerns will determine the range of disciplines and experience required.
Resources	The amount of money and personnel will determine the number and scope of environmental projects and the range of disciplines and the organization of the team.
Innovations	The range of alternatives to be studied, the limitations of the advancement of technology, and the need to develop a project procedure to weld a team interdisciplinary approach will require increased creativity and cooperation and experimentalism from each member of the team and the total team product.
Programs	The nature of the historical basis used to develop established programs, the need to establish new programs, and the limitation in program receptivity owing to politics or budgets will require variations in the team.
Education	The types of traditional courses, degrees, educational philosophy, and educational institution structure and their failure to produce qualified scientists with environmental orientation, or environmentalists with the comprehensive management scope of education, limit the types of teams available to problem solving and require the promotion of education for environmental teamwork.
Application	Regardless of the statutory requirement, the receptivity of the interdisciplinary team depends on proven examples of successful projects. The building of the team will depend directly on these proven cases.

the lure of more prestigious and higher salaried jobs in the so-called management and administrative levels is forcing our scientists and technologists to spend less time in broadening their professional knowledge and more time in attending management conferences, preparing budgets, and performing other tasks removed from the actual practice of their professions.

Corollary to this is the growing lack of confidence and respect between the technologist who stays with his technology and the one who moves into the administrative arena; each views the other's ability and motives with distrust.

If environmentalists of the future are to overcome the complex environmental problems, they must be aware of the need for a broadened interface approach and the structuring of interdisciplinary resources and teams, and they must work to overcome any stigma associated with the less glamorous environmental roles. For example, sanitary engineering has never been attractively competitive with civil engineering, although they are similar expressions of basic intellectual talents and knowledge. It is often difficult to convince the skilled worker to stay at a sewer plant or garbage-handling facility when he can work in a nonwaste facility, even though the work may be very similar.

TABLE 3. DIGEST OF ENVIRONMENTAL ROLES, PLANNING AND IMPLEMENTATION

Professionals	The experienced professional will comprise the main body of broadly trained professionals available to work on environmental teams. Older and some newer graduates will have a broad-based education of comparable scopes for successful comprehensive implementation.
Technicians	Few professionals support the training of technicians to back up professional practices. However, the technician level of training is very important to the success of future environmental planning and implementation, and back-up technicians must be trained and accepted.
Personnel	Environmental planning must include increased production of implementors if the current and expanded environmental goals are to be feasible.
Training	Environmental planning must provide more training programs that are complete in scope if successful implementation is to be a reality.
Waste-Management	The stigma of the low prestige attached to handling wastes or operating technological equipment for waste handling purposes is largely a matter of public attitude. The environmental planner must work to upgrade the public image of this work and the rewards for performing it to guarantee successful implementation of environmental programs.

Table 3 is a digest to serve as a check list for identifying the special environmental roles and employment problems, the reservoir for acquiring present and future implementors of environmental projects, and the need to alter educational and employment attitudes and practices to develop a climate for successful implementation and operation of environmental projects.

TOTAL PROFESSIONAL PRACTICE OR SECONDARY STATUS

The environmental planner, administrator, and the supporting public have many responsibilities to fulfill if improvement of the quality of the environment is to become a reality. The planner and administrator, who may be basically a geologist, will be largely responsible for devising ways to solve the physical problems; then the public must be convinced that those projects are important before funds are provided for their implementation.

In order for this mix to become effective, both the professional and the public must be aware of the *tangible* intricacies of environmental problems and technological solutions. They both must recognize also the *less tangible* aspects of the broad ramifications of the systematic relations of the various parts of the environment, the special problems and advantages related to the application of many different types of training and experience upon a commonly shared problem, and the special problem-solving needs to attain the most improvement over the shortest amount of time.

Only when the geologic profession, practicing in educational institutions, government resource agencies, and private industry, grasps the concept of relevant and timely data gathering and supports the multiprofessional, interdisciplinary approach to problem-solving will environmental mapping or any other product or service of the profession establish geology as a dynamic, primary force in our society and the world. When we obtain the skills and accept the responsibility of making technological, financial, and general administrative interpretations and decisions based on wide information, of which geology is one component, then and only then will individual geologists and the geologic profession realize its full potential. Failure to work toward a role of co-ordination, integration, and leadership such as this will not leave a vacuum for long. Planners, engineers, sociologists, and other professionals with the necessary desire and drive to expand their capabilities and with the courage to accept the additional professional risk, can and will fill the gap.

REFERENCES CITED

Bates, R. L., 1972, Geology and the printed word: Geol. Soc. America, Abs. with Programs (Ann. Mtg.), v. 4, no. 7, p. 446.

Berg, R. R., 1972, Future education of professional geologists: Geol. Soc. America, Abs. with Programs (Ann. Mtg.), v. 4, no. 7, p. 450–451.

Easterbrook, D. J., 1972, Environmental geology of Whatcom County, Washington–A possible model for applied environmental geology: Geol. Soc. America, Abs. with Programs (Ann. Mtg.), v. 4, no. 7, p. 495.

Everett, A. G., 1972, The geologist in environmental and resource decision-making: Geol. Soc. America, Abs. with Programs (Ann. Mtg.), v. 4, no. 7, p. 501.

Garner, L. E., 1972, Environmental geology of the Austin, Texas area: Geol. Soc. America, Abs. with Programs (Ann. Mtg.), v. 4, no. 7, p. 515.

McKee, E. M., 1972, The public wants and needs to know how geology affects both individuals and communities: Geol. Soc. America, Abs. with Programs (Ann. Mtg), v. 4, no. 7, p. 591.

Montgomery, H. B., 1963, What comes before the construction of a dam: Civil Engineers Jour., v. 33, no. 4, p. 56–57.

Parizek, R. R., 1972, Agnes, aftermath and the future: Geol. Soc. America, Abs. with Programs (Ann. Mtg.), v. 4, no. 7, p. 621.

Pessl, F., Jr., 1972, Earth-resource data for land and water management in the Connecticut valley: Geol. Soc. America, Abs. with Programs (Ann. Mtg.), v. 4, no. 7, p. 626.

Reid, R., and Montgomery, E. L., 1972, Geologic hazards pilot study, Flagstaff, Arizona: Geol. Soc. America, Abs. with Programs (Ann. Mtg.), v. 4, no. 7, p. 634.

Skehan, J. W., 1972, The geology and ecology of the Madison Avenue gold mine: Geol. Soc. America, Abs. with Programs (Ann. Mtg.), v. 4, no. 7, p. 668.

Stead, F. L., 1972, The practical aspects of professional ethics: Geol. Soc. America, Abs. with Programs (Ann. Mtg.), v. 4, no. 7, p. 675.

Thomas, J. L., 1972, A different approach to integrating field study into the earth science curriculum: Geol. Soc. America, Abs. with Programs (Ann. Mtg.), v. 4, no. 7, p. 687.

Twiss, R. H., 1972, Use of new-generation geologic information in regional planning: Geol. Soc. America, Abs. with Programs (Ann. Mtg.), v. 4, no. 7, p. 695.

U.S., Congress, House, Committee on Interior and Insular Affairs, 1972, National Land Policy, Planning, and Management Act: 92d Cong., 2d sess., H.R. 7211, Rept. 92–1306.

U.S., Congress, Senate, Committee on Interior and Insular Affairs, 1972a, Land Use Policy and Planning Assistance Act: Congressional Record, 92d Cong., 2d sess., Rept. 92–869, p. 1, 2.

—— 1972b, National Resource Lands Management Act: Congressional Record, 92d Cong., 2d sess., Rept. 92–1163, p. 2.

Zeizel, A. J., 1972, Communication–Keystone between geological science and metropolitan problem solving: Geol. Soc. America, Abs. with Programs (Ann. Mtg.), v. 4, no. 7, p. 713.

MANUSCRIPT RECEIVED BY THE SOCIETY JUNE 25, 1973

Regional Land Use Analysis and Simulation Models: A Step Forward in the Planning Process

CHARLES R. MEYERS, JR.
Regional Planning Consultant, Regional Environmental Systems Analysis
Oak Ridge National Laboratory, Oak Ridge, Tennessee 37830

Land use planning has the potential of playing an important part in man's relation to his physical and cultural environment. However, planners and policy makers are hampered in implementing large-scale planning decisions by their inability to anticipate or visualize the many relations that must be considered in developing rational, comprehensive planning alternatives. To meaningfully control our dynamic environmental system, of which land use can be considered a subsystem, we must have the ability to predict land usage and associated environmental impacts both in an assumed steady state and in the presence of real or imagined perturbation.

In response to the availability of appropriate research tools and the clear need for land use simulation and prediction, the 1960s spawned a proliferation of computer-based simulation models for use in transportation studies. Yet, with some notable exceptions, these sophisticated modeling efforts have not made a significant impact on planning and decision-making–primarily because of their narrow focus and inflexibility.

In large measure, the designers of these models failed to recognize that the real worth of land use simulation modeling lies in the applicability and usefulness of the simulation to real-world planning and decision processes. Potential applications include evaluating effects and implications of alternative land use patterns on service, utility, and transportation systems and measuring environmental impacts on natural systems within the region. Anyone who seeks to manage or plan a region's growth must be able to ask the question, What would happen if . . . ?

Land use simulation must attempt to explain and predict human behavior; therefore, it lacks the precision of more traditional simulation (for example, space flight orbital trajectories). To reduce the complexity involved in modeling human behavior, past planning models have described only one process in the total system and, as a result, have not been put into a context that permits approximation of the total system (for example, including feedback loops between economic, demographic, social, ecological, and political components). Recognizing the shortcoming of not including all relevant processes places one in the dilemma pointed out many times in past modeling experience; that is, in the assimilation of a complex system into a feasible, operational empirical model, the model must be theoretically "elegant" while at the same time operationally "feasible" and must produce accurate projections into the future (Lowry, 1967; Seidman, 1966b). It was recognized in the study made by the Staff, Center for Real Estate and Urban Economics (1968) that major criticism of any land use model can be anticipated to be of two types: (1) the model is too simple to be acceptable on theoretical grounds, or (2) the model is too complex to be operational.

Ian McHarg (1969) has very aptly stated that "The biosphere does not consist of a pyramid of organisms but of ecosystems in which many different creatures coexist in interdependence, each with its own process, apperception, roles, fitness, adaptations and symbioses. . . . They (the process) can be identified not only in degrees of value but of tolerance and intolerance." Only from such understanding of the interacting processes can the land manager incorporate the analysis into the planning process while, in parallel, creating an early warning system to test land use changes before implementation.

A LAND USE MODEL SUPPOSITION

In conceptualizing a regional land use model, one must resolve a number of philosophical modeling questions. The most fundamental of these questions has to do with the mathematical description of land use change. A land use model cannot seek an optimum solution to land use allocation in the mathematical sense, as it is impractical to develop an explicit mathematical function describing land use with our present limited understanding of the land-change process. An approach to this dilemma is a two-level macro/micro or subregion/cell model that iterates on individual incremental allocations such as employment, commercial, and residential categories in a predetermined sequence and attempts to simulate processes observed in the real world by allocating these activities as incremental changes. This then represents an attempt to reconcile the microprocesses of the individual decision unit with the macrobehavior of the system. The changes individually do not upset the steady-state condition of the system, and since the changes are allocated in a prescribed sequence (basic employment, direct residential, and support trade and service, indirect residential, and so forth), the system slowly adjusts to the aggregated change predicted for each iteration. Mathematical rigor can be introduced into the model through the calculated probabilities of land use allocation occurring in each spatial unit (that is, subregion, census tract cell or parcel). The probabilities are used to select the most likely spatial assignment for each allocation. This approach has the advantage of giving an operational framework in which to improve our understanding of the land-conversion process by means of indices that define conversion probability. By improving these indices through time, there will be an evolution toward quantitative and analytical rigor.

This allocation modeling approach and the resulting output requirements lead to a consideration of scale unique to spatial processes. The model recognizes that land use can be modeled at different levels of scale and that each level possesses emergent properties that must be described in combinations of variables unique to that scale. Thus, the model incorporates a hierarchy of scale levels in which larger areal units are aggregates of smaller units. The data classification, cell storage units, and allocation algorithms must reflect this organization. The effect of hierarchy can be viewed as a "zoom" capability wherein one can focus down to the level of detail needed for specific tasks.

Hierarchical structure is forced on a spatial modeler by both the needs of the model and the user. The most important considerations involve the following:

1. Most planners and regional decision makers are location oriented, and their modeling problems reflect this orientation. Regional questions must be answered at a much larger scale than site-specific questions, but the model must be capable of answering *all* questions addressed to it by planners. For example, extensive undulating physiography and hard bedrock close to the surface will hamper construction and increase costs associated with the construction of a new sewer system. These costs can be reflected in emerging land use patterns if the model can process data at appropriate scales of resolution.

2. If activity site selection is accomplished by rank-ordering all possible sites, the site size must be compatible with the land use being allocated, thus requiring many cells to describe the whole region. Although site specificity permits furnishing users with land use decision alternatives at the scale with which they feel comfortable, it has the disadvantages of large data bases and increased processing time. This problem can be alleviated by adoption of a data hierarchy that permits allocating an activity to much larger aggregate site areas before making individual site decisions.

3. Most environmental models require both distributed and point source information on pollutant concentration. Thus, allocation of land use activity is small-scale site specific for point sources and large-scale watershed- or airshed-related for distributed sources. The model accounting system must incorporate both point and distributed descriptors of land use activities.

MODEL ATTRIBUTES MEETING THE NEEDS OF PLANNERS

Ideally, a regional planner needs the capability to assess the present status of the regional system and then to predict the future status in light of alternative plans, management practices, growth patterns, and so forth. Thus, the definition of "model" must be broad and general and yet embrace a myriad of specific factors that influence man and his environment or are influenced by them. To accomplish this, land use models must possess the following characteristics:

1. The model must be *comprehensive.* Many diverse effects of planning (that is, sociological, economic, ecological, political, and demographic) would be anticipated in many diverse fields. If the model does not embrace all these aspects, its focus is limited and it fails in delineating the significant interconnected relations that comprise the system.

2. The model should be *easy to modify*. We are in the

horse-and-buggy era of understanding regional processes, and as new relations become apparent, it should be possible to incorporate new or modified algorithms and to accept updated or modified data.

3. The model should be *exportable*. Each region has its own unique characteristics; however, the general concepts and structural framework of the model should be applicable to other regions. Data and algorithms could be changed in the model, and the basic structure could be utilized for different regions.

4. The model should be capable of making both *baseline and conditional forecasts* by accepting existing conditions or trends and planning scenarios from the planner or model user and simulating their effects in the region. At some level the planner should be able to project his working knowledge into the system during model runs. In the simple example of regional growth simulation, the planner should be able to stop and change a land use or increase migration rates during a run or at any period of time in the future.

5. The model should have a *multipurpose data base*. The data base should serve as a regional information system as well as be usable to the model. Such a data base should provide rapid, accessible, and inexpensive data to a large number of users.

6. The model should be *diverse*. The more use a model receives, the more heterogeneous the users become. Thus, its elements should lend themselves to being organized easily to answer different questions or to handle different decision variables and conditional forecasts.

7. The model should be *reliable*. Reliability should be well established in order to reduce the amount of verification needed. This is achieved by having the model track different events over time before projections are made into the future. If the model begins with historic data and produces results compatible with present events, there is a reasonable assumption that its forecasts may be accurate.

8. The model must have *comprehensible output*. The computer output results of the model run should answer the need of the user; that is, the skilled professional who wants technical output as well as the nonskilled user who wants simplified graphic output.

There are many practical limitations involved in the construction and use of comprehensive land use models. The most serious is lack of knowledge about many of the interacting relations that affect a region over time. Another serious drawback is the unavailability of much basic information and, in many cases, the gaps and inaccuracies in data that do exist. In addition, computer capability is not always present in the regional organization. Possibly the greatest deterrents to widespread use of comprehensive models of this type are that they are costly, new, unproven, and difficult to implement within present institutional structures. The complexity and scope of a comprehensive regional model is such that extensive development and testing are needed before usable models are available to planners for real-world problem-solving applications.

EPISTEMOLOGICAL QUESTIONS OF LAND USE MODELS IN THE PLANNING PROCESS

There appears to be a natural evolution of acceptance of computer models by planners and regional decision makers. This begins with the simple use of data stored and queried in a spatial information system. Utilizing this extension of "seat-of-the-pants" analysis, experimentation that begins with basic variables leads to the understanding of relations not readily apparent. Statistical analysis and simplified submodel development follow until both a working acceptance and use of the model take place. Understanding of and confidence in both the data and the model algorithms become imperative if planners are going to apply model output to real-world problems.

We come now to a fundamental question of model building. How can the land use model and its output be tested to assess the worth of the projections? Clearly, the only true test of a land use model is comparison of *actual* change in land use with *predicted* change. Since this is impossible over the short run, other validation tests are required. In most past cases, land use models have used historic data and regression analysis to develop, tune, and validate the model output. Information most urgently needed by planners concerning the manner in which the future will differ from and not merely replicate relations of the past is missing in this approach. The assumption that somehow past spatial relations will continue into the future is pure folly; extrapolation from the past does not reveal differences between the immediate past and the future. Rather, with a historical perspective, one can delineate changes in relations between variables rather than regularities. This understanding of the relations then serves as a qualitative yardstick to evaluate the reasonableness of the land use model forecasts.

THE FUTURE

Regional models have not yet begun to reveal their potential as planning tools; most of those available are limited, costly, and inadequately reflective of reality. Reliable models for specific purposes are being utilized in many individual cases. For example, housing market models of small geographic areas are gaining rapid acceptance among developers and real estate speculators. Perhaps one of the most important implications of a regional model is that it may lead planners to a greater degree of humility. As we probe deeper and deeper into the complex system of which we are a part, we become increasingly aware of the labyrinthian nature of human motivations, choices, and actions that make up the system.

Models augment and amplify the planners' experience and intuition and can act as an early warning device, indicating needs for change or corrective action that may lie ahead. Most important, however, the models respond to second-, third-, and fourth-order effects when asked the question, What if . . . ?

SELECTED BIBLIOGRAPHY

Alonso, W., 1958, The core of regional science: Regional Sci. Assoc. Papers and Proc., no. 4, p. 3–7.

Behrens, W. W., 1971, The dynamics of natural resource utilization: System Dynamics Group M.I.T., Cambridge, Mass.

Dickinson, R. E., 1947, City, region, and regionalism–A geographical contribution to human ecology: Kegan Paul, Trench, Trupner and Co., Ltd., London, England.

Fuller, R. B., and McHale, J., 1963, Inventory of world resources, human trends and needs: World Res. Inventory, Southern Illinois Univ., Carbondale, Ill.

Lankford, P. M., 1969, Regionalization: Theory and alternative algorithms: Geog. Analysis, no. 1(2), p. 196–212.

Lowry, I. S., 1967, Seven models of urban development: A structural comparison: Highway Research Board, Conf. on Urban Devel. Models.

McHarg, I. L., 1969, Design with nature: Nat. History Press, Garden City, N. Y.

Ringenoldus, J. C., 1970, Water resources and regional land-use plan: Am. Soc. Civil Engineers Proc., Jour. Sanitary Eng. Div., no. SA6, p. 1311–1320.

Seidman, D. R., 1966a, The present and futures of urban land-use models: Proc. 4th Ann. Conf. Urban Planning Inf. Systems and Programs, Univ. California, Berkeley.

—— 1966b, A decision oriented model of urban growth: Proc. 4th Ann. Conf. Operations Research, John Wiley & Sons, Inc., N. Y.

Simmons, J. W., 1965, Descriptive models of urban land use: Canadian Geographer, no. 9(3), p. 170–174.

Sonenblum, S., 1968, The uses and development of regional projections: Issues and Urban Economics, The Johns Hopkins Press, Baltimore, Md.

Staff, Center for Real Estate and Urban Economics, 1968, Jobs, people, and land: Bay Area Simulation Study (BASS): Inst. Urban and Regional Devel., Univ. California, Berkeley.

Taylor, J. L., and Maddeson, R. N., 1968, A land-use gaming simulation: A design of a model for the study of urban phenomena: Urban Affairs Quart., no. 3(4), p. 37–51.

Traffic Research Corporation, 1964, Reliability test report: ENPIRIC, land-use forecasting model: Boston Regional Planning Proj., Boston, Mass.

MANUSCRIPT RECEIVED BY THE SOCIETY JUNE 25, 1973

RESEARCH SPONSORED BY THE NATIONAL SCIENCE FOUNDATION RANN PROGRAM UNDER UNION CARBIDE CORPORATION'S CONTRACT WITH THE U.S. ATOMIC ENERGY COMMISSION

Geonatural Resources Planning

SAMUEL J. MELTZ
Director of Planning, Albany-Dougherty County Planning Commission
Albany, Georgia 31702

ABSTRACT

Geonatural Resources Planning, a comprehensive program specifically initiated to conduct a detailed inventory and analysis of the Earth's natural resources, is relatively new to the field of planning. The composite view obtained allows more objective and logical development alternatives to be derived. In turn, decision makers are provided with greater validity of opinions on which they can base their decisions.

The term "geonatural resources" was derived by combining the Greek word "geo," meaning "of the Earth," to the words "natural resources." Specifically, the *geonatural* or the physical aspects of man's surroundings will refer to the biogeochemical systems and elements, including ecological interrelations, that exist and are part of the life support zone (biosphere) of the Earth. The philosophical foundations of geonatural resources planning are based upon notions of man's behavior, man's use and value of the land, human holding capacities of the land, and the limitation of economic growth.

The major areas of concern are the elements of the defined ecosystem, effects of land use decisions on the ecosystem, economic factors of land use decisions, and the dynamics of the time dimension within the various component relations over time. The conceptual framework consists of three components: (1) to systematically define the elements to be investigated, potential problems, and options; (2) to discuss the tools required to conduct and implement comprehensive programs; and (3) to discuss some of the probable limitations. Emphasis is placed upon the use of engineering geology as an integral element of geonatural resources planning.

INTRODUCTION

The concept that I present in this paper utilizes the map as a basic medium of communication. It is presently being field tested, and I hope will place us in a little better position to cope with some of the overwhelming environmental problems that plague us today. I am not a geologist, I am a generalist—a generalist who is working in the capacity of advisor to city and county elected governmental officials and the overall leadership; in other words, the decision makers of the community.

In this advisory capacity, one of the lessons that I have learned is the need to become sensitive to the practical considerations of effective decision making. These considerations are (1) the ability to render informed decisions, (2) the art of the possible, and (3) public accountability. This logically implies, although scoffed at in certain quarters, that the most elaborate and detailed plans formulated are sometimes the most useless and impractical creations when confronted with the acid test of the real world—public endorsement and support.

When dealing with the complexities of man's relation to his physical surroundings (ecology, ecosystems, and environment, to mention a few), the job of presenting alternatives has become even more complex. Additionally, this new concern stimulated the formulation of several conflicting philosophies: the "so-whaters," or people who do not perceive any problems; the pessimists—prophets of doom; the irrational emotionalists; the "smokestack builders"; and the "holders-of-the-truth."

The results of these lay public philosophies and attitudes toward solutions to environmental problems are predictable and are now observable: *they have led to the lack of political action or action that is more expedient than sound. This expedient political reaction has resulted in scattered, ad hoc, piecemeal attempts at problem solving.* These gestures, at most, can only hope to treat the symptoms of environmental problems; they cannot and will not provide a foundation for achieving realistic solutions to these problems. One of the main obstacles to achieving sound environmental planning and policies is that the American public and their legitimate leaders are hedged in by the single-purpose effects of general apathy, the indecisions of despair, irrational emotionalism, and industrial economics. The leadership has found that this situation has created a very difficult atmosphere in which to formulate and implement rational and logical comprehensive environmental policies.

In my opinion, we are confronted with a professional void in the form of a lack of a comprehensive conceptual philosophy, tested methodologies, valid hypotheses, and sound theories. I further believe that we are in this predicament partly because of the rapid developments in technology that created over-specialized technicians who could only talk to one another.

To overcome these problems and gain the needed comprehensive view, we must involve all the major disciplines and employ operations research. This interdisciplinary approach will be required if practical and workable solutions to these complex environmental problems are to be found. However, it will not suffice merely to incorporate an interdisciplinary approach to complex problem solving; the approach must be founded upon cooperation and coordination if it is to be effective and successful.

GEONATURAL RESOURCES ELEMENTS, PROBLEMS, AND OPTIONS

Three components provide the basis for defining the scope of geonatural resources planning. They also establish areas of concern for geonatural resources utilization and management decision making.

Geonatural Resources Elements

These elements are the living organisms and nonliving entities that are found in the biosphere. It is the interaction between the living organisms and nonliving entities within the biosphere that forms the dynamics of the Earth's ecosystem, and upon which man's existence depends. The quality and quantity of these elements are directly related to man's physical and economic development. In many respects, they act as the independent variable in the equation of man's existence. The geonatural resources elements are

1. Air
2. Water
   a. Surface
   b. Subsurface
3. Land
   a. Urban
   b. Suburban
   c. Rural
   d. Coastal
   e. Natural
4. Minerals
   a. Industrial
   b. Manufacturing
   c. Energy fuels
5. Forest
   a. Wildlife
   b. Recreation
   c. Timber
6. Marine
   a. Estuary
   b. Ocean

Geonatural Resources Problems

This aspect of geonatural resources planning is related to situations that have resulted from a basic lack of understanding and misuse, by man, of the dynamic interrelations of the biosphere. These situations, in most cases, have created a disagreeable state of affairs that is adversely affecting man's well-being. The geonatural resources problems are

1. Pollution
   a. Air
   b. Water
   c. Solid waste
   d. Noise
   e. Pesticides
   f. Radiation
   g. Thermal
2. Human supportive capacity
   a. Population density
   b. Physical hazards
   c. Migration
   d. Population size
   e. Population growth rate
3. Human needs
   a. Food
   b. Clothing
   c. Shelter

Geonatural Resources Option Alternatives

In terms of the geonatural resources elements, there are certain option alternatives that must be considered if decisions for comprehensive policies and actions are to effectively rectify existing problems. These decisions should also incorporate provisions that will assure the sound use of geonatural resources elements if the needs and well-being of human society are to be satisfied and maintained. The option alternatives to be considered are

1. Atmospheric
   a. Meteorology
   b. Weather modification
2. Water
   a. Water supply
   b. Water quality
   c. Saline water conversion
   d. Flood control
3. Land
   a. Public lands
   b. Private lands
   c. Forest lands
   d. Range lands
   e. Wetlands and estuaries
   f. Mined lands
4. Marine
   a. Oceanography
   b. Commercial fishery
   c. Minerals
5. Agricultural
   a. Production of food and fiber
   b. Soil conservation
6. Urban
   a. Residential
   b. Commercial
   c. Industrial
   d. Recreation
   e. Governmental services
   f. Facilities
7. Rural
   a. Agricultural production
   b. Recreation
   c. Revitalization areas
   d. New communities
8. Transportation
   a. Vehicular
   b. Air
   c. Rail
   d. Spaceport
   e. Marine
   f. Pipelines
   g. Transmission lines
9. Mineral
   a. Exploitation and inventory
   b. Research and development
   c. Recycling
   d. Reclamation
10. Energy
    a. Thermoelectric
    b. Hydroelectric
    c. Nuclear
    d. Solarelectric
    e. Energy distribution
11. Recreation
    a. Public
    b. Wilderness
    c. Preservation
    d. Green belt–Open space
    e. Hunting, fishing, and camping
    f. Commercial

TOOLS REQUIRED

The following tools are necessary if the total concept is to be implemented:

1. Aerial photography, remote sensing, and cartographic analysis
2. Bioclimatic analysis
3. Demographic analysis
4. Soils surveys information and analysis
5. Geological inventory and analysis (general)
6. Geomorphological inventory and analysis (specific)
7. Ecological inventory and analysis
8. Spatial analysis
9. Systems analysis

One of the main problems that is preventing the development of effective geonatural resources planning programs is the lack of a synthesis of these tools into the formulation of a working methodology. It is highly doubtful that a unified, coordinated, and cooperative program of geonatural resources planning will be achieved until this specific task is accomplished.

CONCEPT IMPLEMENTATION

Although we began to implement this concept only one year ago, the results appear to be very promising.

We have entered into a multi-year contract with the Earth and Water Division of the Georgia Department of Natural Resources, and we have requested the following:

1. The compilation of detailed geologic and landform maps;

2. An analysis of surface drainage types and conditions;

3. An identification, description, and interpretation of existing and potential geologic hazards (to development);

4. A comprehensive analysis of the aspects of geologic research in the metropolitan area that relates to land use planning;

5. Mineral resource evaluation of the work area;

6. Location and description of geologically important areas where school science classes can observe the bedrock, framework, and earth resources of the area; and

7. Liaison with the National Aeronautics and Space Administration (NASA) to obtain 1:62,500 scale remote sensing data relating to urban and rural planning in the area.

We applied for and were selected as a pilot area by NASA to receive 1:4,800 and 1:24,000 scale remote sensing data.

Recently, I was able to organize a Geonatural Resources Advisory Task Force to the Planning Commission. The task force is interagency and interdisciplinary. The members include geologists, soil scientists, foresters, biologists, and engineers.

We are attempting to develop an information system and construct what I like to call a "public affairs strategy room."

CONCLUSIONS

One overriding point in all of the above is that without assistance from the geologist, particularly the engineering geologist, some of these activities would be difficult if not impossible; however, there is one area that we continually have to work on—that of communication. The reason for this is simple and straightforward: Unless the information received is in a form that can be readily used and understood, it will be of no use to the decision maker or the lay public. This means that the reports, maps, and other information must be submitted in a language that can be understood in practical and not academic terms.

The following points should also be considered:

1. If the proposed National Land Use Policy Act is adopted, most states will not have the capability in either support staff or required resource information to comply with the intent of the Act. This kind of vacuum always results in creating "expert consultants"; the HUD 701 planning programs for communities are a prime example of this.

2. With the current interest in establishing a rural development policy, there is a potential impact on our natural resources base. If this impact is not carefully controlled, we may find ourselves faced with a crisis beyond anything we can now envision. Will we have the time, capability, and resources to guide the impact on our natural resources and prevent rural communities from becoming rural slums?

3. Although some academic institutions are considering establishing environmental planning curriculums, how can these institutions produce the kind of professionals needed when the agencies themselves have not determined specifically what kind of professionals will be needed? If academic institutions begin producing environmentalists as they did engineers, the planning and environmental field may experience the same negative effects that the engineering profession (particularly in the aerospace industry) and teaching profession are presently experiencing.

4. If geonatural resources planning and management are to evolve into a real interdisciplinary field, there is a strong need to develop (a) a conceptual framework and philosophy, (b) a sound methodology, and (c) sound approaches. Further, if this field is to provide effective service, then it must be oriented toward political realities and the pragmatics of public affairs.

5. Many of our academic institutions can provide valuable service functions to the environmental field in basic and applied research capabilities, support staff services, and practitioner-professor roles, to mention a few. However, unless there is an agreement as to the purpose and intent of such service functions, it will be difficult to develop and achieve these new roles for universities and colleges.

SELECTED BIBLIOGRAPHY

Baird, J. V., Bartelli, L. J., Heddleson, M. R., and Klingebiel, A. A., eds., 1966, Soil surveys and land use planning: Madison, Wis., Soc. Agronomy.

Bartelli, Lindo J., 1960, Soil survey information for suburban development: Am. Assoc. Adv. Sci. (Ann. Mtg.).

—— 1962, Use of soils information in urban-fringe areas: Soil and Water Conserv., v. 17, no. 3, p. 99–103.

—— 1966, General soil maps: A study of landscapes: Soil and Water Conserv., v. 21, no. 1, p. 3–6.

Beatty, M. T., Brovold, A. J., and Yanggen, D. A., 1966, Use of detailed soil surveys for zoning: Soil and Water Conserv., v. 21, no. 4, p. 123–126.

Berry, Brian J. L., 1968, Theories of urban location: Assoc. Am. Geographers Resource Paper no. 1.

Berry, Brian J. L., and Marble, Duane F., eds., 1968, Spatial analysis: A reader in statistical geography: Englewood Cliffs, N. J., Prentice-Hall, Inc.

Caldwell, Lynton K., ed., 1967, Environmental studies: Papers on the politics and public administration of man-environment relationships: Bloomington, Ind., Inst. Public Adm. Indiana Univ., 4 vols.

—— 1970, Environment: A challenge for modern society: Garden City, N. Y., The Natural History Press.

Cheatum, E. L., 1971, The natural resource base and land use planning: Jacksonville, Fla., Assoc. Southern Agricultural Workers.

Christiansen, E. A., ed., 1970, Physical environment of Saskatoon, Canada: Natl. Research Council Canada Pub. no. 11378.

DeWiest, Roger J. M., 1965, Geohydrology: N. Y., John Wiley & Sons, Inc.

Doxiadis, Constantinos A., 1968, Ekistics: An introduction to the science of human settlements: N. Y., Oxford Univ. Press, Inc.

Dury, George H., 1969, Perspectives on geomorphic processes: Assoc. Am. Geographers Resource Paper no. 3.

Flawn, Peter T., 1970, Environmental geology: Conservation, land-use planning, and resource management: N. Y., Harper & Row, Pubs.

Georgia State Planning Bureau, 1968, Status of natural resources planning in Georgia: Atlanta, Ga., State Planning Bureau Rept.

Gilbert, Douglas L., 1967, Public relations in natural resources: Minneapolis, Minn., Burgess Pub. Co.

Goldstein, Harold, Wertz, James, and Sweet, David, 1969, Computer mapping: A tool for urban planners: Columbus, Ohio, Battelle Memorial Inst.

Goodenough, Ward H., 1963, Cooperation in change: N. Y., Russell Sage Found.

Gould, Peter R., 1969, Spatial diffusion: Assoc. Am. Geographers Resource Paper no. 4.

Graham, Edward H., 1944, Natural principles of land use: N. Y., Oxford Univ. Press, Inc.

—— 1957, Land use principles and needs, *in* Callison, Charles H., ed., America's natural resources: N. Y., The Ronald Press Co., p. 164–172.

Hackett, James H., and McComas, Murray R., 1969, Geology for planning in McHenry County: Illinois Geol. Survey Cir. 438.
Haggett, Peter, 1965, Location analysis in human geography: London, England, Edward Arnold, Ltd.
Hills, George A., 1961, The ecological basis for land-use planning: Ontario Dept. Lands and Forests Research Rept. no. 46.
—— 1970, Developing a better environment: Ontario, Canada, Ontario Econ. Council.
Isard, Walter, 1956, Locations and space-economy: A general theory relating to industrial location, marker areas, land use, trade, and urban structure: Cambridge, Mass., The M.I.T. Press.
—— 1960, Methods of regional analysis: An introduction to regional science: Cambridge, Mass., The M.I.T. Press.
Johnson, Philip L., ed., 1969, Remote sensing in ecology: Athens, Ga., Univ. Georgia Press.
Kaliser, Bruce N., 1969, Geology for planning: Bear Lake area, Rich County: Utah Geol. and Mineralog. Survey Inv. Rept. 40.
Kellogg, Charles E., 1937, Soil and people: Assoc. Am. Geographers Annals, v. 27, no. 1, p. 142–147.
—— 1964, Planning soil use for both individual and public goals: Soil and Water Conserv., v. 19, no. 1, p. 3–6.
Lutzen, Edwin E., 1968, Engineering geology of the Maxville quadrangle, Jefferson and St. Louis Counties, Missouri: Missouri Div. Geol. Survey and Water Resources Eng. Geology Ser. no. 1.
Lutzen, Edwin E., and Williams, James H., 1968, Missouri's approach to engineering geology in urban areas: Assoc. Eng. Geologists Bull., v. V, no. 2, p. 109–121.
Mayer, Harold M., 1969, The spatial expression of urban growth: Assoc. Am. Geographers Resource Paper no. 7.
McHarg, Ian L., 1969, Design with nature: Garden City, N. Y., The Natural History Press.
Meltz, Samuel J., 1969, Soil and geologic mapping: Phase I - Proposed guidelines for the pilot mapping program: St. Clair County, Illinois, and St. Louis County, Missouri: East St. Louis, East-West Gateway Coordinating Council Rept. no. 38.07.0.
—— 1971, Geonatural resource development: A working paper presenting notions and considerations oriented toward achieving positive action solutions to environmental problems: Limited publication through the cooperation of the Georgia Center for Continuing Education, University of Georgia, Athens, Georgia: Samuel J. Meltz, 1971.
Montgomery, Hugh B., 1969, Environmental analysis in local development planning: Appalachia, p. 1-11 (also this volume).
Moynihan, Daniel P., 1970, The concept of public policy in the 1970's: Paper presented at Hendrix College, Conway, Arkansas, April 6, 1970.
Olgyay, Victor, 1963, Design with climate: Bioclimate approach to architectural regionalism: Princeton Univ. Press.
Olson, Gerald W., 1965, Soil surveys for city citizens: Soil and Water Conserv., v. 20, no. 3, p. 83–85.
Powell, Mel D., Winter, William C., and Bodwitch, William P., 1970, Community action guidebook for soil erosion and sediment control: Washington, D.C., Natl. Assoc. Counties Research Found.
Quay, John R., 1963, Lake County uses soil survey in planning its urban areas: Soil Conserv., p. 99–102.
—— 1968, Using the facts to find the plan: Soil Conserv., v. 33, no. 2, p. 27–29.
—— 1970, Remote sensing: With special reference to agriculture and forestry: Washington, D.C., Natl. Acad. Sci.
Robinson, Arthur H., 1966, Elements of cartography (2d ed.): N. Y., John Wiley & Sons, Inc.
Schlicker, Herbert G., and Deacon, Robert J., 1967, Engineering geology of the Tualatin Valley region, Oregon: Oregon Dept. Geology and Mineral Industries Bull. 60.
Sheaffer, John R., Davis, W. Ellis, and Spieker, Andrew M., 1970, Flood-hazard mapping in metropolitan Chicago: U.S. Geol. Survey Circ. 601-C.
Smith, Douglas C., 1970, Urban highway design teams: Washington, D.C., Highway Users Federation for Safety and Mobility.
Smith, John T., Jr., ed., 1968, Manual of color aerial photography: Falls Church, Va., Am. Soc. Photogrammetry.
Spieker, Andrew M., 1970, Water in urban planning, Salt Creek Basin, Illinois: U.S. Geol. Survey Water-Supply Paper 2002.
Steinitz, C., Murry, T., Sinton, D., Way, D., 1969, A comparative study of resource analysis methods: Cambridge, Mass., Dept. Landscape Architecture Research Office, Graduate School of Design, Harvard Univ.
Theobald, Robert, 1970, Communication to build the future environment: Main Currents, v. 26, no. 4, p. 105–111.
Thornbury, William D., 1954, Principles of geomorphology: N. Y., John Wiley & Sons, Inc.
Tudor, L. Bryant, 1966, Use of soil surveys in urban planning [Master's thesis]: Georgia Inst. Technology.
Tufts, James H., 1918, The ethics of cooperation: N. Y., Houghton Mifflin Co.
U.S., Congress, House, Committee on Education and Labor, 1970, Environmental Education Act: 91st Cong., 2d sess., H. Rept. 1362.
—— 1970, Environmental Quality Education Act of 1970: Hearings on H.R., 14753 before a select subcommittee of the Committee on Education and Labor, 91st Cong., 2d sess.
U.S., Congress, House, Committee on Governmental Operations, 1970, The environmental decade (action proposals for the 1970's): Hearings before a subcommittee of the Committee on Governmental Operations, 91st Cong., 2d sess., H. Rept. 1082, Govt. Printing Office.
U.S., Congress, House, Committee on Merchant Marine and Fisheries, 1971, Administration of the National Environmental Policy Act: Hearing before a subcommittee of the Committee on Merchant Marine and Fisheries, House of Representatives, on Federal agency compliances with sec. 102 (2) (c) and sec. 103 of the National Environmental Policy Act of 1969, 91st Cong., 2d sess., doc. 41, pt. I-II.
U.S., Congress, House, Joint Economic Committee, 1970, The economy, energy, and the environment: A background study: Prepared by the Environmental Policy Div., Legislative Reference Service, Libr. of Cong., joint comm. print, 91st Cong., 2d sess., Govt. Printing Office.
U.S., Congress, Joint House-Senate Colloquium to Discuss a National Policy for the Environment, 1968, Hearing before the Committee on Insular and Interior Affairs, Senate, and the Committee on Science and Astronautics, House of Representatives, 90th Cong., 2d sess., Govt. Printing Office.
U.S., Congress, Senate, 1970, An Act to Amend the Water Resources Planning Act (79 Stat. 244), To add a new title IV–A National Land Use Policy S. 3354: 91st Cong., 2d sess.
U.S., Congress, Senate, Committee on Government Operations, 1969, Intergovernmental Cooperation Act of 1969 and Related Legislation: Hearings before a subcommittee of the Committee on Governmental Operations on S. 60, S. 2035, and S. 2479, 91st Cong., 1st sess.
U.S., Congress, Senate, Committee on Interior and Insular Affairs, 1971, Congress and the Nation's environment: Environmental affairs of the 91st Congress: Committee print, 92d Cong., 1st sess., Govt. Printing Office.

U.S., Congress, Senate, Committee on Public Works, 1970, National Environmental Policy Act Relative to Highways: Hearings before the Subcommittee on Roads of the Committee on Public Works, 91st Cong., 2d sess., Govt. Printing Office.

U.S., Congress, Senate, 1970, A definition of the scope of environmental management: Committee print, 91st Cong., 2d sess., Govt. Printing Office.

U.S., Congress, Senate, 1970, Joint committee report on the environment: S. Rept. 1033, 91st Cong., 2d sess.

U.S., Council on Environmental Quality, 1970, Environmental quality: The 1st Ann. Rept. of the Council on Environmental Quality together with the President's message to Cong., Govt. Printing Office.

U.S., Department of Agriculture and Department of Housing and Urban Development, 1968, Soil, water, and suburbia: Conf. proc., Govt. Printing Office.

U.S., Department of Health, Education and Welfare, Office of Education, 1971, Environmental education: Education that cannot wait: Dept. of Health, Education, and Welfare.

U.S., Department of Housing and Urban Development, 1970, Airport environs: Land use controls: Environmental planning paper, Dept. of Housing and Urban Development.

U.S., Environmental Protection Agency, Water Quality Office, 1971, Guidelines water quality management planning: Environmental Protection Agency.

U.S., President, Message, 1971, Program for a better environment: 92d Cong., 1st sess., doc. 46, Govt. Printing Office.

U.S., President, Message, 1971, A program to save and enhance the environment: 92d Cong., 1st sess., H.R. doc. 46, Govt. Printing Office.

Watson, James E., 1969, What is community development?: Joint pub. Inst. Community and Area Devel. and Georgia Center for Continuing Education, Athens, Ga., Georgia Center for Continuing Education.

—— 1970, Comprehensive community planning and development: Organizing, setting your public goals, gaining support, and creating your new realities: Joint pub. Inst. of Community and Area Devel. and Georgia Center for Continuing Education, Athens, Ga., Georgia Center for Continuing Education.

Watzlawick, Paul, Beavin, Janet H., and Jackson, Don D., 1969, Pragmatics of human communication: N. Y., W. W. Norton and Co., Inc.

Wildman, William, ed, 1969, Soils and land use planning: Conf. proc., Davis, Calif., Univ. California.

Witner, David B., 1965, Soil survey use in a suburban county: Soil and Water Conserv., v. 20, no. 3, p. 88–90.

Woodward, H. B., 1912, The geology of soils and substrata: London, Edward Arnold.

MANUSCRIPT RECEIVED BY THE SOCIETY JUNE 25, 1973

Engineering-Geological Maps for Urban Development

ROBERT F. LEGGET
531 Echo Drive, Ottawa 1, Ontario, Canada

The cities of the world will probably at least double in size before the end of the century, due to world population increases and the apparently irreversible trend from country to city living. The success of the vast building program that this increase will necessitate will depend, ultimately, upon adequate knowledge of the subsurface conditions of all the ground that will be covered by this urban expansion.

Coordination of existing records with the results of new investigations and facts revealed by all new excavation is essential for every urban area. This can be done only in close association with local geology. Engineering-geological maps provide the best means of recording this vital information. Some cities have made a good start at the preparation of such maps; examples are given. All cities must have such maps as basic tools for all future planning if inevitable expansion is to be saved from costly errors. Broad guidelines for this essential public service are presented.

This paper is a summary of my oral presentation at the 1972 Annual Meeting of The Geological Society of America held in Minneapolis, and is based on material in *Cities and Geology* (Legget, 1973), which was in press at the time of the meeting.

The current rate of growth of the cities of North America is little short of phenomenal, as a glance at the Twin Cities area will clearly show. This rapid expansion is a reflection of the increase in population growth since the end of the second world war, assisted to a degree by the corresponding increase in the standard of living. It is but a beginning of a growth pattern that will probably see urban development throughout the world doubled by the year 2000.

If this seems to be just alarmist talk, consider this statement from the official press release (No. 19849-69) issued on 29 October 1969 jointly by the Secretaries of the Department of the Interior and of Housing and Urban Development: "by the year 2000, major urban regions will cover about 340,000 square miles as compared to the present 200,000 square miles. Additional areas will be affected by smaller individual centers." The increased urban area, thus officially estimated, of 140,000 sq mi is difficult to visualize; it is roughly equivalent to the combined areas of England, Scotland, Wales, and Czechoslovakia. And this estimated growth of urban areas by the year 2000—little more than 25 years away!

This prospective growth of cities makes the current tag that "The United States of America paves one million acres a year" a lamentable understatement. At that rate, if we assume by the word "paving" the covering up of natural ground for the physical development of cities, it would take a full century to cover 140,000 sq mi; yet we face this prospect in the next 30 years.

The magnitude of the construction job that is involved in this urban expansion is, in itself, monumental, but we are concerned with the necessary and essential planning that must be done before any construction starts. Such planning cannot be properly done unless it is well and firmly founded upon full knowledge and appreciation of the geology underlying the sites to be developed. Engineering geologists and many civil engineers know this, but it is doubtful whether many others who are engaged in the planning process are aware of it.

In some instances, this has had lamentable results; in other instances, the absence of geological studies has not been noticed because fortune has been kind, and planning based purely on topographical information just happened to be satisfactory also in relation to the underlying geology. Chances like this, however, must not be taken in the future, if only because of the magnitude of the task of planning well for future urban development. An added factor is that many sites previously thought to be unsatisfactory will now have to be used for developmental purposes, thus making their geological study imperative before any decisions are made as to the erection of structures.

When things go well, we tend to forget that the essential stability of every man-made structure depends ultimately upon the ground on which it rests, and in this context the word "structure" includes all facilities constructed for the use of man—buildings, bridges, dams, roads, and airport runways. This very familiarity is probably responsible for the neglect of geology in so much of the literature of planning. Even so eminent a guide as the *Urban Planning Manual* of the American Society of Civil Engineers (in all other respects a notable and most useful volume) gives scant attention to geology.

Fortunately, there are some exceptions to this general neglect of geology in our planning literature. Ian McHarg (1969), a distinguished professor of landscape architecture and planning, has given full attention to local geology as a prerequisite for all planning in his fine book *Design with Nature*. The suddenly increasing interest in what is called "environmental geology" is encouraging, despite the semantic inaccuracy of the name.

If, then, it can be assumed that the vital need for geology in all urban and regional planning is within sight of general acceptance, how can the necessary information best be made available? A general overview of the local geology for any one region usually can be obtained from existing reports of the local, state, or provincial governments, or even from the national geological survey. This information provides a good starting point, but by itself, it is not enough. The geology must be interpreted in terms that will be useful and meaningful to planners. In most cases, much more detailed information will be necessary for urban planning in particular, and even for many cases of regional planning.

For all cities, there exists a vast amount of such information, but it is usually locked up in the private records of local engineering organizations, including those of the cities themselves. Most tunneling beneath cities, for example, is carried out by some public agency. Invaluable information about the local geology revealed by the bore is thus made available to those on the tunneling operation, but only rarely is this information published. (The major tunnels beneath the city of Boston constitute a notable exception to this statement through the work of Professor Marland Billings and his associates, with the cooperation of the local public agencies.)

During the last few decades, it has become common practice to put down deep borings at the sites of all major buildings in order to determine exactly the subsurface conditions upon which their foundations will rest. This is the work of geotechnical engineers, always fully recorded, generally in the form of detailed reports to the designing engineer or architect. Supplementary reports are made recording the actual materials revealed when excavation is carried out. This information is always (or should be) incorporated in the engineering "As-constructed Drawings." But in almost every case, the foundation is completed, the superstructure is erected, the building or bridge put into use, the foundations forgotten about, and the records of the underlying geology (unique though this information will usually be) put away in filing drawers, never to be looked at again unless needed for future changes to or near the structure.

If, for every city, this collection of invaluable information could be made publicly available and, when so available, expertly assessed against the generally known overall picture of the local geology, it would then be possible to prepare detailed engineering-geological maps for all urban areas. These would be of inestimable use in all planning work. They would provide a sound basis for all proposals for the physical development, or redevelopment, of urban areas. They would have the further great advantage of assisting in the planning of all future geotechnical investigations in urban areas by showing clearly what information was already available around any new site that had to be investigated.

The collecting, assembling, and assessment of all such urban subsurface information is clearly a public responsibility. It may be suggested that it could be, for all cities, part of the responsibility of the city engineer, assisted by requisite geological advice. The service to the public that such urban geological information would provide would far exceed the minor costs involved.

Do any cities have such a service available? They have, in a variety of ways. The city of Warsaw, Poland, had a set of four engineering-geological maps made available to it as early as 1936 through the work of two pioneering professors. The city of Prague in Czechoslovakia, with singularly interesting geology beneath its ancient streets, has an excellent system for recording all local geological information. The city of London, England, has the advantage of having always been the headquarters of the Geological Survey of Great Britain. Voluntarily, subsurface information has been given to this organization in such good measure that even before the second world war an excellent series of Six-Inch Maps for the entire area of the great city was available, showing details of key borings throughout the city from which the underlying geology could be deduced.

The Geological Survey of Northern Ireland has recently produced an excellent map of the geology underlying the city of Belfast, with full explanation of its engineering significance printed on the back. The city of Johannesburg in

South Africa has produced a similar map, based on the initial studies of an interested graduate student; a strong committee is now further developing this notable work. Another individual effort started the significant urban geological mapping that is now available for all major Japanese cities, work on the subsurface of Tokyo undertaken individually creating such favorable response that the national government is now assisting financially with a country-wide similar program for the main cities of Japan.

In Canada, The Geological Survey of Canada has recently been engaged on a special winter-works program that has started work of the type described in a number of Canadian cities. An excellent atlas of the *Physical Environment of Saskatoon* (Christiansen, 1970) is already colloquially known as the "Saskatoon Folio." The provincial government of Quebec has recently published a masterly report on the geology of Montreal, prepared by Professor T. H. Clark, in which tribute is paid to the steady accumulation of subsurface information in Canada's metropolis.

For the United States, Boston has already been mentioned appreciatively. Not only has a notable series of papers on the geology of Boston tunnels been published in the *Journal of the Boston Society of Civil Engineers*, but this society has had, since the inception of its journal, a committee responsible for collecting and publishing records of all holes in the city area. It has been humorously observed that the subsurface of Boston has been more probed than that of any other city in the United States, but this remark is really only an indication of the fact that the records of the borings of Boston are better known than those for other cities, since they have been so well assembled and published for general public benefit. Similar work has been done in other cities, although not so extensively; for most United States cities, there are publications of use that do record some useful information about the local subsurface.

The U.S. Geological Survey is now engaged on a number of major programs of studying local urban geology, the results of which will be of wide public benefit. This, however, is only an extension of earlier work. Especially notable were the fine engineering-geological maps published for Oakland and San Francisco in the late fifties. Many of the reports published by the U.S. Geological Survey summarizing long-term studies of the geology of urban regions—such as one on Portland, Oregon—must have proved of great assistance to local engineers, architects, and planners.

If all the examples that could possibly be collected from around the world were to be listed, we would see that only a beginning has been made. If you now consider the "explosion of cities" mentioned at the outset of this paper, then you will recognize the urgency of an immediate and great increase in activity in this field of urban geology if planning for the cities of the immediate future is to be well done. Every geologist would do well to become interested and find out what is being done in his own city about the geology beneath its streets. His support for any local effort in the direction of making local subsurface information publicly available will be, in itself, a minor public service. It will carry with it the distinct possibility that purely scientific information about the local geology will be revealed in this applied work, thus demonstrating yet again the joint engineering and geological benefits to be gained from due attention to urban geology.

REFERENCES CITED

Christiansen, E. A., ed., 1970, Physical environment of Saskatoon, Canada: Saskatchewan Research Council and the Natl. Research Council of Canada, 68 p.

Legget, R. F., 1973, Cities and geology: McGraw-Hill Book Co., N. Y., chap. 3, 624 p.

McHarg, Ian, 1969, Design with nature: Nat. History, 198 p.

MANUSCRIPT RECEIVED BY THE SOCIETY JUNE 25, 1973

Geologic Environment: Forgotten Aspect in the Land Use Planning Process

CHRISTOPHER C. MATHEWSON
and
ROBERT G. FONT*
Department of Geology, Texas A&M University, College Station, Texas 77843

ABSTRACT

Federal, state, and local political pressures are demanding that urban planners and zoning commissions manage all land use to protect the environment. Uncontrolled building practices cause home owners to suffer as a consequence of floods, foundation failures, and numerous other earth processes. Land use planning, as exemplified by the Housing and Urban Development (HUD) 701 program, sadly lacks a sound geologic base. It must be the role of the engineering geologist, versed in both civil engineering and geology, to formulate a geologic base for planning with the physical environment. He must satisfy two fundamental needs: (1) establish engineering geology criteria for specific land use suitability and (2) communicate his findings to the public, a public that includes persons who may be knowledgeable or ignorant about engineering, geology, and (or) planning.

INTRODUCTION

The environmental clamor has subsided to some extent because public decision makers are being forced by new environmental protection laws to produce environmental impact statements and to consider *all* alternatives prior to any decision. These laws, some passed only to placate the source of this clamor, may be a deterrent to the development of prudent land use programs, because the technical basis for decisions is often not considered. Comprehensive urban plans, required by Federal law, are often made by consulting architectural and planning firms who usually do not employ trained geologists or engineers. Some planning contracts, granted under the HUD 701 planning program, state "the Land Use Study will concern itself with the geographic, geo-physical or geologic, and environmental limitations of the area"; whereas others require the planner to "assess those environmental factors which will: (a) minimize or prevent undue damage, unwise use, or unwarranted preempting of natural resources and opportunities; (b) recognize and make prudent allowance for major latent environmental dangers or risks, such as floods, mud slides, earthquakes, air and water pollution; and (c) foster the human benefits obtainable from use of the natural environment by wise use of the opportunities available (e.g. use of natural drainage systems for park and recreational areas)."

These contract requirements, however, are noticeably lacking in a great majority of the HUD 701 urban plans, and where they do exist, they are often very limited. For example, geologic consideration in the environmental assessment portion of the *Development Plan 1970–1990 Lakeside, Texas*, is limited to only the following paragraph, which appears to be based upon the U.S. Department of Agriculture's *Soil Survey*:

> The PRELIMINARY PLAN is a sound one for connecting open space and considering community uses. However, it is implied that Lakeside may permit housing development north of FM 1886. This development should be discouraged because the top-soil east of the TESCO power line is extremely fertile and is already showing signs

*Present address: Department of Geology, Baylor University, Waco, Texas 76703.

of sheet erosion. Stands of broom-weed are a good indicator of high stress. This area is a valuable long term resource to the year 2040 if it remains undeveloped and prudently grazed or not grazed; in addition, by the turn of the century Lakeside will be adjacent to the north to intensive development and this area could well function as a significant buffer against noise, air pollutants and an aquifer recharge region. The walnut soils of this area are rather permeable and probably would function well to maintain water table and contribute to the underlying Paluxy sandstone. If it is developed, every advantage listed above would be virtually lost.

A study of 150 HUD 701 plans to determine the use of geologic and engineering geologic information is summarized in Figure 1. This was generated by reviewing each HUD 701 plan; specifically, listings were made for each geologic or engineering subject incorporated in the plan, and then the depth of each technical study was established.

Up to this point, it has only been implied that valuable data and information are missing and that wise land use management is therefore impossible, except by error. This implication soon becomes fact once one reads the newspaper headlines of flood disasters, landslide and mass-movement catastrophies, home foundation failures, earthquake losses, and similar acts of God—the Rapid City flood, California landslides, and the like.

PLANNING PROCESS

The planning process provides the basic framework through which necessary geologic and engineering information can flow to policy and decision makers. This process is a dynamic analysis and control mechanism—dynamic in that it is the application of a continuous planning methodology to develop solutions for urban problems. The planning process must consist of four phases: (1) the formulation of applicable goals and objectives, (2) the identification of both constraint and opportunity areas, (3) the making of policy recommendations, and (4) the implementation of programs designed to meet the goals and objectives (see Fig. 2). To be both dynamic and effective, the process re-

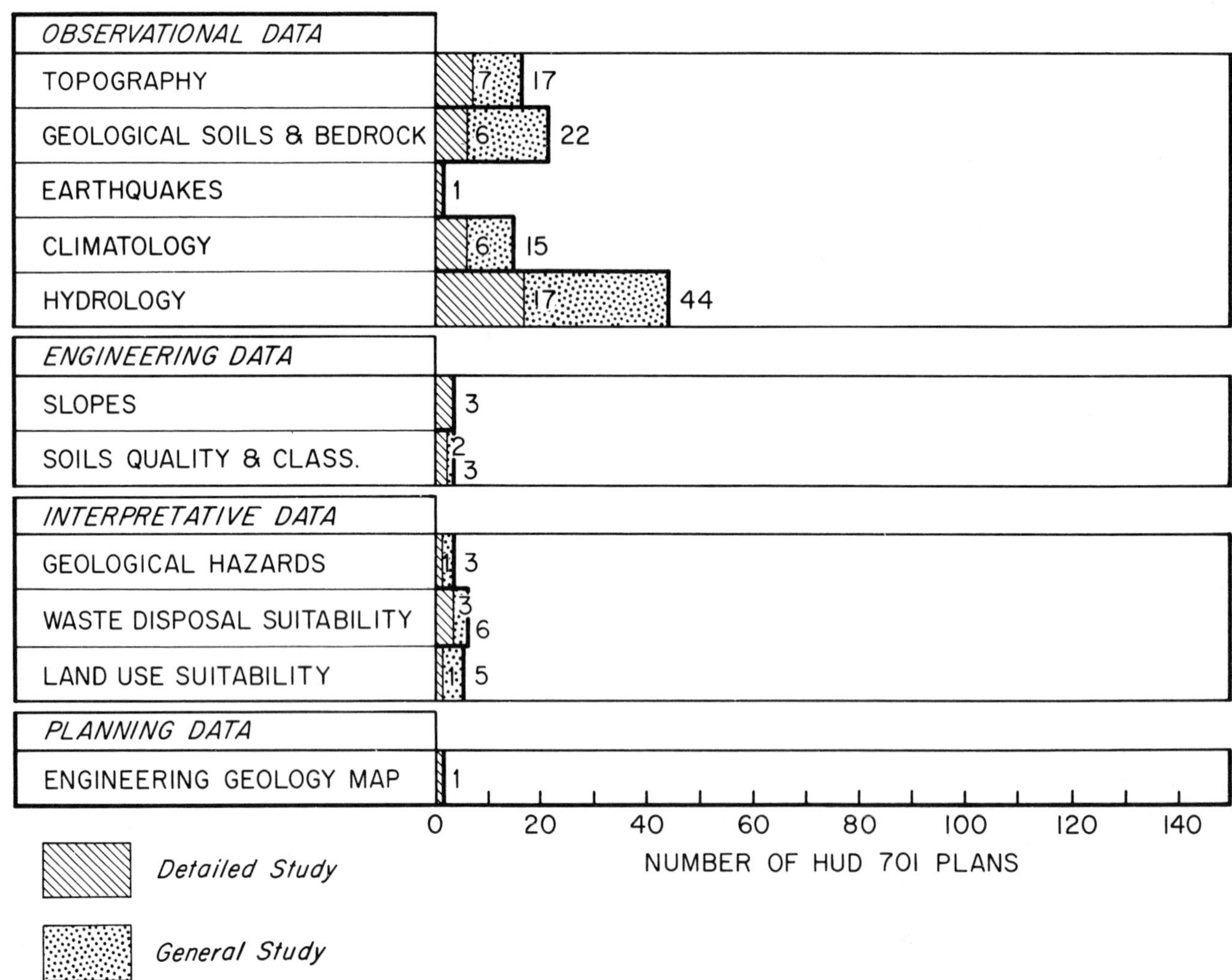

Figure 1. Bar diagram showing geologic information utilized in 150 HUD 701 plans.

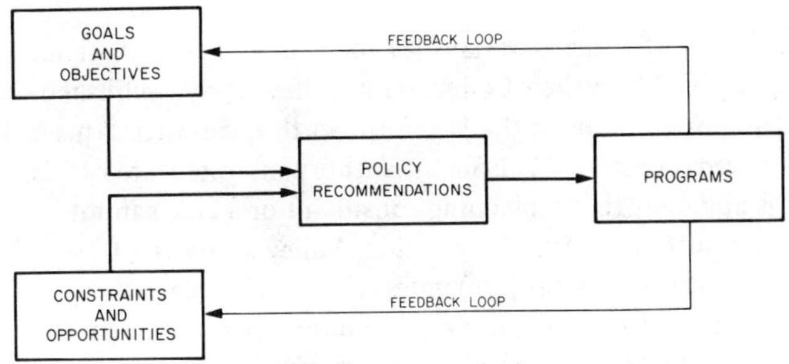

Figure 2. Schematic diagram of the planning process.

quires the active participation of the public, planners and technical specialists, and elected officials in each of these phases (especially in the implementation phase).

The *goals and objectives* phase of any urban plan establishes the intent and direction of the plan. The goal is the idealized optimum final product, while the objectives are the more specific detailed parts of the plan. The role of geology and engineering geology is established during the formulation of the goals. The fundamental geologic goal is to protect the health, welfare, and safety of the population from undue losses caused by geologic processes. The objectives are more specific and could include the following projects: (1) to map and define areas where soil, rock, and water conditions are unsuitable for safe development; (2) to locate, evaluate, and identify valuable resources areas; and (3) to locate, test, and evaluate suitable land areas for safe and clean sites for waste disposal.

The goals and objectives are inhibited or promoted by the *constraints and opportunities* that exist within the planning area. Constraints, as used by planners, are factors that limit the urban plan; opportunities are factors that enhance the plan. These factors are generally divided into two basic classes: (1) supply and (2) structural. Supply includes such things as natural resources; skilled manpower, especially in the local government; and the availability of finance. Structural are those that are built into the physical, socioeconomic, and political framework of every urban community. Two structural constraints, related to the geologic profession, probably have contributed to the lack of geologic information in the HUD 701 plans:

1. The geologist has often remained professionally distant from the political world; as a result geologic input to planning is either not required or not sought by planners and public officials.

2. Geologic solutions to environmental problems are often politically impossible to implement because publications, papers, and presentations are often only statements of existing problems without solutions, or they are unintelligibly written in such geologic jargon that the planner cannot or will not try to understand them.

Land use management depends entirely upon the establishment of correct planning policies. *A set of good policies is the core of planning.* Geologically based policies should strive to (1) define and develop resource potential; (2) preserve environmental quality within the limits of the local constraints; (3) maximize the efficient use of the land; and (4) minimize the loss of life and property caused by geologic processes. *Policies* are strategies for achieving the desired goals and objectives. They are formulated by both the public and private sectors of each community, and they are dependent upon the community's physical environment. Thus, national or state policy intent can be formulated, but it must be adapted or rejected by each community during its planning process. Some generalized policy statements are given below as examples:

1. Geologic hazards. Urbanization in hazardous areas should be prohibited or severely limited by the introduction of high safety and environmental standards.

2. Home site suitability. City government should require that builders and realtors show the prospective buyer the site suitability of his new home. Building codes should be established such that each home structure is designed to fit the geologic environment of the site. Such information should be filed with each plat submitted for approval.

Programs, the end result of policies, are the mechanisms through which the goals and objectives of the urban community are fulfilled. The continuation of these programs is the responsibility of the citizens, their elected and appointed officials, and the local government employees.

CONSTRAINTS TO GEOLOGIC PLANNING

Both supply and structural constraints affect geologic planning for any urban area. The strongest supply constraint is financial, despite the federally funded HUD 701 planning program. A generalized breakdown of the distribution of planning funds for a medium-size consulting firm is given in Table 1. The Land Use Element of the plan is budgeted for about 20 percent of the total grant, and probably represents the most costly part of any HUD 701 grant. It is often the profits from the other elements that make up for the loss in this element. The high cost of the Land Use Element is the result of the necessary field surveys, drafting, and mapping of the *existing* man-made land uses. A minor portion of this money is used to identify or approximate flood-

TABLE 1. DISTRIBUTION OF HUD 701 PLANNING FUNDS

Element	Percent
Base studies	15
Population estimates	10
Land use	20
Community facilities	15
Utilities	16
Capital improvements	12
Administrative controls	6
Comprehensive plan	6

prone areas, unstable slopes, foundation conditions, and the like. These engineering geology characteristics usually are identified through discussions with local residents and city officials.

The geologic profession, in an effort to get geology into planning, has created its own constraint, through the emphasis on a new science of environmental geology. This field has unfortunately tended to attract a number of environmentally aware students who are looking for a relevant education. The environmental geology, and now environmental geoscience, programs appear to be designed along the classical lines of descriptive geology, rather than with a sound background in the applied sciences and engineering mechanics. As a result of this descriptive approach, numerous environmental geology publications present statements of existing problems *without* solutions. The following quote from a widely circulated environmental geology publication exemplifies this problem as it pertains to rock-cut stability:

> Rock lithology, structure, and geometry of the rock mass (description and orientation of bedding, joints, and faults), quantity and direction of ground-water flow, plus other minor geological factors contribute to the development of a stable or unstable rock mass on any man-made slope. As our population grows, there is increasing pressure to develop land with steeper slopes.

The unfortunate aspect of such publications is that the urban planner, planning consultant, elected, appointed, and employed city officials, and the general public do not understand the significance of the statements. Their reaction is one of confusion, or perhaps a feeling that acquiring the necessary geologic information to implement the plans would cost too much. The same occurs when geologic studies are presented in a complex manner that draws upon our professional jargon. As an example of this "over-expert" constraint, another publication states:

> Geologic units include interdistributary muds, barrier-strandplain-chenier swales, abandoned channel-fill muds, overbank fluvial muds, mud-filled coastal lakes and tidal creeks.

This explanation, like the previous example, leaves the non-geologist planner and decision-maker confused and unenlightened.

Another significant constraint is developed when the "McHargian method," "supermapping," and "geocomputerology" try to mix with the financial limitations placed upon the planning process. The McHargian method is the application of a series of transparent overlay maps of a variety of factors arranged in such a way that the most suitable site for a specific land use can be determined (McHarg, 1969). This system is similar to the basic biology approach to the anatomy of a frog in which plastic overlays of skin, circulation system, lungs, and so forth are added together to form one frog (this requires that the user knows a frog). To operate, of course, the method requires a great number of maps, printed on transparent plastic, of various geologic factors in the mapped area, and a trained geologist to evaluate the map overlays for their significance. Pessl and others (1972) in their Connecticut Valley study, published a total of 18 maps of the Hartford North quadrangle; this involved a total of 11 different authors and one state office. It is apparent that a planning consultant or a city cannot afford such maps (the Connecticut Valley study is a U.S. Geological Survey project) unless the costs are carried by the Federal Government. Geocomputerology, the science of applying the McHargian method and geologic mapping to a computer, runs into the same problems as supermapping projects: they are beyond the budget limits of most users. This method has gained favor in the university community where numerous "cheap" workers (commonly known as graduate students) and extensive computer facilities are available.

In contrast to these examples, the recent engineering geology publication *Engineering Characteristics of the Rocks of Pennsylvania* (McGlade and others, 1972) summarizes the geologic, hydrologic, and engineering characteristics of each rock unit shown on the state geologic map. A paragraph of the rock description, bedding, fracturing, weathering, topography, drainage, porosity, ground water, ease of excavation, cut-slope stability, foundation stability, construction materials, and rock test data is presented in easily understood terms. For example, the cut-slope stability of the Conemaugh Group (Pc) is listed as: "Most cuts require moderate angle of slope; disintegration of claystone and shale under resistant sandstone and limestone beds cause rock falls, slumps and landslides; cuts in massive sandstones can be near vertical." The foundation stability of the limestone conglomerate (Trlc) is listed as: "Good quality foundation for heavy structures; sinkholes observed and detailed investigation should be undertaken."

RESPONSIBILITY OF THE GEOLOGIC PROFESSION

The problem, therefore, is how to get the most geologic benefit into *land use management* programs at the *least* cost. The planner needs (1) to receive geologic information in the terms of a "use suitability" and not in terms of geologic processes or descriptions, and (2) this information must be in a form that he can apply directly to his land use management policies. Three diverse professions–planner, geologist, and civil engineer–are necessary to generating solutions for the physical environment portion in any land use management program. However, financial constraints immediately restrict the size of this professional staff. The existing profession of engineering geology combines civil engineering (soil, fluid, and rock mechanics, hydrology, cost analyses, design, construction, and planning) with geology (physical, structural, ground water, geochemistry, sedimentology, and stratigraphy) and therefore offers a suitable technical expert. The specific role of the engineering geologist in land use management analyses is to evaluate the land's suitability for the safest and most economic use. The results are presented in terms of the limitations of the physical environment that are then evaluated in terms of the socioeconomic environment by the planner.

The most effective way in which this information can be transmitted is in map form as shown in Figure 3. All first-order (observational) maps show the technical results of a study. Each factor must be considered and its influence on the community determined. Second-order (engineering) maps are designed for the use of the engineer, developer, or contractor who deals directly with the local earth materials in construction projects. Each parameter considers a basic problem that may be significant in the local area. For a particular region, it may not be necessary to compile all nine maps. It must be stressed, however, that all nine factors are usually encountered and usually influence a community and its changes. Thus, an engineering geology study would not be complete unless the influence or noninfluence of each one of these nine parameters is identified.

Observational and engineering maps are used in synthesizing the third-order (interpretative) maps. Each third-order map shows the suitability of an area for a specific land use purpose or delineates existing and potential hazards. This is the most essential contribution of an engineering geology survey; if it lacks the interpretative portion, it has *failed* in its purpose. First- and second-order maps are of little use to a person untrained in engineering geology or

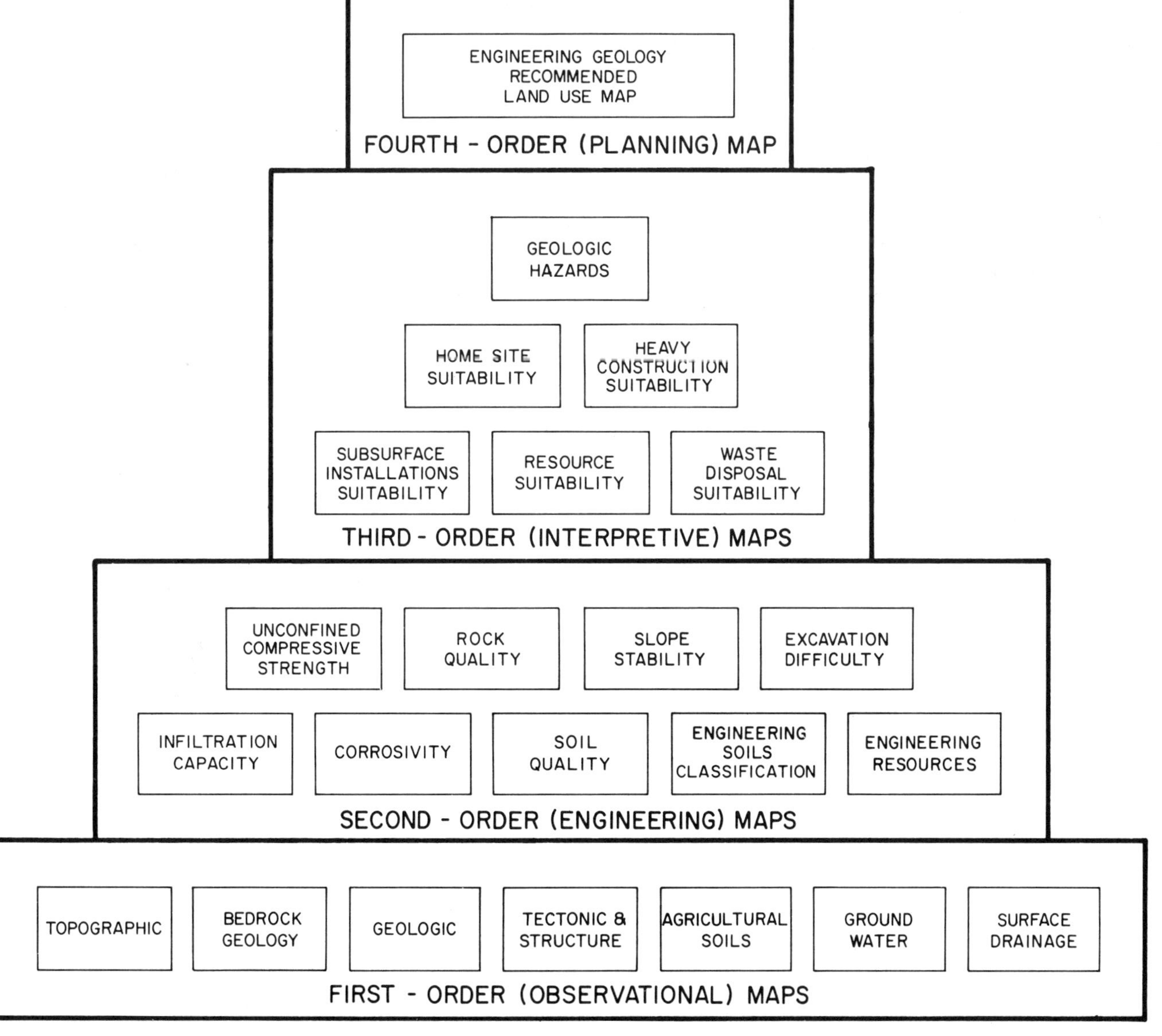

Figure 3. Diagram showing the ordering of engineering geology information for land use decision-making.

geotechnical engineering; to him they are of limited value. In contrast, the third-order (interpretative) maps are designed for direct use by the general public and must be made available to the public.

Finally, the fourth-order map, the engineering geology recommended land use map, is the end result of a thorough study. It combines all the previous information and is therefore designed for the land use planner. It gives a basis of choice by outlining from an engineering geology viewpoint the wisest use of any region. This map is the one from which the architect, the realtor, and the public will benefit the most. It is the ultimate contribution that the engineering geologist can make to the decision-maker.

The advantage of this system of ordered maps is that few maps (third-order) depict the site suitability of an urban area. These maps can be printed black on white and sold to the public at a reasonable cost, in contrast to the McHargian method (McHarg, 1969) which requires many overlay maps of specific information with a corresponding increase in both cost and complexity. Since the third-order maps depict site suitability, they are controlled to a large degree by building technology. For example, a site having a high shrink-swell clay would be defined as unsuitable or marginal for the construction of shallow concrete slab foundations using No. 6 reinforcing mesh. However, the introduction of a special post-tensioned concrete reinforcing method (new technology) designed for swelling soils would affect the delineation of unsuitable home building sites only on the Home Site Suitability Map, since neither the geology nor engineering characteristics of the area change.

This system has a high initial cost because the geologic and engineering maps must be prepared in detail; however, these maps are only required in a limited edition. Once this information is prepared, it acts as a continuous technical base for future restudies to update the third- and fourth-order maps. Thus, maintenance costs are kept at a minimum and could possibly be recovered through the sales of higher order maps.

ACKNOWLEDGMENTS

We express our appreciation to R. A. Cook, Alan Jones, and Ronald Tanenbaum for their reviews and comments of this paper. The graduate students in urban planning kept us honest. A portion of this work has been supported through a grant from the Texas Real Estate Research Center.

REFERENCES CITED

McGlade, W. G., Geyer, A. R., and Wilshusen, J. P., 1972, Engineering characteristics of the rocks of Pennsylvania: Pennsylvania Geol. Survey Bull. EG–1, 200 p.

McHarg, Ian, 1969, Design with nature: Doubleday/Nat. History Press, 197 p.

Pessl, Fred, Jr., Langer, William H., and Ryder, Robert B., 1972, Geologic and hydrologic maps for land-use planning in the Connecticut Valley with examples from the folio of the Hartford North quadrangle, Connecticut: U.S. Geol. Survey Circ. 674.

MANUSCRIPT RECEIVED BY THE SOCIETY JUNE 25, 1973

REVISED MANUSCRIPT RECEIVED SEPTEMBER 19, 1973

Environmental Analysis in Local Development Planning

HUGH B. MONTGOMERY
122 Mt. Pleasant Boulevard, R.D. 6, Irwin, Pennsylvania 15642

We need help in fighting the disease that now threatens our planet—a sort of cirrhosis of the environment.
Conservationist David Brower

As the nation approaches a 300-million population and a trillion-dollar economy and as our strip cities, industrial complexes, and communities grow, the time remaining to prevent the destruction of our natural environment and erosion of our resources needed to support life is becoming dangerously short.

As each man-made change occurs, it has effects on the environment which we, through shortsightedness, frequently fail to anticipate. And since our physical resources are limited and interdependent, the misuse of each resource alters the availability or usefulness of the others.

The nation is rapidly coming to realize that if action is not taken *now* on the national, regional, and local levels to preserve our environment, we will soon lack the environment and resources necessary to sustain our civilization and perhaps human life itself.

Where will more and more people live, work, and play? It may soon be true that in many metropolitan areas the nearest open space for a picnic within 25 mi of the downtown area will be the city dump.

Where will we dispose of the waste products of our increasing population and industrialized society?

The conservation of wildlife, wilderness, and recreation land and the pollution of our water, land, and air have finally become matters of public concern.

In Appalachia, where the mountainous environment has historically been an inspiration to poets and musicians but an impediment to planners, we have a unique opportunity to demonstrate the reciprocal relationship between resource conservation and economic progress.

Although many of Appalachia's environmental and resource problems are regional in character, it is often only at the local level that the planned use of our natural resources can be effective.

It has long been recognized in the region that resources analysis is an important step in certain types of local planning. In determining the location of an industrial plant or industrial park, for example, analyses of water supply and topography have become an important part of standard planning procedure.

Local planners are now becoming aware, however, that their efforts are more effective if a systematic analysis of the total environment is undertaken. In fact, many have discovered that only by incorporating environmental planning in the overall economic and social planning can development be successful (Moser, 1969).

These local planners are seeking answers to such questions as:

1. Why is environmental analysis important, and why should it be carried out during the very early stages of local development planning?

2. What are the various aspects of the environment that should be inventoried and assessed in overall planning?

3. What specific steps must be taken to incorporate environmental factors in any plan for local economic development?

4. What resources exist or are needed to provide local planners with necessary information on environmental factors?

IMPORTANCE OF EARLY ANALYSIS AND PLANNING

On a Sunday in mid-September, a square block area of Scranton, Pennsylvania, pivoted slightly and settled a few inches toward the collapsed tunnels of an abandoned coal mine 90 ft below, leaving front doors that would not open and 2 in. cracks in kitchen walls.

In other areas, automobiles have fallen into gaping holes in city streets because of the collapse of subsurface mines.

The Federal Bureau of Mines estimates that 2 million surface acres have already experienced mine subsidence, and that nearly a million more will sink by the year 2000.

In March 1964, an earthquake in Anchorage, Alaska, caused 14 percent of the city to slide into the Pacific, killing nine people and causing property damage valued at more than $300 million. The land- and mud-slide was caused by the fact that the city was built on clay subsoil which broke away when subjected to the tremors of the earthquake.

In Columbiana, Alabama, in August and September of 1968, major cracks appeared in the concrete filter-plant water reservoir; the water storage tank leaned southeast, out of plumb by 13½ in.; commercial buildings and walls and sidewalks in a federal housing project were damaged; water lines were broken; and the tracks of the Louisville and Nashville Railroad were severed. The cause of the damage was a particular form of subsidence which is common in areas underlain by limestone. The flow of ground water erodes a subsoil layer which lies between the limestone and the topsoil, causing surface depression, tension cracks, and sinkholes.

In May of 1967, rock slides closed off Route 28 in Pennsylvania's O'Hara Township. Some of the boulders were as large as buses, and three days were required to clear the blocked road of rocks and debris. The force of the slide, which extended for 100 ft, also damaged tracks of the Pennsylvania Railroad's main lines through the upper Allegheny Valley.

These are only a few examples of cases where public and private projects have been figuratively—and in some cases, literally—undermined by the lack of thorough environmental analysis. (See Figs. 1 and 2.)

Less spectacular, but equally significant, are areas where the interrelation between the environment and man's activity offers opportunities for improvement of economic and social life.

Knowledge of the type and location of various mineral resources permits hospitals and medical centers to plan staffs capable of treating and undertaking research on the health problems associated with mineral extraction. The Environmental Health Centers established by the Public Health Service have done pioneer work in this field.

Limestone terrains with open-solution channels provide both wells for water supply and adjacent wells for waste disposal. The work of engineers planning water and sewerage systems is facilitated by resource surveys showing where this type of terrain exists.

If soil is relatively impermeable, use of certain human waste disposal systems will result in malodorous and unhygienic conditions. Planners of public and private housing should avoid this unhealthy pitfall.

If highways are built through highly erodable land, erosion control is essential to conserve neighboring soil and streams.

If highways are built over land that contains valuable mineral resources, right-of-way becomes expensive, and the resources are unavailable for economic development.

If landscapes are scarred, archaeological treasures inundated, or virgin land and water defiled, the damage is irreparable and the loss unmeasurable, both in aesthetic and commercial terms.

Local areas may suffer loss of potential mineral land tax revenue because precise information is lacking as to the location and extraction potential of mineral resources within their taxing jurisdictions.

Foresighted resource analysis and environmental planning can help the local planner make wise investment decisions and facilitate his application for various types of funds that are available to finance activities in his area.

The planning activities of Washington County, Pennsylvania, furnish an excellent example of these benefits. Traditionally a coal-mining economy, the county was urged to consider making large-scale investments in reservoirs that would serve as the center of future recreational activites. Before making the decision, this county obtained professional advice on mine drainage, surface subsidence, and the value of minerals in the area that would be affected by the reservoirs. As a result of these studies, it was decided that while construction of some of the reservoirs was feasible, others should not be built because of the negative overall effects of mining on the area. Time was saved, and unwise decisions were avoided.

At another time, when federal agencies were inventorying mine drainage sources in part of the same county, county officials hired their own personnel, arranged to have them trained by federal scientists, and took the initiative in completing an inventory of the entire county. When mine drainage pollution control funds became available from the state government, Washington County was ready to file an immediate application, complete with all necessary detailed backup information.

The importance of environmental planning in connection with specific projects is being recognized more frequently

Figure 1. Improper drainage under a parking lot caused two landslides in two years at a shopping plaza east of Pittsburgh. (Photo by Nick Knezevich)

by governmental bodies at all levels. For example, in 1968 Congress added to the federal highway legislation a clause requiring that in submitting plans for a federally aided highway project, a state highway department must certify that it has considered the "economic and social effects of the location of the project, its impact on the environment and its consistency with the goals and objectives of such urban planning as has been promulgated by the community." The additional words reflect the growing concern of lawmakers that our environment be conserved and the use of our natural resources planned. (See Fig. 3.)

There is also a growth of interest in environmental planning at the state and local levels. The June-July 1969 issue of *Appalachia* included an article (Moser) describing the first environmental geology investigation that has been completed. Stanley Munsey, director of the Muscle Shoals Development District, the area covered by the Study, says that his district will find the study "invaluable." He stated that the industrial development committees of the district will use the information in brochures on industrial sites. The district staff expects to rely on it heavily in determining where facilities, highways, and other investments can best go—in fact, in all future land use planning. It will also furnish up-to-date geological information which technical and vocational schools will incorporate into courses of study.

ASPECTS OF ENVIRONMENT IMPORTANT TO PLANNING

Environmental resources are of two basic types: *biological* or living (plants and animals) and *physical* (air, water, and land). Ecology is the science that explains the nature of the living resources in terms of the physical environment within which they are found and then describes the effects that the biological and physical resources have on each other. In this broad sense, every element of the environment is important in creating a plan for economic development and understanding its long-term effects.

But in developmental planning, some aspects of the environment are more important than others. Those aspects of the *physical* and geological environment that affect other economic and social development projects most are the following:

Figure 2. Unsightly mine subsidence and a landslide caused by heavy rains depict environmental problems. (Photo by Anthony Kaminski)

Air

Man can live only a few seconds without this essential resource. Yet we are polluting it with exhausts from motor vehicles, airplanes, and ships; smoke and gases from residential heating; wastes from industrial plants (especially steel, oil refining, and coal-fired steam power); waste from mining; and certain construction materials, such as silts and sands, which are light and easily blown about, and which are frequently improperly stored. Public Health Service figures show that 72 million tons of carbon monoxide, 26 million tons of sulfur dioxide, and 12 million tons of uncaptured airborne particulate matter were dumped into U.S. air in 1965. This pollution causes corrosion of materials, human illness—much of it serious—and personal discomfort. (See Fig. 4.)

Water

Water comes from two sources—streams and underground flows.

Streams. The quality of stream water is affected by two major factors: the biological systems that inhabit the streams (whose tenants range in size from bacteria to mammals as large as beaver) and the chemical processes that take place in the stream water. Any human activity that affects either of these factors will determine whether the water will be usable and at what expense. Prime examples are the discharge of human and industrial waste. Stream water *quantity* is also a matter of public concern. Planning can assure that future water supplies will be sufficient to meet total needs, and streams with unique qualities (for example, the "white water" rapids which attract adventurous tourists) can be preserved for special priority uses. *Flow* of streams can also be controlled to prevent damaging torrents and floods.

Underground Flow. Underground water performs two important functions: when it comes to the surface, it is a major source of supply, sustaining the flow and quality of streams; if it remains underground, it influences the performance of soils, rock strata, and mineral resources. Underground water is "recharged"—and may be purified—by surface water which seeps down through the soil, subsoil, and rock. If these upper layers are contaminated, however, the underground water is also contaminated by the recharging process. In order to maintain needed supplies of high-quality underground water, its use must be carefully planned, and there must be control of waste disposal which takes place in the subsurface rock strata and on the surface areas where recharge of underground water occurs.

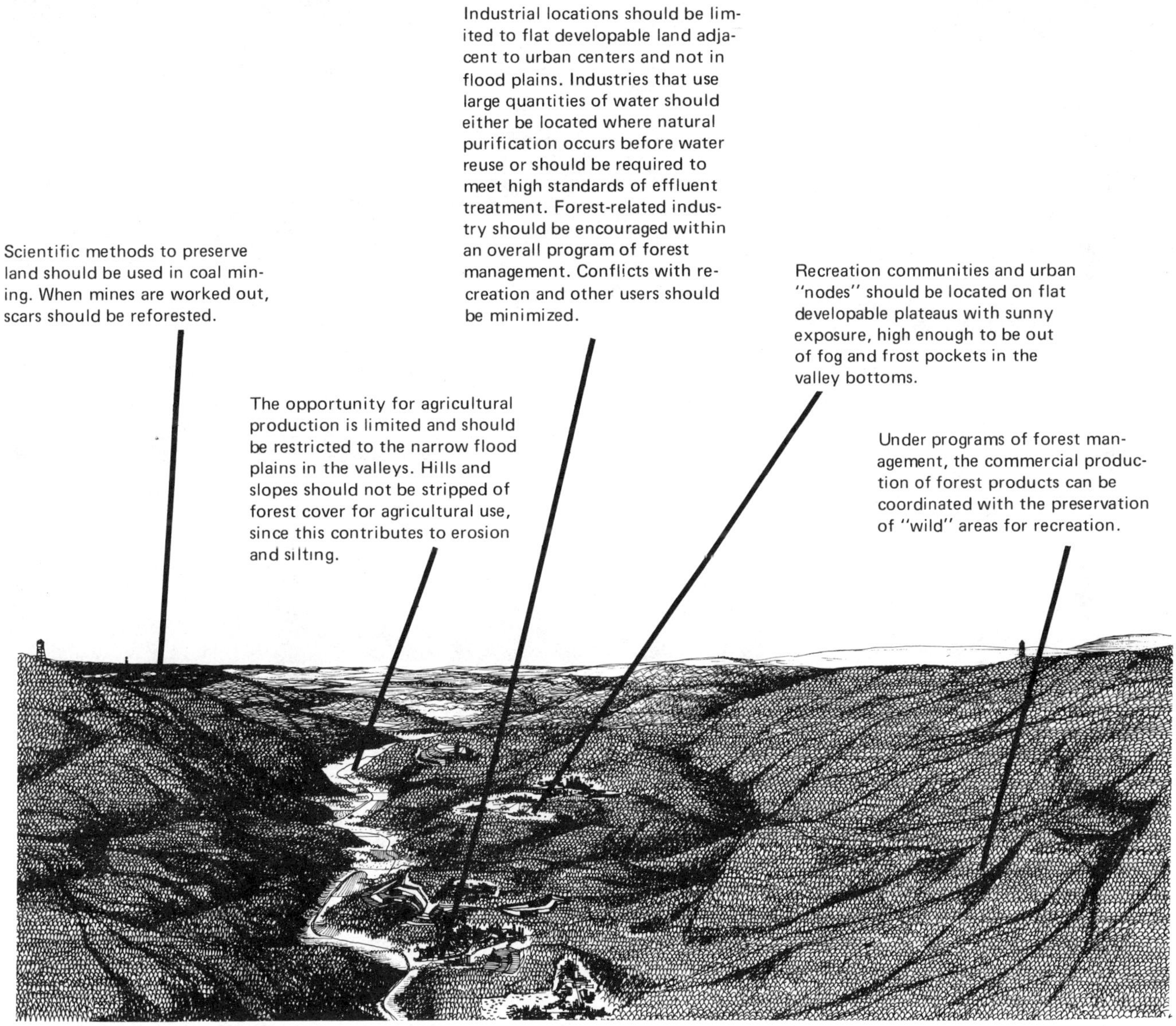

Industrial locations should be limited to flat developable land adjacent to urban centers and not in flood plains. Industries that use large quantities of water should either be located where natural purification occurs before water reuse or should be required to meet high standards of effluent treatment. Forest-related industry should be encouraged within an overall program of forest management. Conflicts with recreation and other users should be minimized.

Scientific methods to preserve land should be used in coal mining. When mines are worked out, scars should be reforested.

Recreation communities and urban "nodes" should be located on flat developable plateaus with sunny exposure, high enough to be out of fog and frost pockets in the valley bottoms.

The opportunity for agricultural production is limited and should be restricted to the narrow flood plains in the valleys. Hills and slopes should not be stripped of forest cover for agricultural use, since this contributes to erosion and silting.

Under programs of forest management, the commercial production of forest products can be coordinated with the preservation of "wild" areas for recreation.

Figure 3. Environmental planning in the Allegheny Plateau. The schematic depiction of Maryland's Georges Creek Valley portrays the various land use patterns of the area—coal mining, agriculture, industry, recreation, and timbering. In a study prepared by the American Institute of Architects and students from the University of Pennsylvania's Department of Landscape Architecture and Regional Planning, the Potomac River basin was divided into four physiographic regions. The Allegheny Plateau area, which is located where the Potomac begins its journey eastward to the sea, is epitomized by the Georges Creek Valley, a 64-sq-mi area south of Frostburg, Maryland. The terrain is rugged. Extensive coal reserves have been largely depleted, leaving a legacy of acid drainage, land subsidence, and erosion from spoil banks. Ample precipitation makes the forests moderately productive; with wise management, they can be used both as a source of timber and as recreational development. The primary objectives of planning, according to the study, should be the reduction of acid effluent during mining operations, rehabilitation of spoil banks, reforestation of cutover forest lands, and further development of the area's recreation potential.

Figure 4. Fumes from this mine fire near Scranton, Pennsylvania, endangered the nearby residential area before reclamation began.

Land

Soil and Topography. Soil is used for farming; it is frequently a mineral resource (as in the case of sand, gravel, and clays); and it plays an important role in the construction of buildings and highways and in the operation of many systems of waste disposal. Since soils vary greatly in composition, their responses to water and gravity also vary. These variations have important effects on the productive capacity of the soil and on its ability to support heavy loads, serve as a medium for waste disposal, or hold its shape and slope after excavation. Variations in topography—hills and valleys, plateaus and ridges—also play an important role in economic development. When we do not know about these differences or fail to recognize their importance in the planning process, the result is irreversible erosion, smelly waste disposal plants, highways that fall apart, and buildings that sink.

Rock. Although we use rock strata in the deeper subsurface layers of the land less frequently than we do the soil, these layers are of critical importance. They are a source of raw materials, including water; they serve as disposal sites; and they give the land "backbone" so it can support heavy structures on the surface. Improper use of this resource results in jeopardized water supplies, ineffective disposal systems, and damaged buildings and highways.

Minerals. The man-minerals relation has written many of the most important pages in history—gold in our American West, diamonds in South Africa, uranium in Canada, oil in the Middle East, and coal in Appalachia. However, the methods used to extract and purify these minerals have frequently degraded our water supplies, contaminated our land, polluted our air, caused subsidence of our land surfaces, and wastefully depleted our supplies of the minerals themselves (Fig. 5). Long-range planning of mineral resource extraction can avoid these problems.

One example of this kind of planning is the multiple land use cycle developed in certain extractive industries where the minerals are found on or near the surface and excavation sites can be filled and then used for other purposes. For example, the phosphate industry in Florida and the sand and gravel industries in New Jersey have transformed excavation sites into artificial lakes, which could be used for recreational purposes. In western Maryland, strip coal mining areas are being filled with city refuse and will be planted over with trees and grass. In the future, these areas may be used as open-space recreational sites.

Historical, Aesthetic, and Archaeological Sites. Certain pieces of land are "where the action was" in our history. Others are superbly scenic, ecologically unique, or geologically unusual. They are finite in number and unreproducible, and should be identified and preserved as part of our visual and inspirational heritage.

STEPS TO INCORPORATE ENVIRONMENTAL FACTORS INTO LOCAL DEVELOPMENT PLANNING

All of the environmental elements described above should be considered by local planners in determining what kinds of development are feasible or preferable in an area.

Information about the natural resources or geologic makeup of an area may be useful in determining overall development goals or objectives or deciding upon priorities among projects. Once these are determined, however, several steps should be taken to assure that information about the environment is incorporated into project planning. Each planner should (1) determine what environmental information is essential to plan a project; (2) survey how much information is readily available from existing sources and obtain it; (3) organize efforts to obtain the remainder of the required information through consultation with professionals or appeals to local, state, and federal agencies for expertise or funds to hire experts; and (4) incorporate this environmental information into plans and project proposals.

Most information about our physical environment is usually presented in map form. Table 1 is designed to help the planner make a decision as to which maps would be most valuable in relation to a specific problem area. The left-hand column shows six major categories of planning assistance decisions: construction (which contains several important subcategories), recreation, mining, cleaning up the environment, preservation of prime agricultural land, and preservation of the heritage. The right-hand column of the table shows the environmental maps that are most useful to planners, arranged according to the major categories of physical resources. The middle column performs a selection, indicating which maps would be of greatest value for each specific type of planning decision.

For example, if a planner is considering the construction of an industrial park, the maps that would be most relevant are the numbered maps listed for this category of construction.

If, in addition, the area under consideration is one where mining has taken place in the past, maps (14) through (17) would be valuable.

Most maps listed in this table have been specifically designed to meet the needs of development planners. Figure 6 illustrates schematically how such maps are prepared. Raw basic data are collected and measured, then organized and presented graphically. The basic data maps are then interpreted and translated into terms that have practical meaning to the planner. In the example shown on this drawing, geologic data on the various types of rock found in an area are ultimately translated into maps that show how much it would cost to pump 1,000 gal of water in various parts of that area.

Following the sequence shown in this schematic presentation. Table 2 shows in detail the steps required to prepare the 18 maps that are valuable in making specific planning decisions—the same 18 maps that were discussed earlier in this article and listed in Table 1.

The left-hand column of Table 2 shows for each resource category the basic environmental data required; the next column lists the basic data maps that can then be prepared. The third and fouth columns show the data interpreted and translated into the final planning maps.

Sources of information for each of the four steps are shown at the bottom of each column.

In order to demonstrate the use of Table 2, let us return to the example of the planner who is interested in developing an industrial park. We found that six maps relevant to his problem (nos. 5, 6, 7, 8, 9, and 10) were in the resource category of *soil*. Table 2 shows that in order to construct these maps the planner would need

1. Raw basic data describing the location, distribution, composition, thickness, and types of soil and the history of origin of the area and its elevation above sea level at various points.

2. Basic data maps (prepared from the above data): maps of soil, contour, and land form.

As indicated in the bottom portion of the table, basic data can be obtained from various state and federal departments and commissions. For most areas, the soil map is available from the State Department of Agriculture and (or) the U.S. Soil Conservation Service. Contour maps may be obtained from the State or U.S. Geological Surveys or the U.S. Army Map Service, and land form maps from the Geological Surveys or the Aeronautical Chart Information Service.

3. Interpreted data maps (prepared from the above information) showing the thickness, fertility, and engineering qualities (such as permeability) of the soil, the degree of land slope, and the location of flood plains.

4. Maps used by planners (compiled from the interpreted data maps), in this case maps (5) through (10), dealing with

Figure 5. Highwalls and landslides left after surface coal mining in Tennessee. (Photo by Donald A. Crane)

such practical concerns as the size of area required for a sewage disposal facility, cost per 1,000 gal for disposal of liquid waste, cost per acre for disposal of solid waste, the volumes of both liquid and solid waste that can be accomodated in different areas, and information showing which areas should be avoided because of the danger of soil slides.

The maps shown in the last two columns of Table 2 are not generally available to planners. They represent specialized analyses of basic data maps enriched by interpretation of other relevant engineering, technical, and economic information, and are usually constructed by teams of natural resource scientists who can furnish expertise in geology, engineering, and economic analysis.

Advanced environmental analysis for planning will need even more efficient literal and graphic shorthand for environmental communication and planning decision-making. An approach is under development to rate subjectively the components of the environment that are quantifiable. These quantified "indicators" can then be combined into a number that conveys a level of quality at a specific time for a particular geographic area. The resulting combination or "index" can pertain to a natural resource, or a pollution characteristic, or any combination of these characteristics, individually or in any grouping appropriate to the decision-making process involved. For example, a rating for water quality could be derived from a number of measurable components (such as acidity) or arbitrarily quantifiable components (such as recreation utility). Each of these indicators could be combined and represented by a single index number that could convey the quality of the component considered for the time and area for which the data measured are applicable. Another index might group two or more of the indicators to convey in a single number a more complex conception of the quality of the area. The index concept can be used to convey still a third level of complexity in environmental analysis by combining characteristic indicators, such as land and water resources and pollution. And finally, at a fourth level of complexity, environmental geology can merge even more intimately with environmental and developmental planning. This is done by merging indicators of the physical and natural environment with similar indicators of demographic and economic indicators. The mapping and other communication possibilities of conveying complex present conditions and future alternatives as unified environmental and developmental concepts are infinite.

How can a planner obtain the maps he needs? The answer to this question depends on the professional competence of his staff in the specialized fields described above. If the staff has the necessary training and experience, the basic geologic data can be obtained from the sources indicated at the bottom of Table 2. Other technical and engineering information can then be gathered and the final tables prepared.

If the planner's staff does not include experts capable of doing this type of analysis, there are several courses of action that can be taken: competent personnel can be added to the staff; local, state, and federal agencies can be stimulated to furnish the required expert assistance; or professional services can be purchased from nongovernmental sources.

The local planning and development district in your area should be able to offer assistance and serve as the coordinator of environmental planning efforts. The following state agencies might be contacted for information or technical advice:

1. Geological surveys
2. Mineral resources agencies
3. Soil surveys
4. Water resource agencies
5. Environmental health divisions of health agencies
6. Environmental conservation practices departments
7. Commerce agencies
8. Public utility commissions
9. Wildlife commissions
10. Forestry departments
11. Regional and state economic development offices

Federal programs of assistance (financial, technical, advisory, and informational) are listed in the *Catalog of Federal Domestic Assistance* (U.S. Office of Economic Opportunity, 1972). A program described in the catalog

TABLE 1. ENVIRONMENTAL MAPS VALUABLE IN LOCAL PLANNING

Types of development	Maps needed
Category of Construction	All maps listed in the next column would be valuable in planning construction. Map 4 and maps 14 through 17 would be particularly important in areas where mining and agriculture are or have been significant activities. Map 18 would be relevant only in areas where special sites are located. Listed below are the maps that would be most useful for each category of construction:
Industrial parks and manufacturing plants	1, 5, 6, 7, 8, 9, 10, 11, 12, 13
Public buildings	1, 5, 8, 9, 10, 11, 12
Public utilities	
Water systems	2, 3, 8, 9, 10, 11, 12
Waste disposal facilities	5, 6, 7, 8, 9, 10, 13, 17
Electric power plants	2, 3, 9, 10, 11, 12
Transportation facilities, including highways	8, 9, 10, 11, 12, 15, 16
Housing developments	5, 6, 7, 8, 9, 10
Development of Recreational Facilities	Map 18 is of critical importance in this category. If construction of a facility is required, the maps listed above would be valuable. If the recreation facility is to be developed on an area where there has been mining activity, maps 14 through 17 are particularly useful. Maps 2 and 8 are of basic importance in development of any recreational facility.
Mining Development, Control and Restoration	All maps would be valuable in this type of planning. Maps 14 through 17 would be particularly important in areas where there has been mining activity. Maps useful in the two categories below are:
Determination of commercial extraction sites	14, 15, 16, 17
Mine waste disposal	6, 7, 8, 9, 10, 13, 17
Cleaning Up the Environment	
(Air, land, and water pollution control)	1, 5, 6, 7, 8, 9, 10, 13, 14, 15, 16, 17
Preservation of Prime Agricultural Land	2, 3, 4, 6, 7, 8, 10, 14
Preservation of the Heritage	2, 8, 18

Types of maps used in planning	
Air	(1) Zones showing ability of areas to tolerate air pollution resulting from new activity
Water	(2) Water yield per well or acre-ft
	(3) Cost per 1,000 gal of water
Land: Soil and Topography	(4) Agricultural products grown, including productivity per acre and dollar value of land
	(5) Size of area required for septic tanks or sewage disposal facilities
	(6) Liquid waste acceptance capacity and cost per 1,000 gal
	(7) Volume of solid waste which can be accommodated per acre and cost per acre of disposal
	(8) Ground water level and yield
	(9) Soils by origin
	(10) Soil slide zones
Land: Rock	(11) Areas requiring intense rock blasting
	(12) Rockfall and landslide zones
	(13) Capacity to accept liquid and solid waste
Land: Minerals	(14) Cost per acre for reclamation
	(15) Subsidence zones
	(16) Underground mine fire areas
	(17) Cost per cu yd or acre for refuse treatment or for quenching or burial of mine fires
Special sites	(18) Location of historic, aesthetic, and archaeological sites

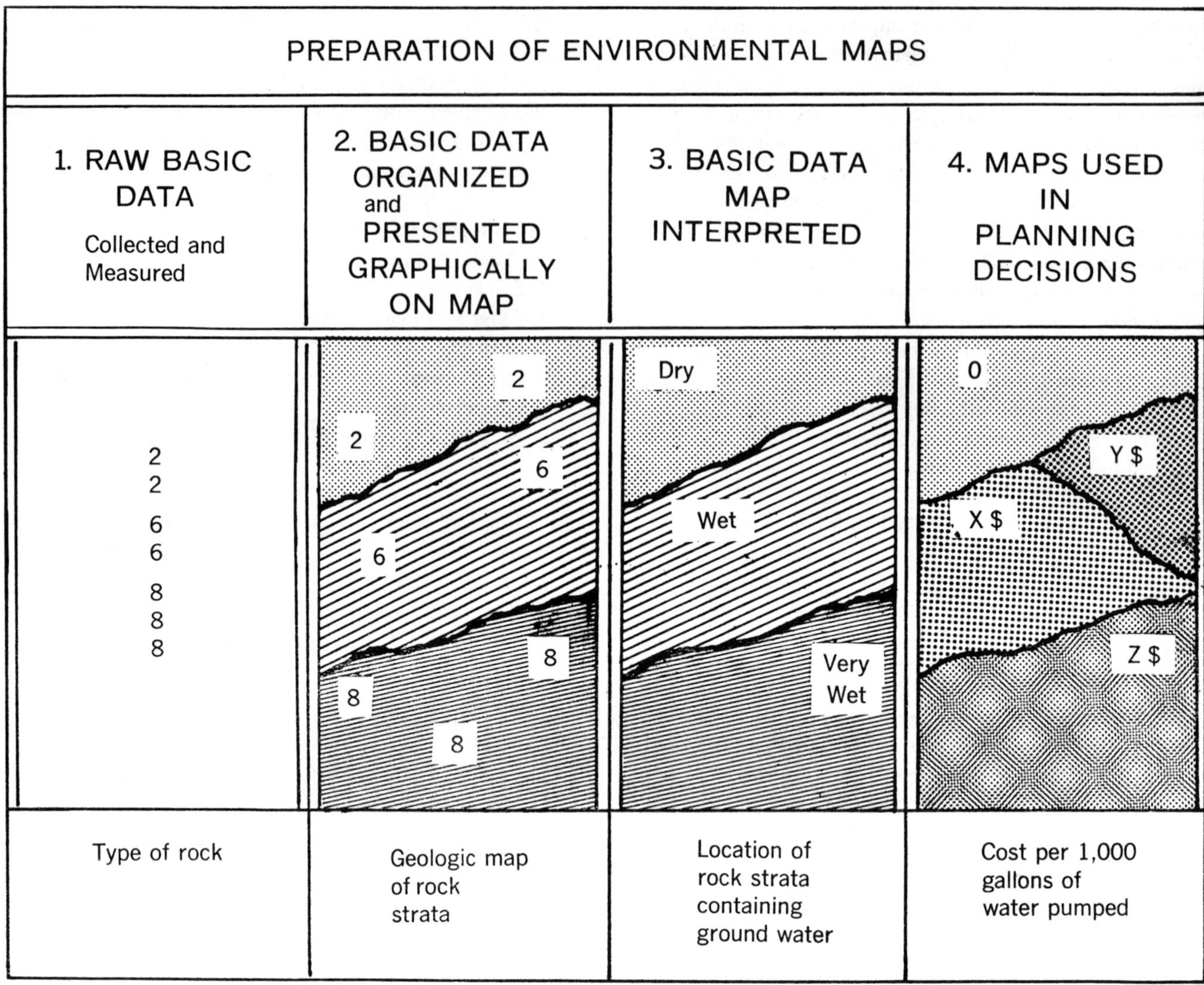

Figure 6. Schematic illustration of the preparation of environmental maps.

may be directly related to a physical resource, as, for example, water supply and development studies or planning for water needs of the future. Other programs support the building and construction of facilities that capitalize on a resource or modify a problem, such as the restoration and preservation of a historic site or the building of a sewage treatment plant. Still other programs, such as training and education in the nature of environmental problems or in skills required for problem-solving, are not directly involved with manipulating the components of the environment but help make the climate more conducive to efficient management of physical resources. Most programs listed are available to state and local governmental bodies of various types, although in some instances public or private educational or research institutions (and in a few cases individuals) are also eligible. Some programs require that applications must be channeled through established working relations between a federal and a state agency. In most instances, the applicant initiates a program, with technical counsel from the federal agency involved. Although most programs that furnish technical counseling and advisory services are handled on a correspondence basis, a few provide for personal consultation. If a planner learns of a program that would be beneficial to his overall plan but which he is not eligible to apply for or to initiate, he may wish to encourage the appropriate body to take advantage of the program.

TOOLS AVAILABLE TO IMPLEMENT LOCAL PLANS

This article has discussed the four steps that need to be taken to incorporate environmental factors into local development planning: (1) determining what environmental information is essential; (2) obtaining whatever is available from existing sources; (3) organizing efforts to obtain the remainder; and (4) incorporating all relevant environmental information into plans and project proposals. The tools used by the planner in these four steps are primarily technical—data, maps, and professional expertise.

Once these four steps have been completed, the final and most important step remains—translating the plans and project proposals into reality. This step requires other types of tools, primarily administrative and institutional in nature.

Legislation

Certain types of local legislation can be effectively used by the planner in implementing his programs: *zoning restrictions* for land, water, and air use; *codes or standards of*

TABLE 2. DATA AND MATERIALS NEEDED TO PREPARE ENVIRONMENTAL MAPS

1. Basic environmental data		2. Maps of basic data	3. Interpreted basic data maps	4. Maps used in making specific planning decisions
Air	Air pressures Moisture content Components Wind velocity	Airsheds map (a)	Patterns of pollutant distribution as affected by time of day and season	(1) Zones showing ability of areas to tolerate air pollution resulting from new industrial activity
Water	Distribution Amount Quality	Ground water geologic map and hydrologic atlas (b) Topographic map showing valleys and streams (c)	Water-supplying strata Ground water recharge areas Ground water quality Surface water flow	(2) Water yield per well or acre-ft (3) Cost per 1,000 gal of water
Land: Soil and Topography	Location Distribution Types Permeability Composition Thickness Elevation above sea level History of origin	Soil map (d) Topographic map (d) Land form maps (b) Orthophoto map (b)	Thickness Fertility Engineering qualities such as permeability Degree of slope Flood plains	(4) Agricultural products grown, including productivity and dollar value per acre (5) Size of area required for septic tanks or sewage disposal facilities (6) Liquid waste acceptance capacity and cost per 1,000 gal (7) Volume of solid waste which can be accommodated per acre and cost per acre of disposal (8) Ground water level and yield (9) Soils by origin (10) Soil slide zones
Land: Rock	Distribution Thickness Composition Engineering qualities	Geophysical map (b) Geophysical map Orthophoto map (b)	Strata engineering qualities	(11) Areas requiring intense rock blasting (12) Rockfall and landslide zones (13) Capacity to accept liquid and solid waste
Land: Minerals	Location Distribution Amount Composition	Resource geologic map (b)	Mineral quality or quantity distribution Depth and type of overlying soil and rock Mineral areas Mine refuse areas	(14) Cost per acre for reclamation (15) Subsidence zones (16) Underground mine fire areas (17) Cost per cu yd or acre for refuse treatment or for quenching or burial of mine fires
Land: Special Sites	Location			(18) Location of historic, aesthetic, and archaeological sites
Basic environmental data may be obtained from: (a) Data sources listed in next column; (b) State and federal departments that carry on construction activities, such as agencies responsible for water resource development, mines and minerals, highways, public buildings, and public construction review bodies. (For information as to whether and where these data can be obtained for a specific area, call or write the STATE GEOLOGICAL SURVEY (or its equivalent); the MAP INFORMATION OFFICE, U.S. Geological Survey, Room 1038, General Services Administration Building, Washington, D.C. 20244; or the GEOLOGIC INQUIRIES GROUP, U.S. Geological Survey, 1625 "I" Street, N.W., Washington, D.C. 20244.) (c) Local and state public utilities commissions for value and services information on water, gas, and coal.		*Types of Maps Published by Government Agencies* available at MAP INFORMATION OFFICE (address at left) include addresses where maps may be ordered. Maps in this column may be obtained from: (a) Not generally available, but currently being developed for some areas by state university meteorological schools and local weather bureau meteorological stations. (b) State geological surveys. General information on U.S. Geological Survey maps, publications, and open-file reports from GEOLOGIC INQUIRIES GROUP (address at left). Maps from WASHINGTON DISTRIBUTION SECTION, 1200 Eads Street, Arlington, Virginia 22202. Copies of reports containing maps from SUPERINTENDENT OF DOCUMENTS, Government Printing Office, Washington, D.C. 20402. Inquiries on water resources to: Mr. James Randolph, Inquiries Unit, WATER RESOURCE DIVISION, U.S. Geological Survey, Washington, D.C. 20244. (c) MAP INFORMATION OFFICE (address at left) and maps from WASHINGTON DISTRIBUTION SECTION (address above). (d) Sources in (b) above and from Commanding Officer, U.S. ARMY TOPOGRAPHIC COMMAND, Attention: 16230, Washington, D.C. 20315. Aerial photographs usually available at state departments of commerce or agriculture. Also write MAP INFORMATION OFFICE (address at left).	These maps are not generally available. For some areas, they may be obtained from sources cited for basic data maps (see column at left). If not available, they can be constructed by a professional geologist from basic data, available data maps, and other relevant engineering, technical, and economic information.	These maps, which are based on special analyses of basic environmental data, are not usually available, but can be constructed by physical resource teams using data and maps described in left-hand portion of this table.

quality, as in the case of construction; and *rules and regulations* (including a system of consistently imposed fines and penalties), which may be passed by local legislative bodies to control various types of activities. In addition, the planner can stimulate interest in state and federal legislation that will encourage environmental planning at all levels.

Scheduling

One of the most important aspects of any plan is the timing of various activities that are scheduled to occur. Planners can use scheduling as a tool to insure that the physical environment is strengthened rather than depleted. For example, if there is need for clean water at a certain location on a stream, it may be more effective to schedule control of upstream industrial waste dumping *before* constructing a water purification plant. After the source of pollution has been brought under control, the need for downstream purification can be reassessed and appropriate measures taken.

Budget and Finance

Working with local legislative bodies, planners can determine where money needs to be spent and then raise part of the money through taxes, user fees, and permit and license fees. These local fiscal tools can be used to reward development activities that conserve our environmental heritage and discourage those that do not. Many resource analysts believe that the use of our public resources, such as air and water, is a privilege that should be paid for, particularly when their use results in their degradation. Analysts also feel that the money obtained from these payments should be used to restore the resources to their original condition. A local use tax on stream water, for example, can be an effective method of financing the purification of water that has been polluted during industrial use.

Organization of Compacts

One of the most effective—and least used—tools available to local planners and legislatures is their ability to organize into compacts. Compacts are of various types. They may be intergovernmental (combining townships, counties, development districts, or states) or interdisciplinary (bringing together experts from various professions). They may be formed to study problems, to propose plans, or to carry out programs. They may deal with many resources or with one, as in the case of watershed compacts, which have been organized because state and local officials have realized the inefficiency and expense of handling water and sewage disposal on a community-by-community basis.

ACKNOWLEDGMENTS

The cooperation and assistance of the Appalachian Regional Commission, and particularly of Elizabeth Moore of the *Appalachia* journal, are appreciated.

REFERENCES CITED

Moser, Paul H., 1969, Alabama investigates resources to determine economic benefits: Appalachia, June–July, Appalachian Regional Commission, 1666 Connecticut Ave., N.W., Washington, D.C. 20235.

U.S., Office of Economic Opportunity, 1972, Catalog of federal domestic assistance: Government Printing Office, Washington, D.C. 20402, 817 p. and appendixes.

MANUSCRIPT RECEIVED BY THE SOCIETY JUNE 25, 1973